INDUSTRIAL TOXICOLOGY

Safety and Health Applications in the Workplace

Edited by

Phillip L. Williams
Senior Research Scientist
Environmental Health and Safety Division
Georgia Tech Research Institute
Georgia Institute of Technology
Atlanta, Georgia

and

James L. Burson
President
J. L. Burson & Associates, Inc.
Marietta, Georgia

VAN NOSTRAND REINHOLD
New York

Library of Congress Catalog Card Number 85-5781
ISBN 0-442-23541-0

Manufactured in the United States of America

Van Nostrand Reinhold
115 Fifth Avenue
New York, New York 10003

Chapman & Hall
2-6 Boundary Row
London SE1 8HN, England

Thomas Nelson Australia
102 Dodds Street
South Melbourne, Victoria 3205, Australia

Nelson Canada
1120 Birchmount Road
Scarborough, Ontario M1K 5G4, Canada

15 14 13 12 11 10 9 8 7 6 5 4

Library of Congress Cataloging in Publication Data

Main entry under title:

Industrial Toxicology.

Includes bibliographies and index.
1. Industrial toxicology. 2. Poisons—Safety measures.
I. Williams, Phillip L., 1952– . II. Williams,
Phillip L., 1952– . [DNLM: 1. Occupational
Diseases—Chemically induced. 2. Environmental pollutants
—Poisoning. WA 465 I437]
RA1229.I54 1985 615.9′02 85-5781
ISBN 0-442-23541-0 (pbk.)

Contributors

James L. Burson, M.S., C.I.H. President, J. L. Burson & Associates, Inc., Marietta, Georgia.

G. Michael Duffell, M.D. Associate Professor of Medicine; Director, Pulmonary Disease Division, Emory University School of Medicine, Atlanta, Georgia.

Daniel R. Goodman, Ph.D. Late Assistant Professor, Department of Pharmacology and Interdisciplinary Toxicology, University of Arkansas for Medical Sciences, Little Rock, Arkansas.

Robert C. James, Ph.D. Assistant Professor, Department of Pharmacology and Interdisciplinary Toxicology, University of Arkansas for Medical Sciences, Little Rock, Arkansas.

Renate D. Kimbrough, M.D. Research Medical Officer, Center for Environmental Health, Centers for Disease Control, U.S. Department of Health and Human Services, Atlanta, Georgia.

Ellen J. O'Flaherty, Ph.D. Associate Professor of Environmental Health, Department of Environmental Health Sciences, College of Medicine, University of Cincinnati, Cincinnati, Ohio.

Joel G. Pounds, Ph.D. Department of Applied Science, Trace Metal Biology and Toxicology Laboratory, Brookhaven National Laboratory, Upton, New York.

Martha Radike, Ph.D. Research Associate, Department of Environmental Health Sciences, College of Medicine, University of Cincinnati, Cincinnati, Ohio.

Robert L. Rietschel, M.D. Associate Professor of Dermatology and Deputy Chairman, Department of Dermatology, Emory University School of Medicine, Atlanta, Georgia.

Woodhall Stopford, M.D., M.S.P.H. Clinical Assistant Professor of Medicine, Division of Occupational Medicine, Duke University Medical Center, Durham, North Carolina.

Christopher M. Teaf, Ph.D. Assistant Director, Hazardous Waste Management Program, Center for Biomedical and Toxicological Research, Florida State University, Tallahassee, Florida.

Phillip L. Williams, M.S., C.I.H. Senior Research Scientist, Environmental Health and Safety Division, Georgia Tech Research Institute, Georgia Institute of Technology, Atlanta, Georgia.

Preface

Purpose of This Book

Industrial Toxicology presents compactly and efficiently the scientific basis of toxicology as it applies to the workplace—in particular to the manufacture, storage, use, and ultimate disposal of industrial materials. The book covers the diverse chemical hazards encountered in the modern work environment and provides a practical understanding of these hazards for those concerned with protecting the health and wellbeing of people at work. Due to its *emphasis on toxicological principles*, we hope that this book will serve the reader both as a learning tool and as a durable professional reference.

Intended Audience

Industrial Toxicology is the direct result of the authors' and editors' involvement in a successful continuing education course presented annually at the Georgia Institute of Technology entitled "Industrial Toxicolgy". The book was written for those health professionals who need toxicological information and assistance beyond that of an introductory text in general toxicology, yet more practical than that in advanced scientific works on toxicology. In particular, we have in mind industrial hygienists, occupational physicians, safety engineers, sanitarians, occupational health nurses, safety directors, and environmental scientists.

Organization of the Book

Industrial Toxicology consists of three parts, which follow a general introductory chapter explaining the relationship between the theory and the practice of toxicology.

Part I establishes the scientific basis of toxicology, which is then applied through the rest of the book. This part discusses concepts such as absorption, distribution, and elimination of toxic agents from the body. Chapters 4 through 9 discuss the effects of toxic agents on specific physiological organs, including the blood, liver, kidneys, nerves, skin, and lungs.

Part II addresses specific areas of concern in the industrial environment—both toxic agents and their manifestations. Chapters 10 through 12 examine toxic effects of metals, pesticides, and solvents. Chapter 13, on Occupational Epidemiology, treats toxic manifestations from a population viewpoint. Chapters 14 through 16 consider areas of great research interest—mutagenesis, carcinogenesis, and reproductive toxicology.

Part III is devoted to the evaluation and control of hazards in the industrial workplace. It provides specific guidelines for developing chemical exposure limits and risk assessment and presents case studies that allow the reader to visualize the application of toxicological principles in industry.

Features

The following features of *Industrial Toxicology* will be especially useful to our readers:

- The book is compact and practical, and the information is structured for easy use by the health professional in both industry and government.
- The approach is scientific, but applied, rather than theoretical. In this it differs from more general works in toxicology, which fail to emphasize the information pertinent to the industrial environment.
- The book consistently stresses evaluation and control of toxic hazards.
- Numerous illustrations and figures clarify and summarize key points.
- *Case histories and examples from industry* demonstrate the application of toxicological principles.
- Chapters include *annotated bibliographies* to provide the reader with additional useful information.
- A comprehensive Glossary of Toxicological Terms is included.

PHILLIP L. WILLIAMS
JAMES L. BURSON

Acknowledgments

A text of this undertaking on the broad topic of industrial toxicology would not be possible except for the contributions made by each of the authors in their field(s) of speciality. In addition, the staff of the Environmental Health and Safety Division at the Georgia Institute of Technology have offered constructive criticism and encouragement during the preparation of the text. In particular, the editors gratefully acknowledge the endless hours of typing and manuscript preparation provided by Rachel McCain and Susanne Keiller, along with reviews of drafts provided by Anthony DeCurtis.

Finally, we would like to thank the management of the Georgia Tech Research Institute (formerly the Engineering Experiment Station) for their support of this project. Their vision of both the course and a textbook have encouraged the editors to accept this challenging assignment.

Contents

Chapter 1

Bridging Theory and Practice in Industrial Toxicology

Phillip L. Williams

During recent years the technical sophistication of industrial processes has increased dramatically. Today, new industries such as those engaged in by "high technology" companies continually develop new materials and chemicals. In the past it was not unusual for chemicals to be introduced into the workplace even when little was known about their effects on man or the environment. Asbestos, vinyl chloride, formaldehyde, and benzene are all examples of materials used for many years before the results of health research indicated that they could be dangerous in the workplace. Workers then began to question the risks sustained through exposure to these and other materials while conducting certain operations.

As a consequence of this concern, a variety of new regulations have come into existence, designed to encourage identification of hazardous chemicals prior to use. Countless hours of research are now spent in determining appropriate ways in which such chemicals can be safely used. This knowledge is then put into practice in the workplace. Much of this work is done by industrial toxicologists.

Industrial toxicology is the subdiscipline of toxicology primarily concerned with evaluating human health effects posed by chemical exposure in the workplace. Understanding the theory of industrial toxicology requires a working knowledge of toxicology in general. Such an understanding is particularly true for individuals responsible for the health and safety of employees.

In an occupational setting, people from several professions may share the task of improving the safety of work environments. An industrial hygienist may be required to recognize, evaluate, and control a variety of health hazards. A safety engineer may be responsible for designing and maintaining a safe work environment. An occupational health nurse may handle the biological monitoring of employees. An occupational physician may be responsible for the overall medical program. An environmentalist may be concerned with the safe disposal of the company's hazardous waste. The product safety and liability engineer may evaluate new products prior to marketing. The risk control manager may be responsible for workmen's compensation coverage. Even line supervisors play a role by overseeing the day-to-day implementation of safety and health policies. All of these people practice industrial toxicology to some extent.

As many health professions overlap and interplay to maintain a safe working environment, the purpose of this book is to bridge the gap between the theory and actual practice of industrial toxicology, and to demonstrate how it is applied to maintaining workers' health and safety.

To judge impending or potential toxicity requires an understanding of chemical actions and interactions, as well as an understanding of biologic mechanisms. Accordingly, the first part of this book explains general principles and examines the basic sorts of physiological disorders that may be caused by toxins in the occupational setting. This includes a discussion of the anatomy and function of the major organs and the types of adverse effects that arise within them.

The book approaches these topics by concentrating on specific organs, because the reverse arrangement—listing chemicals and describing their effects—would be an endless process. New chemicals are continually being introduced. But organs are biological constants. New information about the body is generated at a much slower pace than that at which new chemicals appear. Therefore, given a general understanding of how the human body operates and how adverse physiological factors manifest themselves, the reader should be able to assess the risks associated with various individual chemicals.

Industrial toxicology is an applied science; its theory is designed for implementation. Therefore, the second part of this book applies theoretical knowledge to specific areas of concern.

The third part gives specific examples of applications in industrial situations.

The book emphasizes the application of information about the toxic effects of chemicals in an industrial setting, and every effort has been made to use work-related examples. In some cases, however, the best illustrations are found in other settings. When these other types of exposure are discussed, the principles applicable to industrial settings will be emphasized.

Some general concerns encountered when evaluating the validity of animal testing data will be considered. However, the book will not describe the detailed protocol followed in animal testing. It is assumed that those reading this book for industrial toxicological information will not be responsible for conducting specific animal tests.

After reading this book, you should be able to design a logical approach for use when evaluating industrial chemical exposures. You may also learn to identify areas in which the data are incomplete, and those situations in which published data about the toxicity of a chemical may be extrapolated to a safe exposure level for humans.

Suggested Reading

Anderson, K., and Scott, R. 1981. *Fundamentals of Industrial Toxicology*. Ann Arbor (Michigan): Ann Arbor Science Publishers, Inc.

Boyland, E., and Goulding, R. 1968. *Modern Trends in Toxicology*. London: Butterworth and Company, Ltd.

Burgess, W. A. 1981. *Recognition of Health Hazards in Industry*. New York: John Wiley and Sons, Inc.

Clayton, G., and Clayton, F. (Eds.) 1978—1980. *Patty's Industrial Hygiene and Toxicology*. 3rd Edition. Three volumes. New York: John Wiley and Sons, Inc.

Fairhall, L. T. 1957. *Industrial Toxicology*. 2nd Edition. Baltimore: Williams and Wilkins Co.

Finkel, A. J. (Ed.) 1983. *Hamilton and Hardy's Industrial Toxicology*. 4th Edition. Littleton (Massachusetts): John Wright, Inc.

Hayes, A. W. (Ed.) 1982. *Principles and Methods of Toxicology*. New York: Raven Press.

The Industrial Environment: Its Evaluation and Control. 1973. U.S. Department of Health, Education, and Welfare, Public Health Service, Center for Disease Control, National Institute for Occupational Safety and Health. Washington, D.C.

Lee, S. D., and Mudd, S. B. 1979. *Assessing Toxic Effects of Environmental Pollutants*. Ann Arbor (Michigan): Ann Arbor Science Publishers, Inc.

Mayers, M. R. 1969. *Occupational Health: Hazards of the Work Environment*. Baltimore: Williams and Wilkins Co.

Proctor, N. H., and Hughes, J. P. (Eds.) 1978. *Chemical Hazards of the Workplace*. Philadelphia: J. B. Lippincott Co.

PART I

Conceptual Aspects

Chapter 2

General Principles of Toxicology

Robert C. James

Introduction

Part I of this book provides a concise description of the basic principles of toxicology, in order to enable readers to make reasonable judgments about potential hazards, risks, and consequences of chemical-biologic interactions within the workplace.

This chapter explains:

- What toxicologists study, and the scientific disciplines they draw upon.
- Dose and response, and the information conveyed through their relation.
- Descriptive toxicology, or the generating of dose-response data.
- How data from animal tests can be extrapolated to human exposure.

2.1 What Toxicologists Study

Toxicology is the study of chemical or physical agents that produce adverse responses in the biologic systems with which they interact. These responses span a broad physiologic spectrum, ranging from something relatively minor, such as irritation or tearing, to a more serious response such as liver or kidney damage, or even permanent disabilities such as lung disease, kidney failure, or cancer. Because these responses occur within such a broad range, toxicologists have defined a number of specialities, as Table 2-1 shows. Every toxicologist, however, performs one or both of the two basic functions of toxicology, which are: (1) to examine the nature of the adverse effects produced by chemical or physical agents, and (2) to assess the probability of their occurrence.

While a great variety of toxicities can result from chemical exposure in the workplace, these toxicities can be classified into a few general categories. For example, it is usually useful to separate adverse responses by the temporal

Table 2-1. Types of Toxicologists.

Type	Focus of work
Clinical toxicologist	Concerned with the effects of chemical (drug) poisoning and the treatment of poisoned people.
Descriptive toxicologist	Concerned directly with the toxicity-testing of chemicals.
Environmental toxicologist	Concerned with the ultimate environmental fate of chemicals and their impacts upon the biological ecosystem and human populations.
Forensic toxicologist	Concerned with applying techniques of analytical chemistry to answer medicolegal questions about harmful effects of chemicals.
Industrial toxicologist	Concerned with disorders produced in individuals who have been exposed to harmful materials during the course of their employment.
Mechanistic or biochemical toxicologist	Concerned with elucidating the biochemical mechanisms by which chemicals exert toxic effects.
Regulatory toxicologist	Concerned with assessing descriptive data with regard to the risk involved in the marketing of chemicals and their legal uses.

duration of their effects:

- Acute toxicity: Generally has a sudden onset for a short period of time (that is, a reversible effect).
- Chronic toxicity: Marked by long or permanent duration, constant or continuous (that is, a permanent or irreversible effect).

Likewise, toxicities may be distinguished by their general sites of action:

- Local toxicity: Occurs at the site of application or exposure, between the toxicant and the biologic system.
- Systemic toxicity: Requires absorption of the toxicant within the body and then distribution of the toxicant via the bloodstream to susceptible organ(s), which are the sites of action.

Carbon tetrachloride, for example, has been used in industry as a degreaser, solvent, and chemical intermediate, and it has the ability to produce all of the above toxicities. At high air concentrations it can produce local effects by inducing irritation to the eyes and throat. The inhaled vapors of carbon tetrachloride can be absorbed to produce an acute, systemic toxicity, such as inebriation or the depression of the exposed person's mental capacity, which quickly wears off if exposure is halted. In exposures of sufficient magnitude or length, carbon tetrachloride might also produce chronic or irreversible effects, such as liver and kidney damage, which could prove both permanent and life-threatening.

Toxicities are also categorized by the response they produce or the organ they affect. Thus, acute or chronic problems might be traceable to hepatotoxins (liver poisons), nephrotoxins (kidney poisons), or neurotoxins (poisons to the nervous system), as well as to immunotoxins (poisons of the immune system), teratogens (poisons causing birth defects), allergens sensitizers (agents that stimulate the production of natural bodily substances that can lead to harmful, allergic-like reactions), or carcinogens (toxicants that cause malignant tumors). Thus, to describe, evaluate, and predict the occurrence of an industrially related toxicity, one should understand aspects of physiology, biology, biochemistry, pathology, mathematics, and other scientific disciplines.

All of these categories of response will be discussed in detail in other chapters of the book.

2.2 The Dose-Response Relationship

The chemical agent or toxicant that can produce an adverse effect in a biological system may do this either through an alteration of normal function or the destruction of life. This definition is broad, because all chemicals are toxic at some dose; that is, all chemicals are capable of altering some function or producing death in some biological organism. Given the broad range of toxicities any substance might eventually invoke, the wisdom of Paracelsus (1493–1541) becomes clear: "All substances are poisons; there is none which is not a poison. The right dose differentiates a poison and a remedy." This statement serves to emphasize the importance of risk assessment, which is the evaluation of those circumstances and conditions under which an adverse effect can be produced (discussed in Chapter 17). All chemicals are toxic and produce harm, but only within prescribed conditions of usage. Consequently, there are no harmless substances, but only harmless ways of using substances.

Since all chemicals are toxic, what judgments determine their use? To answer this, one must understand the use of dose-response relationships in toxicology.

First, let us define some terms used in discussing the dose-response relationship.

Defining Dose and Response

A particular toxicity test is said to exhibit a dose-response relationship when a consistent mathematical relationship describes the proportion of test organisms responding for a given dosage interval for a given exposure period. For example, the number of mortalities might increase consistently as the amount of the chemical introduced into the organism is increased; or, for therapeutic agents, adverse effects might decrease consistently as the dosage is increased.

The design of any toxicity test incorporates:

1. The selection of a test organism.
2. The selection of a response to monitor or measure.
3. An exposure period or test duration.
4. A series of doses to test.

Possible test organisms range from isolated cellular material or selected strains of bacteria through higher-order plants and animals. The response or biological endpoint can range from subtle changes in organism physiology or behavior to death of the organism, and exposure periods may vary from a few hours to several years. Clearly, those tests are sought for which the response is not subjective and can be consistently determined; that are conclusive even when the exposure period is relatively short; and in which test species respond in a manner that relates to the likely human response. However, some tests are selected because they yield indirect measurements or special kinds of responses that are useful because they correlate well with another response of interest; for example, the determination of mutagenic potential is used as a measure of carcinogenic potential.

Frequency-Response and Cumulative-Response Graphs

Not only does response to a chemical vary with different species of test organisms, response also varies within a group of test organisms of the same species. Experience has shown that typically this intraspecies variation follows a normal (Gaussian) distribution when a plot is made relating the frequency of response of the organisms and the magnitude of the response for a given dose (see Figure 2-1, a). Well-established statistical techniques exist for this distribution and reveal that two-thirds of the test population will exhibit a response within one standard deviation of the mean response, while approximately 95 percent and 99 percent, respectively, lie within two and three standard deviations of the mean. Thus, a relatively small number of experimental groups can be tested; and then statistical techniques can be used to define the probable response of the average organism to a given dose (the mean) and the range within which most test organisms will respond (±one or two standard deviations about the mean).

Typically, frequency-response curves are not used. Instead, cumulative dose-response curves are used, which depict the summation of several frequency-response curves for a range of dosages. Graphically, the results are depicted as a point (average response) with bars extending above and below it

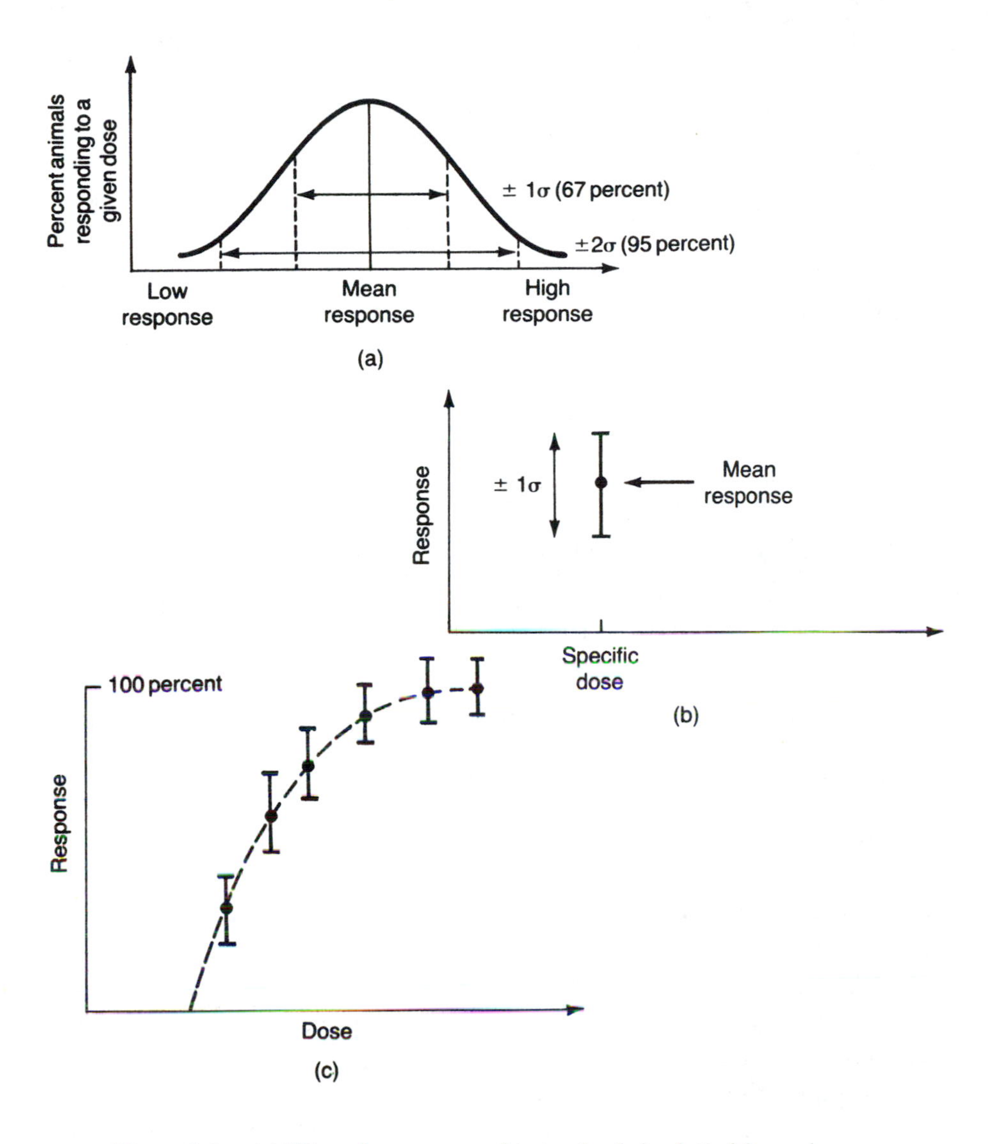

Figure 2-1. (a) When the response of test animals is plotted for a given dose, we see that some may show a minimal effect while others are more affected by the same dose. Plotting the percent of animals showing a particular magnitude of response gives a bell-shaped curve about the mean response. One standard deviation in either direction from the mean should encompass the range of responses for about two-thirds (67 percent) of the animals. Two standard deviations in both directions encompasses 95 percent of the animals. (b) The probable response for a test animal can therefore be easily predicted by testing *n* animals at a dose. By plotting the average of the *n* values as a point bracketed by one standard deviation, the probable response of an animal should fall within the area bracketed about the mean at least two-thirds of the time. (c) By plotting the cumulative dose-response (the probable responses for various doses), we generate a curve that is representative of the probable response for any given dose.

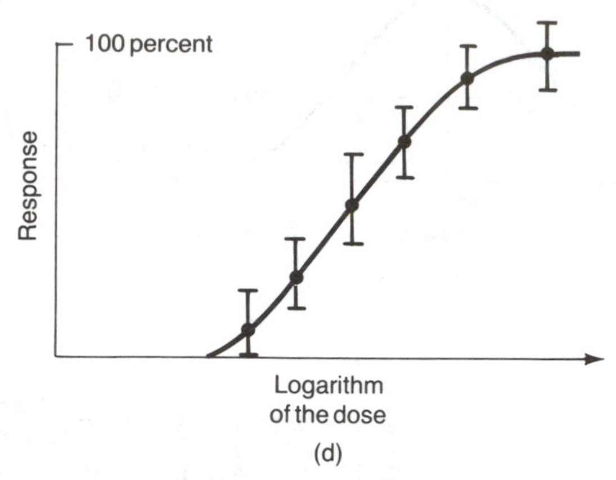

Figure 2-1. continued. (d) By plotting the cumulative dose-response curve, using the logarithm of the dose, we transform the hyperbolic shape of the curve to a sigmoid curve. This curve is nearly linear over a large portion of the curve, and it is easier to see or estimate values from this curve.

to exhibit one standard deviation greater and less than this average response (see Figure 2-1, b). A further refinement is made by plotting the cumulative response in relation to the logarithm of the dose, to yield plots that are typically linear, from which several basic relationships can be readily identified (Figure 2-1, c, d).

Ways in Which Dose-Response Data Can Be Used

Dosages are often described as lethal (LD), in a test where the response is mortality; effective (ED), in tests where the response is a desirable effect; or toxic (TD), where the response is an undesirable toxicity other than death. Construction of the cumulative dose-response curve enables the identification of doses that affect a given percent of the exposed population. For example, LD_{50} is the dosage lethal to 50 percent of the test organisms (see Figure 2-2), or one may choose to identify a less hazardous dose, such as LD_{10} or LD_{01}.

Dose-response data allow the toxicologist to make several useful comparisons or calculations. As Figure 2-2 shows, comparisons of LD_{50} for toxicants A, B, and C indicate the potency (toxicity relative to the dose used) of each chemical. Knowing this potency allows comparisons to be made against other chemicals, which is particularly informative when done against substances one is already familiar with. In this way one may approximate the human risk or safety to a specific exposure by extrapolating the animal test data (i.e., relative potency between chemicals) to doses established for more familiar substances, and thereby approximate a safe exposure level for man to the new chemical. For toxic effects one may assume man is as sensitive to the toxicity as the test species is, and not more susceptible; then the test dose used (mg/kg) multiplied by the average human weight (about 70 kg for a man and 60 kg for a

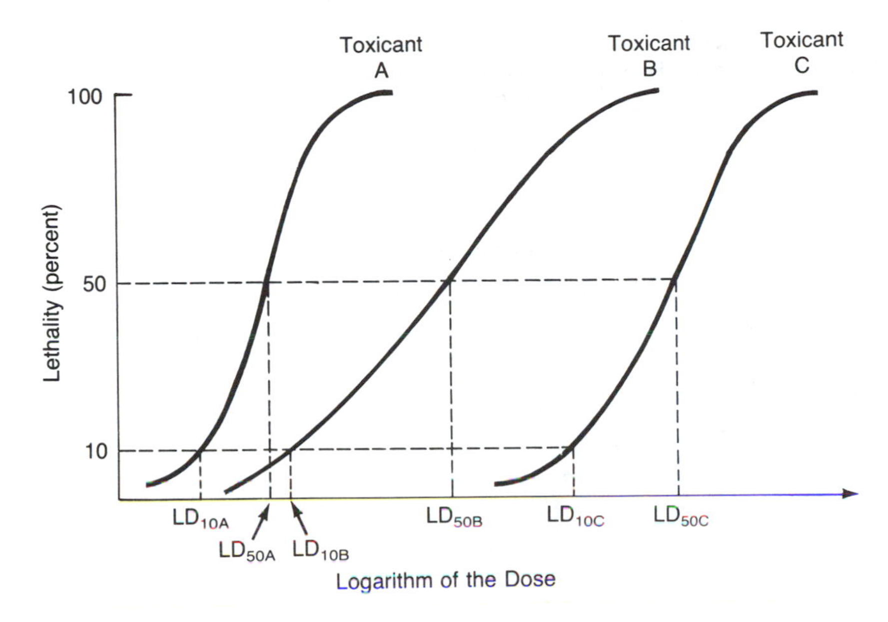

Figure 2-2. By plotting the cumulative dose-response curves (log dose), one can identify those doses of a toxicant or toxicants that affect a given percentage of the exposed population. Comparing the values of LD_{50A} to LD_{50B} or LD_{50C} ranks the toxicants according to relative potency for the response monitored.

woman) will give the toxic human dose. A relative ranking system developed years ago can then be used to categorize the acute toxicity of the chemical (see Table 2-2).

Dose-response curves are also useful because often the dose-response curve for a relatively safe acute toxicity such as odor, tearing, or irritation occurs at much lower doses than more severe toxicities such as coma or liver injury, and at much lower doses than fatal exposures. This situation is shown in Figure 2-3, and it can be easily seen that understanding the relationship of the three dose-response curves might allow the use of the symptoms represented by the ED curve to prevent overexposure and the occurrence of more serious toxicities. In fact, the difference in dose between the toxicity curve and a safely monitored effect represents the margin of safety. Typically, the margin of safety is calculated, as in Figure 2-4, by dividing LD_{50} (the higher, lethal dose) by ED_{50} (a lower, innocuous dose). The higher the margin of safety, the safer the use of the nontoxic (nonharmful) dose-response curve as an indicator and limit of exposure. One may also want to use a more subjective definition of the margin of safety (for example, TD_{10}/ED_{50} or TD_{01}/ED_{10}) depending upon the circumstances of the substance's use and the ease of identifying and monitoring the safe response; the seriousness of the toxicity produced; or if the toxic dose-response curve overlaps some portion of the non-toxic response curve.

Table 2-2. A Relative Ranking System that Can Be Used to Categorize the Acute Toxicity of a Chemical. (Reproduced with Permission ofthe American Industrial Hygiene Association JOURNAL.)

	Probable oral lethal dose for humans	
Toxicity rating or class	Dose	For average adult
1. Practically nontoxic	> 15,000 mg/kg	More than 1 quart
2. Slightly toxic	5,000–15,000 mg/kg	Between pint and quart
3. Moderately toxic	500–5,000 mg/kg	Between ounce and pint
4. Very toxic	50–500 mg/kg	Between teaspoonful and ounce
5. Extremely toxic	5–50 mg/kg	Between 7 drops and teaspoonful
6. Supertoxic	< 5 mg/kg	Less than 7 drops

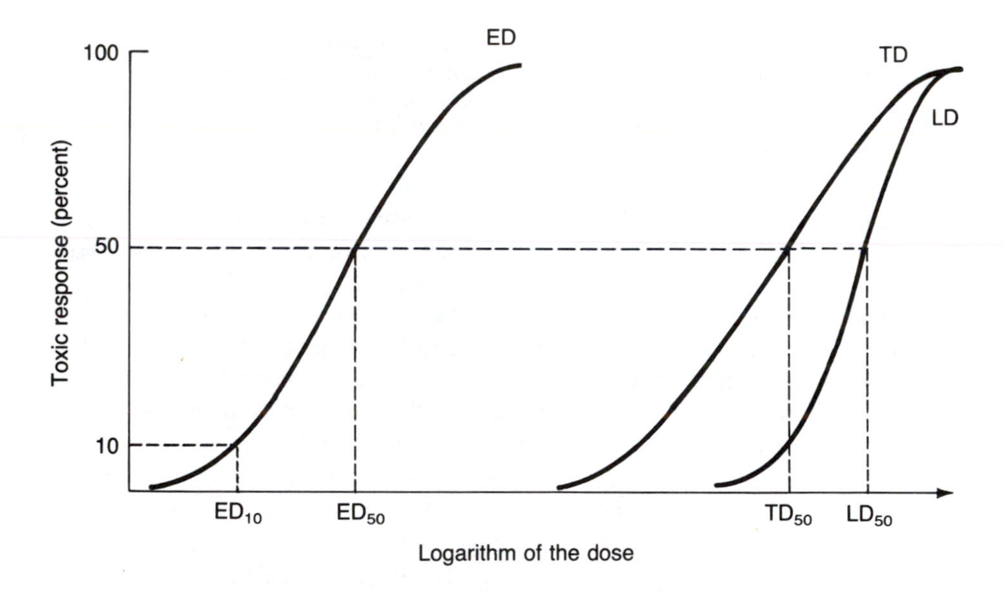

Figure 2-3. By plotting or comparing several dose-response curves for a toxicant, one can see the relationship existing for several responses the chemical might produce. For example, the effective response (ED curve) might represent a relatively safe acute toxicity, such as odor or minor irritation to the eyes or nose. The toxic response (TD curve) might represent a serious toxicity, such as organ injury or coma. The lethal response (LD curve), of course, represents the doses producing death. Thus, finding symptoms of toxicity in a few people at effective response (ED_{10}) would be sufficient warning to prevent a serious or hazardous exposure from occurring.

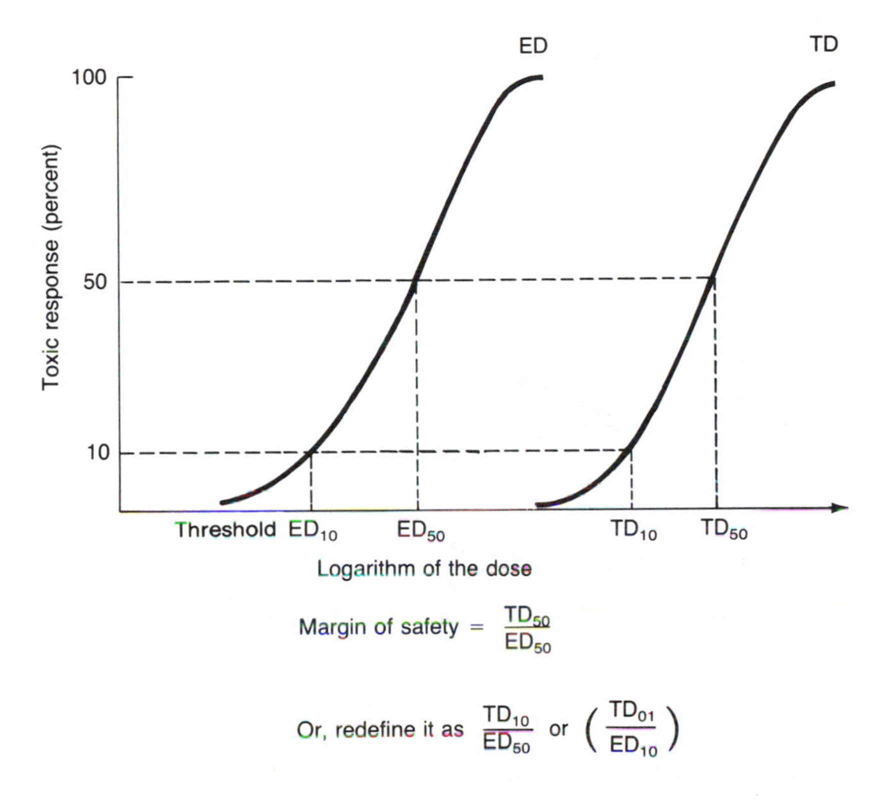

Figure 2-4. By plotting an ED curve (safe reversible response) against a lethal-dose or toxic-dose curve, one can calculate the margin of safety that exists between seeing the safe response and not eliciting the undesirable response.

Finally, the use of dose-response curves allows for the calculation or identification of the threshold dose or exposure. The threshold is the lowest point on the dose-response curve, that dose below which an effect by a given agent is not detectable. Thus, all doses and exposures less than the threshold dose represent are safe doses for all the exposed subjects.

Avoiding Incorrect Conclusions from Dose-Response Data

While the dose-response relationship can be determined for each toxicity of a toxicant, one must be cognizant of certain limitations when using these data. First, if only single values from the dose-response curves are available, as in the case of toxicants A and B in Figure 2-5, it must be kept in mind that single values will not provide any information about the slope of the curve. Thus, while A would appear to be the more toxic chemical, this is not true at lower doses. Toxicant B has a lower threshold and is more potent at lower doses. Once someone is exposed to a toxicant, the width of the dose-

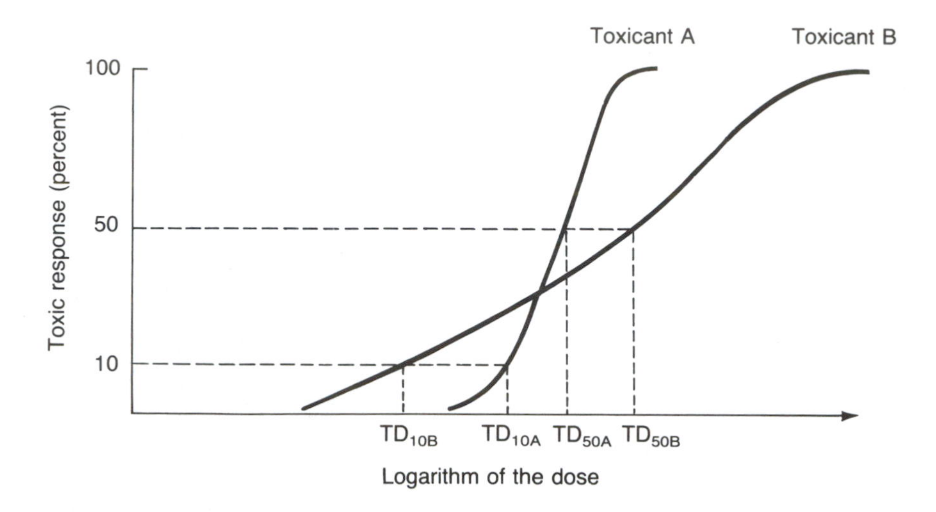

Figure 2-5. The shape of the dose-response curve is important. If one found the LD_{50} values for toxicants A and B from a table, one would erroneously assume that A is (always) more toxic than B. The figure demonstrates that this is not true at low doses.

response interval is as important as when toxicity begins (threshold). Actually, in this regard toxicant A is of greater concern, not necessarily because of its lower LD_{50} and LD_{100}, but rather because of the sleeper dose-response curve once persons are overexposed (exceed the threshold) there is much less variability among individuals, and more persons are affected with subsequent increases in dose. In other words, once the toxic level is reached, the margin of error for substance A decreases more rapidly than for substance B, because each incremental increase in exposure greatly increases the percent of individuals affected.

Second, acute toxicity, which is most often generated in tests because of the savings in time and expense, may not accurately reflect chronic toxicity dose-response relationships. The type of adverse response generated by a substance generally differs with acute and chronic exposures. Chronic toxicities, being irreversible, can rarely be the same as acute or reversible adverse responses. For example, both toluene and benzene cause depression of the central nervous system, and for this acute effect toluene is the more potently toxic of the two compounds. However, benzene is of greater concern to those with chronic, long-term exposures to it, because it is carcinogenic while toluene is not.

Third, you usually have little information to guide you in deciding what animal data will best mimic the human response. (Is your test species less sensitive or more sensitive than humans?) Unfortunately, physiologic differences vary among animal species. These differences may cause changes in the disposition and biotransformation of the toxicant; with other factors differing in animals, this constitutes an area of paradox in toxicological research. We use animals as models to study the toxic mechanisms of many chemicals. Yet the proper selection of the animal to serve as the test system ideally requires prior knowledge of which animal species most closely resembles man with respect to the chemical interaction of interest. Thus, the toxicologist is almost always faced with a paradoxical situation. The goal of the toxicologist's study is the prediction of chemical effects on humans by using animal studies. However, selection of the right animal for that study requires a prior knowledge of the fate and effects of the chemical in man (the goal), as well as its fate and effects in various animals. Thus, once data are generated in a test species there are always inherent limitations to extrapolating the observed effects to humans. This is especially problematic when, as sometimes happens, one of the species tested is susceptible to a very undesirable effect, such as cancer or birth defects, yet several other species show no such effects. In that situation, determining or choosing which species represents the human response most accurately of course has a great impact upon the estimated risk.

Variables Influencing Dose-Response Curves

Characteristics of the test species or the human population may alter the dose-response curve or limit its usefulness. The following variables should be considered when extrapolating toxicity data:

Route of Exposure. How a substance enters the body determines how much of it enters (rate of absorption) and which organs are exposed to the largest concentration of the substance. Both of these variables may in turn determine rates of metabolism and excretion. For example, the amount of chemical that is orally toxic may not be toxic, or may be more toxic, when applied to the skin.

Sex. Gender characteristics may affect the toxicity of some substances. Women have a larger percent of fat in their total body weight than men; women also have different susceptibilities to reproduction-system disorders or teratogenic effects. Some cancers and disease states are sex-linked. Large sex-linked differences may also be present in animal data.

Age. Older people have differences in their musculature and metabolism, which change the disposition of chemicals within the body and therefore the levels required to induce toxicity. At the other end of the spectrum children have higher respiration rates and different organ susceptibilities (they are less sensitive to central nervous system (CNS) stimulants and more sensitive to CNS depressants), differences in the metabolism and elimination of chemicals, and many other biological characteristics that distinguish them from adults in the consideration of risks or chemical hazards.

Effects of Chemical Interaction (Synergism, Potentiation, and Antagonism). Synergists are chemicals with the same toxicity that, when combined, cause a greater than additive effect. Potentiation occurs when a chemical that does not produce a specific toxicity nevertheless increases the toxicity caused by another chemical. Antagonists are chemicals that, when combined, diminish each other's measured effect. The following chart compares these effects:

Effect	Relative toxicity (hypothetical)
Additive	2 + 3 = 5
Synergistic	2 + 3 = 20
Potentiation	2 + 0 = 10 (*Example*: Alcohol potentiates the acute toxicity of many chlorinated hydrocarbons.)
Antagonism	4 + 6 = 8, or 4 + (−4) = 0, or 4 + 0 = 1

Modes of Chemical Interaction. There are four ways in which chemical interactions can be increased or decreased.

1. Functional: Both chemicals affect the same physiologic function.
2. Chemical: The chemical interaction between the two compounds affects the toxicity of one of the chemicals.
3. Dispositional: The absorption, metabolism, distribution, or excretion of one of the chemicals is altered by the second chemical.
4. Receptor-mediated: When two chemicals bind to the same tissue receptor, the second chemical, which differs in activity, competes for the receptor and thereby alters the effect produced by the first chemical.

Genetic Makeup. Unfortunately or fortunately we are not all born physiologically equal; this provides both advantages and disadvantages. For example, people deficient in glucose-6-phosphate dehydrogenase (G6PD deficiency) are more susceptible than others to the hemolysis of blood by aspirin or certain antibiotics. On the other hand, some individuals metabolize such drugs as isoniazid faster than other people and are therefore less susceptible to their serious side effects.

Factors Influencing the Toxicity of a Chemical

We have seen that many factors may affect the predicted response. But there are also additional factors influencing the toxicity of the chemical in question.

1. Physical and chemical properties of the toxic substance may affect its absorption or alter the probability of exposure: Chemical composition (salt, free base, anion, etc.), physical characteristics (particle size, liquid

or solid, etc.), chemical properties (volatility, solubility, etc.), presence of impurities (may alter absorption or toxicity), and stability of products in the mixture are examples.

2. Exposure conditions may affect the amount of toxicant an individual is exposed to: Concentration, amount, type of exposure (skin, oral, inhalation, etc.), and duration (acute or chronic) are examples.
3. Heterogeneous makeup of persons exposed may alter the toxic response: Genetic status, immunologic status, nutritional status, hormonal status, age, sex, body type, health, and concurrent diseases are examples.
4. Environmental conditions may heighten or lessen the effect of exposure or the toxicity: How the toxic substance is carried (air, water, soil, food), presence of additional chemicals (synergism, antagonism), temperature and air pressure (volatility, ventilation, etc.), safety protection equipment and methods for handling toxic chemicals, medical facilities, and training are examples.

2.3 Descriptive Toxicology: Testing Adverse Effects of Chemicals and Generating Dose-Response Data

Since the dose-response relationship aids both basic tasks of toxicologists—namely, identifying the hazards associated with a toxicant, and assessing the conditions of its usage—it is now appropriate to summarize toxicity testing, or descriptive toxicology. While a number of tests may be used to assess toxic responses, each test rests on two assumptions.

- The effects produced by the toxicant in the laboratory test are assumed to be qualitatively similar to the effects produced in man. Therefore, the test species or organisms are useful surrogates for humans.
- The exposure of animals to high doses of toxic agents is necessary because this uncovers all possible toxic effects and is more economical than attempting to reproduce the pattern of lower exposures among larger populations that is characteristic in human groups.

Which tests or testing scheme to follow depends upon a chemical's use and the likelihood of human exposure. In general, part or all of the following scheme might be applied as the required testing program.

Level I: Testing for acute exposure (for the hazard of handling known toxicants)

a. Plot dose-response curves for lethality and possible organ injuries.
b. Test eyes and skin for irritation.
c. Make a first screen for mutagenic activity.

Level II: *Testing for subchronic exposure* (*to characterize the toxicities of a chemical*)

a. Plot dose-response curves (for 90-day exposures) in two species; the test should use the expected human route of exposure.
b. Test organ toxicity; note mortality, body weight changes, hematology, and clinical chemistry; make microscopic examinations for tissue injury.

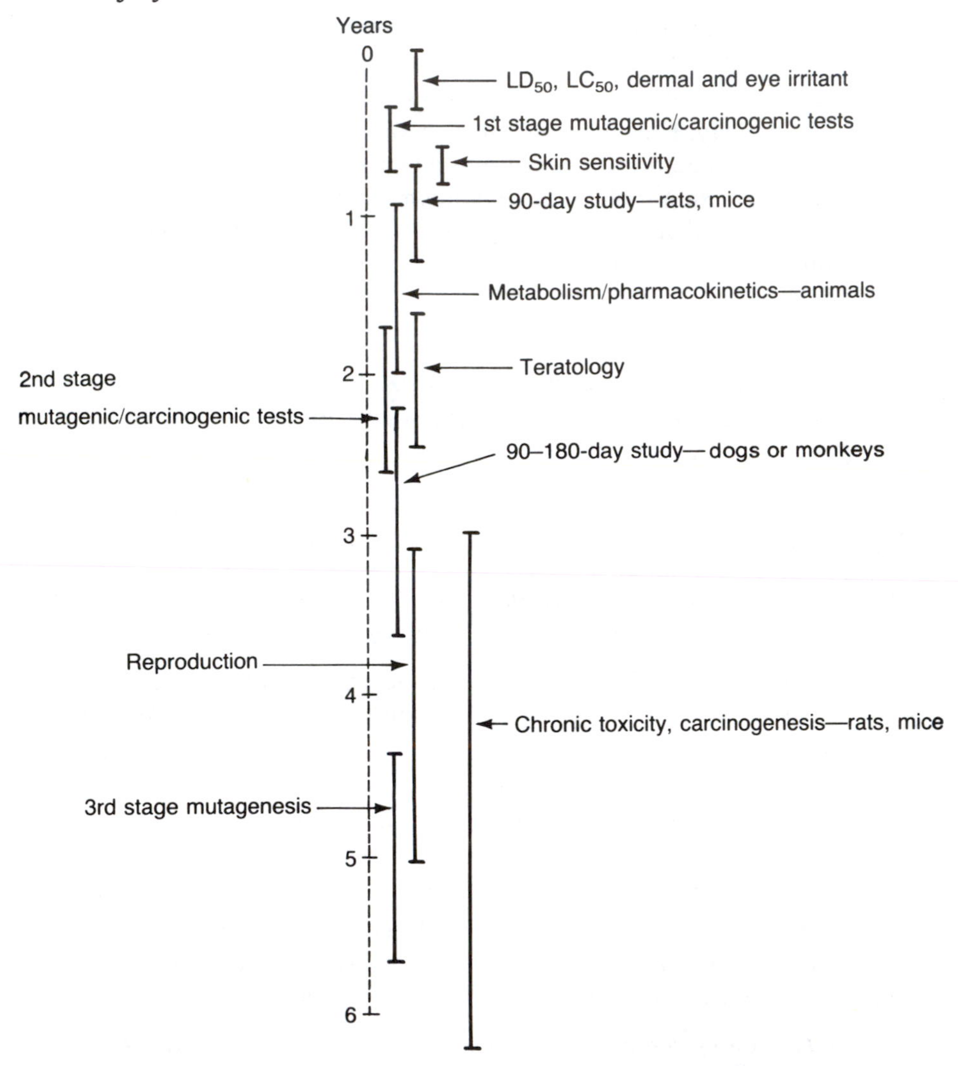

Figure 2-6. A time line showing the approximate time that it might take to test a chemical having a broad exposure to the human population. The bars represent the approximate time required to complete the tests and suggest when testing might be initiated and completed.

c. Make a second screen for mutagenic activity.
d. Test for reproductive problems and birth defects (teratology).
e. Examine the pharmacokinetics of the test species: the absorption, distribution, metabolism, and elimination of chemicals from the body.
f. Conduct behavioral tests.
g. Test for synergism, potentiation, and antagonism.

Level III: *Test for chronic exposure (final assessment of the probable human impact)*

a. Conduct mammalian mutagenicity tests.
b. Conduct a two-year carcinogenesis test in rodents.
c. Examine pharmacokinetics in humans.
d. Conduct human clinical trials.
e. Compile the epidemiologic data of acute and chronic exposures.

Establishing the safety and hazard of a chemical is a costly and time-consuming effort. For example, the rodent bioassay for carcinogenic potential requires from two to three years to obtain results, at a minimum cost of $500,000, and positive results in the end may severely limit or prohibit the use of the chemical in question, which entails additional costs. Figure 2-6 outlines the approximate time required to test and develop the safety of chemicals assumed to have widespread human impact.

2.4 Extrapolation of Animal Test Data to Human Exposure

Several models can be used to extrapolate the human risk of chemical exposure from tests of toxicity in animals. The model chosen is primarily determined by the health hazard of most concern. In general, however, only two types are used. The first type consists of extrapolating the human risk directly from the dose for which there was no observable animal toxicity. This method can be applied to most toxicities or health hazards (except cancer), since thresholds are assumed for these hazards. The second type of model is generally used to assess the risk associated with carcinogens. Since many scientists assume no identifiable threshold for this toxicity, any exposure involves risk. This concept dictates that mathematical models be used to extrapolate to exposures far below the dosages that induce observable responses in the test animal population.

For noncancer-causing toxicants (those with threshold toxicity), the models for extrapolating risk are relatively simple and similar to the methods that have been suggested or used by the National Academy of Sciences and various governmental agencies such as the Food and Drug Administration or the Environmental Protection Agency. These go by such names as Suggested No

Adverse Response Level (SNARL), Lowest Observable Effect Limit (LOEL), and No Observable Effect Limit (NOEL). Basically this type of calculation assumes that humans are as sensitive as the test species used. Therefore, the amount of a chemical ingested by the test animal that gives no toxic response is considered the safe upper limit of exposure for humans; that is, it represents the human threshold dose.

Calculating Threshold Safety: A Safe Human Dose

The calculation of a safe human dose essentially makes an extrapolation on the basis of the size differential between humans and the test species. Usually this is a straightforward body-weight extrapolation, but sometimes the surface area differences are used. The calculation is similar to the following:

$$\text{Safe human dose (SHD)} = \frac{\text{ThD}_{0.0}(\text{mg/kg/day}) \times 70\ \text{kg}}{\text{SF}} = \#\ \text{mg/day},$$

where $\text{ThD}_{0.0}$ denotes the threshold, or no-observable-effect dose in the test species, and SF denotes a safety factor, which depends on the reliability of the data used for extrapolation.

Typically, the safety factor used varies from 10 to 1000, depending on the availability of animal data and whether or not there are any human data to substantiate the reliability of the number. Of course, the number calculated should use chronic exposure data if chronic exposures are expected. This type of model calculates one value. Exposure at or below this value is considered safe.

Routes of Exposure and SHD

Once the safe human exposure has been estimated it may be necessary to convert that amount to the corresponding safe exposure level for the expected route of exposure. That is, while some exposure (mg/day) may be the safe daily intake for compound X, the level of contamination will differ depending upon whether the person is exposed via water, food, or air.

Exposure by Inhalation. Inhalation is usually a major route for occupational exposures and safe levels are determined by the comparison of airborne concentrations to established standards. For converting the safe daily intake to a safe air concentration the following formula may be used:

$$\text{Dosage} = \frac{(\alpha)(\text{BR})(\text{C})(t)}{\text{BW}} = \#\ \text{mg/kg},$$

where

α = percent of the chemical absorbed by the lungs (if not known, considered to be 100 percent);

BR = breathing rate of the individual (which, for a normal worker, can be estimated as two hours of heavy breathing at 1.47 m^3/hr and as six hours of moderate breathing at 0.98 m^3/hr), depending upon the size and physical activity of the individual;

C = concentration of the chemical in the air (mg/m^3);

t = time of exposure in hours (usually considered to be 8 hr);

BW = body weight in kilograms (considered to be 70 kg for men and 60 kg for women).

Thus, using the animal data, the preceding formula can be converted to calculate the safe air concentration if the SHD is known:

$$C = \frac{\text{SHD}}{(\alpha)(\text{BR})(t)}. \quad (\textit{Note}\text{: SHD = dosage adjusted for safety factor, } \times \text{BW})$$

This type of calculation can be used in two important ways:

- To predict a safe occupational airborne concentration for a chemical when there are no established airborne standards.
- To compare an established occupational airborne standard (such as the TLV or threshold limit value established by the ACGIH, or an OSHA standard) to newly derived animal toxicity data.

For many environmental exposures it may be assumed that $\alpha = 100$ percent, and that (BR)(t) is 30 m^3 for a 24-hour period. Thus, for a constant daily exposure the formula reduces to:

$$C = \frac{\text{SHD (as calculated previously)}}{30\ \text{m}^3/\text{day}} = \frac{\#\ \text{mg}}{\text{m}^3}.$$

Should it be desirable to express the safe air concentration in parts of toxicant per million parts of air, the value of C may be converted to a ppm level by the following relationship:

$$\text{ppm} = \frac{(\#\ \text{mg/m}^3) \times 24.5}{\text{MW}},$$

where MW is the molecular weight of the chemical (g/mol), and 24.5 is the amount (liters) of vapor per mole of contaminant at 25°C and 760 mm Hg.

Exposure by Ingestion. If the chemical is ingested with water or food through contamination (e.g., hands to mouth during smoking or eating), then the level of contamination is merely the safe human dose divided by the average daily consumption of the food source in question. For example, two liters of water per day is usually accepted as the daily rate of human water consumption. Therefore, the acceptable safe water concentration of a chemical would be:

$$\text{Water concentration} = \frac{\text{SHD}}{2\ \text{liter/day}} = \#\ \text{mg/liter.}$$

Testing for Carcinogens

The second type of risk, which is applied to carcinogens, can also be calculated according to several methods or models. However, all differ in their basic assumptions or in the mathematical expressions they use. Hence, at the low exposures in question, estimations of risk can vary dramatically. How to define "threshold" is an important toxicologic and regulatory consideration for mutagenic/carcinogenic chemicals. Current theory suggests that for genotoxic carcinogens (initiators) there is no threshold and that the rate of cancer is a continuous response with a decreasing probability all the way down until zero exposure is met (see Figure 2-7). This means that every exposure theoretically carries some risk (some mathematical probability of an adverse reaction) with it, and this assumption underlies the extremely low levels of allowable exposure proposed for carcinogenic substances. In a currently favored multistage model, the risk is linearly proportional to the exposure at low doses. While each exposure is considered to carry some risk, the acceptable risk or safe dose is usually suggested as the exposure range for a 10^{-5} to 10^{-7} risk. This means

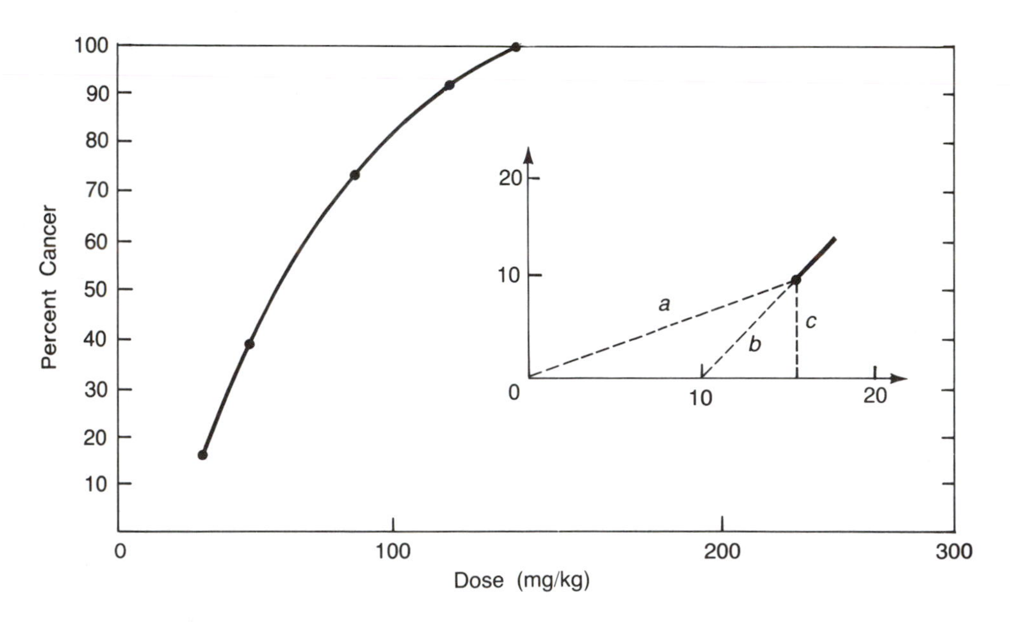

Figure 2-7. A dilemma is faced when extrapolating test results to a safe dose (threshold dose) when you only have high-dose (high-response) data. Does curve a, b, or c best represent the low-dose response?

that a lifetime of daily exposure to that dose would increase cancer by 1 person in 100,000 (10^{-5}) or by 1 person in 10,000,000 (10^{-7}). These concepts are further discussed in Chapters 14, 15, and 17.

Summary

Toxicology is a broad scientific field that utilizes basic knowledge of many other scientific disciplines.

- A toxicologist must understand these disciplines in order to discover and examine the variety of adverse effects produced by any toxicant.
- A toxicologist must utilize an understanding of each particular poison so as to develop antidotal therapies and guidelines for risk prediction and prevention.
- A toxicologist uses dose-response relationships as a basic means of identifying the potency (toxic doses) and toxicities (responses) that determine a chemical's relative hazards.

There are many types of toxicity tests, and many factors can affect the outcome of a test or create uncertainty about its extrapolation to a heterogeneous human population.

- Often the inherent toxicity of a compound cannot be altered; then the only way to lower the risk is to lessen the exposure.
- Likewise, when unknown compounds are suspected of posing a hazard, or confidence in the estimate of their toxicity is poor, the only way to limit the risk and its liability is to limit exposure.

References and Suggested Reading

Albert, A. 1968. *Selective Toxicity*. 4th Edition. London: Methuen & Co., Ltd.

Bhatnagar, R. S. (Ed.) 1980. *Molecular Basis of Environmental Toxicity*. Ann Arbor (Mich.): Ann Arbor Science Publishers, Inc.

Deichmann, W. B., and Gerarde, H. W. 1969. *Toxicology of Drugs and Chemicals*. New York: Academic Press.

Doull, J., Klaassen, C. D., and Amdur, M. O. (Eds.) 1980. *Casarett and Doull's Toxicology: The Basic Science of Poisons*. 2nd Edition. New York: Macmillan Publishing Company.

DuBois, K. P., and Geiling, E. M. K. 1959. *Textbook of Toxicology*. New York: Oxford University Press.

Guthrie, F., and Perry, J. (Eds.) 1980. *Introduction to Environmental Toxicology*. New York: Elsevier.

Hamilton, A. 1943. *Exploring the Dangerous Trades*. Boston: Little, Brown & Co.

Hayes, W. (Ed.) 1982. *Principles and Methods of Toxicology*. New York: Raven Press.

Hayes, W. I. (Ed.) 1972. *Essays in Toxicology*. New York: Academic Press, Inc.

Hodgson, E., and Guthrie, F. (Eds.) 1980. *Introduction to Biochemical Toxicology*. New York: Elsevier.

Paget, G. E. (Ed.) 1970. *Methods in Toxicology*. Philadelphia: F. A. Davis Co.

Stolman, A. (Ed.) 1969. *Progress in Chemical Toxicology*. 4th Edition. New York: Academic Press.

Chapter 3

Absorption, Distribution, and Elimination of Toxic Agents

Ellen J. O'Flaherty

Introduction

This chapter identifies fundamental principles of toxicology, and discusses:

- The relationship between the concerns of the toxicologist and the occupational health specialist.
- The broad principles that govern transfer of molecules across membranes.
- The factors that influence absorption of foreign compounds from the GI tract and the lung, and across the skin.
- Simple kinetic models that describe disposition (distribution and elimination).
- Mechanisms of elimination (biotransformation and excretion).
- Mechanisms of action.

3.1 Toxicology and the Safety and Health Professions

Occupational health specialists work with humans in industrial settings, while toxicologists usually work with rats or mice. Sometimes it may seem that no relationship exists between the two kinds of activity. Yet the working relationship between the toxicologist and the occupational health specialist can and should be a productive one. Each is able to do something the other cannot, and the activities of each complement the activities of the other.

If effects observed in workers can be reproduced in a laboratory animal, it becomes possible to investigate the mechanisms that might reasonably be expected to produce such effects. On the other hand, shedding some light on the mechanism by which a stipulated effect is produced in a test animal species may make it easier to find ways to prevent such effects from occurring in humans. Such an understanding may also help to identify subtle or delayed effects that have not been observed in workers, but which health professionals should be aware of. For these reasons, the toxicologist concerned with industrial exposure should work closely with industrial safety and health professionals.

The uncertainties associated with converting test results in small animals to data relevant to humans are frequently stressed. Quantitative differences among species do exist. Occasionally, these differences can be so great that they make us fail to keep fundamental similarities in mind. Physiological and biochemical attributes characteristic of a particular species can shift patterns of absorption, distribution, metabolism, excretion, or effect, in significant ways. Yet the basic principles that control these processes are the same for all mammalian species. These principles will be surveyed in this chapter.

Toxic Agents and the Body

Figure 3-1 will be referred to several times during this chapter. It presents an overview of the behavior of any foreign compound as it enters the body, is transported to the target tissue, exerts its effect, and is eliminated. Toxicant exterior to the body (T) is absorbed into the body and then into the blood (T_B). From the blood, it is both eliminated (T_E) and distributed to various tissues, including the target tissue (T_{TT}). The target tissue is the tissue on which the toxicant exerts its effect (E).

$$
\begin{array}{c} \quad\quad T_E \\ \quad\nearrow \\ T \rightarrow T_B \rightarrow T_{TT} \rightarrow E \end{array}
$$

Figure 3-1. An overview of a foreign compound (T) being absorbed into the blood (T_B), from which it is both eliminated (T_E) and distributed to the target tissue (T_{TT}), where it exerts its effect (E).

Generally, it is considered that a toxicant must be absorbed in order for it to have an effect, but this is not always true. Some toxicants are locally toxic or irritating. For example, acid can cause serious damage to the skin even though it is not absorbed through the skin.

Although in Figure 3-1 a distinction is made between the target tissue and blood, in some instances the blood itself represents the target tissue. Carbon monoxide, for example, combines with hemoglobin to form carboxyhemoglobin, whose presence in the blood reduces the availability of oxygen to the s. Hemolytic agents such as arsine are also active in the blood compartment, and blood is their target tissue (see Chapter 4). But most often the target tissue is a tissue other than the blood.

The Significance of the Target Tissue

The target tissue or target organ is not necessarily the tissue in which the toxicant is most highly concentrated. For example, over 90 percent of the lead in the adult human body is in the skeleton, but lead exerts its effects on the kidney, the central and peripheral nervous systems, and the hematopoietic system. It is well known that chlorinated hydrocarbons tend to become concentrated in body fat stores, but they are not known to exert any effects in these tissues. Whether distribution and/or storage processes such as these are actually protective—that is, whether they act to lower the concentration of toxicant at its site of action—is not certain. There is experimental support for the idea that certain highly localized and specialized sequestration mechanisms, such as incorporation of lead into intranuclear lead inclusion bodies or binding of cadmium to the tissue protein metallothionein, do indeed function as protective mechanisms. Whatever the case with regard to their function, however, the existence of sequestration mechanisms for many compounds means that bulk movement of a toxicant through the body or its kinetic behavior, as reflected in total plasma and tissue concentrations, must be interpreted with care. The concentration of the biologically active toxicant within the target tissue controls the action—the dynamic behavior—of the toxicant.

3.2 Transfer across Membrane Barriers

Every compound that reaches the systemic circulation and has not been intravenously injected has had to cross membrane barriers. Therefore, the first topic to be considered is the membrane itself and what enables a toxicant to cross it.

All membranes are similar in composition and structure. They consist of a phospholipid bilayer, toward the interior of which are positioned the long hydrocarbon or fatty acid tails of the phospholipids, and toward the outside of which are the more polar and hydrophilic portions of the phospholipid molecules. The fatty acid tails align themselves in the interior of the membrane to form an interior structure that is relatively fluid at body temperatures. The polar portions of the phospholipid molecules maintain a relatively rigid outer structure. Proteins embedded throughout the lipid bilayer have specific functions that will be considered later.

There are three principal mechanisms by which molecules can traverse membranes:

- Passive transfer or diffusion.
- Facilitated diffusion.
- Active transport.

Passive Transfer

Passive transfer does not involve the participation of any membrane proteins. Two factors determine the rate at which passive diffusion takes place across a membrane: (1) the difference between concentrations of the compound on the two sides of the membrane, and (2) the ease with which a molecule of the compound can move through the lipophilic interior of the membrane. Three major factors affect ease of passage: lipid solubility, molecular size, and degree of ionization.

The Partition Coefficient. The lipid solubility of a compound is frequently expressed by its partition coefficient. The partition coefficient is defined as concentration in an organic phase divided by concentration in the aqueous phase at equilibrium. The organic phase is often chloroform, hexane or heptane, or octanol. The partition coefficient is determined by shaking the chemical with water and the organic solvent, and measuring the concentration of the chemical in each phase when equilibrium has been reached.

Although the partition coefficient does not have much meaning as an absolute value, it is very useful in measuring relative lipophilicities of a series of compounds. It is the rank order that is meaningful in most cases. For example, it has been shown that the partition coefficients of the nonionized forms of several series of representative drugs can be correlated with their rates of transfer across a number of biological membrane systems—from intestinal lumen into blood, from plasma into brain and into cerebrospinal fluid, from lung into blood, etc. In general, as lipophilicity increases, the partition coefficient increases also, and so does speed of movement through the membrane.

Molecular Size. The second important feature of the molecule determining ease of movement across a membrane is molecular size. As the cylindrical radius of the molecule increases, with lipophilicity remaining approximately constant, rate of movement across the membrane decreases. This is because transfer of larger molecules is slowed by frictional resistance and, depending on the structure of the molecule, may also be slowed by steric hindrance. Very small molecules, in contrast, may move across the membrane more rapidly than would be predicted on the basis of their partition coefficients alone. Small molecules are likely to be more water-soluble than their larger homologs. If this is the case, they may be able to move through membrane pores.

Pores are features of all membranes. Their size varies with the nature and function of the membrane. Cell membranes will not allow passage of water-soluble molecules larger than about 4×10^{-4} μm in diameter, while blood capillary walls allow passage of water-soluble molecules up to about 30×10^{-4} μm in diameter. This size range excludes plasma proteins, which thus are retained within the plasma fluid volume. Even within this size range of water-soluble compounds, however, rate of transcapillary movement is inversely proportional to molecular radius.

Degree of Ionization. The third important feature of the molecule in determining ease of movement through membranes is its degree of ionization. Electrolytes are ionized at the pH values of body fluids. With the exception of very small ionized molecules that can pass through membrane pores, only the nonionized forms of most electrolytes are able to cross membranes. The ionized forms are generally too large to pass through the aqueous pores, and are insufficiently lipophilic to be transferred by passive diffusion. The rate of diffusion therefore will depend not only on the amount of an electrolyte present in the nonionized form, but also on the ease with which the nonionized form of the molecule can cross the membrane: that is, on its molecular size and lipophilicity.

All ionizable acids and bases have a pK_a value related to the dissociation constant K. The dissociation constant is always expressed for either acids or bases as an acid dissociation constant, K_a:

$$\text{For acids, } K_a = \frac{(H^+)(A^-)}{(HA)}.$$

$$\text{For bases, } K_a = \frac{(H^+)(B)}{(HB^+)}.$$

Thus, pK_a is the negative logarithm of the acid dissociation constant. If, for example, the acid dissociation constant K_a is 10^{-3}, then pK_a is 3.

The degree of ionization in body fluids depends on the pH of the medium as well as on the pK_a of the acid or base. This relation can be expressed by the Henderson-Hasselbalch equations:

$$\text{For acids, } pK_a - pH = \log\frac{\text{(nonionized form)}}{\text{(ionized form)}} = \log\frac{(HA)}{(A^-)}.$$

$$\text{For bases, } pK_a - pH = \log\frac{\text{(ionized form)}}{\text{(nonionized form)}} = \log\frac{(HB^+)}{(B)}.$$

When the pH is equal to the pK_a, half of the acid or base is present in the ionized form and half in the nonionized form. At pH less than the pK_a, acids are less completely ionized. At pH greater than the pK_a, bases are less completely ionized. Another way of stating these relationships is that at a given pH, acids having large pK_a (weak acids) are not as fully ionized as strong acids, while bases having large pK_a (strong bases) are more fully ionized (associated with a hydrogen ion) than weak bases.

When Schanker and his co-workers (Hogben *et al.*, 1959) studied the intestinal absorption of a series of acids and of a series of bases in the rat, they found that the percentage of the drug absorbed depended on its pK_a and on whether it was an acid or a base. Weak acids and bases, largely nonionized at the intestinal surface pH of 5.3, were readily absorbed, while strong acids (pK_a less than 3) and strong bases (pK_a greater than 7) were not. When the pH of the intestinal contents was increased from about 4 to about 8 by dissolving the test compounds in strongly buffered solutions rather than in water, the percentage absorption of the acids was decreased and that of the bases was increased.

Facilitated Diffusion

The second type of passage across a membrane is facilitated diffusion. Facilitated diffusion is sometimes thought of as a special case of active transport. It requires participation of a carrier protein molecule. Because the number of carrier molecules is limited, facilitated diffusion has a maximum rate. It also can be inhibited selectively, either competitively or noncompetitively, by materials that are similar to the diffusing material and therefore compete for binding sites on the carrier protein, or that affect the carrier in some other way. If equilibrium is reached, concentrations on both sides of the membrane are equal. In other words, with regard to percentage of substance absorbed, facilitated diffusion gives the same net results as passive diffusion, but facilitated diffusion is faster.

Several carrier protein systems have been studied in reasonable detail (Guidotti, 1976). Although many of the proteins that are incorporated into membranes appear at only one membrane face, all membrane carrier proteins that have been studied are known to span the membrane and to surface at both faces. It is probable that carrier proteins undergo a conformational change, associated with the binding of the diffusing molecule to the carrier, that facilitates their transport across the membrane. The requirement that the diffusing molecule must be able to bind to the carrier protein confers a certain degree of specificity on this mechanism. As might be expected, facilitated mechanisms have evolved to transport essential nutrients across membrane barriers. For example, the transport of essential nutrients such as sugars and amino acids into red blood cells and into the central nervous system takes place by facilitated diffusion.

Active Transport

Active transport is the third general process by which molecules can traverse cell membranes. In addition to its requirement for a carrier molecule, active transport requires controlled energy input. It maintains transport against a concentration gradient so that, when equilibrium is reached, the concentrations on the two sides of the membrane are not equal. Adenosine triphosphate (ATP) is the source of the energy required to maintain this concentration gradient.

Active transport processes are critical to conservation and regulation of the body's supply of essential nutrients. These are important functions of the kidney and the liver. Another function of these organs is excretion of toxic compounds and their metabolites. Accordingly, there are at least two active transport processes in the kidney for excretion into the urine (one for organic acids and one for organic bases) and at least three in the liver for elimination into the bile (one for acids, one for bases, and one for neutral compounds). It is also possible that there is an active transport process in the liver for the elimination of certain metals into the bile. Thus, active transport is a key mechanism for the excretion of toxicants from the body.

Of course, other processes may operate simultaneously with active transport processes. Facilitated diffusion can occur along with active transport, and passive diffusion will take place in the presence of a concentration gradient whenever physical factors are sufficiently favorable.

Specialized Transport Processes

There are also specialized processes that transfer molecules across membrane barriers. Phagocytosis takes place in the alveoli of the lung and also in the reticuloendothelial system of the liver and the spleen. In phagocytosis, and another process, pinocytosis, the cell membrane surrounds a particle to form a vesicle that detaches itself and moves into the cell interior. For macromolecules such as proteins, phagocytosis is particularly important. It is thought to be biologically significant in interiorizing certain kinds of specialized enzyme systems. Phagocytosis is also important because it is the mechanism used by the alveolar macrophage in scavenging particulates in the alveoli of the lung.

3.3 Absorption

For most practical purposes, we consider absorption to be absorption into the systemic circulation. The leftmost arrow in Figure 3-1 therefore represents absorption. Figure 3-2 is a schematic diagram of the human or animal body. It is drawn so as to show the lumen of the gastrointestinal (GI) tract and its contents exterior to the animal body. Epithelial tissue lines the GI tract and absorption takes place across the epithelium.

As illustrated in Figure 3-2, toxicants may be absorbed from the gastrointestinal tract, from the lung, or through the skin. In experimental studies,

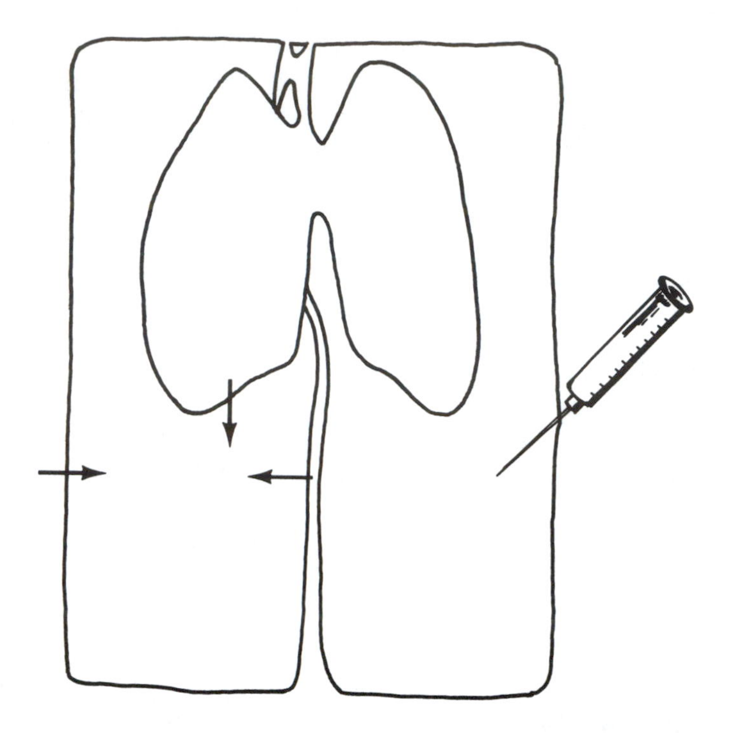

Figure 3-2. Schematic diagram demonstrating entry of a chemical into the human body.

toxicants may also be injected as indicated in Figure 3-2. Injections are commonly given intravenously, intraperitoneally, subcutaneously, or intramuscularly.

Gastrointestinal Tract

The GI tract is a very important route for absorption of toxicants. Toxicants may be present in food or in drinking water or, if they have been inhaled but are relatively large particulates, they may have been collected in the nasopharyngeal area, swallowed, and then subjected to transfer and transport processes in the GI tract. Essentially only one cell thick, the wall of the GI tract is specialized not only for absorption but also for elimination.

Absorption from the GI tract is strongly site-dependent, since the pH varies from the very acidic range of about 1 to 3 in the stomach (depending upon the amount and quality of the food and when it was eaten) to around 5 to 8 in the small intestine and colon (depending on the location, the food, and the intestinal microflora). The intestinal contents can therefore be neutral or even slightly basic.

Absorption of Organic Acids and Bases. Application of the Henderson-Hasselbalch equation, discussed above, to organic acids, which have pK_a values of 3–5, suggests that they should be relatively well absorbed from the acidic pH of the stomach. Salicylic acid is shown as an example in Figure 3-3. Its pK_a is about 3. The efficiency of its transfer across the gastric mucosa is dependent on the concentration gradient of the nonionized form across the mucosa as well as on the physical features of salicylic acid that control its rate of diffusion. As Figure 3-3 shows, in the stomach there are 100 nonionized molecules of salicylic acid for every salicylate ion. On the plasma side of the mucosal cell, however, there is relatively little salicylic acid; salicylate ion is overwhelmingly the dominant species. These calculations were carried out for steady-state conditions. In fact, once salicylic acid molecules have entered the plasma, they will be both ionized to a large extent and carried away from the absorption site by the plasma flow. These factors combine to promote efficient absorption of organic acids from the stomach.

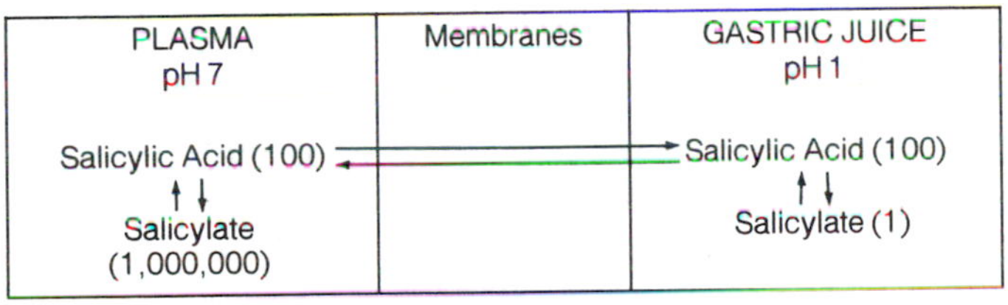

PLASMA pH 7	Membranes	GASTRIC JUICE pH 1
Salicylic Acid (100)	⇄	Salicylic Acid (100)
↑↓		↑↓
Salicylate (1,000,000)		Salicylate (1)

Figure 3-3. Partitioning of salicylic acid across the gastric mucosa. The numbers in parentheses are the numbers of molecules of the ionized or un-ionized species present on either side of the membrane. (Reproduced with permission from O'Flaherty, *Toxicants and Drugs: Kinetics and Dynamics*. New York: John Wiley and Sons, Inc. 1981. Figure 3-9.)

Organic bases, on the other hand, are largely ionized at the pH of the stomach contents, and so are more efficiently absorbed from the intestine.

Determinants of GI Absorption. A number of other factors are important in determining whether, and how rapidly, a compound will be absorbed from the GI tract. The physical factors, such as lipid solubility and molecular size, that determine the rate of diffusion of nonionized species have already been mentioned. Diffusion is also favored by the presence of villi and microvilli in the intestine. These greatly increase the surface area available for diffusion. Thus, even though absorption may not be particularly efficient per unit surface area, the very large total surface area helps to promote absorption in the intestinal portion of the GI tract.

Facilitated and active transport mechanisms present in the GI tract provide specialized transport for essential nutrients and electrolytes, including sugars, amino acids, sodium, and calcium. There is every reason to believe that a toxicant that mimics the molecular size, configuration, and charge distribution of an essential nutrient sufficiently well will be transported by the carrier process that is already in place for absorption of that nutrient, but few examples of such behavior have been identified. 5-Fluorouracil has been shown to be absorbed by a pyrimidine transport mechanism (Schanker and Jeffrey, 1961). Interaction among metal ions with respect to their use of common transport mechanisms has been somewhat more fully documented (Hamilton and Smith, 1978; Toraason *et al.*, 1981). A recent study has shown that there is a maximum absorption rate for lead in the rat at very high concentrations (Aungst *et al.*, 1981).

There are other considerations regarding absorption in the GI tract. Compounds that are unstable at the acidic pH of the stomach will not even reach the intestine to be absorbed there. Other compounds are susceptible to alteration by the actions of intestinal microflora or of gastrointestinal enzymes. For example, the formation of carcinogenic nitrosamines from secondary amines by intestinal flora is very well known.

Another factor in gastrointestinal absorption is the rate at which foodstuffs pass through the GI tract. If the rate of passage is slowed, the length of time during which the compound is available for absorption is increased. Other important factors include the chemical and physical makeup of the compound, its solubility, and its interactions with other compounds present in the GI tract. Age and nutritional status of the individual may also affect absorption from the GI tract.

Skin

The second major pathway for absorption is the skin. The skin is a very effective barrier to absorption, primarily because of the outermost keratinized layer of thick-walled epidermal cells, the stratum corneum, which in general is not very permeable to toxicants, although its permeability varies from location to location. There may be slight absorption through sweat glands or hair follicles, but these structures represent a very small percentage of the total surface area and are not ordinarily important in the process of absorption through the skin.

The stratum corneum is generally rate-limiting in the process of absorption through the skin. Compared with the total thickness of the epidermis and dermis together, the thickness of the stratum corneum is relatively slight, but this barrier is very important inasmuch as many compounds are unable to pass through it readily. All toxicants that penetrate the skin appear to do so by passive diffusion.

There are mechanisms by which absorption through the skin can be increased. Abrasion, which damages or removes the stratum corneum, greatly increases the permeability of the damaged area. The skin is normally partially hydrated; further increases in the degree of hydration increase permeability and promote absorption. Certain solvents, such as dimethyl sulfoxide (DMSO), also increase skin permeability and facilitate absorption of toxicants.

Certain toxicants can produce systemic injury by absorption through the skin. Hydrocarbon solvents, such as hexane, can produce a peripheral neurotoxicity by percutaneous absorption, and carbon tetrachloride can produce liver injury. Certain insecticides have caused deaths in industrial and field workers after absorption through the skin.

The Lung

The third major site of toxicant absorption is the lung. In occupation-linked toxicology, the lung is a very important route. Gases and vapors such as carbon monoxide, sulfur dioxide, and volatile hydrocarbon vapors are absorbed through the lung, and liquid or particulate aerosols, such as sulfuric acid aerosols or silica dust, are also deposited and/or absorbed in the lung. With solid and liquid particulates, the site of deposition is critical to the degree of absorption of a compound.

Solid and Liquid Particulates. The lung can be thought of as consisting of three basic regions: the nasopharyngeal region, the tracheobronchiolar region, and the distal or alveolar region. Particles that are roughly 5 μm or greater in diameter are generally deposited in the nasopharyngeal region. If they are deposited very close to the surface, they can either be sneezed out, blown out, or wiped away. If they are deposited slightly farther back, they may be picked up by the mucociliary escalator system and moved back up into the nasopharyngeal region where they may be swallowed and absorbed through the GI tract in accordance with their solubility and absorption characteristics. Particles that fall into the size range of 2 to 5 μm generally reach the tracheobronchial area before they impact the lung surface. Most of these particles are also cleared by the mucociliary escalator back up to the nasopharyngeal region, where they are either eliminated directly or swallowed and absorbed or excreted in the GI tract. Particles smaller than one μm in diameter may reach the alveolar regions of the lung. Absorption in the lung, if it takes place at all, will most likely take place in the alveolar region, although there may be some absorption in the tracheobronchiolar region, particularly if the material is soluble in the mucus.

Size is probably the most important single characteristic determining the efficiency of particulate absorption in the lung. Size determines the region of the lung in which the aerosol is likely to be deposited; even within the range of very small particles that reach the alveolar region and may be absorbed there, size is inversely proportional to the magnitude of particle deposition.

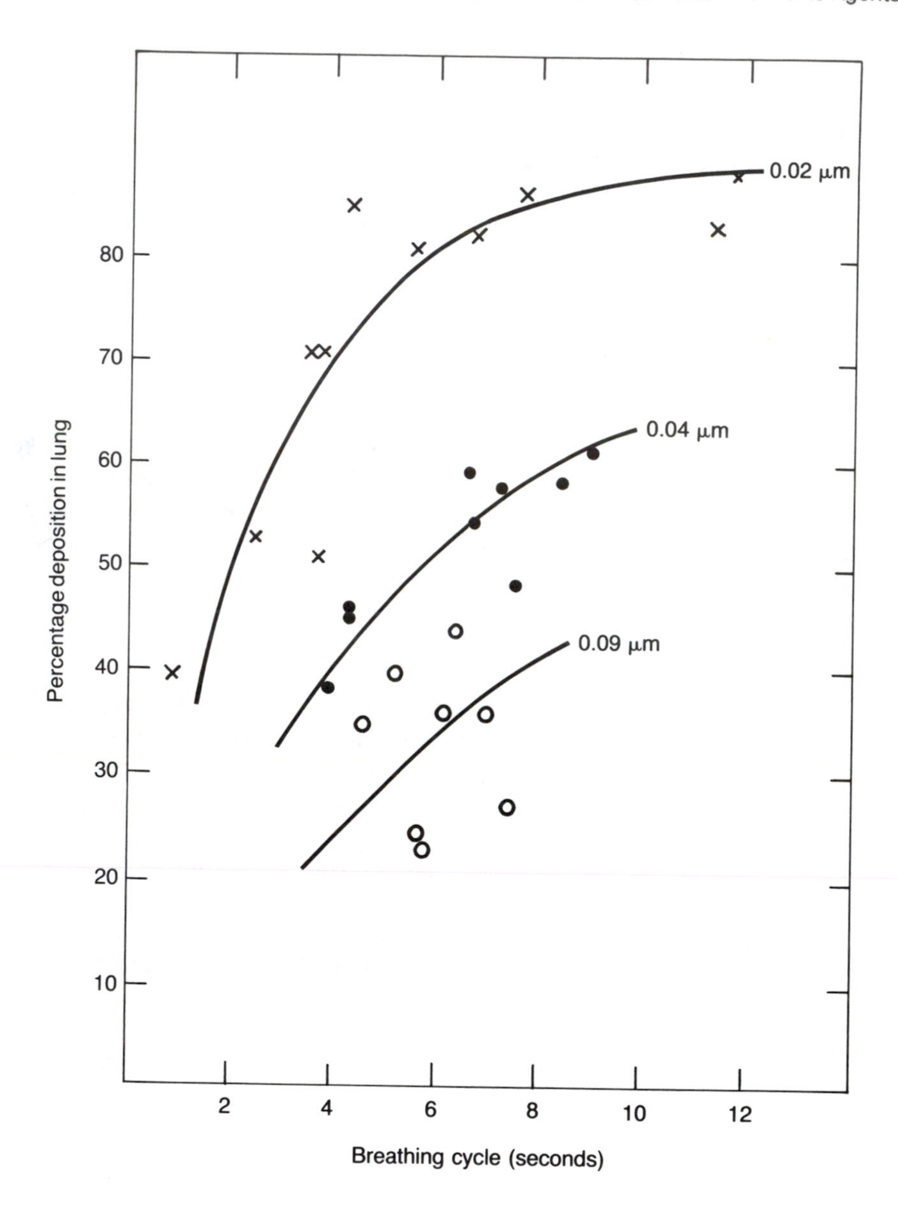

Figure 3-4. Deposition in the lung of lead particles of various sizes. (Reproduced with permission from Chamberlain *et al.*, *Investigations into Lead from Motor Vehicles*. Harwell (England): AERE. Publication No. R9198. 1978. Figure 5.3.)

Figure 3-4 shows the dependence of lead deposition in the human lung on the size of the lead particles in an artificially generated lead sesquioxide aerosol that was inhaled by the subjects (Chamberlain *et al.*, 1978). Size is expressed as diffusion mean equivalent diameter (DMED), a measure of mean particle diameter; the amount deposited was calculated as the difference between the amount of lead that entered the lung and the amount that the subject exhaled. Thus, the regions of the lung in which the particles were actually deposited were not identified. However, DMED's for all three aerosols were less than 1 μm. For standard man, with a breathing cycle of about 4 sec, lung deposition of lead varied from about 24 percent for particles with a DMED of 0.09 μm to 68 percent for very small particles with a DMED of 0.02 μm.

After being deposited in the alveolar region, particulates may be dissolved and absorbed into the blood stream, reaching the systemic circulation directly. If they are not readily soluble, they may be phagocytized by alveolar macrophages and then either transferred directly to the lymphatic system, where they may remain for a considerable time period, or moved together with the macrophage to the mucociliary escalator for clearance by that route. They may also occasionally remain in the alveolus for an extended period of time. Absorption of particulates tends to be slower than absorption of gases and vapors, and appears to be controlled primarily by the solubility and membrane-transfer characteristics of the particulate.

Gases and Vapors. Absorption of gases and vapors in the lung depends on solubility. Compounds with very high solubility in blood may be almost completely cleared from the inhaled air in one respiration. For such compounds, increasing the rate of blood flow makes very little difference. Nearly all the gas is transferred to the blood during each respiration. The only way to increase absorption would be to increase the rate of respiration, that is, to increase ventilation. Thus, absorption of these compounds is said to be ventilation-limited. If these compounds are also lipid-soluble, they will find their way rapidly to the lipid depots of the body.

Chloroform is a good example of such a compound. It is very highly lipid soluble, and is readily cleared from inspired air. As the blood circulates through the body, the chloroform is transferred to fat, so that the blood is also effectively cleared and during its next pass through the lung is able to pick up more chloroform. The absorption of chloroform is ventilation-limited.

For gases that are poorly soluble in the blood, there is rather limited capacity for absorption. Often these are not compounds that are readily cleared from the blood. The blood may become saturated quickly, and the only way to increase absorption then is to increase the rate of perfusion of the lung. Such compounds are said to be flow-limited in their absorption characteristics. Of course, there is a range of transition between these two extremes of absorption behavior.

3.4 Disposition — Distribution and Elimination

Unlike absorption, disposition consists not just of one kind of process but, rather, of a number of different kinds of processes taking place simultaneously. Disposition includes both distribution and elimination, which occur simultaneously in almost all cases. Elimination is also made up of two kinds of processes, excretion and biotransformation, which usually take place simultaneously.

Distribution and elimination are often considered independently of each other. While it is necessary to do this in discussing them, it is important to remember that they occur at the same time, as illustrated in Figure 3-1. If a substance is effectively excreted, it will not be distributed into peripheral tissues to any great extent. On the other hand, wide distribution of the compound may impede its excretion.

Kinetic Models

Before going into the specific mechanisms for distribution, excretion, and biotransformation, it is useful to consider some simple kinetic disposition models. Rates of distribution are related to kinetic distribution constants. Rates of metabolism, or biotransformation, and of excretion are related to kinetic constants of elimination. It is possible to integrate all the essential information about distribution, metabolism, and excretion of a compound into a single kinetic model. Such models can be used to help formulate predictions about the behavior of a toxicant under different exposure conditions and, if they are used correctly, they can sometimes also help to suggest a toxicant's site of action. We will not explore kinetic models in great depth, but you should be aware that they are now widely used and will continue to be widely used in the future.

One-Compartment Open Model. The most basic kinetic model is the one-compartment open model (Figure 3-5), in which the compound is assumed to have been introduced instantaneously into the body, distributed instantaneously and homogeneously, and eliminated at a rate that is at all times directly proportional to the amount left in the body, or "first-order" rate. The constant of proportionality between the rate of elimination and the amount in the body is the elimination rate constant. For this model, the logarithm of concentration in the blood is a linear function of time, as shown in Figure 3-6. The elimination rate constant is the absolute value of the slope of this line.

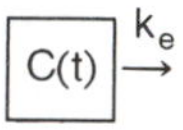

$$\boxed{C(t)} \xrightarrow{k_e}$$

Figure 3-5. The linear one-compartment open model. $C(t)$ is the concentration, which is a function of time; and k_e is the elimination rate constant. (Reproduced with permission from O'Flaherty, 1981. Figure 3-13.)

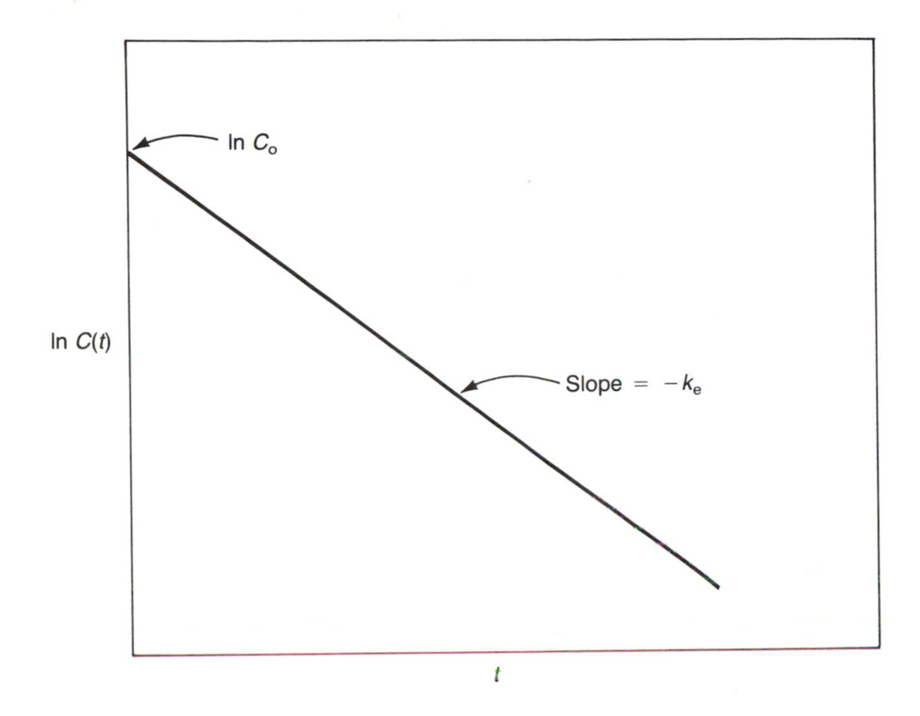

Figure 3-6. Plot of $\ln C$ versus t for the linear one-compartment open model. C_0 is the concentration at time $t = 0$, assuming instantaneous distribution. (Reproduced with permission from O'Flaherty, 1981. Figure 3-15(a).)

Half-Life. Half-life is the time required for half the compound to be lost from the plasma. The nature of the half-life is such that it can be measured at any point on a concentration-versus-time curve. Thus, it is a constant value that characterizes the behavior of the compound in the blood. The half-life is inversely proportional to the elimination rate constant.

Two-Compartment Open Model. Few compounds obey straightforward, first order, one-compartment kinetics. For most compounds it is necessary to consider at least a two-compartment model (Figure 3-7). In the two-compartment model, the compound is assumed to have entered the first compartment, which includes the blood, to have been distributed instantaneously and homogeneously throughout this compartment, and then to have been simultaneously eliminated and distributed to another compartment, from which it can return to the first compartment. All of the distribution and elimination steps are presumed to be first-order (passive).

Concentration in the first compartment declines smoothly as a function of time. Concentration in the second or peripheral compartment rises, peaks, and subsequently declines (Figure 3-8). A half-life may also be calculated for a

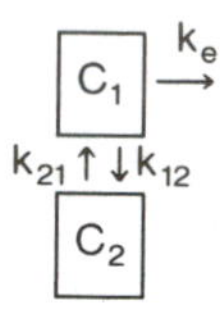

Figure 3-7. The linear two-compartment model. C_1 and C_2 are the concentrations in the first and second compartments, respectively; and k_{12} and k_{21} are the rate constants for transfer between the two compartments. (Reproduced with permission from O'Flaherty, 1981. Figure 3-22.)

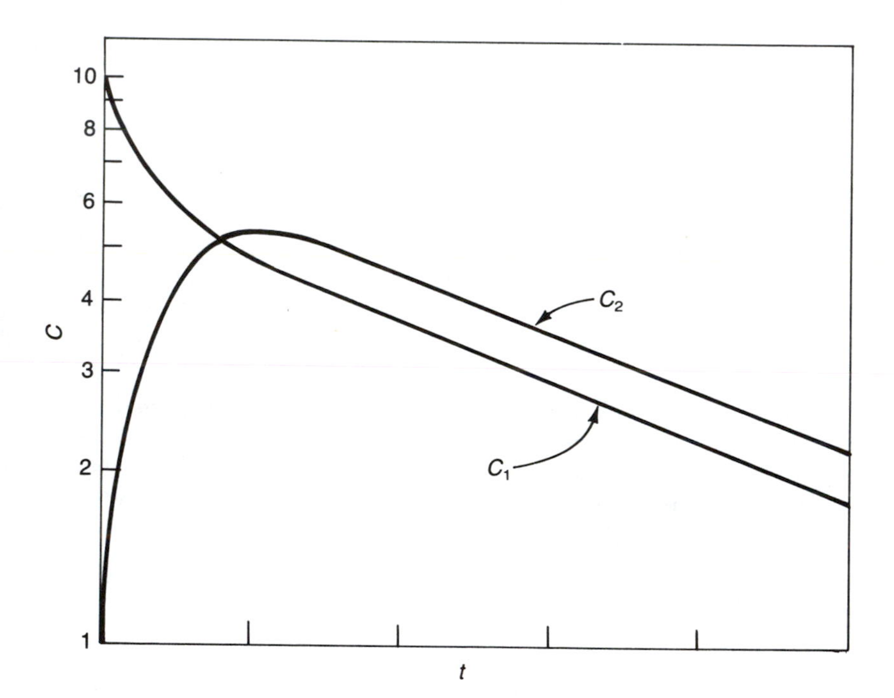

Figure 3-8. Concentration-versus-time plot for the linear two-compartment open model, showing $\ln C$ and $\ln t$ for the central (C_1) and peripheral (C_2) compartments. (Reproduced with permission from O'Flaherty, 1981. Figure 3-24(b).)

compound whose kinetic behavior fits a two-compartment model. Calculated from the terminal slope of a plot of the logarithm of the concentration in the first compartment as a function of time, this half-life is designated the biological half-life. It is the parameter most frequently used to characterize the *in vivo* kinetic behavior of an exogenous compound.

Other Kinetic Models. Other features of the compound's kinetic behavior or of its mode of administration may be incorporated into the model as appropriate. For example, there may be more than one peripheral tissue compartment; or absorption, which is never truly instantaneous even for intravenous injection, may be first-order instead (e.g., a single oral exposure, in which case the rate of absorption is directly proportional to the amount in the GI tract). If, on the other hand, exposure is roughly constant and continuous (e.g., a contaminant is widely dispersed in ambient air), the important group of models discussed in the next subsection is developed.

Mention should also be made of the important group of models that incorporate non-first-order elimination. While absorption and distribution are generally considered to be passive processes and therefore first-order, often elimination is not. Frequently this is because metabolism or excretion is saturable, or capacity-limited, due to a limitation on the number of active transport sites in organs of excretion or on the number of metabolizing enzyme active sites. When all active elimination sites are occupied, elimination is said to be saturated, or zero-order. Elimination by a saturable process is often referred to as elimination by Michaelis-Menten kinetics. Kinetic models incorporating Michaelis-Menten elimination have been developed (Wagner, 1973).

Chronic Exposures. Most industrial or other environmental exposures are not acute. Acute exposures do occur, but chronic exposures are much more frequent in the industrial setting. When exposure is constant and continuous over a long period of time, a steady state or plateau level will be reached in all tissues. As long as elimination processes remain first-order (typical, for example, of excretion by glomerular filtration in the kidney, or of metabolism as long as it is not saturable in the concentration range), this steady state should be directly proportional to both the magnitude of exposure and the biological half-life.

If exposure were truly constant, the plateau level would be constant also. More commonly, exposure is intermittent, in which case blood concentrations at steady state will cycle in a way that reflects the absorption and elimination characteristics of the compound as well as the exposure pattern (Figure 3-9). However, this cycling will take place about a constant mean that, like the plateau level, is predictable from the equivalent constant exposure rate and the biological half-life. This is one of the reasons why biological half-life is such an important attribute. Together with exposure rate, it determines mean steady-blood level irrespective of whether exposure is continuous or intermittent.

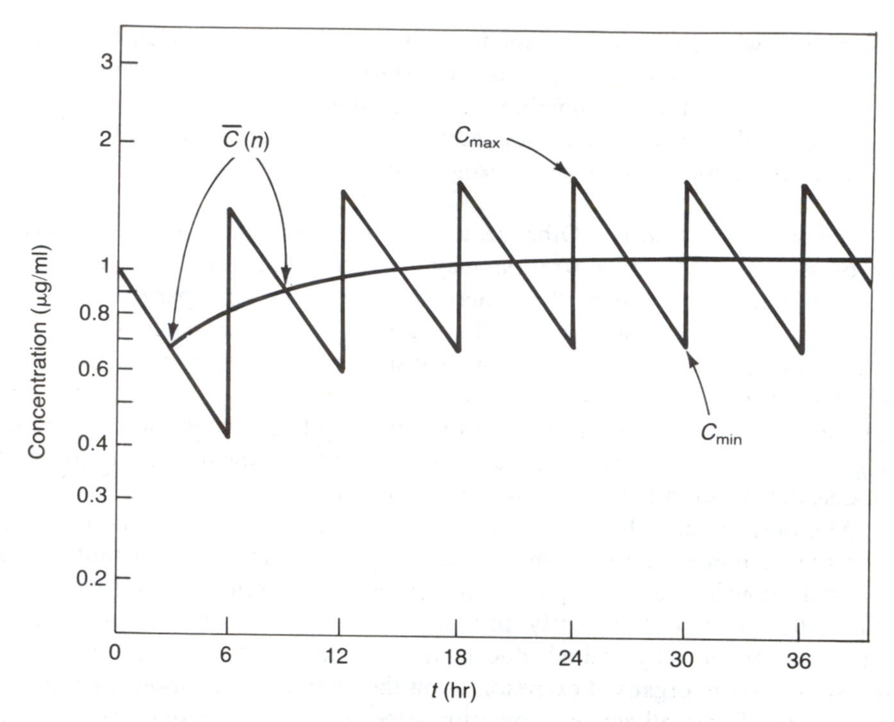

Figure 3-9. The relationship between average concentration, $\overline{C}(n)$, calculated for repetitive administration, and the time course of concentration change during continuous administration of a hypothetical compound. C_{max} and C_{min} are the maximum and minimum concentrations in each time interval between doses, assuming instantaneous distribution of each successive dose. (Reproduced with permission from O'Flaherty, 1981. Figure 5-4.)

However, the individual exposed to large amounts of a substance at intervals will experience greater peak concentrations in blood and tissues following each new exposure than will an individual exposed to the same total amount given as frequent small doses. If the large peak concentrations are associated with toxicity or with saturation of elimination processes, then it becomes important to consider the pattern of administration as well as the equivalent exposure rate.

Biotransformation

Biotransformation is the first mechanism for the elimination of toxicants and drugs that will be discussed. Biotransformation reactions in general can be divided into two types: phase I and phase II reactions.

Phase I reactions are catabolic or breakdown reactions (oxidation, reduction, and hydrolysis). Phase II reactions are synthetic reactions in which an additional molecule is covalently bound to the parent or the metabolite in order to make the conjugate product more water-soluble.

In general, all biotransformation mechanisms are directed toward production of metabolites that can be more readily eliminated from the body than the parent compound. Phase I reactions may occasionally produce metabolites that are somewhat less water-soluble than the parent. However, such metabolites are presumably more susceptible to phase II transformations, which result in greater water solubility and excretability.

Oxidation. A large fraction of exogenous compounds is metabolized by oxidative mechanisms. The most important of the oxidative enzyme systems is the group of monooxygenases containing cytochrome P-450. This system consists always of two enzymes: NADPH (nicotinamide adenine dinucleotide phosphate) cytochrome P-450 reductase, and the cytochrome P-450 itself. The two enzymes together insert one atom from molecular oxygen into the substrate while the other oxygen atom is reduced to water in combination with two protons.

Cytochrome P-450 is not a single enzyme but rather a group of enzymes with different but frequently overlapping specificities. Cytochrome P-450 enzymes have been found in the endoplasmic reticulum of the liver of all species examined for their presence. They occur also in the endoplasmic reticulum of the lung, the wall of the intestinal tract, the kidney, the skin, and the brain, as well as in other tissues. Although cytochrome P-450 monooxygenase activity is greatest in the liver, the presence of these enzymes in other tissues may be of concern if a toxic metabolite is generated *in situ.*

One important group of metabolites formed by the cytochrome P-450 monooxygenases is the epoxides, particularly epoxides of aromatic compounds. The epoxides of aromatic hydrocarbons are believed to be responsible for the mutagenic or carcinogenic effects of these hydrocarbons. This is an instance in which the biotransformation system, in producing a compound that is more polar than the parent compound, produces a metabolite that is significantly more toxic than the parent. Aromatic epoxides are not particularly stable and will rearrange spontaneously to form phenols. However, they may be present in tissues long enough to exert mutagenic or carcinogenic action.

Monooxygenases containing cytochrome P-450 catalyze a variety of reactions in addition to epoxidation, including N-hydroxylation; N-, O-, or S-dealkylation; aromatic and aliphatic hydroxylation; sulfoxidation; and desulfuration. Amine oxidase does not contain P-450, but like the P-450 system it can oxidize and hydroxylate nitrogen atoms. Amine oxidase is also present in the endoplasmic reticulum of cells. It appears that the structure of an amine—more specifically, the pK_a of the nitrogen atom—along with the relative abundance of the two enzymes in the tissue in which the amine is metabolized, largely determines whether it will be oxidized by amine oxidase or by the P-450 system.

Another important oxidizing and detoxifying enzyme is epoxide hydratase. Epoxide hydratase transforms an epoxide into a dihydrodiol, which is not toxic and which, being polar, is readily excreted. Therefore, the epoxide, once formed, can rearrange spontaneously to phenol form or can be attacked by epoxide hydratase to form a dihydrodiol. Epoxide hydratase is present in many of the same tissues as cytochrome P-450 monooxygenase and is also associated with the endoplasmic reticulum. It can act rather quickly on an epoxide once the epoxide is formed.

The dehydrogenases, particularly alcohol and aldehyde dehydrogenases, are familiar oxidizing enzymes. In this case, the aldehyde may or may not be less toxic than the parent alcohol. However, the action of aldehyde dehydrogenase will produce the corresponding acid, which usually is less toxic and also will be readily excreted.

Reduction. Reduction of azo and nitro compounds to the corresponding amines is catalyzed by the same cytochrome-P-450-containing monooxygenase system that catalyzes many of the oxidation reactions. Some of these reductions can be catalyzed directly by only a part of that system, NADPH cytochrome P-450 reductase. However, reduction of other compounds in this group requires the complete system.

Hydrolysis. Restricted to esters and amides, hydrolysis is catalized by a number of esterases and amidases. There are esterases in practically all tissues, including the blood. Most are fairly nonspecific with regard to their substrate requirements.

Esterases have been broadly classified as arylesterases (hydrolyzing aromatic esters); carboxyl esterases (hydrolyzing aliphatic esters); cholinesterases (hydrolyzing esters in which the alcohol moiety is choline); and acetylesterases (catalyzing compounds in which the acid moiety of the ester is acetic acid). There is, however, considerable overlap of substrate specificity from one class of esterase to another. Amidases also are rather nonspecific. Hydrolysis of amides tends to take place more slowly than hydrolysis of esters.

Phase II Reactions (Conjugations). Phase II reactions generally involve the synthesis of a molecule by conjugation of the parent or metabolite with a compound that renders the product significantly more polar than the substrate. There are several different kinds of conjugation reactions. Probably the most important, at least in humans, is glucuronidation. Glucuronides are synthesized by means of the enzyme glucuronyl transferase, with uridine-5′-diphospho-α-D-glucuronic acid as the source of glucuronic acid. The enzyme is found in the endoplasmic reticulum of many tissues, but most especially in the liver. It has a very broad specificity. Thus, the glucuronic acid can be conjugated to a number of compounds, including aliphatic and aromatic alcohols, mercaptans, certain acids, and primary and secondary aliphatic and aromatic amines. As a result, glucuronidation is a particularly important mechanism for rendering toxicants more polar and more readily excreted.

A second phase II mechanism is the formation of mercapturic acid derivatives. This conjugation is catalyzed by glutathione *S*-transferases in the presence of the molecule glutathione, a tripeptide. Glutathione is first attached to the substrate by a thioether bond. Then two of the amino acids are lost from the conjugate. Finally, an acetyl group is added to form the mercapturic acid derivative. This is a particularly important mechanism, not so much quantitatively as because the requirement of glutathione *S*-transferase for an electrophilic carbon on the substrate means that it will attack those compounds, such as epoxides, capable of reacting with and altering or damaging key cell components. Thus, glutathione *S*-transferase is capable of rendering some of the more highly toxic metabolites harmless.

Other enzymes present in animal tissues can catalyze a variety of other conjugations. Foreign compounds may be conjugated to one of the amino acids glutamine and glycine, or they may be sulfated, methylated, or acetylated. The enzymes that catalyze these transformations are variously located in the endoplasmic reticulum, the mitochondria, or the cytoplasm of the cell.

Methylation is an example of a phase II reaction that does not necessarily result in increased polarity. It is, however, quantitatively of very minor importance. Acetylation is another phase II reaction that may occasionally result in decreased polarity and increased toxicity.

Factors Affecting Biotransformation

A number of factors influence the contribution of biotransformation processes to the overall disposition of a compound. These factors can be classified in two categories: factors that affect the rate of the biotransformation process itself, and factors that influence the rate of delivery to the biotransformation site but may not actually affect the rate of the biotransformation process once the compound has arrived at that site.

Factors Affecting the Rate of the Biotransformation Process

Species and Strain. Variation in species and strain may influence the rate of biotransformation. For example, the amount of phenylacetic acid given to a series of primates and nonprimates and the amount excreted in a 24-hr period after conjugation either to glutamine or to glycine were compared in one study (Williams, 1974) (Table 3-1). The percentages conjugated to glutamine or to glycine varied widely among the species. In man, nearly all of it was found to be conjugated to glutamine. In nonprimates, none was conjugated to gluta-

Table 3-1. Conjugation of Phenylacetic Acid in Primates.

Family	Species	Percent of 24-hr excretion as conjugated with	
		Glutamine	Glycine
Anthropoids			
1. Hominidae	Man	90	0
2. Hylobatidae	—	Not tested	
3. Cercopithecidae	Rhesus monkey Cynomolgus monkey Green monkey Red-bellied monkey Mona monkey Mangabey Drill Baboon	30–90	0.1–1.0
4. Cebidae	Capuchin monkey	64	10
	Squirrel monkey	75	2
5. Callitrichidae	Marmoset	80	1
Prosimians			
Lemuridae	Bushbaby	0	80
	Slow loris	0	87
Nonprimates			
Ten species		0	80–100

Source: Adapted from Williams, R. T., "Interspecies Scaling." *In* Teorell *et al*. (Eds.), *Pharmacology and Pharmacokinetics*. New York: Plenum Publishing Corporation. 1974. Table IV.

mine: between 80 and 100 percent was conjugated to glycine. In nonhuman primates, the percentage conjugated to glutamine varied between 30 percent and 90 percent.

With respect to strain, marked differences in metabolism in inbred laboratory strains of rats and mice have been demonstrated. Such differences may be less important in other, less inbred species.

Another species difference is worth noting. Many studies of metal toxicity have been carried out in the cat, a furred animal. Since metals can be excreted via the hair, it is difficult to extrapolate from observations made of metal disposition in furred animals to humans, for whom that particular mechanism of excretion is not quantitatively as important.

Other species differences may not so much affect rates of biotransformation as delivery, but they are nonetheless worth mentioning in this context. For example, proportions of body fat to total body weight differ. For compounds sequestered in fat depots, the amount of body fat may be an important factor controlling access to biotransformation sites. Plasma proteins differ in kind and amount from species to species (see below). And, finally, different populations of gut flora can produce qualitatively and quantitatively different metabolites in different species.

Age and Sex. Marked sex-related effects on biotransformation rate are usually limited to hormone-related differences in the cytochrome P-450 enzymes. Biotransformation is well known to be generally less efficient in infants than in adults, but biotransformation capacity develops rapidly after birth. The activities of certain enzymes continue to fluctuate with age, especially during adolescence, but tend to stabilize during adulthood.

Nutritional Status. Nutritional status is important for biotransformation. Good nutrition supplies the metals required for biotransformation processes and prevents protein deficiencies that would reduce the ability of the body to synthesize key enzymes. It is known that deficiencies of copper, zinc, calcium, or protein can decrease the activity of metabolizing enzymes.

Enzyme Induction or Inhibition. Metabolizing enzymes may be induced or inhibited, either by the substrate itself or by simultaneous exposures to some other compound. Over three hundred exogenous compounds are known to induce enzyme activity. Phenobarbital and 3-methylcholanthrene are two well-known compounds used by toxicologists to induce certain of the cytochrome P-450 enzymes. Induction can be extremely rapid. In fact, one dose of many compounds will cause significant induction of metabolizing enzymes. This effect may persist for quite a long time after cessation of exposure. Enzyme activities have been observed to be greater in test subjects than in control subjects for almost a year after exposure to some compounds.

Some compounds can directly inhibit the activity of cytochrome-P-450-containing monooxygenases. This group of compounds includes SKF 525-A, as well as drugs that may be given therapeutically.

Other Factors. Circadian rhythms in metabolizing ability may be correlated with variations in endocrine function associated with the light/dark cycle in experimental animals.

The presence of disease in the metabolizing organ will reduce metabolic efficiency if damage is sufficiently severe. The liver, however, has considerable reserve capacity for its normal functions. It is possible for some liver damage to occur and for no evidence of impaired biotransformation to appear.

One factor not always taken into consideration is the individual's genetic makeup. Several instances are known of specific individual genetic differences in humans that have affected important metabolizing functions.

Factors Affecting Delivery

Organ Perfusion. The factors influencing the rate of delivery of a compound to the biotransformation process are somewhat fewer. One of them is the rate of organ perfusion. An increase in the rate of blood flow naturally presents more of a dissolved compound to the metabolizing site per unit of time. As long as the capacity of the organ to extract and metabolize the compound has not been exceeded, therefore, an increase in the rate of organ perfusion will lead to an increase in biotransformation rate. However, it is possible for the rate of perfusion to exceed the maximum capacity of the organ to extract and metabolize the compound. As blood flow continues to increase beyond this point, the relative efficiency of extraction decreases. Rates of organ perfusion may be altered by temperature, by activity level, and by certain drugs.

Competing Processes. Another factor influencing the rate of toxicant delivery to the biotransformation site is the efficiency of competing processes. Metabolism occurs simultaneously with excretion and with distribution to other organs. If those processes are unusually efficient, metabolism may be relatively less important.

Protein Binding. The third factor is protein binding. It is often stated that protein binding reduces the availability of the bound compound for metabolism. This is sometimes, but not always, the case. Protein binding is an equilibrium process. As soon as some of the unbound chemical has been removed from the plasma, the equilibrium will reestablish itself. Some of the chemical will be released from its protein-bound form and become unbound and available. If the metabolizing enzyme system is extremely efficient, it can gain access to both bound and unbound chemicals.

Therefore, the effect of protein binding depends on whether the metabolizing enzyme system is capable of extracting and metabolizing protein-bound as well as free toxicant in the course of a single passage of plasma through the metabolizing organ. If it is, then the effect of protein binding is analogous to that of the organ perfusion rate, discussed above. Protein binding can reduce the half-life of the compound by providing very efficient delivery to the biotransformation site. If, on the other hand, protein-bound toxicant is not available to the metabolizing enzyme system, then protein binding will increase the half-life by sequestering the toxicant and thus delaying its access to the biotransformation site.

Excretion

Excretion takes place simultaneously with biotransformation and, of course, with distribution. The kidney is probably the single most important excretory organ in terms of the number of compounds excreted, but the liver and lung are of greater importance for certain kinds of compounds. The lung is active in excretion of volatile compounds and gases. The liver, because it is a key biotransforming organ, may excrete metabolites before they have a chance to reach the systemic circulation.

Excretion in the Kidney. About 20 percent of all dissolved compounds of less than protein size are filtered by the kidney in the glomerular filtration process. Glomerular filtration is a passive process; it does not require energy input. Filtered compounds may be either excreted or reabsorbed. Passive reabsorption in the kidney, as elsewhere, is a diffusion process. It is governed by the usual principles. Thus, lipid-soluble compounds are subject to reabsorption after having been filtered by the kidney. The degree of reabsorption of electrolytes will be strongly influenced by the pH of the urine, which determines the amount of the compound present in a nonionized form.

It is to be expected that some control could be exerted over the rate of excretion of weak acids and bases by adjusting urine pH. This type of treatment can be used very effectively in some cases. For example, alkalinization of the urine by administration of bicarbonate has been used to treat salicylic acid poisoning in humans. Alkalinization causes the weak acid to become more fully ionized; the ionized molecule is excreted in the urine rather than reabsorbed.

There are also active secretory and reabsorptive processes in the tubules of the kidney. These processes are specialized to handle endogenous compounds; active reabsorption helps to conserve the essential nutrients, glucose and amino acids. These pathways can also be used by exogenous compounds, provided that the compounds have the structural and electronic configurations required by the carrier molecules.

Renal clearance can be defined as rate of excretion divided by concentration in the plasma. Its units are volume of plasma per unit time. Thus, it represents a hypothetical plasma volume "cleared" of solute during the specified period of time. A compound such as creatinine that is filtered but not secreted or reabsorbed is cleared in humans at a rate of about 125 ml/min. Compounds that are reabsorbed as well as filtered have clearances that are less than the creatinine clearance. Compounds that are actively secreted can have clearances as large as the renal plasma flow, about 600 ml/min.

It should also be noted that the presence of disease in the kidney can affect the half-life of a compound eliminated via the kidney, just as the presence of disease in the liver can affect the half-life of a compound that is biotransformed to a significant extent.

Excretion in the Liver. The liver is both the major metabolizing organ and a major excretory organ. When a toxicant that is absorbed from the gastrointestinal tract is eliminated in the liver by metabolism or excretion before it can reach the systemic circulation, it is said to exhibit a "first-pass effect." In addition, metabolites formed in the liver from compounds already present in the systemic circulation may be excreted into the bile before they themselves have had a chance to circulate. Although it does not excrete as many different compounds as the kidney does, the liver is in an advantageous position with regard to excretion, particularly of metabolites.

There are at least three active systems for transport from liver into bile: one for organic acids, one for organic bases, and one for neutral organic compounds. There probably is also a transport system for metals. These transport processes are efficient and can extract protein-bound as well as free compounds. It is not known precisely what characteristics cause a compound to be excreted either into the bile or into the urine. In general, it seems that compounds with molecular weights greater than about 300 or 325 are more frequently found in the bile, while compounds with smaller molecular weights are more frequently found in the urine.

Once a compound has been excreted by the liver into the bile, and thereby into the intestinal tract, it can either be excreted in the feces or reabsorbed. Most frequently the excreted compound itself, being water soluble, is not likely to be reabsorbed directly. However, glucuronidase enzymes of the intestinal microflora are capable of hydrolyzing glucuronides, thereby releasing less polar compounds that may then be reabsorbed. The process is termed enterohepatic circulation. It can result in greatly extended retention of the compounds recycled in this manner. Techniques have been developed to interrupt the enterohepatic cycle by introducing an adsorbent that will bind the compound and carry it through the gastrointestinal tract.

Certain factors influence the efficiency of liver excretion. Liver disease can reduce the excretory as well as the metabolic capacity of the liver. On the other hand, a number of drugs increase the rate of hepatic excretion by increasing bile flow rate. For example, phenobarbital produces an increase in bile flow that is not related to its ability to induce metabolizing enzymes. Whether the increased rate of bile flow will increase the rate of elimination of a compound that is both metabolized and excreted by the liver depends on whether the rate-limiting step is the enzyme-catalyzed biotransformation or the transfer from liver to bile. Therefore, if the transfer from liver to bile is the rate-limiting step, then enhancement of the rate of bile flow will enhance the rate of excretion.

Excretion in the Lung. The third major organ of elimination is the lung, the key organ for the excretion of volatile liquids and gases. Pulmonary excretion, like pulmonary absorption, is by passive diffusion. For example, the rate of transfer of chloroform out of pulmonary blood is directly proportional to its concentration in the blood. Essentially, pulmonary excretion is the reverse of the uptake process, in that compounds with low solubility in the blood are perfusion-limited in their rate of excretion, whereas those with high solubility are ventilation-limited. Highly lipophilic compounds that have accumulated in lipid depots may be present in expired air for a very long time after exposure.

Other Routes of Excretion. Minor routes of excretion are skin, hair, sweat, nails, and milk. It has already been mentioned that hair can be a significant route of excretion for furred animals, and indeed the amount of metal in hair, like the amount of a volatile compound in exhaled air, is frequently used as an index of exposure in laboratory animals. Hair is not quantitatively an important route of excretion in humans, although it can be used to estimate magnitude of exposure. Sweat and nails are only rarely of interest as routes of excretion, simply because loss by these routes is quantitatively so slight.

Milk may be a major route of excretion for some compounds. Milk has a relatively high fat content, three to five percent or even higher, and therefore compounds that are lipophilic may be excreted in milk to a significant extent. Some of the toxicants shown to be present in milk are the highly lipid-soluble chlorinated hydrocarbons: for example, the polychlorinated biphenyls (PCB's) and DDT. Certain heavy metals may also be excreted in milk. Lead is thought to be excreted into milk by the calcium transport process.

3.5 Mechanisms of Action

Simultaneous operation of absorption, distribution, biotransformation, and excretion processes determines whether an exogenous compound will reach a target tissue, as shown in Figure 3-1. The target tissue is that tissue in which the compound exerts its effect. The mechanism by which a toxicant produces an effect is symbolized by the rightmost arrow in Figure 3-1. What is the relationship between concentration in the target tissue and the effect of the compound?

The most widely accepted theory of mechanism of action is the receptor theory, in which it is assumed that the compound must combine with a receptor site in order to act. The receptor site is not an imaginary construct. It may be a binding site on a transport protein, for example, or part of a membrane, or the active site of an enzyme. Whatever the nature of the receptor site, its occupancy by an active molecule initiates an effect. The effect is presumed to be reversible. As soon as the toxicant is dissociated from the receptor site, the effect ceases. Therefore, receptor theory does not describe the mechanism of action of compounds that act irreversibly, such as carcinogens.

Most pharmacologically active exogenous compounds, and many that are toxicologically active, act by binding to important sites on enzymes or on regulatory macromolecules. The site to which the exogenous compound binds is one already specialized to accept an endogenous compound, either a metabolic intermediate or a regulatory molecule such as a neurotransmitter or other hormone. Thus, an exogenous compound must meet certain size, configuration, and electronic charge requirements before it can be bound to a receptor site. Highly selective receptor sites can be activated only by exogenous compounds that resemble the natural activator, or agonist, very closely.

Occupancy of a receptor site by a toxicant can result either in inappropriate stimulation of an effect or in inhibition of the action of the natural agonist. In either case, stimulation or inhibition is limited by availability of receptor sites, as shown in Figure 3-10; there, concentration of a toxicant at the receptor sites, X, is plotted against a fraction, expected effect over maximum effect, θ/θ_m. The curve in Figure 3-10 is a hyperbola. This relationship is more frequently expressed, however, in the form shown in Figure 3-11, in which concentration is transformed by conversion to a logarithmic scale. In Figure 3-11, the dose-effect curve is sigmoid; notice that an approximately linear midsection can be fitted along the curve, from 20–80 percent of maximum effect.

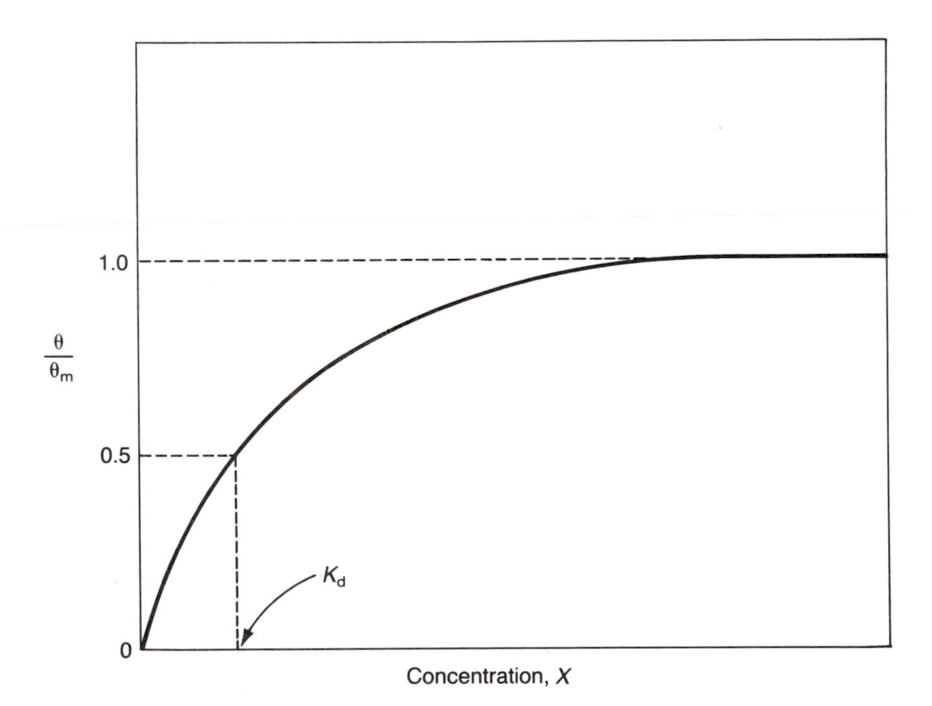

Figure 3-10. The dependence of fraction of maximum effect on effector concentration. K_d is the dissociation constant for binding of the toxicant to the critical receptor site. (Reproduced with permission from O'Flaherty, 1981. Figure 6-4(a).)

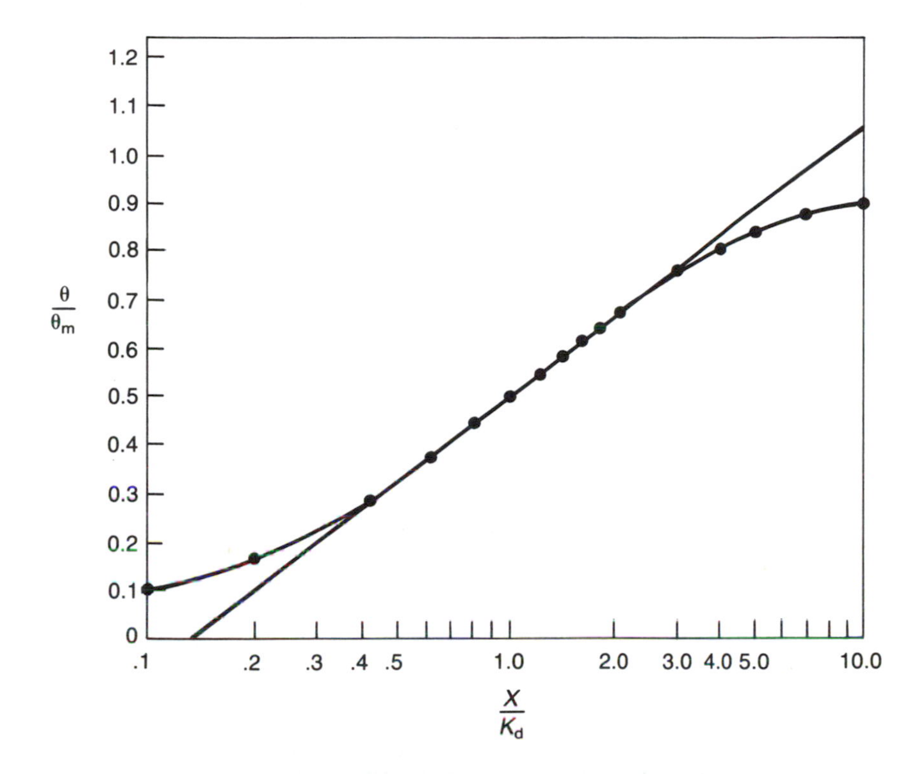

Figure 3-11. The theoretical log dose–fraction of maximum effect relationship. Both effect and concentration are expressed in dimensionless form, as fraction of maximum effect, θ/θ_m, and as fraction of the dissociation constant, X/X_d, respectively. The slope at the midpoint is also shown. (Reproduced with permission from O'Flaherty, 1981. Figure 6-10.)

As predicted by receptor theory, the relationship between the logarithm of concentration (or dose) is frequently observed to be linear over the effect range of interest. For example, Figure 3-12 (Carruthers *et al.*, 1974) shows that the mean reduction of exercise heart rate in patients being treated with the β-adrenergic-blocking drug practolol is directly related to the logarithm of mean blood practolol concentration.

Summary

This chapter has conveyed some of the general biochemical and physiological principles that govern absorption, distribution, and elimination of toxic agents,

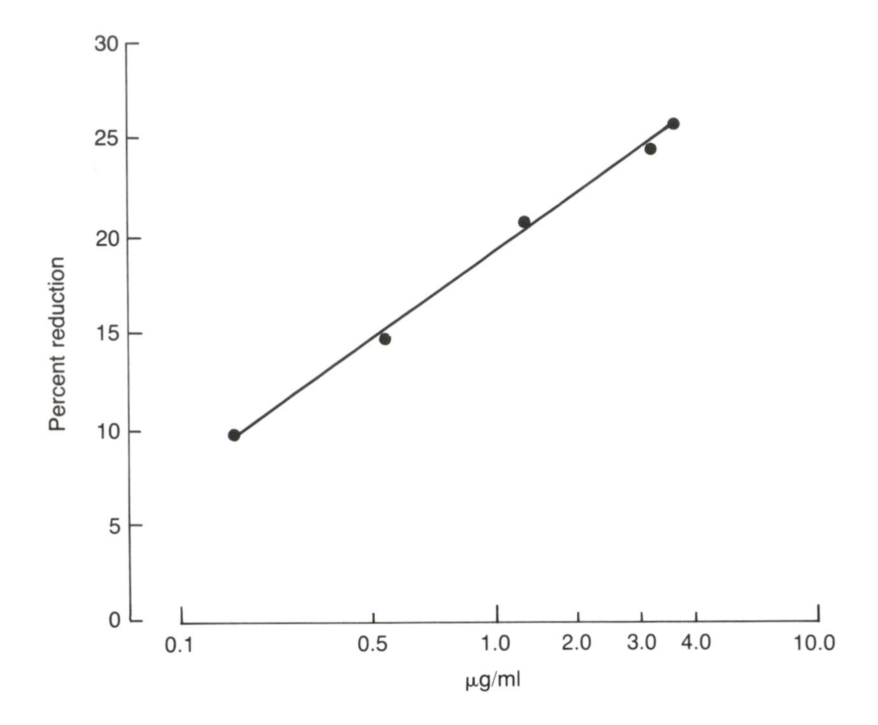

Figure 3-12. Correlation of the mean reduction of exercise heart rate with the logarithm of the mean blood practolol level. (Reproduced with permission from Carruthers *et al*., "Blood Levels of Practolol after Oral and Parenteral Administration and Their Relationship to Exercise Heart Rate." *Clin. Pharmacol. Ther., 15* (1974): 497–509. Figure 6.)

in particular:

- The importance of lipid solubility, molecular size, and degree of ionization to the rate at which a molecule moves through a membrane by passive transfer or diffusion.
- The characteristics of other transfer processes such as facilitated diffusion, active transport, phagocytosis, and pinocytosis.
- Absorption from the gastrointestinal tract with particular emphasis on the importance of pH as a determinant of absorption of ionizable organic acids and bases as well as on compound-specific and host-related factors such as lipid solubility and molecular size, the presence of villi and microvilli in the intestine, the possibility that the compound can be absorbed by facilitated or active transport mechanisms, and the action of gastrointestinal enzymes or intestinal microflora.
- The structure of the skin, and factors controlling the rate of diffusion across the skin.

- Absorption of solid and liquid particulates and of gases and vapors in the lung.
- Simple kinetic models describing disposition (distribution, metabolism, and excretion).
- Biotransformation, including phase I (oxidation, reduction, and hydrolysis) and phase II (conjugation) reactions.
- Factors affecting rate of biotransformation: species, strain, age, sex, nutritional status, induction or inhibition of enzymes, circadian rhythms, the presence of disease in the metabolizing organ, and genetic makeup.
- Factors affecting rate of delivery to the site of biotransformation: rate of organ perfusion, protein binding, and the relative efficiency of competing processes such as distribution (sequestration) and excretion.
- Excretion from kidney, liver (including enterohepatic circulation), lung, and by less universal routes such as skin, hair, sweat, nails, or milk.
- Receptor theory of action.

References and Suggested Reading

Aungst, B. J., Dolce, J. A., and Fung, H. L. 1981. "The Effect of Dose on the Disposition of Lead in Rats after Intravenous and Oral Administration." *Toxicol. Appl. Pharmacol., 61*: 48–57.

Carruthers, S. G., Kelly, J. G., McDevitt, D. G., and Shanks, R. G. 1974. "Blood Levels of Practolol after Oral and Parenteral Administration and Their Relationship to Exercise Heart Rate." *Clin. Pharmacol. Ther., 15*:497–509.

Chamberlain, A. C., Heard, M. J., Little, P., Newton, D., Wells, A. C., and Wiffen, R. D. 1978. *Investigations into Lead from Motor Vehicles*. Harwell (England): AERE. Publication No. R9198.

Guidotti, G. 1976. "The Structure of Membrane Transport Systems." *Trends in Biochem. Sci., 1*:11–13.

Hamilton, D. L., and Smith, M. W. 1978. "Inhibition of Intestinal Calcium Uptake by Cadmium and the Effect of a Low Calcium Diet on Cadmium Retention." *Environ. Res., 15*:175–184.

Hogben, C. A. M., Tocco, D. J., Brodie, B. B., and Shanker, L. S. 1959. "On the Mechanism of Intestinal Absorption of Drugs." *J. Pharmacol. Exp. Therap., 125*:275–282.

O'Flaherty, E. J. 1981. *Toxicants and Drugs: Kinetics and Dynamics*. New York: John Wiley & Sons, Inc.

Schanker, L. S., and Jeffrey, J. J. 1961. "Active Transport of Foreign Pyrimidines across the Intestinal Epithelium." *Nature, 190*:727–728.

Toraason, M. A., Barbe, J. S., and Knecht, E. A. 1981. "Maternal Lead Exposure Inhibits Intestinal Calcium Absorption in Rat Pups." *Toxicol. Appl. Pharmacol., 60*:62–65.

Wagner, J. G. 1973. "Properties of the Michaelis-Menten Equation and Its Integrated Form Which Are Useful in Pharmacokinetics." *J. Pharmacokinet. Biopharmaceut., 1*:103–121.

Williams, R. T. 1974. "Interspecies Scaling." *In* Teorell, T., Dedrick, R. L., and Condliffe, P. G. (Eds.) *Pharmacology and Pharmacokinetics*. New York: Plenum Press. p. 108.

Chapter 4

Hematotoxicity: Toxic Effects in the Blood

Robert C. James

Introduction

This chapter discusses:

- The origin, formation, and differentiation of blood cells.
- Oxygen transport and interferences.
- Chemically induced blood disorders.
- Specific chemicals that adversely affect the blood.
- Clinical tests used to evaluate hematotoxicity.

4.1 Basic Hematopoiesis: The Formation of Blood Cells, and Their Differentiation

Origin

All blood cells originate from undifferentiated mesenchymal cells. From these stem cells, originating in the bone marrow, clones of cells differentiate and ultimately appear as mature (committed) cells with a specific purpose. These differentiated cells form specific elements of whole blood, such as red blood cells (erythrocytes), which provide oxygen transport; platelets (thrombocytes), which provide clots against hemorrhage; or white blood cells (leukocytes), which protect against foreign material and scavenge debris. Stimulation of the stem cell pool is carried out by the "poietins," each blood cell type having its own poietin or stimulating factor. Anemia is a decreased number of red blood cells; thrombocytopenia refers to a decreased number of platelets; leukopenia is a decreased number of white blood cells.

Red Blood Cells (RBC) — Erythrocytes

Erythropoiesis is the term defined as that process by which red cells are formed. Renal erythropoietic factor (REF) is released by the kidneys and converts proerythropoietin, a blood protein released by the liver, to erythropoietin. Release of REF is increased by the conditions of hypoxia and anemia, and by the presence of cobalt, while it is decreased by polycythemia (high number of red blood cells) or hyperoxia. Erythropoietin, in turn, converts stem cells to proerythroblasts, which ultimately mature into red blood cells.

In the fetus, red blood cells are made by the liver, spleen, and lymph nodes. However, during the last part of gestation, the bone marrow becomes the sole red-blood-cell-producing organ and usually remains so for the rest of a person's life. Demands for increased oxygen transport, such as are occasioned during chronic hypoxia, can lead to a reactivation of red-blood-cell formation in the liver and spleen, but this reserve capacity is not enough to sustain life by itself. Therefore, the bone marrow is the organ of concern when dealing with a chemical-induced change in the number of circulating blood cells.

Anemia, as previously stated, is a decreased number of red blood cells. Such a decrease, if it becomes severe enough, is of grave concern because the resultant decrease in the oxygen content of arterial blood can lead to anemic hypoxia (oxygen deprivation) in important tissues, such as the brain. There are four general types of anemia to consider, each with a specific and different cause for the decrease in red blood cells.

- Microcytic hypochromic anemia: An anemia that results when red blood cells have a low hemoglobin content. This occurs when red cells are produced rapidly to help the body recover from the rapid loss of blood, as through hemorrhage. Frequently, a person does not absorb iron from the intestines fast enough to produce adequate hemoglobin in the red cells. Iron deficiency in the diet will also produce this anemia.
- Hemolytic anemias: A loss of red blood cells owing to their abnormal fragility. Examples are hereditary spherocytosis; sickle cell anemia (hemoglobin crystalizes at low O_2 concentrations); thalassemia (inability to properly synthesize hemoglobin); and erythroblastosis fetalis (Rh-positive baby, Rh-negative mother). Certain diseases (malaria), chemicals, and autoimmune disorders also lead to hemolytic anemia.
- Megaloblastic microcytic anemia: A defect in DNA synthesis possibly secondary to a deficiency in folic acid or vitamin B_{12}.
- Aplastic anemia: A decrease in red blood cells caused when there is loss or dysfunction of the bone marrow.

Platelets — Thrombocytes

Thrombocytes, the smallest blood cells and the first line of defense against blood loss, are derived from the largest stem cell in the bone marrow—the megakaryocyte. Platelets accumulate almost instantaneously at the site of any rupture within the blood vessels. Within seconds the platelets, which are normally nonsticky, begin adhering to the subendothelial surfaces exposed by trauma and to the exposed collagen fibers. The platelets react to this adhesion by undergoing degranulation and releasing ADP (adenosine 5′-diphosphate), which in turn causes the adhesion and aggregation of new platelets. Simultaneously, other platelet factors increase thrombin formation, which transforms fibrinogen to fibrin and provides the webbing of the platelet clot.

Thus, platelets serve two purposes. First, they act as physical plugs to help seal a ruptured vessel, and second, they participate in fibrin formation, which enables the size of the clot to grow and eventually encompass the compromised area.

Thrombocytopenia is a state of decreased blood levels of platelets. Chemicals (drugs) are the most common cause of thrombocytopenia, especially anticancer drugs. Certain diseases or chemicals may also hamper clot formation by preventing fibrin formation (e.g., warfarin), or, like the common medicine, aspirin, by inhibiting platelet aggregation.

White Blood Cells (WBC) — Leukocytes

Leukocytes have the most complex function of the various types of blood cells but, in general, they defend the body against foreign bodies. They fall into two basic defense mechanism categories: phagocytic cells and immunocytic cells.

Phagocytic cells—that is, the granulocytes (neutrophils, eosinophils, and basophils) and the monocytes—physically engulf and destroy foreign particles. Granulocytes spend less than a day in circulation before they attach to vascular cell walls, the vascular epithelial cells, and/or pass through the blood vessels to become deposited in various tissues, to form a somewhat protective barrier. Specific chemicals released from inflammatory lesions may cause a migration of granulocytes to the injured area. Monocytes exist in the blood from three to four days and then migrate to reticuloendothelial tissues, such as the liver, spleen, and bone marrow. There they stay for several months, protecting against foreign bodies and destroying senile red blood cells.

Granulocytosis, or a high blood titer of granulocytes, is the term indicating a count greater than 10,000/mm^3. This condition is usually triggered by infection. Granulocytopenia, or a low granulocyte count, is the term used when the count is less than 1,000/mm^3 and recurrent infection is likely. This may be caused by chemically induced bone marrow damage seen with such drugs as the phenothiazines, antithyroid drugs, etc. Leukemia, a cancerous proliferation of white blood cells, is indicated when the count is greater than 30,000/mm^3. Chemical inducers of leukemia include benzene, chloramphenicol, and phenylbutazone. Neither monocytosis nor monocytopenia appears to be related to a specific chemical injury. However, they are seen as a generalized phenomenon of bone marrow damage or change.

Immunocytes are produced in the thymus (T cells) and bone marrow (B cells). Immunocytic defense mechanisms are of two types. T cells inhibit the migration of macrophages from an area of infection or a foreign body (antigen), secrete cytotoxic factors, recruit other T cells into the affected area, and activate B cells against the antigen. B cells react to T cells and antigens by secreting antibodies. Antibodies act by direct attack on the invader or via activation of the complement system and cause a lysis or destruction of the foreign cell membranes. Complement may digest the cell membrane of the antigen; attach to the cell surface and increase phagocytosis by body macrophages; agglutinize the antigens by forming chains or clots of complement (antibody) and antigen; or increase inflammatory responses.

4.2 Oxygen Transport

Hemoglobin

Hemoglobin (Hb), the oxygen-carrying protein of the red blood cells, consists of four separate peptide chains (two alphas and two betas). The protein chains (globin) have irregularly folded conformations that enclose the heme group in a hydrophobic pocket which forms the oxygen binding site. The active site of the heme group is an iron molecule bound by four coordination bonds to the nitrogen molecules of a porphyrin ring. Of the two remaining coordination bonds, one is associated with an imidazole residue from the globin chain and the remaining bond is available for reversible binding with oxygen.

Oxygenation of the iron molecule of the heme causes changes in the tertiary structure of the hemoglobin molecule, which in turn affects the oxygen-binding affinity. This change in the oxygen-binding affinity of hemoglobin is known as cooperativity. Because hemoglobin consists of four globin molecules, the net result of cooperativity is four different dissociation constants as four different oxygen molecules leave the heme group. This results in the characteristic sigmoid shape of the hemoglobin-O_2 dissociation curve (see Figure 4-1). The release of the first oxygen molecule triggers a cooperative change that greatly facilitates the release of the second, which when released triggers a change facilitating the release of the third oxygen molecule. The total oxygen content of normal blood is about 20 ml O_2/100 ml blood; the release of about one fourth (one O_2 molecule per Hb) of the blood's oxygen capacity is equivalent to 5 ml O_2/100 ml blood). Such a change requires about a 60-mm-Hg-drop in the blood's partial pressure for oxygen (P_{O_2}); the release of the second O_2 molecule (next 5 ml O_2/100 ml blood) requires only a 15-mm-Hg-drop in pressure; the third only a 10-mm-Hg-drop in pressure. Through cooperativity, when tissues are in greatest need of O_2 (in circumstances of low oxygen pressure and low tissue-oxygen content), hemoglobin loses its affinity for O_2; thus O_2 is released for use by the tissues more easily in those areas of greatest need.

Physiologic regulators are known that can shift the position of the hemoglobin-O_2 curve, to increase (a shift to the left) or decrease (a shift to the right) the P_{O_2} necessary to half-saturate the hemoglobin. Both the hydrogen ion content (pH) and 2,3-diphosphoglycerate shift the curve to the right and thus decrease the affinity of hemoglobin for O_2; and a rightward shift in the curve would appear theoretically to benefit ischemic diseases, because oxygen would

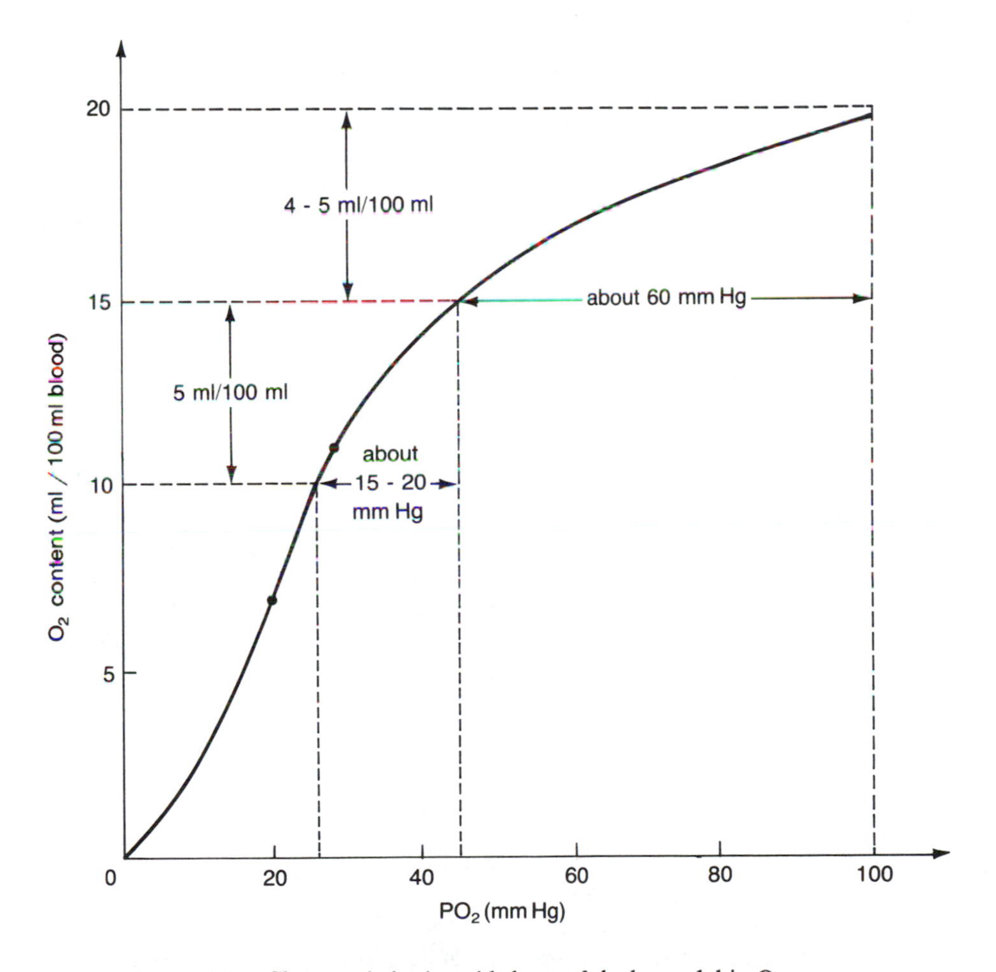

Figure 4-1. Characteristic sigmoid shape of the hemoglobin-O_2 dissociation curve. To liberate the first 4–5 ml of oxygen, the partial pressure of oxygen within the blood must drop about 60 mm Hg. The second 5 ml of oxygen per 100 ml of blood is liberated with a drop in pressure of only 15 to 20 mm Hg.

be released more easily, but no clinical benefits have yet been shown. A leftward shift, as caused by carbon monoxide, is generally regarded as undesirable because less O_2 would be released by the blood as the oxygen content of tissue decreased.

4.3 Hematotoxins That Induce Hypoxia

Hypoxia and Chemical Asphyxiants

Hypoxia is synonymous with anoxia and refers to any condition in which there is an inadequate supply of oxygen to the tissues. The first types of hypoxia we will be concerned with are those characterized by an insufficient oxygen supply, even though the rate of blood flow is normal. This type of hypoxia, asphyxial hypoxia or anoxic hypoxia (a somewhat redundant terminology), can be caused by inadequate respiration (i.e., respiratory paralysis, suffocation, asphyxiation) or by chemical interference with the normal oxygen transport by hemoglobin.

Carbon Monoxide

Carboxyhemoglobin. Carbon monoxide is perhaps the best-known example of a chemical agent that decreases the oxygen transport of the blood and produces an anoxic or asphyxial hypoxia. The mechanism whereby carbon monoxide elicits this toxic effect results from the fact that carbon monoxide is a stronger ligand for hemoglobin than is oxygen, and therefore has a stronger binding affinity. This means that carbon monoxide molecules compete more successfully than oxygen molecules for the hemoglobin binding sites that normally carry oxygen. The chemical equation describing the competition for binding is known as the Haldane equation, which is

$$\frac{[\mathrm{HbCO}]}{[\mathrm{HbO_2}]} = \frac{M[P_{\mathrm{CO}}]}{[P_{\mathrm{O_2}}]},$$

where M denotes the carbon monoxide binding-affinity.

In humans, M is reported to be anywhere from 210–245. That is, carbon monoxide binds the hemoglobin more than 200 times more tightly than oxygen. Thus, at equal concentrations of the two gases, the blood would contain some 200 times more carboxyhemoglobin (HbCO) than oxyhemoglobin (HbO_2).

Another way of examining the competition for the hemoglobin binding site is to calculate the concentration of carbon monoxide necessary to bind to and inactivate 50 percent of the body's total hemoglobin:

$$[P_{CO}] = \frac{[P_{O_2}]}{M}.$$

Or, since normal air is 21 percent oxygen, then the content of carboxyhemoglobin is equivalent to oxyhemoglobin content when

$$[P_{CO}] = \frac{1}{210}[P_{O_2}] = \frac{21 \text{ percent oxygen}}{210} = 0.1 \text{ percent}.$$

Thus, in normal ambient air, 50 percent of a person's hemoglobin is inactivated by carbon monoxide when the air concentration of carbon monoxide approaches 0.1 percent. Carbon monoxide also affects the hemoglobin-oxygen dissociation curve and shifts it to the left. Thus, there is not only less hemoglobin-carrying oxygen when carbon monoxide is present, but the oxygen is bound tighter and less is released in conjunction with a drop in the P_{O_2}. It is similar to being limited to the top portion of the hemoglobin-oxygen dissociation curve, as the benefit of cooperativity is lost. That is, the decreased affinity for O_2 (which causes hemoglobin to more readily release the next O_2 molecule) is lost because the tight binding by carbon monoxide decreases cooperativity and diminishes the amount of O_2 released by any drop in P_{O_2}. Thus, the binding of hemoglobin by carbon monoxide decreases the amount of O_2 carried by red blood cells and reduces their release of oxygen to oxygen-deprived tissues (Figure 4-2).

The carbon monoxide binding-affinity differs among species. One of these differences is the source of a misunderstanding inherent in a safety procedure of former times. Before many of the modern hygienic tests were known, miners took caged canaries into the shafts with them as an early warning device of oxygen deprivation or carbon monoxide accumulation. Small birds were useful indicators because their metabolic rates and, hence, respiration rates are much faster than those of humans. Consequently they achieve any change in equilibrium much more quickly than people do. Generally speaking, any conditions causing a fatal hypoxia in both humans and birds would become critical to a bird much sooner than for a man, and thus changes in the bird's condition served as an early warning to leave the tunnel.

However, the carbon monoxide binding affinity of hemoglobin for the canary is only 110; the canary therefore contains less carboxyhemoglobin at steady state than a human contains for any given carbon monoxide concentration. Thus, at high carbon monoxide concentrations—that is, those toxic to the bird, which are well above those toxic for humans—the bird dies first and warns the miner. However, at lower carbon monoxide concentrations, a steady state is attained in the bird first, but the amount of hemoglobin deoxygenated in the bird is not as high as that deoxygenated in humans, and the miner expires first (warning the *bird* to leave, as it were). The decisive dividing line is about 0.15 percent carbon monoxide. At concentrations of 0.05 percent–0.15

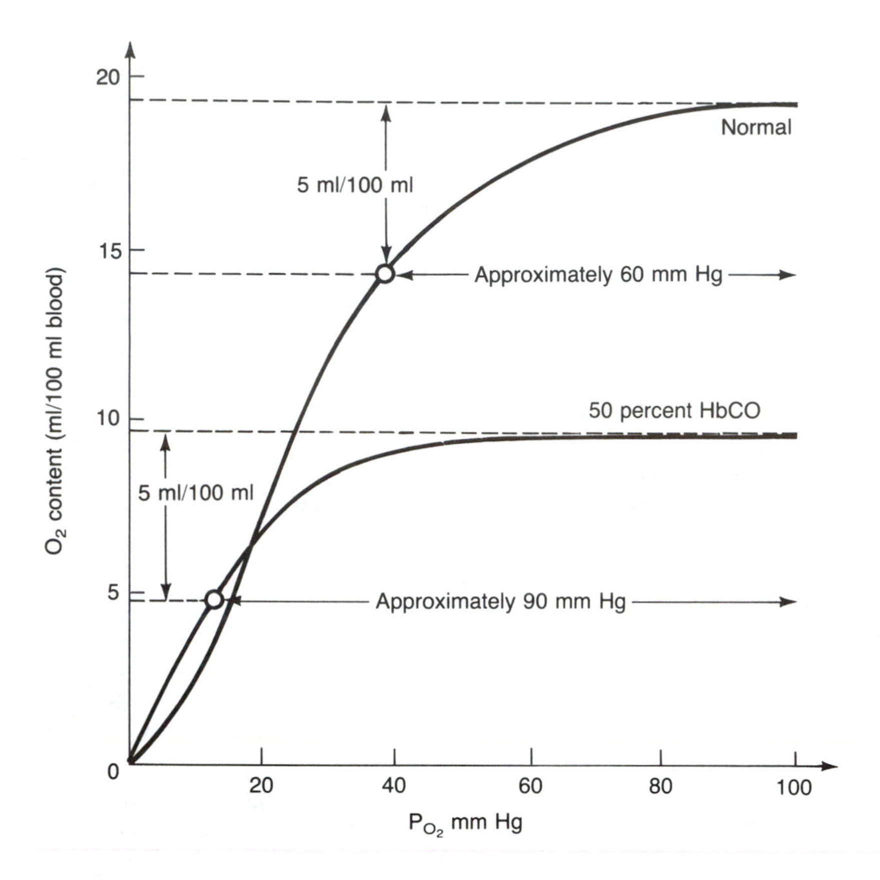

Figure 4-2. The decreased oxygen-binding caused by carbon monoxide decreases the blood's capacity, and requires a larger drop in pressure before oxygen is released.

percent, the miner dies first and at 0.20 percent or greater the bird dies first (see Figure 4-3).

The lethal concentration of carbon monoxide is small and does not significantly decrease the amount of oxygen. Thus, chemoreceptors that normally detect drops in the oxygen concentration are not triggered and respiration is not increased to match the decrease in the blood's oxygen-carrying capacity. Peripheral vasodilation of blood vessels, which occurs in response to the slowly developing hypoxia, may reduce blood pressure until cardiac output does not suffice. As the blood pools in peripheral tissues and less is returned to the lungs and heart, fainting becomes a possibility. Other symptoms include headache, weakness, nausea, and dizziness. Carboxyhemoglobin is a cherry red color, like fully oxygenated blood, and the venous circulation may impart a slight flush to the skin in carbon monoxide poisoned persons.

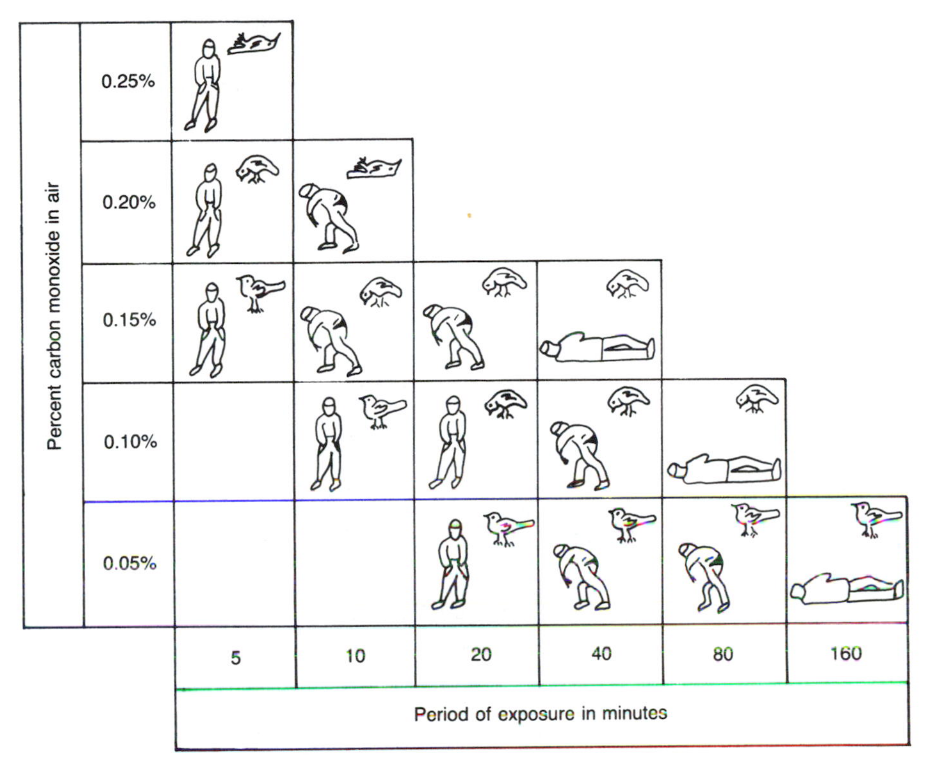

Figure 4-3. Schematic representation of how the carbon monoxide concentration in the mine air would affect the behavior of the canary or miner. Graphs are approximations of time to toxicity and lethality. (Based on Spencer, T. D. "Effects of Carbon Monoxide on Man and Canaries." *Ann. Occup. Hyg.* 5: 1961. Figure 3.)

Nitrites, Nitrates, and Aromatic Amine Compounds

Methemoglobin. The iron in the heme of hemoglobin is normally in the ferrous state (Fe^{++}), but it can be oxidized by certain chemicals to the ferric state (Fe^{+++}). The resulting pigment—methemoglobin—is greenish brown to black in color and does not bind oxygen or carbon monoxide. The chemicals capable of causing methemoglobin (MetHb) formation include nitrites, aromatic amines, nitro compounds, and chlorate salts.

A very poor correlation exists between methemoglobin levels and signs of hypoxia, because a chemical that causes methemoglobin may have additional side effects. For example, inorganic nitrites and organic aliphatic nitrite or nitrate chemicals not only cause methemoglobin formation but also relax blood vessels (cause vasodilation), which results in blood pooling or stagnant hypoxia, and this decrease in blood flow causes the hypoxia already present to be all the more harmful.

The nitrate anion is not capable of oxidizing hemoglobin itself; instead, it causes methemoglobin, because bacteria in the gut reduce it to nitrite, which is its toxic form. Normal levels of methemoglobin in the blood are around 2 percent. Clinical features of methemoglobin, like cyanosis, become evident when levels reach 15 percent or greater.

Diaphorase (methemoglobin reductase) is the major enzyme responsible for reducing methemoglobin (Fe^{+++}) back to hemoglobin (Fe^{++}). This enzyme requires reduced nicotinamide adenine dinucleotide (NADH) as a cofactor that supplies the electrons or reducing equivalents. A dormant system is also

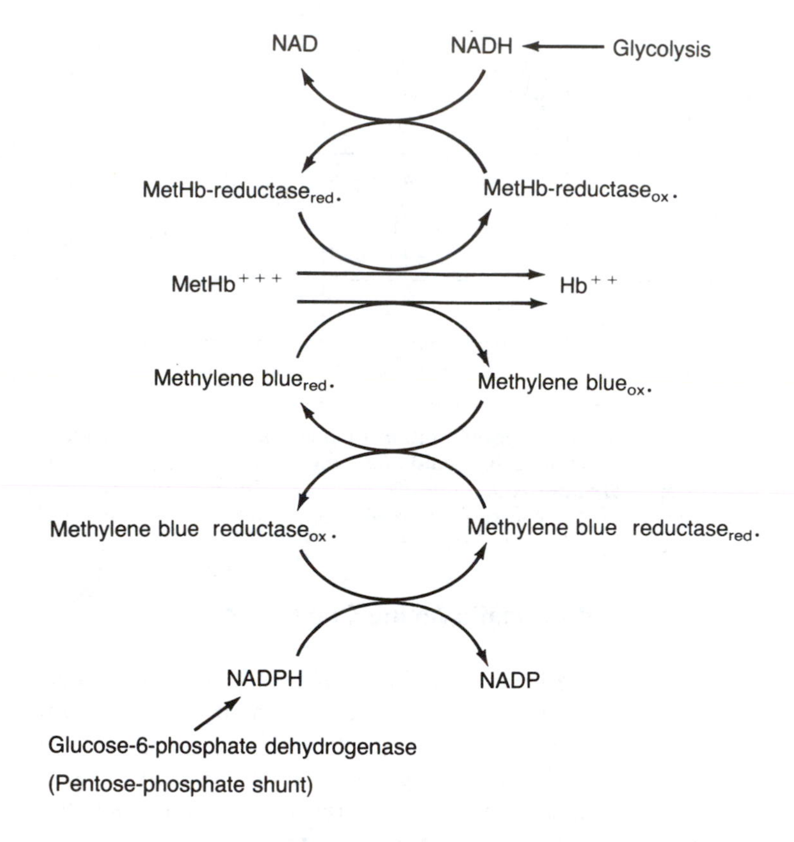

Figure 4-4. The spontaneous (NADH) and the dormant (NADPH) methemoglobin reductase systems. Methemoglobin (MetHb) reductase is active in intact red cells in the presence of substrates that can provide for NAD reduction. The NADPH system requires intact red cells, glucose or its metabolic equivalent, a functioning pentose-phosphate shunt, and methylene blue. Methylene blue reductase reduces methylene blue, which in turn nonenzymatically reduces MetHb. (Based on Doull *et al.* (Eds.) *Casarett and Doull's Toxicology: The Basic Science of Poisons*. 2nd Edition. New York: Macmillan Publishing Company. 1980.)

present within the body, but requires exogenously added electron carriers such as methylene blue to transfer electrons to Fe^{+++} (see Figure 4-4). Ascorbic acid (slow reaction) also appears to be capable of reducing MetHb and is useful to those persons genetically deficient in diaphorase.

Red blood cells deficient in glucose-6-phosphate dehydrogenase have a decreased pentose-phosphate shunt activity and, therefore, have difficulty in maintaining stores of NADPH. In such cells, methylene blue fails to accelerate methemoglobin reduction. Similarly, glutathione reductase cannot maintain red cell stores of glutathione without NADPH. Thus, these red blood cells are more susceptible to endogenous oxidizers, such as peroxide or superoxide, as well as being less capable of combatting methemoglobin-forming-toxicants.

Sulfhemoglobin and Heinz Bodies

Sulfhemoglobin is probably a mixture of oxidized and partially denatured hemoglobins; although they persist for the life of the red blood cells, they have never been generated in life-threatening concentration *in vivo*. Sulfhemoglobin is often part of a more serious triad produced by oxidative stress on the red blood cell that also includes Heinz bodies and hemolysis. Heinz bodies, after staining, appear to consist of clumps of denatured hemoglobin covalently bound to the interior of the red cell membrane. Some chemicals that produce the Heinz-body-hemolytic anemia in normal red blood cells are chlorates, naphthalene, arsine, phenylhydrazine, and a methylene blue overdose, but all of these in lower doses, and many other agents as well (fava beans, primaquine, aspirin), produce this effect in cells deficient in glucose-6-phosphate dehydrogenase.

4.4 Cytotoxic Hypoxia

Cytotoxic hypoxia results from an interference of the utilization of oxygen during cell metabolism in the presence of an adequate oxygen supply and blood flow. In this situation, tissue oxygen concentration may be normal or higher than normal because the chemical creates an inability to utilize O_2 during continued transport.

Cyanide (CN) Poisoning

Cyanide inhibition of cytochrome oxidase halts electron transport, oxidative phosphorylation, and aerobic glucose metabolism, resulting in a lactic acidemia and high concentrations of HbO_2 in the venous return (thereby bringing a flush to skin that could be confused with HbCO). About 20 to 40 percent of the population is congenitally unable to detect the bitter almond odor of CN. The lethal dose by mouth in an adult is about 100 mg of sodium or potassium cyanide and death because of central respiratory arrest is rarely delayed more than an hour.

Complex formation between the cyanide ion and methemoglobin constitutes the basis for an effective method of treatment and also for the most widely used method for determining MetHb levels in blood. Treatment involves the administration of amyl nitrite by inhalation and the intravenous injection of $NaNO_2$ to generate an effective but safe level of methemoglobin,

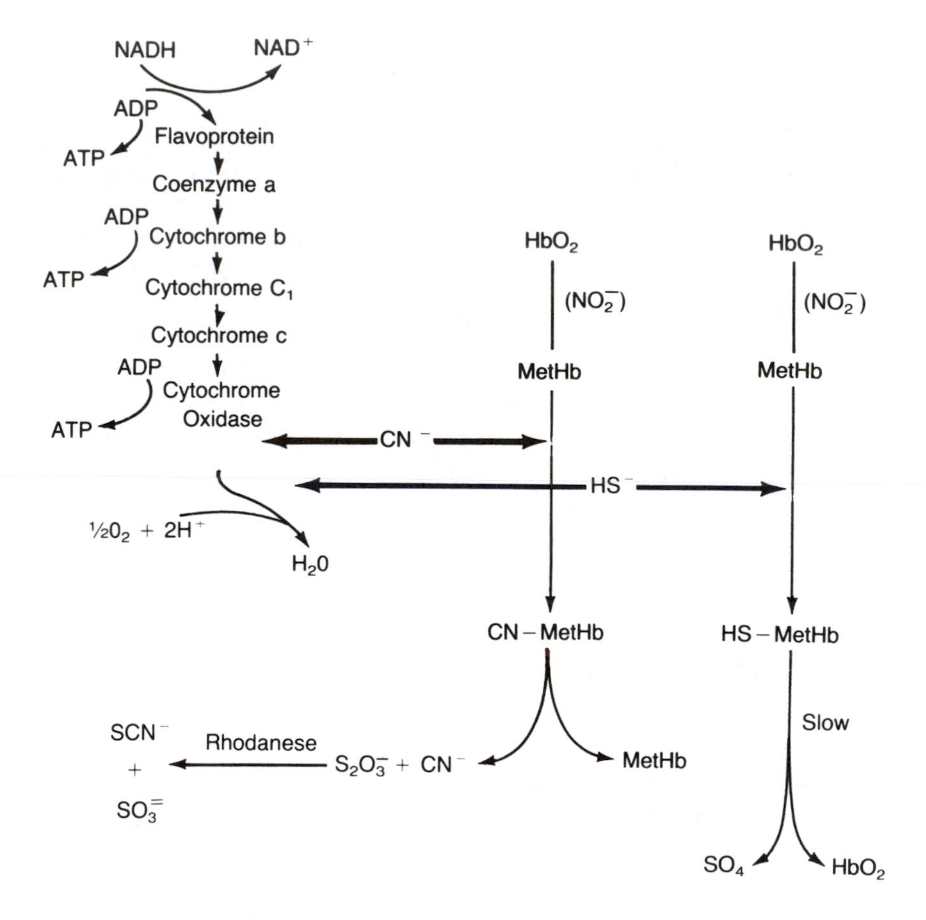

Figure 4-5. Schematic depiction of the electron transport chain through which the oxidation of NADH derived from sugar metabolism generates ATP. Both the cyanide (CN^-) and hydrogen sulfide (HS^-) anions bind to and inhibit cytochrome oxidase. However, both anions also bind the Fe^{+++} ion methemoglobin (MetHb) formed by the oxidation of hemoglobin with nitrate (NO_2^-). CN-MetHb denotes cyanmethemoglobin; HS-MetHb denotes sulfmethemoglobin.

which removes free cyanide and shifts the cyanide concentrations away from cytochrome oxidase (see Figure 4-5). This is, however, at best a stopgap measure because:

1. Cyanmethemoglobin (CNMetHb) cannot transport oxygen.
2. Although very stable, CNMetHb will eventually dissociate to yield free CN.
3. The total amount of CN in the body may be so large that any tolerable level of MetHb will be quickly saturated.

Actual detoxication is achieved by the administration of thiosulfate, which, under the influence of sulfurtransferase (hepatic rhodanese), reacts with cyanide to form thiocyanate (SCN^-), a relatively nontoxic substance readily excreted in the urine:

$$Na_2S_2O_3 + CN^- \xrightarrow{\text{sulfurtransferase}} SCN^- + Na_2SO_3$$

The final reaction is slowly reversible through the action of thiocyanate oxidase. Therefore, if symptoms of poisoning appear or recovery is slow, the sodium nitrite and sodium thiosulfate treatment should be repeated.

The spectrophotometric determination of MetHb is based on the fact that an absorption maximum at 635 nm is abolished by the addition of cyanide (formation of cyanmethemoglobin). Thus, one cannot accurately measure MetHb levels in cyanide-poisoned patients.

Hydrogen Sulfide Anion (HS^-) Poisoning

HS^- is generated by H_2S gas, and is as potent an inhibitor of cytochrome oxidase as the cyanide anion. H_2S is a gas common to sewers and oil fields. The signs and symptoms of HS^- poisoning are similar to cyanide poisoning, except that H_2S is an irritant gas that may produce conjunctivitis of the eyes or pulmonary edema.

The treatment for hydrogen sulfide poisoning is the same as that used for cyanide poisoning, except that no specific treatment is used to degrade the HS^- bound to MetHb. The hydrosulfide anion is inactivated by oxidized glutathione or other simple disulfides. The sulfide so generated may then be metabolized to sulfates.

4.5 Chemical-Induced Blood Disorders

Dyscrasias

There are a number of chemicals, both occupational and medicinal, that are capable of inducing blood dyscrasias (blood disorders). There are numerous

Table 4-1. Chemicals Reported to Have Caused Thrombocytopenia.

Acetaminophen (Tylenol)	Diazepam (Valium)	Phenobarbital
Aminopyrine	Diethylstilbestrol	Phenylbutazone
Aspirin and salicylates	Digitoxin	Potassium iodide
Benzene	Dimercaprol	Quinidine
Bismuth	Disulfiram	Quinine
Chloramphenicol	Insulin	Stilbestrol
Chlordane	Isoniazid	Tetracycline
Corticosteroids	Lindane	Toluene dyisocyanate
Dextropropoxyphene (Darvon)	Mercurials	Trinitrotoluene

dyscrasias that are named for or defined by the element or elements within the blood that are affected. The variety of blood dyscrasias is large, the number of chemicals involved is very large, and in most instances the mechanism of action is not understood. For example, there are some 1066 drugs that might be responsible for one or several types of blood disorders. There are 182 different drugs suspected or known to cause thrombocytopenia. Other disorders and the number of drugs involved include aplastic anemia, 127 drugs; agranulocytosis, 112 drugs; hemolytic anemia, 96 drugs; leukopenia, 90; neutropenia, 71; and many more could be listed. Since the nature and causes of these disorders are so diverse and not always well understood, this chapter will only attempt to define some of these disorders and list some of the suspected agents. It is hoped that this information will alert the reader to the potential for chemical-induced diseases of the blood and the need for routine blood analysis when accidental or chronic chemical exposure occurs.

Table 4-2. Chemicals Reported to Have Induced Agranulocytosis.

Acetazolamide	Cyclophosphamide	Nitrogen mustards
Amidopyrine	Desipramine	Paracetamol
Ampicillin	Diazepam	Phenylbutazone
Arsphenamine	DDT	Procainamide
Barbiturates	Dinitrophenol	Propylthiouracil
Benzene	Ethacrynic acid	Quinidine, quinine
Busulfan	Hydroxyquinalone	Salicylates
Cephaloridine	Indomethacin	Sulfa drugs
Chloramphenicol	Mercurial diuretics	Thiourea
	Nitrofurantoin	Trinitrotoluene

Thrombocytopenia

Thrombocytopenia is any condition in which there is an abnormally low number of platelets (thrombocytes) in the circulating blood. Because of their decreased platelet count, persons with this ailment are susceptible to hemorrhage (bleeding) of all causes. Agents toxic to thrombocytes could decrease the number of circulating platelets via a mechanism consisting of either a direct toxic action or by some immunological effect. Additionally, little is known about these chemical-induced toxicities and whether or not they affect the "poietins." Thus, for thrombocytopenia or any other blood cell disorder, many possible mechanisms for chemical-induced toxicity exist. Some of the chemicals reported to have caused thrombocytopenia are shown in Table 4-1.

Agranulocytosis

Agranulocytosis is a condition characterized by a reduction in white blood cells with a great reduction in polymorphonuclear leukocytes. Persons afflicted with agranulocytosis are therefore susceptible to severe or fatal infections. Agents reported to have the potential for inducing this condition are shown in Table 4-2.

Aplastic Anemia and Pancytopenia

The term aplastic refers to a condition of hypocellular or acellular bone marrow. Therefore it rarely, if ever, is a condition in which only the erythrocytes (red blood cells) are affected (aplastic anemia). Therefore, pancytopenia, which is a decrease in all blood elements (i.e., red blood cells, white blood cells,

Table 4-3. Chemicals Reported to Have Induced Aplastic Anemia and Pancytopenia.

Alkylating agents	Chloroquine	Oxyphenbutazone
Amitriplyline	Colchicine	Phenylbutazone
Ampicillin	Diazepam	Potassium perchlorate
Arsenicals	Gold compounds	Propylthiouracil
Arsphenamine	Hydrochloroquine	Quinacrine
Aspirin	Insecticides	Quinidine
Benzene	Isoniazid	Salicylates
Carbon tetrachloride	Lindane	Streptomycin
Chloramphenicol	Meprobamate	Sulfa drugs
Chlordane	Methimazole	Tetracycline
		Trinitrotoluene

Table 4-4. Chemicals Reported to Have Induced Immune-Related Hemolytic Anemia.

Amidopyrine	Penicillin
Antazoline	Phenacetin
Cephalosporins	Quinidine, quinine
Insecticides	Rifampicin
Insulin	Salicylates
Isoniazid	Stibophen
	Sulfa drugs

and platelets) is usually a more accurate description of a disease associated with aplastic changes in bone marrow. The seriousness of these blood disorders is obvious. Some of the agents reported to have the potential for inducing such effects are shown in Table 4-3.

Hemolytic Anemia

Hemolytic anemia is a decreased number of red blood cells caused by the lysis or destruction of these cells. There are two general forms of hemolytic anemia other than genetically acquired structural defects in red blood cells. First, there is an immune-related hemolytic anemia, in which the chemical probably serves as a hapten and converts the red blood cell to an antigen. Some of the chemicals reported to induce this type of hemolytic anemia are shown in Table 4-4.

Table 4-5. Chemicals Reported to Have Induced Hemolytic Anemia in Persons Deficient in Glucose-6-Phosphate Dehydrogenase.

Acetanilide	Pamaquin, pantaquin
Aminopyrine	Phenacetin
Aspirin	Phenylhydrazine
Chloroquine	Potassium perchlorate
Dapsone	Primaquine
Dimercaprol	Probenecid
Furazolidine	Quinidine, quinine
Mepacrine	Salicylates
Methylene blue	Sulfa drugs
Naphthalene	Toluidine blue
Nitrofurantoin	Vitamin K (water-soluble analogs)

Table 4-6. Chemicals Reported to be Capable of Hemolyzing Red Blood Cells.

Arsine	Methyl chloride
Benzene	Naphthalene
Butyl cellusolve	Nitrobenzene
Carbutamide	Phenylbutazone
Chloramphenicol	Phenylhydrazine
Chlorpromazine	Primaquine
Dimercaprol	Quinacrine
Lead	Streptomycin
Mephenytoin	Tolbutamide
	Trinitrotoluene

The second general form of hemolytic anemia results from an enzyme deficiency in the glycolytic pathway of the red blood cells. When cells are defective in the pathway converting glucose to adenosine triphosphate (ATP), or energy, they have difficulty maintaining the proper ionic balance and in maintaining glutathione, which protects the cells against peroxidative attack. The most common enzyme defect in the glycolytic pathway is glucose-6-phosphate dehydrogenase deficiency. It affects millions of persons throughout the world; more than 100 variants of the trait have been identified. Persons with the deficiency are particularly sensitive to red cell hemolysis induced by chemicals. Some of the chemicals producing this anemia are shown in Table 4-5.

Several chemicals are capable of hemolyzing red blood cells in all individuals. Some of these chemicals are shown in Table 4-6.

4.6 Evaluation of Hematotoxicity

Blood cell counts and a number of other hematologic measurements are routinely analyzed when a patient is admitted to a hospital and are part of most medical surveillance programs. Most of these analyses are done easily and inexpensively on automated instruments. The analyses include:

1. Hematocrit: The hematocrit is the percentage of red blood cell mass (volume) within whole blood. It is measured as the height of the red blood cells in a column compared to the height of the original whole blood, after centrifugation is used to pack the red blood cells. Normal values are about 40–54 percent for males and 37–47 percent for females. The hematocrit is usually three times the hemoglobin volume and is an indicator for anemia.

2. *Hemoglobin*: Hemoglobin is usually measured along with the hematocrit, as anemia may be a problem in a person having a normal number of red blood cells that are low in hemoglobin (the O_2 carrier). Normal values are usually expressed as grams per hundred milliliters of blood and range at 14–18 g/100 ml for males and 12–16 g/100 ml for females.

3. *Red Cell Count*: The normal number of red blood cells is 5.4 ($\pm$0.8) $\times 10^6/\mu l$ in males and 4.8 ($\pm$0.6) $\times 10^6/\mu l$ in females.

4. *White Cell Count*: The normal range in adults is from 4 to 11 $\times 10^3/\mu l$. Children have higher counts; these diminish between the ages of seven and puberty. Often a differential white blood cell count is made to ensure that there is a normal distribution of the various white blood cells. The following is the approximate count expected: segmented neutrophils, 3800/μl (51 $\pm$ 15 percent); band neutrophils, 620/μl (8 $\pm$ 3 percent); eosinophils, 200/μl (2.7 percent); lymphocytes, 2500/μl (34 $\pm$ 10 percent); monocytes, 300/μl (4 percent).

5. *Platelet Count*: The normal range is 1.5 to 4.5 $\times 10^5/\mu l$.

Summary

Blood cells consist primarily of three cell lines:

- Erythrocytes or red blood cells.
- Thrombocytes or platelets.
- Leukocytes or white blood cells.

Red blood cells transport oxygen to all tissues.

- Any compromise of this transport function or any chemical-induced loss of red blood cells has serious implications, especially for the nervous system, which cannot maintain itself via anaerobic metabolic pathways.
- One must always protect against high concentrations of any gas, inert or otherwise, because all may lead to asphyxiation at sufficiently high concentrations.
- The reader should now be aware of those chemical asphyxiants and cytotoxic poisons that interfere with either the transport or utilization of oxygen and thereby lead to a general tissue hypoxia.

Platelets function primarily as clotting cells to plug ruptured vessels.

- Besides aspirin, which prevents platelet aggregation (i.e., plug formation), or oral anticoagulants (e.g., warfarin and other coumarins), which prevent the liver's synthesis of clotting factors, a number of chemicals can lower the level of platelets within the blood.

- Platelet deficiency can render its victim extremely susceptible to fatal internal bleeding, and persons with a significantly low platelet count should not work under hazardous conditions where injury or bruises could lead to hemorrhage.

The white blood cells protect the body against infection and foreign substances.

- As with the red blood cells and platelets, a number of chemicals insidiously compromise the immune defense system by lowering the white cell count.

The chemicals that are toxic to the blood elements or lower them are numerous and diverse in mechanism. It is not possible to discuss all of the chemicals that affect the blood system, but this chapter can alert the reader to the nature of hematotoxicity, and to the necessity of monitoring and preventing it.

References and Suggested Reading

Brown, B. A. 1980. *Hematology: Principles and Procedures*. Philadelphia: Lea & Febiger.

Dougherty, W. M. 1976. *Introduction to Hematology*. 2nd Edition. St. Louis: C. V. Mosby.

Girdwood, R. H. 1976. "Drug-Induced Hematological Disorders." *In* Israes, M., and Delamore, I. (Eds.) *Hematological Aspects of Systemic Disease*. Philadelphia: B. Saunders. Chapter 19.

Hartmann, P. M. (Ed.) 1980. *Guide to Hematologic Disorders*. New York: Grune & Stratton.

Isselbacher, K. J., Adams, R. D., Braunwald, E., Martin, J. B., Petersdorf, R. G., and Wilson, J. D. 1980. *Harrison's Principles of Internal Medicine*. 9th Edition. New York: McGraw-Hill Book Company.

Smith, R. 1980. "Toxic Responses of the Blood." *In* Doull, J., Klaassen, C. D., and Amdur, M. O. (Eds.) *Casarett and Doull's Toxicology: The Basic Science of Poisons*. 2nd Edition. New York: Macmillan Publishing Company.

Valeria, C. 1975. "Drugs, Hormones, and the Red Cells." *In* Surgenor, D. (Ed.) *The Red Blood Cell*. New York: Academic Press. Chapter 31.

Williams, W. J., Beutler, E., Erslev, A., and Rundles, R. W. 1983. *Hematology*. 3rd Edition. New York: McGraw-Hill Book Company.

Wintrobe, W. M. (Ed.) 1981. *Clinical Hematology*. 8th Edition. Philadelphia: Lea & Febiger.

Chapter 5

Hepatotoxicity: Toxic Effects in the Liver

Robert C. James

Introduction

This chapter will familiarize the reader with:

- The basis of liver injury.
- Normal liver functions.
- The role the liver plays in certain chemical-induced toxicities.
- Evaluation of liver injury.
- Specific chemicals that are hepatotoxic.

5.1 The Physiologic and Morphologic Bases of Liver Injury

Physiologic Considerations

The liver, the largest gland in the body, is often the target organ of chemical-induced tissue injury, a fact recognized for over 100 years. While the chemicals toxic to the liver and the mechanisms of their toxicity are numerous and varied, several basic factors underlie the liver's susceptibility to chemical attack.

First, the liver maintains a unique position within the circulatory system. As Figure 5-1 shows, the liver receives a large portion of the venous return. It effectively "filters" the blood coming from the lower body, kidneys, spleen, and gastrointestinal tract before this blood is pumped through the lungs for

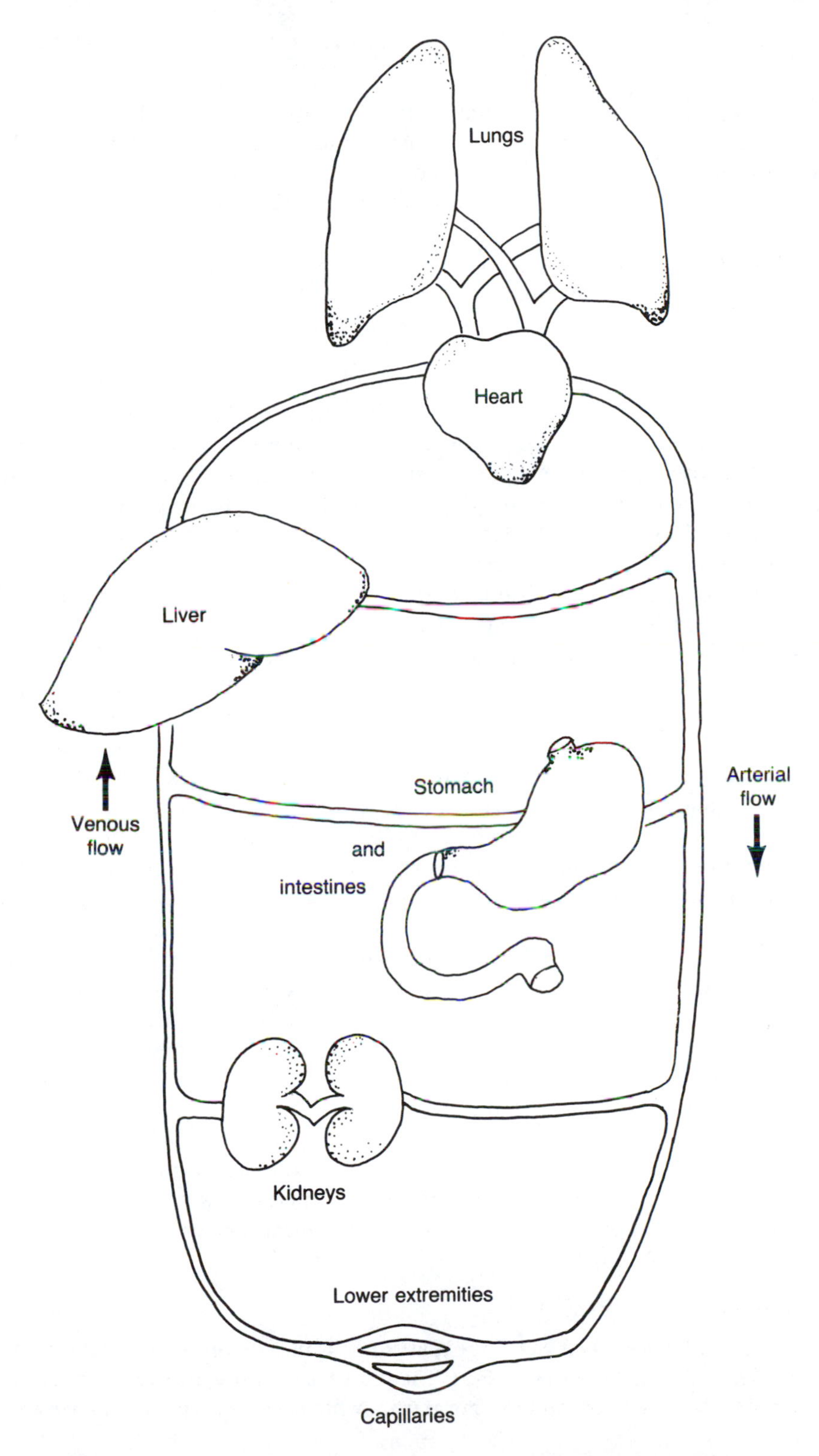

Figure 5-1. The liver maintains a unique position within the circulatory system.

reoxygenation. This unique position in the circulatory system aids the liver in its normal functions, which are: (1) carbohydrate storage and metabolism, (2) metabolism of hormones, endogenous wastes, and foreign chemicals, (3) synthesis of blood proteins, (4) urea formation, (5) metabolism of fats, and (6) bile formation.

The liver "sees" a large portion of the blood supply, and it is the first organ perfused by nutrients absorbed in the gut. Thus, the liver removes foreign chemicals that reach the bloodstream, regardless of the route of administration or exposure.

The liver's prominence causes it to have increased vulnerability to toxic attack, however. The liver can particularly affect or be affected by those chemicals ingested orally or administrated intraperitoneally (i.e., into the abdominal cavity), as many tested substances are, because it is the first organ perfused by the blood containing the chemical. The liver removes and metabolizes almost all blood-transported substances. If removal and metabolism of a chemical is done rapidly and extensively by the liver, then this "first pass" effect upon chemicals administered orally or intraperitoneally can dramatically decrease blood levels of the chemical before the chemical can reach other organs. Thus, a chemical toxic to the liver, or one activated by the liver to a toxic form, may be more toxic when given intraperitoneally than when absorbed through the lungs or skin, because disposition of the chemical to other tissues, which follows absorption, will lower the concentration of the chemical in the blood before the blood reaches the liver, and will lengthen the time necessary for the liver to clear the chemical from the body.

A second reason for the liver's susceptibility to chemical attack is that it is the primary organ for the biotransformation of chemicals within the body. In general, the desired net result of the biotransformation process is to alter the metabolized compound so that: (1) it is no longer biologically active within the body, and (2) it is more polar and water-soluble and, consequently, can be excreted from the body. Thus, for most instances the liver acts as a "detoxification" organ. It lowers the activity and blood levels of a chemical that might otherwise accumulate to toxic levels within the body. For example, it has been estimated that the time required to excrete one half of a single dose of benzene would be about 100 years if the liver did not metabolize it. The primary disadvantage to the liver's role as the main organ metabolizing chemicals, however, is that toxic or reactive chemicals can be formed during the biotransformation process. Of course, the liver, as the generator, is usually the organ most often affected by these "bio-activated" chemical species.

Morphologic Considerations

The liver can be described as a large mass of cells packed around vascular trees of arteries and veins (see Figure 5-2). The most basic unit within the liver is the group of cells lying between the central vein, which drains away cell wastes and

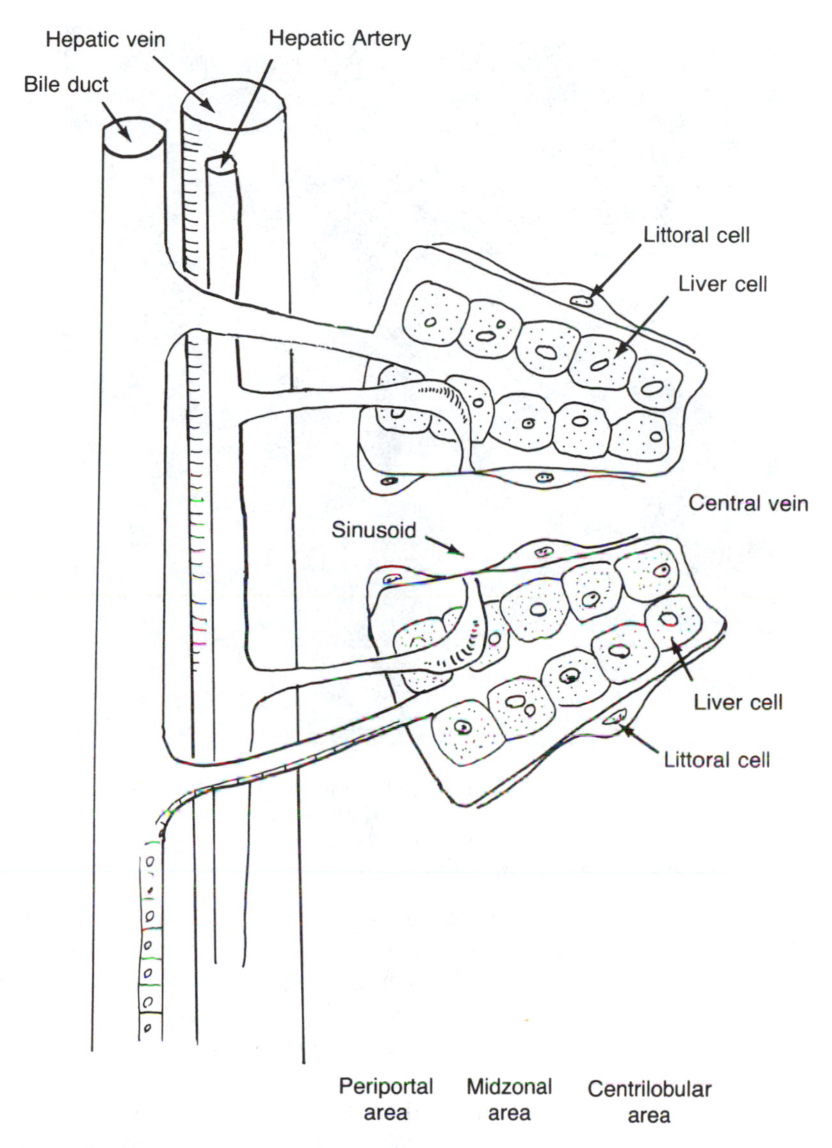

Figure 5-2. A schematic representation of a liver lobule.

products, and the hepatic artery/portal vein (i.e., periportal) system, which supplies oxygen and nutrients. This basic functional unit is termed the liver lobule (sometimes referred to as a liver acinus). The human liver lobule is some one to two millimeters in diameter and several millimeters long. The human liver contains 50,000 to 100,000 such lobules. Since the central vein drains several liver lobules, in cross-section they radiate out from the central vein in triangular groups of cells, somewhat like the spokes of a wheel (Figure 5-3).

Figure 5-3. A cross-sectional demonstration of several liver lobules radiating out from the central vein in triangular groups of cells. (1) Periportal area; (2) Midzonal area; and (3) Centrilobular area. (Courtesy of Dr. Michael Franklin, University of Utah.)

The cells of the lobule closest to the artery are termed the periportal hepatocytes or cells within the periportal area of the lobule. The cells closest to the central vein are termed centrilobular hepatocytes or are said to lie in the centrilobular area of the lobule. The cells in the middle, between both areas, are sometimes considered separately and are then termed midzonal hepatocytes.

The importance of knowing about the lobule structure is that cells within the periportal area, by virtue of being closest to the arterial supply, are exposed to the highest concentration of oxygen, nutrients, and chemicals (Figure 5-3). Also, the cells within the periportal and centrilobular areas—the two main regions—differ greatly in enzymatic activity and sometimes in the enzymes present within the cells. For example, cytochrome P-450, an enzyme capable of oxidizing most chemicals, is found to be largely concentrated in the centrilobular area. On the other hand, the periportal cells contain higher concentrations of glutathione and transaminase enzymes. The net result of this regional distribution of certain enzymes within the lobule is that some toxicants may only be toxic to specific portions of the lobule, while others are nonspecific in their effect. This in turn leads to differences in the histopathological changes characteristic of different hepatotoxins.

5.2 Biotransformation: A Basic Liver Action upon Exogenous Chemicals

The word metabolism, especially in reference to the liver, has two meanings. Metabolism is used by biochemists to refer to those biochemical reactions that serve to generate cellular energy and thereby maintain cellular life. Chemical metabolism as used by toxicologists or pharmacologists refers to biotransformations (i.e., biochemical alteration or transformation) of xenobiotics (foreign compounds) that alter the structure of the xenobiotic compound. In this chapter, then, biotransformation and chemical/xenobiotic metabolism are to be considered synonymous terms and may be used interchangeably. We have previously discussed the primary objectives of biotransformation, which are:

- To biochemically alter a chemical substance so as to alter its biologic effects.
- To biochemically transform the chemical to a more polar, and therefore more water-soluble species, which favors an increased elimination from the body.

Biotransformation pathways and byproducts may be categorized in several different ways. One classification that is routinely used is to classify the pathway as a phase I or phase II type or reaction (Figure 5-4). Phase I reactions are oxidation/reduction reactions. These reactions alter the chemical by attaching some functional chemical group on the molecule—for example, an alkyl group, hydroxyl group, sulfur or nitrogen atom, double bond, etc.—and oxidize or reduce that chemical group to another oxidation state. Phase II reactions alter the chemical by attaching a new prosthetic group to a functional portion of the xenobiotic. For example, phase II enzymes attach sugars, peptides, amino acids, acetyl, methyl, or sulfate groups to primarily alcoholic, carboxylic, sulfhydryl, or amine groups on the chemical being metabolized.

Phase I Reactions

The Cytochrome P-450 Enzymes. The most important enzyme system involved in phase I reactions is that mediated by the cytochrome P-450 enzymes, which is sometimes referred to as the mixed-function oxidase system (MFO system). This system comprises two kinds of enzymes. One is a flavoprotein, NADPH cytochrome *c* reductase (i.e., sometimes called NADPH cytochrome P-450 reductase), which oxidizes NADPH to $NADP^+$, thereby deriving electrons or energy to drive a reaction. This reductase is coupled to other enzymes known as cytochrome P-450, each of which is a heme-containing enzyme that binds the xenobiotic and has the catalytic site responsible for

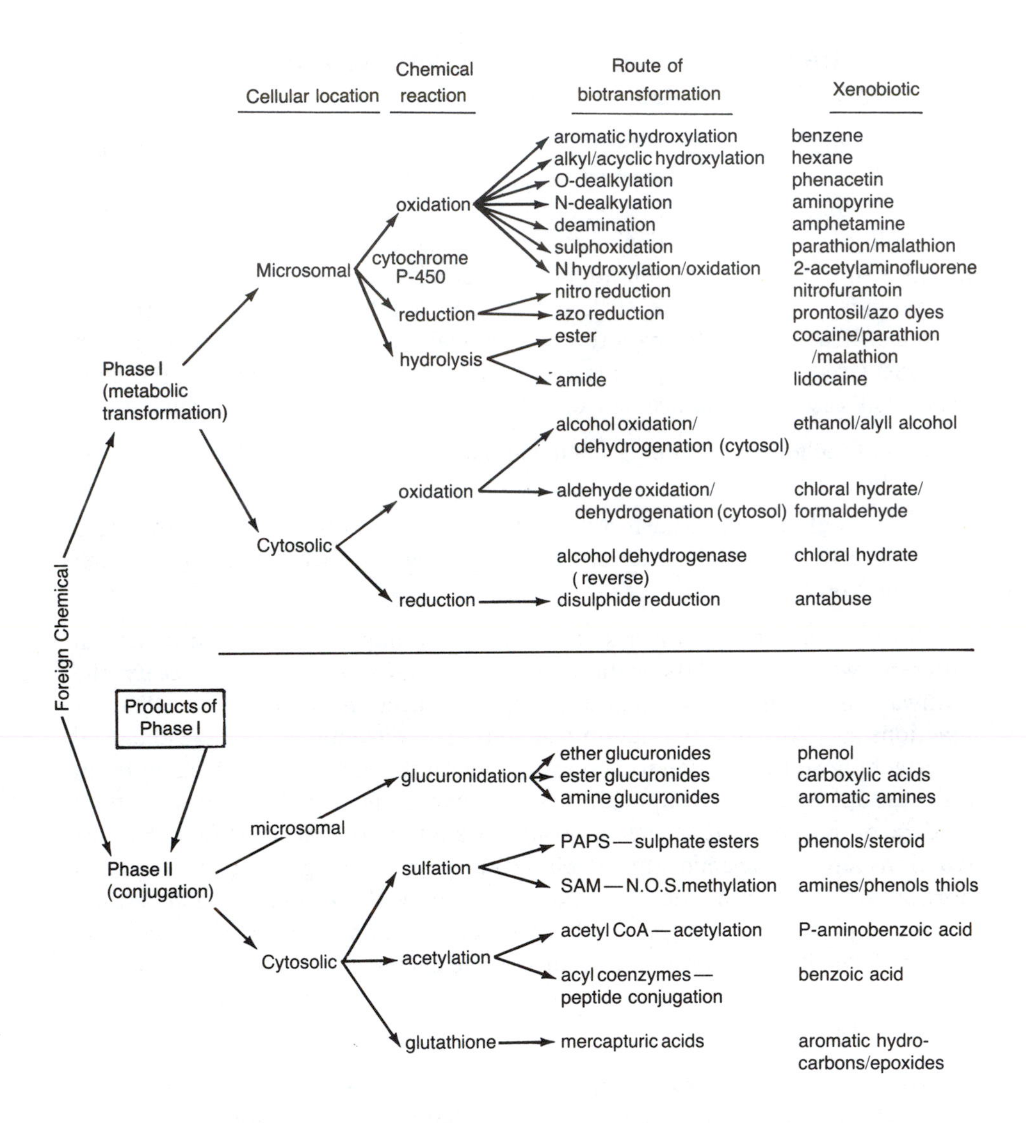

Figure 5-4. Xenobiotic biotransformation in the liver. (a) Phase I reactions. (b) Phase II reactions. (Courtesy of Dr. Michael Franklin, University of Utah.)

the oxidation/reduction being performed. In the process, the chemical being metabolized interacts with an activated oxygen molecule of which one atom ends up as part of a water molecule and the other is inserted into or cleaves off a portion of the chemical (see Figure 5-5). The cytochrome P-450 system is therefore actually a family of different cytochrome P-450 enzymes with different reactions to catalyze. These are very versatile enzymes, and the route of biotransformation that a chemical might undergo varies. Some of the reactions catalyzed by the various cytochrome P-450 enzymes are:

- Aromatic or aliphatic hydroxylations (RH → ROH).
- Dealkylations from functional groups such as nitrogen, sulfur, and oxygen ($R{-}N{-}CH_3 \rightarrow R{-}NH + CH_3OH$).
- Arene oxide formation.
- Epoxidation ($R{-}C{=}C \rightarrow R{-}\overset{O}{\widehat{C{-}C}}$).
- Desulfuration and sulfoxidation ($R{-}S{-}R \rightarrow R{-}\underset{\underset{O}{\|}}{S}{-}R$).
- Deamination and *N*-hydroxylation (R—NH → R—N—OH)
- Azo, nitro, and hydroxylamine reductions ($R{-}NO_2 \rightarrow R{-}NH_2$).
- Carbonyl reductions ($R{-}\overset{\overset{O}{\|}}{C}{-}R \rightarrow R{-}\overset{\overset{OH}{|}}{C}{-}R$).

As can be seen above, many reactions are possible when a chemical is metabolized by cytochrome P-450, and often many different metabolites are generated for a given chemical. For example, methamphetamine may be

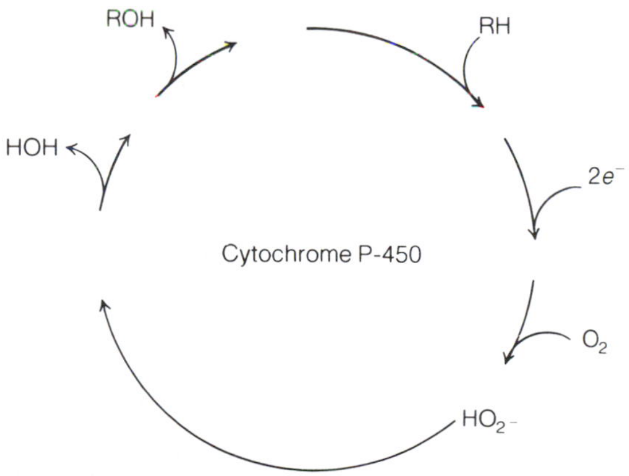

ROH
RH
HOH
$2e^-$
Cytochrome P-450
O_2
HO_2^-

Figure 5-5. Proposed scheme for the metabolism of substrates by the monoxygenases containing cytochrome P-450.

dealkylated at the nitrogen, deaminated, or hydroxylated on the alkyl side-chain or on the benzene ring:

Methamphetamine

CH_3OH

HO → ⟨benzene ring⟩ — C — C — N — C ↗

↑ (under first C) ↓ (under N)

OH

Hydroxylation Deamination Dealkylation

This metabolite pattern may be dose-dependent, and increasing or decreasing the dose may alter the percentage of the chemical that is metabolized by each pathway. Thus, at high concentrations, when the normal pathways are saturated, the chemical may in some instances be metabolized by a new pathyway, which may generate a less common, toxic metabolite.

Tissue Distribution of the Cytochrome P-450 Enzymes. The cytochrome P-450 enzymes are localized primarily within the endoplasmic reticulum of the hepatocyte, although small amounts have also been identified in the mitochondria and in the nuclear membrane. Thus, when the liver is homogenized, the microsomal fraction contains the cytochrome P-450 system. As was stated earlier, the liver is the primary organ for metabolism, but because cytochrome P-450 is such a useful and versatile family of enzymes for metabolizing chemicals, these enzymes enjoy a wide tissue distribution and are found throughout the body. The lungs and kidneys are secondary organs of biotransformation for many chemicals and have approximately 10–30 percent of the liver's capacity to metabolize a compound, depending upon the substrate.

Microsomal preparations from other tissues have also demonstrated that cytochrome P-450 exists in these tissues as well, and although these organs are generally smaller than the liver, their microsomal activity bears approximately the following percentage relationship in comparison to liver microsomes:

Organ or tissue	Capacity for metabolizing compounds (percent relative to liver microsomes)
Gut	10
Adrenal cortex	50–75
Testes	10–20
Spleen	5
Heart	3
Muscle	1
Brain	1
Placenta	1
Skin	1

Increasing and Decreasing the Rate of Synthesis of Cytochrome P-450. Many factors can alter the rate at which chemicals are metabolized in the body. Again, these changes are primarily concerned with liver metabolism. In 1964 scientists noticed that the presence in the liver of a second chemical might alter the rate of metabolism of a chemical already present there. Those compounds that increase metabolism are called "inducers," because they induce the synthesis of more cytochrome P-450 or of a specific type of cytochrome P-450 that can more efficiently metabolize the compound in question. Hundreds of chemicals have been identified as inducers of xenobiotic metabolism. Some examples of inducing agents are listed in Table 5-1.

There are also many chemicals that inhibit cytochrome P-450. Compounds such as the formulation called SKF-525A, piperonyl butoxide, chloramphenical, cobaltous chloride, aminotriazole, α-naphthyl isocyanate, carbon disulfide, carbon tetrachloride, bromobenzene, and many others can inhibit metabolism. There are basically three mechanisms of inhibition: (1) competive binding to and metabolism by cytochrome P-450, (2) inhibition of synthesis of heme or cytochrome P-450, and, (3) agents that inactivate or destroy cytochrome P-450 or the endoplasmic reticulum.

Table 5-1. Some Compounds Reported to Induce Biotransformation.

Drugs	Industrial chemicals
Aminopyrine	Alcohols
Amphetamine	Aldrin/dieldrin
Barbiturates	Chlordane
Chloral hydrate	Chloroform
Chlordiazepoxide (Librium)	DDT, DDD
Chlorpromazine	DMSO
Diazepam (Valium)	Heptachlor
Diphenhydramine	Ketones
Ethanol	Lindane
Ethanol pyridione	PCB compounds
Glutethimide	Piperonyl butoxide
Halothane	Pyrethrum
Imipramine	Toxaphene
Meprobamate	
Morphine	
Nicotine	
Phenylbutazone	Polyaromatic hydrocarbons
Phenytoin	Benz(a)pyrene
Promazine	Dibenzanthracene
Propoxyphene (Darvon)	3-Methylcholanthrene
Steroids	
Sulfanilamide	
Thalidomide	
Trimethadione	
Urethane	
Zoxazolamine	

Other Factors Affecting Biotransformation by Cytochrome P-450 Enzymes. Besides induction and inhibition there are additional factors to consider that can affect biotransformation. Many of the following factors have been shown in animal studies to alter rates of metabolism, and many or all may be extrapolated to human metabolism.

- Diet: A low protein diet decreases xenobiotic metabolism; fasting enhances or inhibits, depending on the particular chemical that must be metabolized.
- Nutrition: Calcium and copper deficiencies reduce metabolism, as do deficiences of zinc or iron; vitamin C, A, or E deficiencies depress metabolism.
- Hormone levels: ACTH (adrenocorticotropic hormone) increases metabolism; growth hormones increase it; thyroxine enchances while thyroidectomy decreases metabolism; diabetes has a mixed effect; glucocorticoids enhance it; anabolic steriods enhance it.
- Age: Metabolism is generally lower in neonates and the aged.
- Sex: Many compounds demonstrate differences in the rates and routes of metabolism between males and females.
- Genetics: Some pathways demonstrate definite genetic variations.
- Diurnal factors: Metabolism seems to be greatest during the portion of the day when food is ingested.
- Pathophysiology: Disease states may alter metabolism by modifying the absorption, distribution, and excretion of chemicals, altering the nutritional states, altering the rate of blood and oxygen delivery to the liver, etc.; liver disease generally decreases cytochrome P-450-dependent metabolism.

Other Phase I Reactions. While this chapter focuses primarily on cytochrome P-450 because of its importance, there are other phase I reactions. Other phase I enzymatic systems that have been identified include: amine oxidase; epoxide hydratase, which cleaves reactive epoxides into dihydrodiols; esterases, which hydrolyze a variety of ester compounds with different basic components; amidases, which hydrolyze amides, and alcohol dehydrogenase.

Phase II Reactions

Phase II reactions always increase the size and molecular weight of the chemical metabolized because they add some chemical group to the molecule (Figure 5-6). Although in general these chemicals are more water-soluble, both acetylation and methylation may significantly decrease the compound's water solubility and thereby slow elimination via the kidneys. Following are several enzyme systems that catalyze phase II reactions.

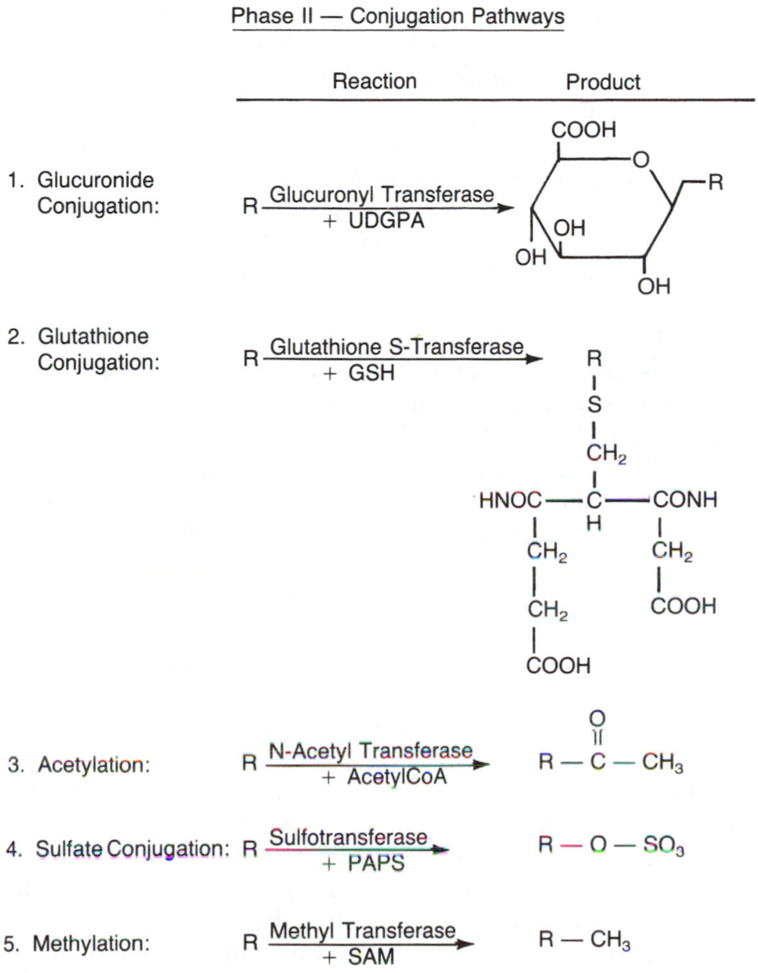

Phase II — Conjugation Pathways

	Reaction	Product
1. Glucuronide Conjugation:	R —Glucuronyl Transferase + UDGPA→	COOH, O, R, OH, OH, OH
2. Glutathione Conjugation:	R —Glutathione S-Transferase + GSH→	R, S, CH_2, HNOC—C—CONH, H, CH_2, CH_2, CH_2, COOH, COOH
3. Acetylation:	R —N-Acetyl Transferase + AcetylCoA→	R — C(=O) — CH_3
4. Sulfate Conjugation:	R —Sulfotransferase + PAPS→	R — O — SO_3
5. Methylation:	R —Methyl Transferase + SAM→	R — CH_3

Figure 5-6. Conjugation pathways for phase II reactions. UDGPA, uridine-5′-diphospho-α-D-glucuronic acid; GSH, glutathione; acetyl CoA, acetyl coenzyme A; PAPS, 3′-phosphoadenosine-5′-phosphosulfate; SAM, *S*-adenosylmethionine.

Glucuronyl Transferase. Glucuronyl transferase is a membrane-bound (endoplasmic reticulum) enzyme that conjugates a sugar molecule to the alcoholic, phenolic, amino, carbamyl, sulfonamide, and thiol groups of chemicals (Figure 5-6). This is perhaps the most common and the most important phase II or conjugation reactions. There is evidence for multiple forms of glucuronyl transferase, and many compounds that induce the cytochrome P-450 system (which also increases the amount of endoplasmic reticulum in the liver) also induce and increase glucuronidation in various organs. The liver is the most important organ by far, but glucuronyl transferase is also found in the microsomal fractions of cells from the kidneys, lungs, intestines, spleen, brain, placenta, and skin. Competing substrates can inhibit this system, as can some of the classical cytochrome P-450 inhibitors like SKF-525A and piperonyl butoxide. Alcohol (ethanol) also inhibits glucuronyl transferase *in vivo* as it shifts the NAD^+ : NADH ratio in the cell in favor of the reduced form. The oxidized form, NAD^+, is required in the synthesis of UDPGA (uridine-5′-diphospho-D-glucuronic acid, the cofactor required by glucuronyl transferase and a high-energy form of the sugar that becomes attached to the chemical being metabolized). Glucuronide metabolites are excreted in the urine and the bile. Generally, compounds with a molecular weight greater than 300 are conjugated and excreted in the bile. In some instances, bacteria in the gut cleave the glucuronide and liberate the compound, whereupon it is reabsorbed by the gut and returns to the liver. This "entero-hepatic circulation," as it is termed, is the reason a compound like phenolpthalein can have a long-lasting laxative effect: its elimination is slowed by "enter-hepatic" recirculation within the body.

Most mammalian cells also contain an enzyme that cleaves glucuronide metabolites or compounds. This enzyme, β-glucuronidase, is found in lysosomes of the cell. It may function to liberate conjugated hormones, thereby liberating the active parent chemical, but the exact physiologic significance of this widespread enzyme remains undefined at the present time.

Glutathione S-Transferase or Mercapturic Acid Formation. The glutathione *S*-transferases are a group of enzymes that conjugate the tripeptide glutathione to a variety of chemicals. Mercapturic acid metabolites may then be formed from the glutathione conjugate by removing the end amino acids of the tripeptide (glutamate and glycine), followed by acetylation of the cysteine residue that remains. The reactions or functional chemical groups of a xenobiotic that may be conjugated are diverse, and include the dehalogenation of alkyl, aryl, and cycloalkanes; reactions with expoxides and double bonds; *N*-hydroxy compounds; and the conjugation and inactivation of reactive/electrophilic compounds in general. The glutathione *S*-transferase enzymes are located in the cytosol of cells and are found primarily in the liver, kidney, gut, spleen, and lungs. These enzymes, like glucuronyl transferase, are induced by many of the compounds that induce cytochrome P-450 (e.g., DDT, phenobarbital, PCB compounds, 3-methylcholanthrene). And like cytochrome P-450, the inducing agents may only increase specific forms of glutathione transferases. Both thyroidectomy and hypophysectomy also increase activity. Glutathione *S*-transferases are inhibited by organic anions (e.g., bilirubin, bromosulophthalein, probenecid, and furosemide, because these bind to ligandin, a liver protein, and thereby inhibit a portion of the transferases) and by chemicals that react with the many sulfhydryl groups (i.e., thiol or SH groups) on the protein. Examples of sulfhydryl-reactive compounds include the mercuric ion, ethacrynic acid, diethyl maleate, and reactive/alkylating metabolites (carbanions, etc.) Like glucuronyl transferase and cytochrome P-450, glutathione *S*-transferase is depressed in the fetus and newborn.

The importance of glutathione *S*-transferase has only somewhat recently been realized. Man is continually exposed to, or generates during metabolism, a host of reactive electrophilic compounds, which, if unchecked, would quickly cause serious, permanent, and fatal damage to many tissues within the body. Because glutathione is a fairly strong nucleophile, the glutathione *S*-transferase system becomes a major inactivating or detoxifying mechanism against these electrophilic substances.

N-Acetyltransferase. *N*-acetyltransferase is another cytosolic (soluble cell fraction) enzyme system that conjugates chemicals by using the acetyl donor, *S*-acetyl CoA. The enzyme conjugates amino, sulfhydryl, and hydroxyl groups of chemicals. It is found primarily in the phagocytic cells of the reticuloendothelial system and is therefore present in liver, gut mucosa, lung, spleen, kidney, and blood tissues. There are multiple forms of this enzyme and the expression of some forms is an often cited example of genetically determined differences in metabolism. There also exist deacetylating enzymes within the body, and the amount of a particular chemical excreted as an acetyl conjugate reflects the balance between the two processes. Certain lipids like tricaprate or steriods like estradiol as well as agents stimulating the reticuloendothelial system induce *N*-acetyltransferase. *In vitro* inhibitors are chloromercuribenzoate, *N*-ethylmaleimide, and Cu^{+}, Zn^{++}, Mn^{++}, or Ni^{++} cation concentrations of at least 0.14 mM.

Sulfotransferase. The conjugation of sulfate to alcoholic, phenolic, and amino groups of xenobiotics is accomplished by a number of cytosolic enzymes known as sulfotransferases or sulfokinases. These enzymes utilize an activated or energy-rich form of sulfate known as PAPS, 3′-phosphoadenosine-5′-phosphosulfate. There are multiple forms of the sulfotransferases, and some of the isoenzymes require magnesium as a cofactor. Sulfation is usually a minor route of metabolism, as the cellular sulfate pool is small and easily exhausted. Sulfotransferases for conjugating xenobiotics are found primarily in the liver, kidney, intestinal mucosa, and placenta. These enzymes are balanced by sulfatases, which remove the sulfate groups they add.

Other relatively minor but general phase II reactions are methylation by the methyltransferases and amino acid conjugation by some enzymes.

Figure 5-4 summarizes the location and variety of biotransformation reactions that a chemical might undergo in the body.

5.3 Biotransformation as a Mechanism of Chemical-Induced Toxicities and Liver Injury

As stated earlier in this chapter, biotransformation, as a general rule, is the process whereby the body biochemically converts a chemical to a more soluble and less active metabolite. As with any generalization, there are exceptions to this rule, and the exceptions produce a variety of interesting consequences. For example, the initial drug metabolism of chloral hydrate, a "mickey finn" ingredient, converts it to the effective sedative-hyparotic compound, trichlorethanol. Several anticancer agents must be metabolized to an active form as well. And perhaps one of the earliest and most important discoveries of the metabolic activation of a chemical to its useful form was the discovery of the sulfa drugs, the sulfonamides. Although the azo dye prontosil was patented in 1932 as the first effective antibacterial agent, by 1935 it was shown that prontosil was actually metabolized in the body to the active chemical species, *p*-aminobenzenesulfonamide, and thus began the synthetic proliferation of a family of important antibiotics, the sulfonamide drugs.

Bioactivation and Detoxification of the Same Chemical

While some of the bioactivations that convert inert or largely nontoxic chemicals to an active form within the body are beneficial, many are not. Moreover, although biotransformation is a useful and necessary process that usually protects the body from an accumulation of the chemicals the body is daily exposed to, the potential for adverse or toxic reactions is always there. The insecticides parathion and malathion are classic examples of biotransformation working to "bioactivate" ineffective compounds to active and toxic form. One biotransformation pathway for the parathion molecule is a desulfuration pathway, which, by replacing the sulfur atom with an oxygen atom, converts parathion to the active anticholinesterase inhibitor paraoxon (see Figure 5-7, a). In addition to this activation pathway, an esterase reaction can convert both parent chemical and active metabolite to inactive byproducts. Since the hydrolysis of paraoxon converts a toxic chemical to nontoxic species, this step is a "detoxification" process. On the other hand, hydrolysis of the essentially nontoxic parathion by the esterase pathway would be considered a nontoxic biotransformation. In Figure 5-7, b, the steps of parathion metabolism have been replaced by the four potential end results of any chemical undergoing metabolism. It can be metabolized to nontoxic metabolites (step 1), it can be metabolized to a toxic/reactive metabolite (step 2), the toxic metabolite may then undergo detoxification, which is the biotransformation to nontoxic metabolites (step 3), or it may interact with a biologic entity to produce a toxic effect (step 4). Thus, many toxicants, particularly hepatotoxins, are chemicals

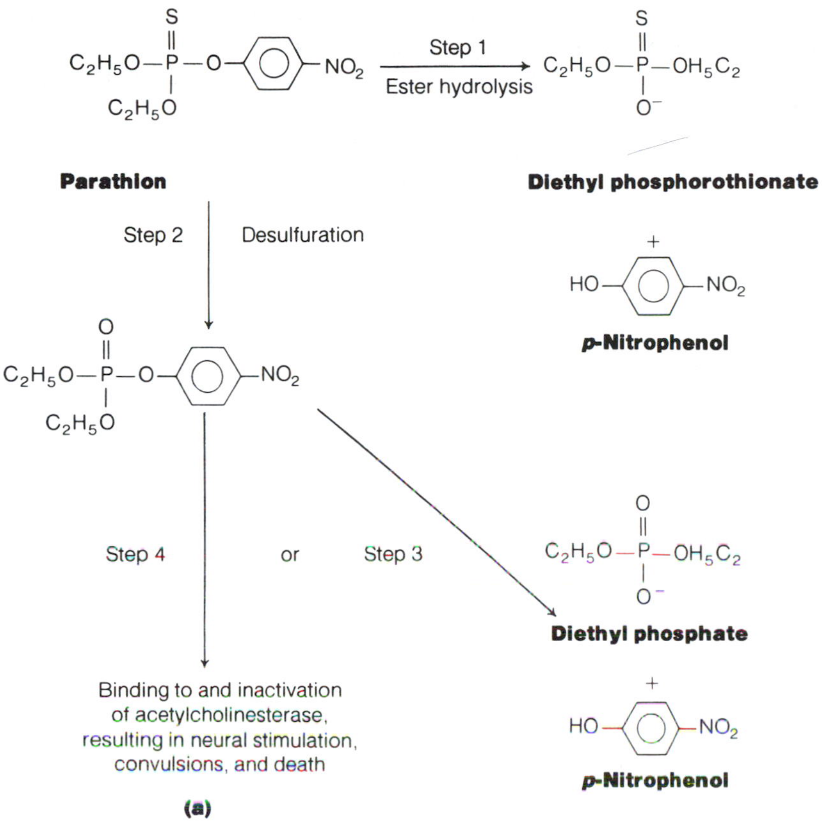

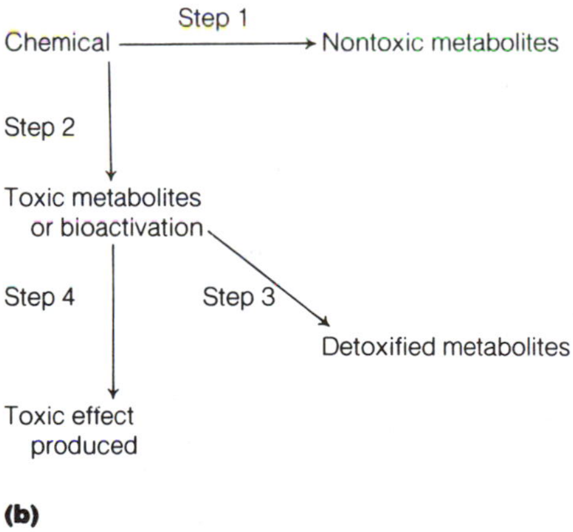

Figure 5-7. (a) Biotransformation of parathion. (b) The four potential end results of any chemical undergoing metabolism.

for which steps 2 and 4 are the predominant or a significant portion of the total biotransformation process. Furthermore, factors that affect the balance between any of the steps (i.e., induction, inhibition, antidote) affects the toxicity of the chemical.

In order to provide a better understanding of biotransformation processes and their role in hepatotoxicity, the bioactivations of two well-known hepatotoxins are explained below.

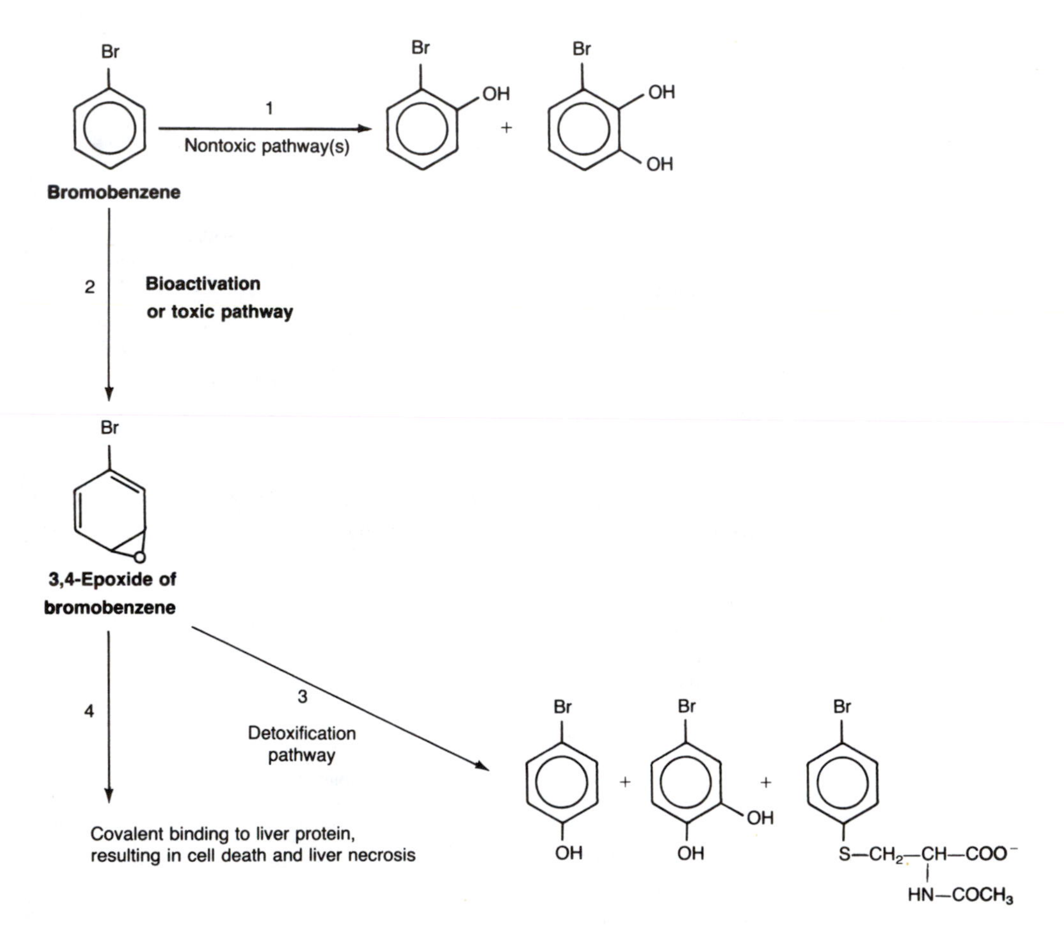

Figure 5-8. Pathways of biotransformation for bromobenzene.

Bromobenzene. Bromobenzene causes a centrilobular necrosis in the liver when administered to rats. This injury is increased by the inducing agent phenobarbital and is decreased by the inhibitor SKF-525A; thus injury is caused by a metabolite of bromobenzene rather than the parent compound. Interestingly, 3-methylcholanthrene, an inducing agent, lowers the toxicity of the compound by increasing the rates of pathways 1 and 3 (Figure 5-8). On the other hand, phenobarbital increases injury by altering the rate of step 2.

Acetaminophen. Acetaminophen (Tylenol) is an over-the-counter medication useful in the relief of minor pain. However, at doses well above the

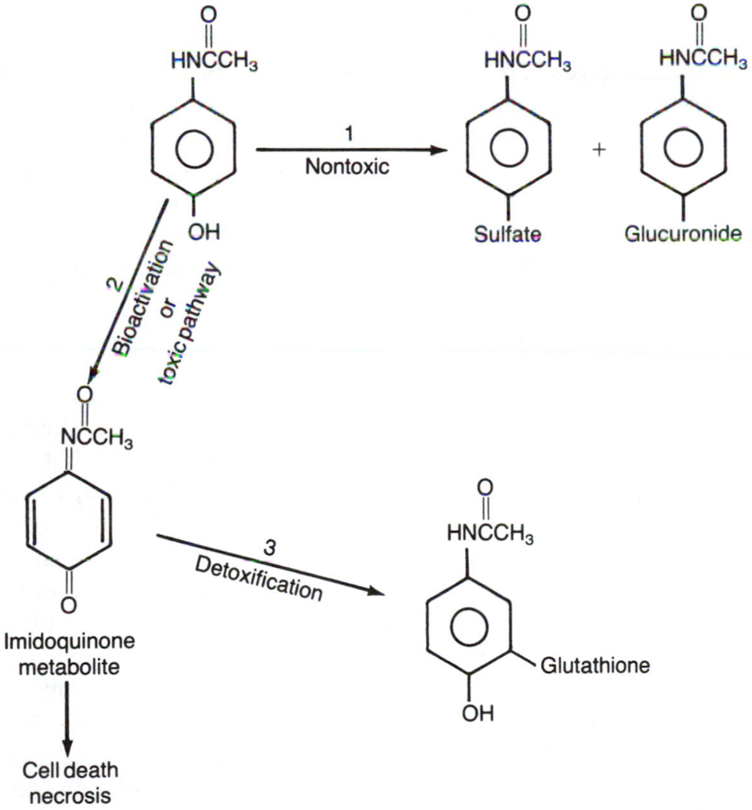

Figure 5-9. Pathways of biotransformation for acetaminophen (Tylenol).

therapeutic regimen it is capable of inducing centrilobular necrosis (Figure 5-9). At high doses the metabolic pathways represented by step 1 are overwhelmed and metabolism via step 2 is also initiated. If the dose is sufficiently high, metabolites formed by the second pathway can not be adequately detoxified by step 3 (inactivation by glutathione conjugation); and step 4—that is, the attack of the cell and cell death—becomes a significant pathway for the toxic metabolites not detoxified by glutathione.

Conjugated Metabolites as Bioactivating Agents. Several other chemicals can also illustrate bioactivation; probable pathways have been proposed for numerous chemicals. Interestingly, while the cytochrome P-450 enzymes, with their oxidative capabilities, are generally thought to be the enzymes responsible for the bioactivation of chemicals, and in many cases they are, studies have also shown that some conjugated metabolites are reactive and toxic as well. For example, sulfation and glucuronidation of the *N*-hydroxy metabolite of phenacetin yields reactive metabolites that can covalently bind protein. The sulfation, glucuronidation, or acetylation of *N*-hydroxy-2-acetylaminofluorene yields reactive, mutagenic, protein-, and nucleic acid-binding metabolites. Even glutathione, which inactivates many electrophilic, toxic chemicals, has been demonstrated to conjugate certain halogenated ethanes to a thiol half-mustard metabolite that is highly reactive and mutagenic. Therefore, even though cytochrome P-450/oxidative metabolism may be regarded as a prime candidate when bioactivation to a toxic intermediate is suspected, potentially any and all pathways of biotransformation may generate a reactive intermediate.

Factors That Influence Bioactivation Also Influence Toxicity

Because bioactivation and detoxification of chemicals arise from the biotransformation of these xenobiotics, the factors that influence biotransformation in general may alter the toxicity of a chemical. Factors such as diet, age, sex, genetics, nutrition, inducing agents, and inhibitory chemicals may all alter one or more of the biotransformation pathways that convert the toxicant in question to some other chemical form. Thus, the outcome may be an increase or decrease in toxicity, but such a change does not necessarily reflect a change in bioactivation alone, and may result from any of a number of possible changes in nontoxic, toxic, and detoxification pathways that significantly alter the balance, and therefore the outcome, of these competing pathways.

Inasmuch as there are numerous species differences in the metabolism of chemicals, both between other animals and between animals and man, while animals are often useful models for predicting the toxicity of a chemical they may not always reflect the actual human condition.

Cellular Damage Due to Reactive Intermediates. Besides generating chemicals, like paraoxon, that are toxic to other tissues, biotransformation has enormous toxicologic significance for the liver itself. Considering the numerous metabolic reactions a chemical can undergo, there is a significant potential for the generation of a reactive chemical intermediate. Reactive chemicals are usually toxic because of their ability to interact with, and thereby inhibit or destroy, important cellular enzymes or vital cellular macromolecules, such as DNA, or cellular organelles like the mitochondria. Some examples of reactive intermediates observed during the bioactivation of chemicals to a toxic species are epoxides, arene oxides, free radicals, and carbanion metabolites. These reactive substances are highly electrophilic and many react with any of the numerous nucleophilic substances normally contained within the cell.

Disease Due to Reactive Intermediates. Besides the possibility for severe and fatal liver injury, the generation of reactive/toxic intermediates during biotransformation carries with it the risk of suffering latent or insidious types of diseases as well. Many nontoxic chemicals (promutagens or procarcinogens) are metabolized to a reactive mutagenic or carcinogenic chemical and with such chemicals comes an increased risk of cancer or birth defects (Figure 5-10). Thus the biotransformation process becomes a critical and integral element in the understanding and prevention of the toxicities produced by these chemicals. In addition to the mutagenic/carcinogenic activity of many of the reactive chemicals the liver is subjected to, the liver may also develop chronic, disabling

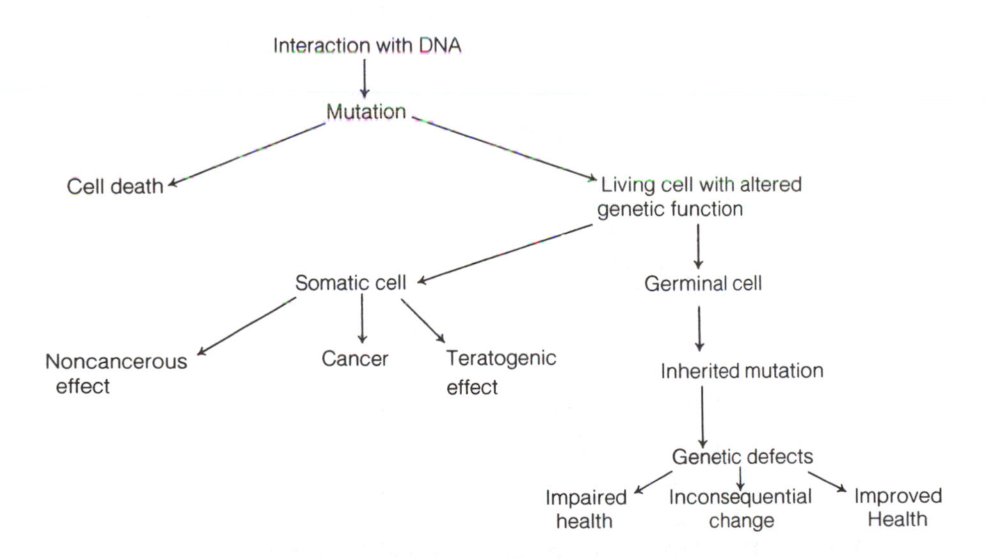

Figure 5-10. Route by which a nontoxic chemical may become a reactive mutagenic or carcinogenic chemical and the potential cellular interactions that may occur which lead to chronic toxicities.

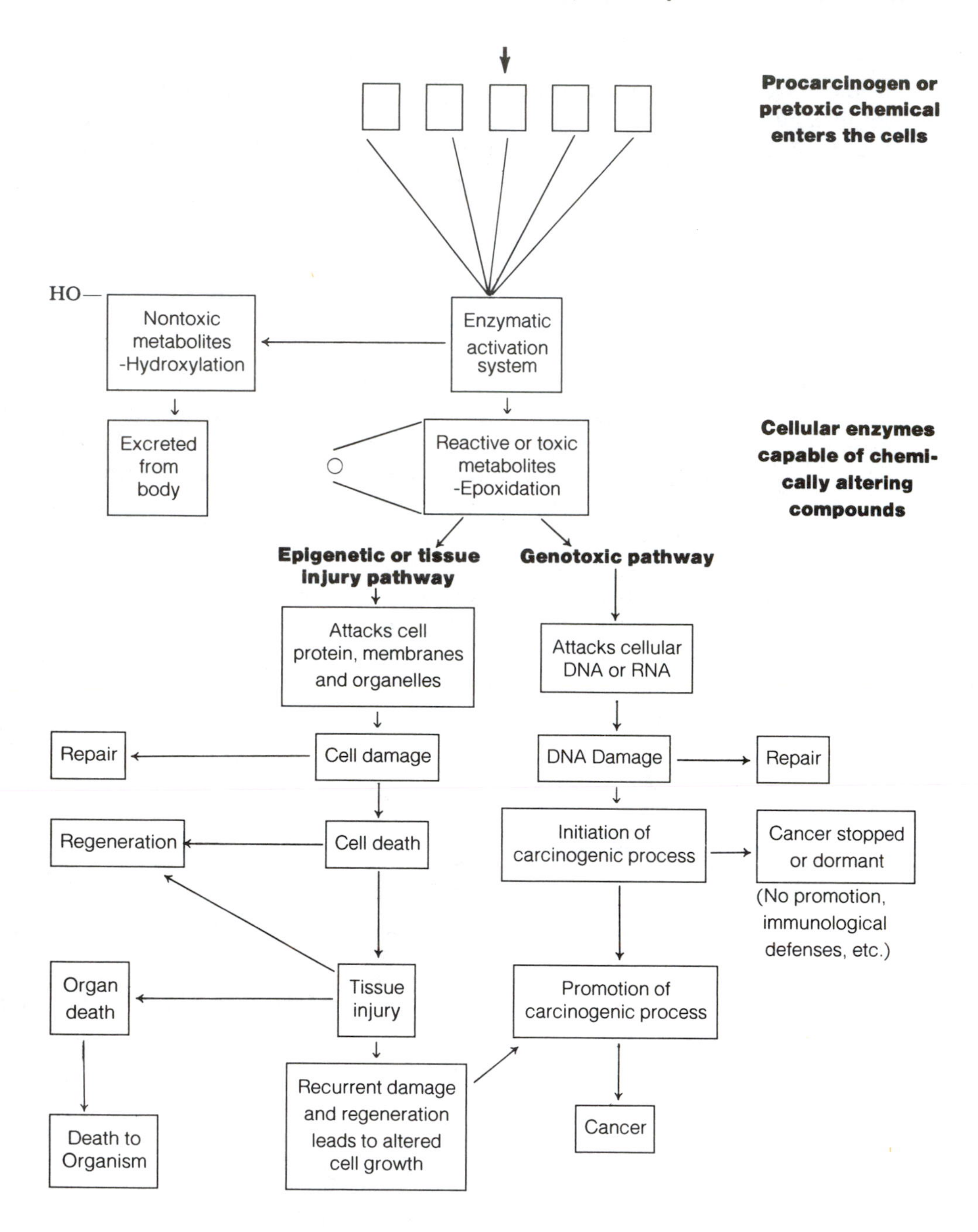

Figure 5-11. Bioactivation of chemicals and the potential cellular interactions that may occur which lead to chronic toxicities.

diseases like cirrhosis or hepatocellular carcinoma, because of the constant recurrent tissue injury it suffers from chronic exposure to a particular hepatotoxin. That is, although an exposure may be kept low enough to prevent acutely severe or fatal liver injury, if it is sustained for a long time, the recurrent injury and regeneration of scar tissue, etc., may lead to chronic disease states (Figure 5-11). Therefore, exposure to hepatotoxins carries both acute and chronic risks, and chronic diseases may result from single or multiple exposures and by a variety of mechanisms.

5.4 Liver Injury

Classifications of Injury

Chemical-induced liver injury does not result in a single type of injury. Rather, the type of lesion observed is dependent upon both the chemical involved, the dose, and the duration of exposure. Thus, hepatotoxic agents may be classified by the type of injury produced and what criteria are used to delineate the injury. For example, hepatotoxins might be classified morphologically. That is, they may be separated or identified by the areas of the liver lobule they injure or cause necrosis to. Table 5-2 contains a partial listing of hepatotoxins illustrating this classification.

Hepatotoxins may be classified as to whether they are cytotoxic (i.e., toxic to hepatocytes), or cholestatic (i.e., damaging to bile flow), as shown in Table 5-3.

Table 5-2. Hepatotoxic Agents and Liver Areas Endangered by Them.

Centrilobular Hepatotoxins	Midzonal Hepatotoxins	Periportal Hepatotoxins
Acetaminophen	Anthrapyrimidine	Acrolein
Aflatoxin	Beryllium	Albitocin
Bromobenzene	Carbon tetrachloride	Allyl alcohol
Carbon tetrachloride	Furosemide	Arsenic
Chloroform	Ngaione	Iron
DDT	Paraquat	Manganese
Dinitrobenzene		Phosphorous
Trichloroethylene		

Table 5-3. Hepatotoxic Agents and the Kinds of Liver Cells They Endanger.

Cytotoxic (Necrotic) Agents	Cholestatic Agents
Acetaminophen	Anabolic steriods
Aflatoxin	Arsphenamine
Allyl alcohol	Chlorpromazine
Bromobenzene	Diazepam
Carbon tetrachloride	Estradial
Dimethylnitrosamine	Mepazine
Phosphorous	Thioridazine
Urethane	

Hepatotoxins may also be classified by the organelles affected within the hepatocyte. For example, compounds damaging the plasma membrane or the endoplasmic reticulum are: carbon tetrachloride, thioacetamide, phallodin, dimethylnitrosamine, allyl alcohol, and many others. Chemicals toxic by virtue of their damage to mitochondria are: hydrazine, ethionine, dichloroethylene, carbon tetrachloride, phosphorous, and others. Chemicals toxic to the nucleus or its components include: beryllium, aflatoxin, galactosamine, ethionine, nitrosamines, and others.

Beside the lobular area of injury, the cells injured, or the organelles injured, hepatotoxins may be separated by their mechanisms or biochemical lesions. For example, some chemicals may induce toxicity by initiating the peroxidation of lipids, others by inhibiting protein synthesis, some by inhibiting nucleic acid synthesis, or by alkylating and inactivating enzymes or other important molecules.

The above discussion should make it clear that many types of injury are produced and that there are a number of ways to classify hepatotoxic agents according to the way the injury induced is classified.

Histopathologic Descriptions of Liver Injury

Acute liver injury can be separated into several distinct categories, as defined below.

- Lipid Accumulation: A number of agents that produce liver injury cause an abnormal accumulation of fats in the liver, primarily triglycerides. However, fatty liver by itself does not necessarily mean liver dysfunction, and the relevance of this response to injury is presently questionable.
- Cholestasis: Hepatobiliary dysfunction resulting in inflammation of the biliary tree and/or decrease of bile flow and excretion.

- Necrosis: Cell destruction or cell death.
- Hepatitis: Inflammation of the liver, usually viral in origin. It is a diffuse cell injury with some spotty or isolated liver cell necrosis.

Chronic liver injury can be separated into the following categories.

- Cirrhosis: A progressive disease of the liver characterized by diffuse damage to hepatic parenchymal cells with nodular regeneration and fibrosis. It is often associated with liver dysfunction, frequently resulting in jaundice and portal hypertension.
- Hepatocellular carcinoma: Characterized by hepatomegaly with multiple scattered nodules and large malignant tumors throughout the liver.

5.5 Evaluation of Liver Injury

Since the liver is the primary site of chemical biotransformation it is often the scene of injury by many chemical toxicants. This is particularly true for the chlorinated aliphatics, like carbon tetrachloride or trichloroethylene, and the persistent chlorinated aromatic compounds, like polychlorinated biphenyls or lindane. Thus, serum tests are often used in industry to screen for overexposures and to detect the liver injury caused by certain chemicals. Basically there are two types of tests: (1) those that measure liver function, and (2) those that measure serum levels of intracellular liver enzymes lost to the blood during cell damage. The function tests assess some normal liver function, such as the clearance from the bloodstream of a substance like bilirubin, which is largely dependent upon the liver for elimination. The enzyme tests measure the serum levels of a protein that is synthesized by the liver and under normal circumstances is at low levels within the blood. The following liver function tests are often used.

1. Dye clearance tests
   a. Bromosulfophthalein (BSP). A dye, only 2 percent of which is excreted by the kidneys and for which there is minimal uptake by muscle tissue. Seventy percent of the BSP cleared by the liver is conjugated by glutathione, while the rest is excreted unchanged. BSP clearance can be impaired during liver injury, and hepatotoxins known to increase the clearance time are the narcotics and the chlorinated hydrocarbons. BSP is also affected by cholestasis.

   b. Indocyanine green (ICG). A water-soluble dye cleared from the bloodstream by the liver and excreted unchanged. Testing ICG clearance in conjuction with the BSP clearance helps differentiate whether the loss of BSP clearance is related to liver uptake or impaired glutathione conjugation.

2. Prothrombin time. The liver produces most of the blood clotting factors; therefore, an increase in the prothrombin time may indicate a delay in clotting caused by a lack of these factors and, thus, liver insufficiency. This test can be misleading if there is a vitamin K deficiency, which would also increase the time it takes to form a fibrin clot—i.e., increase the prothrombin time.
3. Serum albumin. An important blood protein synthesized by the liver. Usually this is a rather insensitive test and serum levels of albumin are usually only decreased when severe, chronic injury is present.
4. Bilirubin. The liver conjugates bilirubin, which is a normal breakdown product of the heme from red blood cells; therefore, clearance of this, and serum levels of bilirubin, indicate the status of liver function.

The functional tests generally have limited use as early indicators of liver damage because they are fairly insensitive, registering only significant damage capable of imparing liver function in grossly measurable ways. Therefore, they are more useful for evaluating the consequences of chronic or permanent injury; often they do not or cannot reveal acute, reparable liver injury.

The hepatocellular enzyme markers provide far more sensitive tests. For example, serum gamma-glutamyl transaminase (SGGT) may be elevated after one drinks alcoholic beverages. These tests are based upon identifying and measuring the serum activity of enzymes present within liver cells. Thus, increased concentrations of these enzymes in serum indicate hepatocellular damage and leakage of these enzymes from liver cells into the bloodstream. These tests include:

1. Aminotransferases. These enzymes transfer the amine group from one amino acid skeleton to another. They detect early membrane permeability. Serum glutamic-oxaloacetic transaminase (SGOT) is found in the liver and in muscle tissue, while serum glutamic-pyruvic transaminase (SGPT) is found mainly in the liver and to a lesser extent in the heart. Thus, SGPT is a better indication of liver damage, while an increase in either enzyme by itself may not be adequate proof of liver injury (as other tissues may be contributing the enzyme to the serum). SGGT is more sensitive an indicator than either SGOT or SGPT and is elevated by both hepatocellular and biliary obstruction.
2. Serum alkaline phosphatase. An enzyme produced by the bones and the liver. That produced in the liver is useful primarily for revealing biliary obstruction; it can have higher than normal levels in children, especially in adolescents during a rapid increase in growth. The liver isoenzyme can be differentiated from the bone isoenzyme, yet it is easier to use the SGPT test or other enzyme markers than to use serum elevations of alkaline phosphatase isoenzymes as confirmation of liver injury.

3. Serum 5′-nucleotidase. A marker for hepatobiliary disease. The physiologic function of this marker is not known. Although the substance is normally increased during liver injury or pregnancy, its levels are not affected by bone disease and therefore it complements the alkaline phosphatase test by eliminating the possibility that alkaline phosphatase elevations might be coming from bone tissue.
4. Others. There are a host of other enzymes released during cell damage that can be used as a test of liver injury. Isocitrate dehydrogenase and lactate dehydrogenase are fairly sensitive indicators but are not specific for damage to liver cells because they are common to other tissues as well. Less commonly used than the above but specific for liver injury are ornithine carbamoyl-transferase, sorbitol dehydrogenase, alcohol dehydrogenase, and glutamate dehydrogenase.

Summary

The liver is a particularly interesting and important organ in the practice of industrial toxicology.

- Its unique position within the circulatory system ensures that it is the first "critical" organ to be exposed to chemicals absorbed from the gut; hence, in some instances it is sensitive to chemical attack for reasons related to chemical distribution.
- Since it is the primary organ functioning to break chemicals down into more water-soluble, less biologically active chemical species, it contains many enzymes that may instead biotransform the chemical into a reactive, more toxic form.

Although the liver performs many important physiologic functions (e.g., carbohydrate and lipid metabolism, synthesis of certain blood proteins, urea formation, etc.), chemical metabolism (i.e., biotransformation) is a major toxicological consideration, because this process impinges on every chemical entering the body, even if sometimes it does so only in a minor way.

- By oxidizing or reducing a portion of a chemical or by adding another molecule to it, the liver alters the chemical's toxicity, its tissue-binding properties, and its distribution and duration within the body.
- While for the most part this chemical change benefits the body, sometimes it leads to a toxic mutagenic or carcinogenic chemical byproduct.
- While the main drive of biotransformation is to protect the body from attaining high chemical levels within the various tissues, the lack of

specificity and predictability in the process sometimes leads to bioactivation, or an increase in a chemical's toxicity.

Metabolism of a toxic chemical has three possible outcomes.

- It may form a nontoxic metabolite.
- It may generate a toxic metabolite that is subsequently detoxified.
- Cellular and tissue injury may result because a toxic metabolite has not been rendered harmless by detoxification.

It is often the balance between these possible occurrences that determines the eventual outcome of the chemical exposure in question.

Many characteristics of a person exposed to a toxic chemical may alter the person's metabolism and, in doing so, alter chemical toxicities.

- These include diet or nutritional status, age, sex, and the genetic makeup of the individual.
- These characteristics may account for the inter-individual variations observed in the responses to chemical exposure.
- Species differences in the biotransformation of a chemical account for many differences that exist in the extent or type of toxicity observed in various species.

Because the liver is susceptible to such a variety of chemically induced toxicities, the occupational health specialist should be aware of the potential for liver toxicity, and should become familiar with clinical serum measurements that reflect the status of liver function, so as to screen or monitor for injury when excessive chemical exposure is suspected.

References and Suggested Reading

Becker, F. 1974. *The Liver: Normal and Abnormal Functions, Part A*. New York: Marcel Dekker, Inc.

Becker, F. 1975. *The Liver: Normal and Abnormal Functions, Part B*. New York: Marcel Dekker, Inc.

Cameron, H., Linsell, D., and Warwick, G. 1976. *Liver Cell Cancer*. Amsterdam: Elsevier.

Jakoby, W. B. (Ed.) 1980. *Enzymatic Basis of Detoxification, Volumes 1 and 2*. New York: Academic Press.

Jenner, P., and Teston, B. 1981. *Concepts in Drug Metabolism, Volumes A and B*. New York: Marcel Dekker, Inc.

Jollow, D., Kocsis, J., Synder, R., and Vainio, H. (Eds.) 1977. *Biological Reactive Intermediates: Formation, Toxicity, and Inactivation*. New York: Plenum Press.

Neal, R. 1980. "Metabolism of Toxic Substances." *In* Doull, J., Klaassen, K. C., and Amdur, M. O. (Eds.) *Casarett and Doull's Toxicology: The Basic Science of Poisons, 2nd Edition*. New York: Macmillan Publishing Co.

Plaa, G. 1980. "Toxic Responses of the Liver." *In* Doull, J., Klaassen, K. C., and Amdur, M. O. (Eds.) *Casarett and Doull's Toxicology: The Basic Science of Poisons, 2nd Edition*. New York: Macmillan Publishing Co.

Plaa., G., and Hewitt, W. (Eds.) 1982. *Toxicology of the Liver*. Target Organ Toxicology Series. Robert L. Dixon, Editor-in-Chief. New York: Paven Press.

Popper, H., Bianchi, L., and Reutter, W. 1977. *Membrane Alterations as Basis of Liver Injury*. Lancaster (England): MTP Press Ltd.

Zimmerman, H. 1978. *Hepatotoxicity*. New York: Appleton-Century-Crofts.

Chapter 6

Nephrotoxicity: Toxic Effects in the Kidneys

Daniel R. Goodman

Introduction

This chapter will give the occupational health professional information about:

- The importance of kidney functions.
- How toxic agents disrupt kidney functions.
- What measurements are done to determine kidney dysfunctions.
- What industrial agents cause kidney toxicity.

6.1 Basic Kidney Functions

The principal excretory organs in all vertebrates are the two kidneys. The primary function of the kidney in humans is excreting wastes from the blood in the form of urine. However, the kidney plays a key role in regulating total body homeostasis. These homeostatic functions include the regulation of extracellular volume, the regulation of calcium metabolism, the control of electrolyte balance, and the control of acid-base balance.

Important Kidney Functions Not Usually Considered as Toxic Endpoints

Renal Erythropoietic Factor. The kidney synthesizes hormones essential for certain metabolic functions. For example, hypoxia stimulates the kidneys to secrete renal erythropoietic factor, which acts on a blood globulin (proerythropoietin) released from the liver to form erythropoietin, a circulating glycoprotein with a molecular weight of 60,000 daltons. The erythropoietin acts on erythropoietin-sensitive stem cells in the bone marrow, resulting in enhanced hemoglobin synthesis and increased production and release of red blood cells from the bone marrow into the circulating blood. By this compensatory mechanism the kidneys help correct for the hypoxia. Thus, in chronic renal failure, anemia usually develops, in large part due to decreased synthesis of erythropoietic factor because of damage to the kidney tissues responsible for its synthesis.

In addition to hypoxia, androgens and cobalt salts also increase production of renal erythropoietic factor by the kidneys. In fact, administration of cobalt salts produces an overabundance of red cells in the blood (i.e., polycythemia) by this mechanism. Polycythemia has been observed in heavy drinkers of cobalt-contaminated beer.

Regulation of Blood Pressure. The kidney apparently regulates blood pressure in several ways. One is by producing renin, a proteolytic enzyme, which cleaves a plasma protein globulin to form angiotensin I. Angiotensin I is converted to angiotensin II, a potent vasoconstrictor. The angiotensin II stimulates release of aldosterone from the adrenal cortex, and aldosterone increases reabsorption of sodium in the kidney, leading to an increase in blood plasma osmolality and an increase in extracellular volume. A decrease in the mean renal arterial pressure is the stimulus controlling the kidney renin production and the compensatory increase in arterial pressure by the above mechanisms. In addition, renal disease and narrowing of the renal arteries are known to cause sustained hypertension in humans. It appears that the kidney produces vasodepressor substances that are thought to be important in the regulation of blood pressure. Thus, changes in the kidney that disturb the renin-angiotensin-aldosterone system and/or secretion of the vasodepressor substances are suspected of playing a key role in the etiology of certain forms of hypertension.

Metabolism of Vitamin D. The kidney also plays a key role in the metabolism of vitamin D, thus performing a vital function in the hormonal regulation of calcium in the body. Vitamin D_3 (cholecalciferol) is relatively inactive. The liver hydroxylates vitamin D_3 to 25-hydroxycalciferol and then the kidney hydroxylates the 25-hydroxycalciferol to 1,25-dihydroxycalciferol, the most potent active form of vitamin D. The kidney is also the key to the metabolism of parathyroid hormone, another hormone important to calcium regulation. If the kidney is damaged, thereby disrupting its role in vitamin D and parathyroid hormone metabolism, the development of a renal osteodystrophy can occur, which is characterized by skeletal disease and hyperplasia of the parathyroid gland.

From the preceding paragraphs it should be clear that the kidney plays an essential role in maintaining a number of vital body functions. Therefore, if a disruption of normal kidney function is caused by the action of a toxic agent, a number of serious sequelae can occur besides a disruption in blood waste elimination. However, for clinical purposes, alterations in the excretion of wastes are the principal endpoints for determining the action of nephrotoxins.

Nevertheless, it must be remembered that changes in the other functions may also be present, even if they are not conveniently or routinely measured as toxic endpoints.

6.2 The Physiologic and Morphologic Bases of Kidney Injury

The adult kidneys of reptiles, birds, and mammals (including humans) are nonsegmental and drain wastes only from the blood (principally breakdown products of protein metabolism). The kidneys are paired organs that lie behind the peritoneum on each side of the spinal column in the posterior aspect of the abdomen. The adult human kidney is approximately 11 cm long, 6 cm broad, and 2.5 cm thick. In adults, human male and female individual kidney weights range from 125–170 g and 115–155 g, respectively. On the medial or concave surface of each kidney is a slit called the hilus through which pass the renal artery and vein (Figure 6-1, b). From each kidney a common collecting duct called a ureter carries the urine posteriorly to where it can be voided from the body.

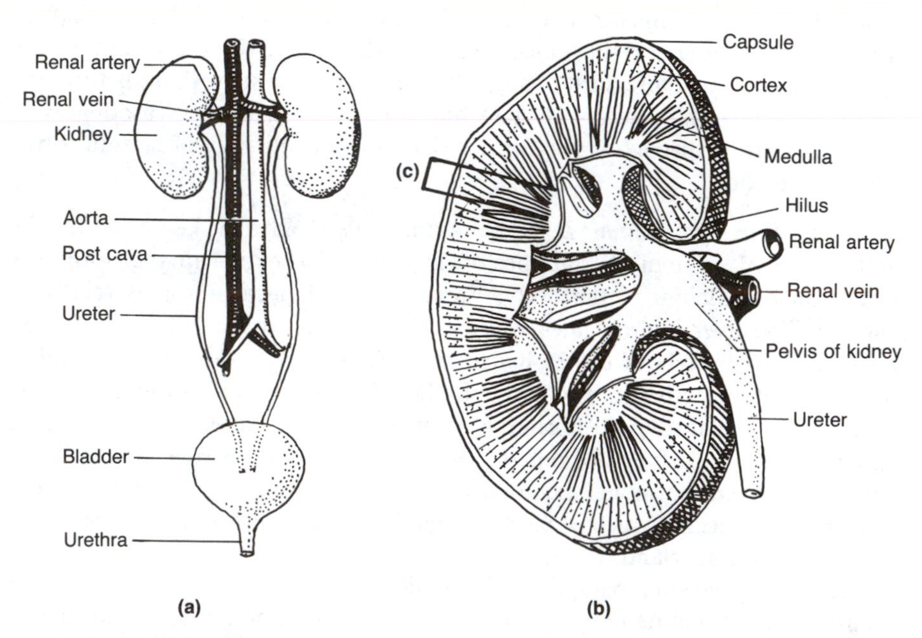

Figure 6-1. The human renal excretory system. (a) The complete excretory system. (b) Cross-section of kidney. (c) Representative section for the enlargement in Figure 6-2.

Blood Flow to the Kidneys

The kidneys comprise approximately 0.5 percent of the total body weight, or approximately 300 g in a 70-kg human. Yet the kidneys receive just under 25 percent of the total cardiac output, which is about 1.2–1.3 l blood/min, or 400 ml/100 g tissue/min. The rate of blood flow through the kidneys is much greater than through other very well perfused tissues, including brain, heart, and liver. If the normal blood hematocrit (i.e., that proportion of blood that is red blood cells) is 0.45, then the normal renal plasma flow is approximately 660 to 715 ml/min. Yet only 125 ml/min of the total plasma flow is actually filtered by the kidney. Of this, the kidney reabsorbs approximately 99 percent, resulting in a urine formation rate of only about 1.2 ml/min. Thus, the kidneys, which are perfused at approximately 1 l/min, form urine at approximately 1 ml/min or 0.1 percent of the perfusion. Because of the great blood flow to the kidneys, a chemical in the blood is delivered in relatively great quantities to this organ. In addition, because of the large mass-balance differential between blood flow and urine formation, the metabolic energy required by the kidney is great. Roughly 10 percent of the normal resting oxygen consumption is needed for the maintenance of proper kidney function. Therefore, the kidney is sensitive to agents that induce ischemia, or a lack of oxygen via a decrease in blood flow.

Barbiturate intoxication is an excellent example of a condition that induces acute renal failure because of a decrease in blood flow resulting in ischemia. Acute intoxication by barbiturates is characterized by severe hypotension (i.e., low blood pressure) and shock. The severe decrease in blood pressure results in a decrease in filtration of the plasma, resulting in a decrease (oliguria) or cessation (anuria) of urine formation. At an early stage this is called prerenal failure, and a reversal in the blood deficit to the kidney will restore normal renal function. However, a time comes when renal sufficiency cannot be restored because of the cell death caused by ischemic anoxia, and the resultant renal failure is irreversible. In this situation, the accumulation in the blood of wastes normally excreted (uremia) results in death. It should be remembered, then, that any agent or physical trauma that causes severe hypotension and shock may produce acute renal failure and eventually death by a similar mechanism.

Kidney Structure

Each human kidney consists of an outer cortex and an inner medulla (see Figures 6-1, b and 6-2). The cortex constitutes the major portion of the kidney and receives about 85 percent of the total renal blood flow. Consequently, if a toxicant is delivered to the kidney in the blood, the cortex will be exposed to a very high proportion. The cortex of each kidney in humans contains approximately one million excretory units called nephrons. Agents toxic to the kidney generally injure these nephrons, and such agents are therefore referred to as

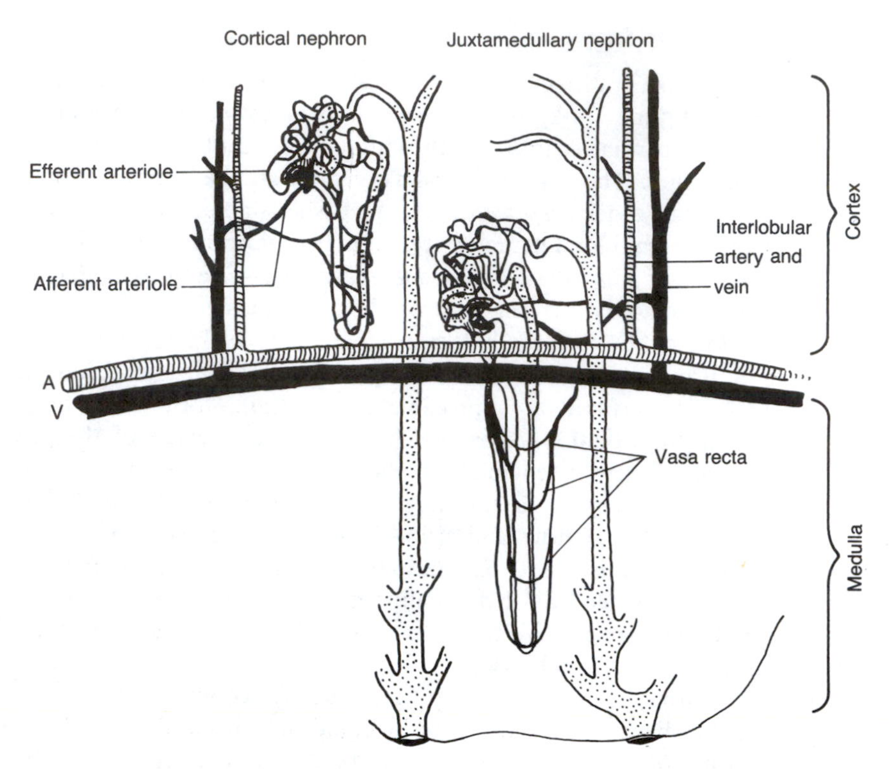

Figure 6-2. Cortical and juxtamedullary nephrons. Enlargement of representative kidney section in Figure 6-1, c. (Based on Brenner, B., and Rector, F., *The Kidney*. Philadelphia: W. B. Saunders Co., 1976.)

nephrotoxins. Degeneration, necrosis, or injury to the nephron elements is referred to as a nephrosis or nephropathy.

An individual nephron may be divided into three anatomical portions: (1) the vascular or blood-circulating portion, (2) the glomerulus, and (3) the tubular element (Figures 6-2 and 6-3). The glomerulus, which is about 200 μm in diameter, is formed by the invagination of a tuft of capillaries into the dilated, blind end of the nephron (Bowman's capsule). The capillaries are supplied by an afferent arteriole and drained by an efferent arteriole. These vascular elements deliver waste and other materials to the tubular element for excretion, return reabsorbed and synthesized materials from the tubular element to the blood circulation, and deliver oxygen and nourishment to the nephron.

The Glomerulus and Glomerular Filtration. The glomerulus behaves as if it were a filter with pores 100 Å in diameter, or about 100 times more permeable than the capillaries in skeletal muscle. Substances as great as 70,000 daltons can appear in the glomerular filtrate, but most proteins in the plasma are still too large to pass through the glomerulus. Therefore, a substance that is, for

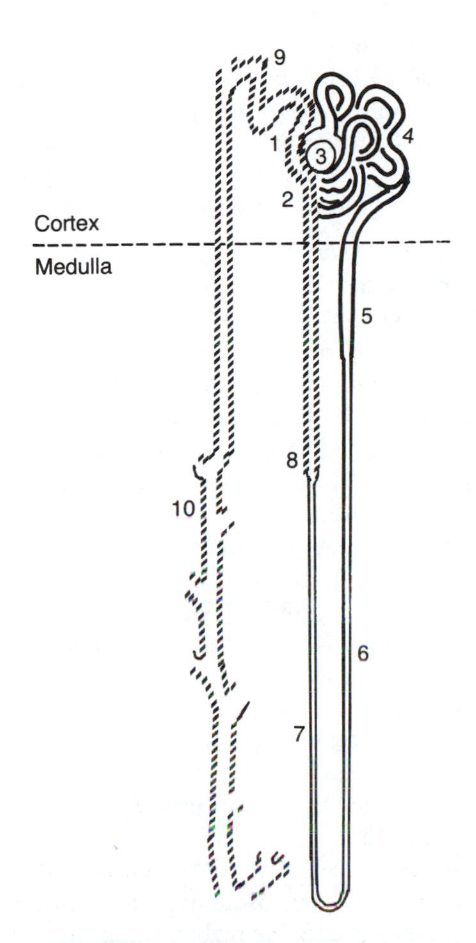

Figure 6-3. Juxtamedullary nephron. (1) Afferent arteriole; (2) efferent arteriole; (3) glomerulus; (4) proximal convoluted tubule; (5) proximal straight tubule (pars recta); (6) descending limb of the loop of Henle; (7) thin ascending limb of the loop of Henle; (8) thick ascending limb of the loop of Henle; (9) distal convoluted tubule; and (10) collecting duct. (Based on Doull, J., *et al.* (Eds.) *Casarett and Doull's Toxicology: The Basic Science of Poisons*. 2nd Ed. New York: Macmillan Publishing Company, 1980.)

example, 75-percent-bound to plasma proteins has an effective filterable concentration of one fourth its total plasma concentration. Small amounts of protein, principally the albumins, which are important chemical-binding proteins, may appear in the glomerular filtrate, but these are then normally reabsorbed.

The glomerular filter can be made more permeable in certain disease states and by actions of certain nephrotoxins, and both circumstances may result in the appearance of protein in the urine (proteinuria). Also, transient but significant proteinuria occurs normally after prolonged standing or strenuous exercise. If damage to the glomerular element is severe, the result is a loss of a large amount of the plasma proteins. If this occurs at a rate greater than the rate at which the liver can synthesize the plasma proteins, the result will be hypoproteinemia (lower than normal levels of proteins in the blood) and a concomitant edema due to the reduction in osmotic pressure. This clinical picture is sometimes referred to as the nephrotic syndrome.

Nephron Tubules and Tubular Reabsorption. The tubular element of the nephron selectively reabsorbs 98 to 99 percent of the salts and water of the initial glomerular filtrate. The tubular element of the nephron consists of the proximal tubule, the loop of Henle, the distal tubule, and the collecting duct (see Figure 6-3). The proximal tubule consists of a proximal convoluted section (pars convoluta) and a distal straight section (pars recta). Substances that are actively reabsorbed in the proximal tubule include glucose, sodium, potassium, phosphate, amino acids, sulfate, and uric acid. Essentially all amino acids and glucose are reabsorbed in the proximal tubule and virtually none normally appear in the urine. Agents toxic to the proximal tubule then will cause amino acids and glucose to appear in the urine (aminoaciduria and glycosuria). Even though 250 grams of glucose normally pass through the kidney daily, no more than 100 milligrams are usually excreted in 24 hours. However, glucose does appear in excess quantities in the urine if high blood-glucose levels produce a glucose load in the filtrate and this exceeds the resorptive capacity of the proximal tubule of the nephrons. This occurs in diabetes mellitus, where excess glucose appears in urine because excessive amounts of glucose in the blood plasma filtrate have overwhelmed the glucose transport system in the nephron.

As solutes are reabsorbed in the proximal tubule, so is water, owing to an osmotic gradient with plasma. Thus, isotonicity is maintained in the proximal tubule even though there is a selective reabsorption of solutes. Approximately 75 percent of the glomerular filtrate fluid is reabsorbed in the proximal tubule. If tubular reabsorption of substances is compromised, then less water is reabsorbed. The result is diuresis (increased urine flow) and polyuria (excess urine production). Toxic agents can cause polyuria by affecting active solute reabsorption.

Tubular Secretion. Active transport of certain organic compounds into the tubular fluid also occurs in the proximal tubule. There are two separate active secretory systems in the proximal tubule, one for anionic (negatively charged) organic chemical species, and a similar but separate system for cationic (positively charged) organic chemical species. The organic anion secretory system is the better studied. Organic cations such as tetramethyl ammonium are actively secreted but this system is not as well studied as the organic anion secretory system.

Not only are both transport systems separate, but both have their own separate competitors and inhibitors. Penicillin and probenecid are actively secreted by the organic anion secretory system. As a consequence, they inhibit the excretion of PAH (*p*-amminohippuric acid) and each other. In fact, probenicid was used in the past to prolong the half-life of penicillin in the blood since it inhibits its secretion into the proximal tubules and its subsequent excretion in the urine. It should be pointed out that these organic anions do not compete for and do not inhibit the secretion of organic cations or vice versa. Remember, then, that substances reabsorbed from the tubular element of the nephron will have a clearance significantly less than the glomerular filtration rate (approximately 125 ml/min), while those secreted into the tubules will have a clearance significantly greater than the glomerular filtration rate in the adult human.

The Loop of Henle. After the glomerular filtrate has passed the proximal tubule in the nephron, it moves into the loop of Henle. A nephron with a glomerulus in the outer portion of the renal cortex has a short loop of Henle, whereas a nephron with a glomerulus close to the border between the cortex and medulla (juxtamedullary nephrons) will have a long loop of Henle extending into the medulla and pyramids (Figures 6-2 and 6-3). Approximately 15 percent of the nephrons in humans are juxtamedullary. As the tubule descends into the medulla there is an increase in osmolality of the interstitial fluid. In the descending limb the tubular fluid becomes hypertonic (high in salt) as water leaves the tubule to maintain isoosmolality with the hypertonic interstitial fluid. However, in the thick segment of the ascending portion of the loop of Henle the tubule becomes impermeable to water, and sodium is actively transported out of the tubule with a decrease in the osmolality of the filtrate and an increase in the osmolality of the interstitial fluid. The sodium transport in the ascending limb is necessary for maintenance of the interstitial fluid-concentration gradient. An additional 5 percent of the glomerular filtrate fluid is reabsorbed in the loop of Henle, making a total of 80 percent of the total water reabsorbed at this point.

Urine Formation. Once the tubular fluid enters the distal convoluted tubule and collecting duct, it is hypotonic (low salt concentration) in comparison to blood plasma, because of the active transport of sodium out of the tubule at the loop of Henle. In the presence of vasopressin, the antidiuretic hormone, the collecting duct becomes permeable to water and the water leaves the tubular fluid in order to maintain isoosmolality. However, in the absence of vasopressin, the collecting duct is impermeable to water, which results in excretion of a large volume of hypotonic urine. Normally, another 19 percent of the original glomerular filtrate fluid is reabsorbed in the last process, making a total of 99 percent reabsorption for the fluid filtered at the glomerulus. Thus, the normal flow of urine is only about 1 ml/min, while in the absence of vasopressin it can be increased to 16 ml/min.

The kidney's ability to concentrate urine is determined by the measurement of urine osmolality. Urine osmolality can vary between 50 to 1400 mOsm/liter.

Certain nephrotoxins compromise the kidney's ability to concentrate the urine. These changes occur early after the exposure to the nephrotoxin and frequently foreshadow graver consequences.

The excretion of urea, a metabolic breakdown product of protein, is a special case. Urea passively diffuses out of the glomerular filtrate of the tubules as fluid volume decreases. At low urine flow more urea has the opportunity to leave the tubule. Under these conditions only 10–20 percent of the urea is excreted. At conditions where the urine flow is high the urea has less opportunity to diffuse through membranes with the water; this results in a 50–70 percent secretion of urea. A second factor in urea excretion is that it accumulates in the medullary interstitial fluid along a concentration gradient. Since the walls of the collecting ducts are permeable to urea fluid where they pass through the medulla, the urea content is higher than it would be if they had passed only through regions with low urea concentration.

It should be noted that passive reabsorption occurs for all nonionic compounds, while for chemicals in an ionic form there is essentially no passive reabsorption. For organic acids, a basic urine is desirable for maximization of excretion, since more of the acid will be ionized at higher pH (Haldane equation, Chapter 4). For organic bases, an acidic urine is desirable for maximal excretion, because more of the basic compound will be ionized.

6.3 Evaluation of Kidney Injury

Determining the excretion rate of certain drugs is a useful clinical procedure for diagnosing the functional status of the kidney. This rate of elimination in the urine is the net result of three renal processes:

- Glomerular filtration.
- Tubular reabsorption.
- Tubular secretion.

The rates of glomerular filtration and tubular secretion are dependent on the concentration of the drug in the plasma; and the rate of reabsorption by the tubules is dependent on the concentration of drug in the urine.

The Glomerular Filtration Rate

The glomerular filtration rate (GFR) can be measured in intact animals and humans by measuring both the excretion and plasma levels of those chemicals that are freely filtered through the glomeruli and neither secreted nor reabsorbed by the kidney tubules. The substance used should ideally be one that is freely filtered, not metabolized, not stored in the kidney, and not protein bound. Inulin, a polymer of fructose with a molecular weight of 5200, meets these criteria. For measuring the glomerular filtration rate the inulin is allowed to equilibrate within the body, and then accurately timed urine specimens and plasma samples are collected.

The following general formula is used to determine the clearance in this procedure:

$$\frac{U_a(V)}{P_a} = \mathrm{Cl},$$

where

U_a = concentration of substance a per milliliter urine;
V = urine volume excreted per unit time;
P_a = concentration of substance a per milliliter of plasma;
Cl = clearance of substance per unit of time.

For clearance of inulin (in), the following values can be used to demonstrate a sample calculation:

$U_{\text{in}} = 31$ mg/ml;
$V = 1.2$ ml/min;
$P_{\text{in}} = 0.30$ mg/ml.

Thus,

$$\frac{(31\ \text{mg/ml}) \times (1.2\ \text{ml/min})}{0.30\ \text{mg/ml}} = 124\ \text{ml/min}.$$

The normal human glomerular filtration rate in adult humans is about 125 ml/min and inulin clearance is routinely used as a measure of glomerular function. The GFR is not only a measure of the functional capacity of the glomeruli, it also indicates the kidney's ability to concentrate urine by removal of water. By comparing the amount (milliliters) of urine voided in one minute to the amount (milliliters) of plasma cleared, information can be gained as to the amount of water reabsorbed during passage through the tubules.

Diseases or nephrotoxins that affect the glomerulus or those that produce renal vascular disease have a profound effect on the glomerular filtration rate. Indeed, any significant renal disease or nephrotoxic compromise can decrease the glomerular filtration rate. It should also be realized that any agent inducing severe hypotension or shock will likewise reduce the glomerular filtration rate.

Measurement of certain natural endogenous substances in the blood can be used to assess glomerular function as well. The measurement of blood-urea-nitrogen (BUN) and plasma creatine are two endogenous compounds routinely measured for the clinical assessment of glomerular function. As glomerular filtration decreases, BUN and plasma creatinine become more elevated. Normal BUN ranges from 5 to 25 mg/100 ml, while serum creatinine ranges from 0.5 to 0.95 mg/ml of serum.

Nephrotoxins may also disrupt the selective permeability of the glomerular apparatus. Normally, the result is an increase in porosity in the glomerulus; protein enters the glomerular filtrate and subsequently the urine. Therefore, if a compound causes an excretion of large amounts of protein into the urine it must be suspected as a nephrotoxin, and measurement of protein in urine, particularly those of high molecular weight, is used to determine which chemicals produce toxic changes to the glomerulus. The normal excretion of protein in humans is no more than 150 mg in 24 hours.

Renal Plasma Flow

Some organic acids, such as *p*-aminohippuric acid (PAH), can be used in clearance studies to obtain information about the total amount of plasma flowing through the kidneys. PAH is transported so effectively that it is almost completely removed from the plasma in a single passage through the kidney (i.e., 80–90 percent). Any chemically induced reduction in the PAH clearance may be caused by either a disruption of the active secretory process or by an alteration of the renal blood flow.

In a clinical setting, measurements can be made of the concentration of PAH per milliliter of plasma (P_{PAH}), of the concentration of PAH per milliliter of urine (U_{PAH}), and of the volume of urine excreted per minute (V). Using the formula that was previously discussed, the clearance of PAH in ml/min can be calculated. This calculation will represent the rate of plasma flow through the kidneys (average renal plasma flow in the normal, healthy adult male is about 650 ml/min).

Excretion Ratio

Another useful calculation for evaluating kidney injury is the excretion ratio:

$$\text{Excretion ratio} = \frac{\text{Renal plasma clearance of drugs (ml/min)}}{\text{Normal GFR (ml/min)}}.$$

If the ratio is less than 1.0 it indicates that a drug has been partially filtered, perhaps also secreted, and then partially reabsorbed. A value greater than 1.0 indicates that secretion, in addition to filtration, is involved in the excretion. A substance that is completely reabsorbed, such as glucose, would have an excretion ratio of 0, and a substance such as PAH that is completely cleared can have a ratio of about 5.

Additional Clinical Tests

Alterations in renal function can be determined by a variety of other tests. A battery of such tests includes urinary pH, measurement of urine volume, and a determination of the excretion of sodium and potassium. An excess of protein or the appearance of sugar in the urine would indicate abnormalities in renal function as would changes in urine sediments. These are all general tests but they can provide information about the changes in total kidney function.

6.4 Nephrotoxic Agents

Cadmium

The kidney is the organ most sensitive to the toxic effects of cadmium. Workers in factories where nickel/cadmium batteries are manufactured, who are exposed to excessive amounts of cadmium oxide, exhibit consistent proteinuria, and proteinuria appears to be the most sensitive indication of the renal toxicity of cadmium. The proteins appearing in the urine are usually of low molecular weight (e.g., 20,000 to 30,000 daltons). The presence of largely low-molecular-weight proteins suggests interference with protein absorption in the nephron at the site of the proximal tubule. Proximal tubule damage of the nephrons caused by cadmium is also evidenced by glycosuria, aminoaciduria, and the diminished ability of the kidney to secrete PAH.

In Japan excessive cadmium intake was also linked to a peculiar form of renal osteodystrophy known as ouch-ouch disease or *itai-itai byo*. It has been proposed that this disease is caused by excessive loss of cadmium and phosphorus in the urine, combined with dietary calcium deficiency.

The kidney naturally accumulates cadmium. Normally cadmium accumulates in the kidney over the lifetime of the individual until the age of 50. About 50 percent of the total burden of cadmium in the body is borne by the liver and kidney, with the kidney having 10 times the concentration of the liver. Cadmium induces synthesis in the liver of metallothionein, a protein with a high binding-affinity for cadmium. While metallothionein acts to protect certain organs, such as the testes, from cadmium toxicity, it may enhance cadmium toxicity in the kidney. Chronic cadmium exposure has also been implicated as a factor in hypertension. However, while the development of hypertension may involve the kidney, the role of cadmium in the etiology of hypertension in humans is far from conclusive.

Mercury

Inorganic mercury (Hg^{++}) is a classical nephrotoxin. It is used as a model compound for producing kidney failure in animals, and massive doses of mercuric ion can cause fatal renal failure. A brief polyuria is followed by oliguria or even anuria. The anuria (kidney failure) leads, of course, to a life-threatening accumulation of bodily wastes and may last many days. If recovery occurs, a polyuria follows, which is probably caused by a decreased sodium absorption in the proximal tubule. Such disturbances in tubular function may last several months.

The part of the nephron most sensitive to mercuric ion toxicity is the pars recta or straight portion of the proximal tubule (Figure 6-3). PAH secretion principally occurs in the pars recta and PAH transport is very sensitive to mercuric ion. In contrast, the mercuric ion has much less effect on glucose absorption, because this occurs principally in the relatively unaffected convoluted portion of the proximal tubule. At higher levels, however, the whole nephron becomes affected.

Acute mercury poisoning is rare. In chronic exposure, mercury-caused renal toxicity is characterized by proteinuria. If severe, a large loss of plasma protein occurs, which causes hypoproteinemia (low amounts of protein in the blood) and edema in the extremities of the body. The mercuric ion appears to cause both tubular and glomerular damage upon chronic exposure in humans.

Lead

Lead is a known nephrotoxin in humans. Lead causes damage principally to the proximal tubule of the nephron. Reabsorption of glucose, phosphate, and amino acids is depressed in the proximal tubule. This leads to glycosuria, aminoaciduria, and a hyperphosphaturia with hypophosphatemia. These changes are reversible upon treatment with a chelating agent such as ethylene-diamine tetraacetic acid (EDTA), but only when the lead exposure has been relatively short. Long-term, prolonged exposure to lead may cause an irreversible dysfunction and morphologic changes. This is manifested by intense interstitial fibrosis accompanied by tubular atrophy and dilation. The glomeruli will be involved in later stages of the disease. Eventually, long-term lead exposure syndrome results in renal failure and death. There has been linkage of the chronic renal damage to saturnine (lead-induced) gout, in which uric acid is increased in the kidney.

Other Toxic Metals

Table 6-1 gives a listing of those metals known to be toxic to the kidneys. As with most nephrotoxicities, the proximal tubule appears to be the most sensitive to toxic effects, with more extensive nephron involvement at higher dosages.

Table 6-1. Metal Nephrotoxic Agents.

Metal	
Cadmium	Metals of principal concern
Lead	Metals of principal concern
Mercury	Metals of principal concern
Arsenic	
Bismuth	
Chromium	
Platinum	
Thallium	
Uranium	

Halogenated Hydrocarbons

Carbon tetrachloride (CCl_4) and chloroform ($CHCl_3$) are nephrotoxins. Again, the proximal tubule appears to be the portion of the nephron most sensitive to damage by these agents. However, lesions are seen in other parts of the nephrons as well. It should be noted that carbon tetrachloride causes severe hepatic necrosis in man, but the ultimate cause of death is kidney failure.

It appears that chloroform and carbon tetrachloride are activated to a toxic chemical species in the kidney by a mixed-function oxidase system similar to that found in the liver. The toxic metabolite convalently binds to tissue macromolecules in the kidney, and this leads to nephrotoxicity (see Chapter 5).

Bromobenzene, tetrachloroethylene, and 1,1,2-trichloroethylene also produce toxic effects to the kidney similar to those of chloroform and carbon tetrachloride.

Methoxyflurane (1,1-difluoro-2,2-dichloromethyl ether) is a halogenated surgical anesthetic that causes renal failure in humans and animals. Its kidney toxicity is characterized with a polyuria-type renal failure—i.e., a net loss of fluid with concomitant increases in serum osmolality, serum sodium, and BUN. Methoxyflurane is metabolized to inorganic fluoride anion and oxalate (as described in the next discussion). The fluoride anion has been shown to be responsible for acting on the collecting tubules; this results in vasopressin resistance and causes polyuria.

Agents Causing Obstructive Uropathies

A number of agents cause nephrotoxicity through physical deposition in the tubular sections of the nephron. Certain chemical agents can be concentrated in the tubular fluid to levels well above their solubility limit in water. The result is that crystals are deposited in the kidney tubules, causing physical damage. Methotrexate and sulfonamide drugs can cause nephrotoxicity by this mechanism.

Acute renal failure of this type is also associated with the ingestion of ethylene glycol. Ethylene glycol is metabolized to oxalic acid by the body; the acid in turn is deposited in the lumen of the tubule of the nephron as well as within the cell of the tubule as insoluble calcium oxalate salt. However, the ethylene glycol additionally appears to cause a nephrotoxicity to the proximal tubule, which is independent of oxalate deposition. The deposition of large quantities of oxalate crystals in the tubular elements of the nehprons probably contributes to the nephrotoxicity observed. Oxalate found in the leaves of rhubarb is of a sufficient quantity that it can cause deposition of oxalate crystals in the tubular elements of the nehpron, and can lead to nephrotoxicity. Part of the nephrotoxicity observed to be caused by methoxyflurane is likewise believed to be caused by deposition of calcium oxalate crystals in the tubular elements of the nephrons.

Agents Producing Pigment-Induced Nephropathies

A number of chemicals can cause the release of certain pigments such as methemoglobin, hemoglobin, and myoglobin into the blood. When this occurs, an associated acute renal failure may develop. Arsine gas causes massive hemolysis of red blood cells, which results in hemoglobinuria and associated renal failure (see Chapter 4 for a general listing of hemolytic agents, methemoglobin formers, etc.).

Heroin overdosage can result in a prolonged pressure on dependent muscles and a lysis of the muscle cell leading to a release of myoglobin into the blood. Heroin may also cause some direct lysis of the muscle cells. The result can be myoglobinuria and ultimately acute renal failure. Aniline dyes are another group of chemicals that have been shown to release methemoglobin, with an associated renal failure.

Therapeutic Agents

Table 6-2 lists a number of therapeutic agents known to cause nephrotoxicity.

Acetaminophen appears to cause nephrotoxicity by being oxidized by the microsomal P-450 oxygenase system in the renal cortex to the form of a toxic metabolite. The microsomal P-450 oxygenase system of the kidney is similar to that of the liver (see Chapter 5).

Cephalosporadine reaches high toxic concentrations in the nephron because the organic ion transport system of the proximal tubule secretes it into the tubule. The nephrotoxicity of cephalosporadine can be diminished by compounds that compete with the organic-anion-secretion system in the proximal

Table 6-2. Therapeutic Agents Known to Cause Nephrotoxicity.

Acetaminophen (analgesic)
Aminoglycoside antibiotics
Gentamycin
Kanamycin
Neomycin
Streptomycin
Amphotericin B (antibiotic)
Cephalosporadine (antibiotic)
Colistimethate (antibiotic)
Polymyxin B (antibiotic)
Tetracyclines (particularly outdated formulations) (antibiotics)

tubule, such as probenicid. The resulting decrease in tubular concentration of cephalosporidine in tubular fluid results in elimination of toxicity.

Another number of therapeutic agents can, in certain individuals, elicit a nephrotoxicity by an allergic type of reaction. However, such nephrotoxicities are usually only rarely encountered.

Summary

The kidney performs a number of functions essential for the maintenance of life:

- Elimination of waste products (particularly nitrogen-containing wastes from the metabolism of proteins) from the blood.
- Regulation of acid-base balance, extracellular volume, and electrolyte balance.

Toxic agents that disrupt these key functions can be life threatening.

The kidney is a highly metabolic organ sensitive to deprivation of oxygen, and any agent that significantly impedes renal flow will cause two adverse sequelae; acute renal failure that can result in death.

- First, less blood plasma will reach the kidney, resulting in a decrease in removal of blood wastes with a resulting increase of wastes in the blood (i.e., uremia).
- Second, if blood flow is compromised long enough, tissue ischemia will result in irreversible organ damage.

Nephrotoxins, agents toxic to the nephron, the principal excretory unit of the kidney, also disrupt key life-preserving functions.

- The glomerulus normally filters out the high-molecular-weight proteins from the blood. However, toxic agents will increase its permeability, allowing these proteins to appear in urine.
- Agents that damage the tubular element of the nephron will compromise its ability to reabsorb solutes such as glucose and amino acids, which are necessary for normal maintenance of the body, or disrupt sodium transport out of the nephron tubule, which could result in diuresis or excess urine formation or an unbalancing of the body's ionic (salt) homeostasis.
- If damage to the nephron is excessive, renal failure can decrease or completely stop urine flow, and cause death by poisoning from the body's own waste products.

Many agents directly toxic to the nephron are commercially or industrially important.

- Mercury, lead, and cadmium are industrially the most important nephrotoxic metals.
- Halogenated hydrocarbons, particularly carbon tetrachloride and chloroform, are nephrotoxic.
- Certain therapeutic agents, such as phenacetin, aspirin, and the aminoglycoside antibiotics are directly nephrotoxic.

Chemicals can cause nephrotoxicity indirectly.

- Some agents deposit crystals in the tubular element of the nephron, resulting in physical damage.
- Hemolytic agents such as arsine gas are capable of pigment neuropathy by releasing hemoglobin into the blood.

References and Suggested Reading

Berndt, W. O. 1982. "Renal Methods of Toxicology." *In* Hayes, A. W. (Ed.) *Principles and Methods of Toxicology*. New York: Raven Press. pp. 447–474.

Brenner, B., and Rector, F. 1976. *The Kidney*. Philadelphia: W. B. Saunders Co.

Ganong, F. 1973. *Review of Medical Physiology*. Los Altos (California): Lange Medical Publications. pp. 510–532.

Hook, J. B. 1980. "Toxic Responses of the Kidney." *In* Doull, J., Klaassen, C. D., and Amdur, M. O. (Eds.) *Casarett and Doull's Toxicology: The Basic Science of Poisons*. 2nd Edition. New York: Macmillan Publishing Company. pp. 232–245.

Pitts, R. 1968. *The Physiology of the Kidney and Body Fluids*. 2nd Edition. Chicago: Year Book Medical Publications.

Porter, G. A., and Bennett, W. A. 1981. "Toxic Nephropathies." *In* Brenner, B. M., and Rector, F. C. (Eds.) *The Kidney*. 2nd Edition. Vol. II. Philadelphia: W. B. Saunders Co.

Ullrich, K., and Marsh, D. 1963. "Kidney, Water, and Electrolyte Metabolism." *Ann. Rev. Physiol., 25:* 91.

Weiner, I. 1967. "Mechanisms of Drug Absorption and Excretion: The Renal Excretion of Drugs and Related Compounds." *Ann. Rev. Pharmacol., 7:* 39.

Chapter 7

Neurotoxicity: Toxic Effects in the Nervous System

Robert C. James

Introduction

The purpose of this chapter is to inform the reader of the effects that workplace exposures to chemicals may have in the nervous system. Often a particular chemical may be better known for producing toxicity or symptoms of acute poisoning in other organs; and yet this same chemical may also be capable of inducing chronic disorders in the nervous system as well. In reading this chapter, bear in mind that:

- The nervous and endocrine systems together act to control the functions of the rest of the body's organs.
- The nervous system is unique in its complexity and its variety of control. It comprises the peripheral nerves and the central nervous system (i.e., spinal cord and brain). Within these two subdivisions are numerous cell types with specific and differing functions and physiology.
- Neurotoxins are diverse compounds that may be toxic to specific regions, specific cell types, and specific cell functions within the nervous system.

The complexity of the nervous system and the selectivity of various neurotoxins together account for the wide variety of neurologic manifestations observed after chemical injury. For example, neurotoxins may selectively impair:

- Protein synthesis.
- The propagation of electrical impulses along nerve axons.
- Neurotransmitter activity.
- The maintenance of the myelin sheath.

Some neurotoxins are capable of fairly selective injury and may damage only hearing or sight, or specific portions of the brain or peripheral nerves.

7.1 The Mechanism of Neuronal Transmission

Transmission of Nerve Impulses

Figures 7-1 and 7-2 depict the important electrochemical changes that occur during the conduction of an impulse down a nerve. The following is a brief description of the physiology of neuronal transmission.

The resting nerve membrane is largely impermeable to both sodium (Na^+) and potassium (K^+) ions. Through active transport it pumps potassium into the cell and pumps sodium out of the cell. Thus, while the membrane is at rest, there is a larger concentration of sodium outside the cell than within it, and a larger concentration of potassium inside the cell than outside it. Although sodium attempts passive diffusion down its concentration gradient and into the cell while potassium attempts the opposite, the selective permeability of the membrane to these ions works with the active transport system to maintain concentration differentials across the membrane. And while the membrane is not completely impermeable, it is less permeable to sodium than to potassium. Thus, at steady state, there is more potassium outside the cell than is counterbalanced by intracellular sodium. This difference in the total concentration of positive ions leads to an electrochemical gradient across the membrane such that there is a net positive charge outside and a net negative charge inside the cell (see Figure 7-1, a).

During stimulus (Figure 7-1, b), a portion of the nerve (the axon) becomes permeable to sodium, and sodium ions rush into the cell, down the established electrochemical gradients. This portion of the membrane becomes depolarized as the positive charge moves into the cell; through a positive feedback cycle, the sodium permeability increases further, leading to a rapid rising phase of the action potential of the cell. As can be seen in parts (c) and (d) of Figure 7-1, the local depolarization affects the adjacent gradient, which in turn increases the sodium permeability of nearby portions of the membrane. This causes a self-sustaining wave of depolarization—a small local current—to traverse the nerve, as the flow of ions across the membrane at each successive point causes a change in the membrane permeability at the next point. In this manner an electrochemical impulse travels down the neuronal axis.

Repolarization begins in the same place the impulse started (see Figure 7-1, d). As the electric potential becomes positive within the cell itself, the action potential is shut off as the sodium permeability is terminated and potassium permeability begins to increase. This allows potassium to flow down its concentration gradient and out of the cell, to reestablish the resting state electrical differential across the membrane. The membrane once again becomes

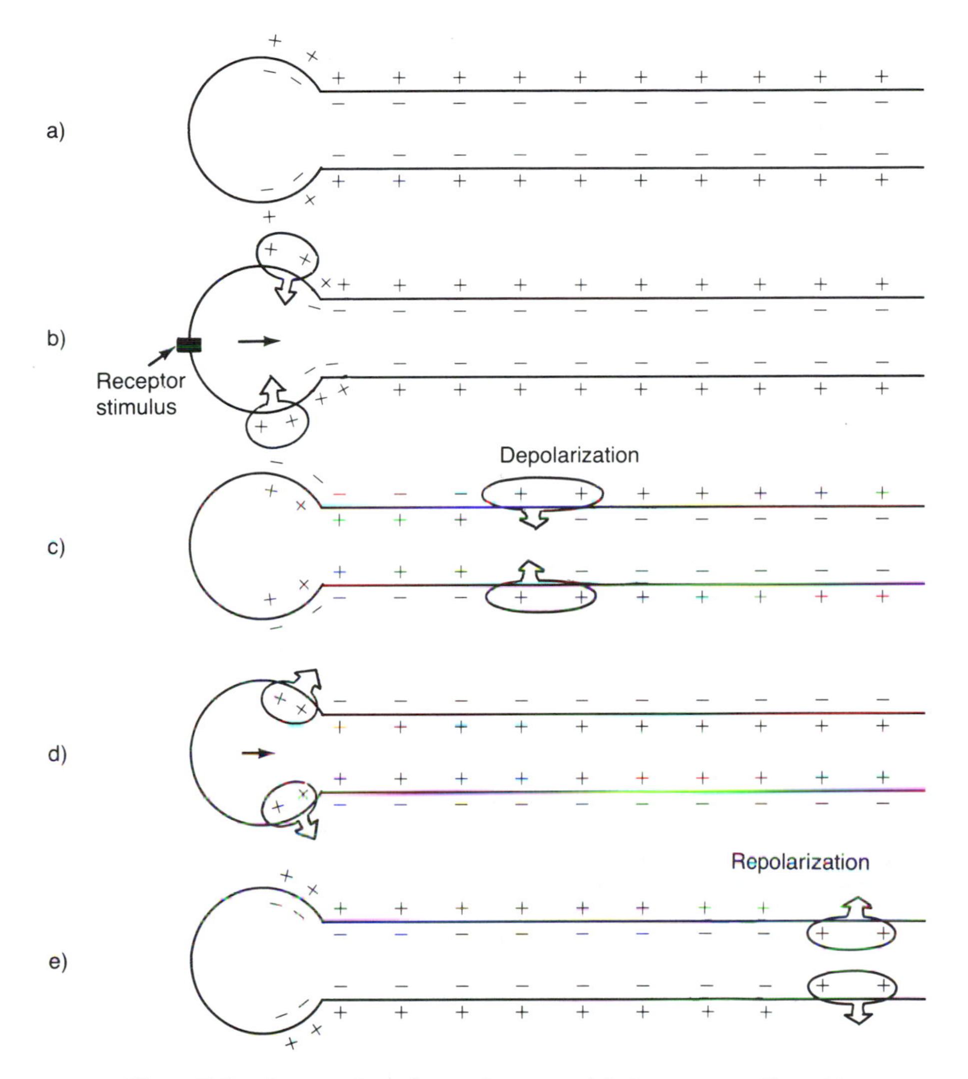

Figure 7-1. Propagation of an action potential along a nerve fiber. (a) The resting electrochemical potential of a nerve. (b) Stimulation alters the sodium permeability of the nerve. (c) As sodium ions rush in, the adjacent gradient begins to depolarize, which increases sodium permeability and allows sodium to enter this part of the nerve as well. This action propagates down the nerve as a small, local current. (d) Repolarization begins in the same place the impulse started. The high positive charge inside the cell increases the potassium permeability. Potassium ions flow out of the cell and reestablish the resting potential. (e) Repolarization travels down the nerve until it is complete.

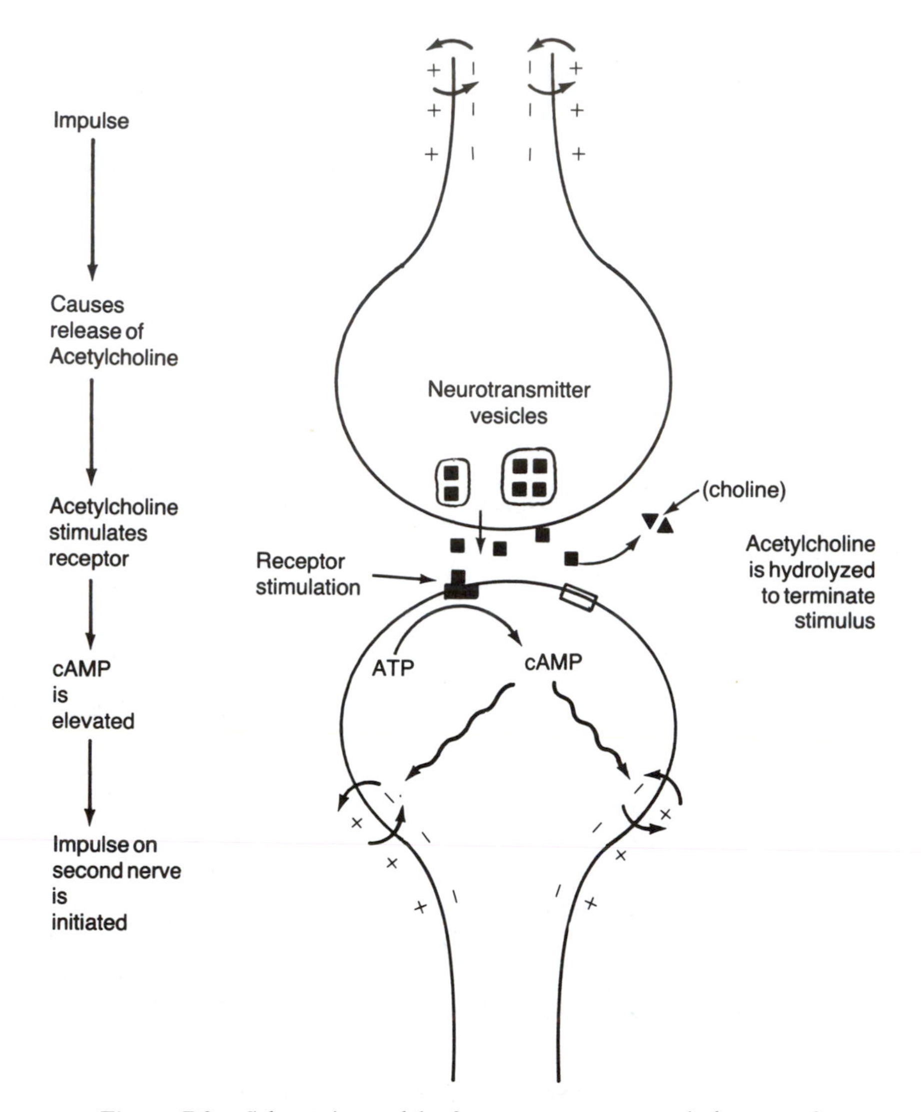

Figure 7-2. Schematic model of a nerve synapse and the use of neurotransmitters to pass the stimulus from nerve to nerve. Neurotransmitters are released and stimulate receptors. Receptor stimulation increases cAMP levels, which affect sodium/potassium ATPase activity and hence the electrochemical gradient, membrane permeabilities, etc. Stimulus is ended either by (1) the breakdown of acetylcholinesterase, or (2) the reuptake of epinephrine in adrenergic nerves.

selectively permeable to both ions, and active transport pumps sodium out of the cell and potassium into it, thus re-establishing the original offsetting concentration gradients.

Neurotransmitter Activity

When the impulse reaches the end of the nerve, it causes the release of neurotransmitters, the chemicals responsible for passing the impulse along to the next nerve. This is accomplished because, as in Figure 7-2, the neurotransmitter (in this case acetylcholine) diffuses from the end of the nerve, transmitting the stimulus across the synaptic junction by binding to and thereby stimulating receptors on the postsynaptic nerve. The stimulation of these receptors then triggers the impulse down the axis of this nerve and along to the next one in the neural chain.

It now may be clear from this short discussion that there are two mechanisms by which toxins can affect neural transmission. One is by altering the ion flux that generates the electrical impulse; the other is by affecting the activity of the neurotransmitter that is released to pass the impulse on to the next nerve. The following sections discuss neurotoxins that work by both mechanisms.

7.2 Agents That Disrupt Neuronal Transmission

Blocking Agents

Blocking agents are toxins that prevent the continuation of the electrical impulse. Blocking agents include:

1. Botulinum toxin. This prevents the release of acetylcholine, the neurotransmitter responsible for continuing electrical activity in the parasympathetic portion of the nervous system and at neuromuscular junctions.
2. Tetrodotoxin (produced in puffer fish) and Saxitoxin (produced in dinoflagellates). These act by blocking the sodium channels and the influx of sodium into the cell, which represents the initial phase or initiating portion of the impulse.

Depolarizing Agents

These agents depolarize the cell or eliminate the electrochemical gradient that normally exists within it.

1. Batrachotoxin (alkaloid secreted by frogs). This increases the membrane permeability to sodium, destroying the sodium gradient and the established electrical potential.

2. Dichlorodiphenyl trichloroethane (DDT). This chemical depolarizes the presynaptic nerve terminal repeatedly by increasing sodium permeability. Each impulse becomes exaggerated and a repetitive firing is induced after the initial stimulus, which eventually leads to convulsant activity.
3. Pyrethrins. These are similar in action to DDT.

Stimulants

Stimulants are agents that increase the excitability of neurons.

1. Strychnine. This increases the level of neuronal excitability in the central nervous system by preventing the activity of inhibitory neurons. Strychnine acts as a competitive antagonist of the inhibitory transmitter (glycine) at postsynaptic sites.
2. Picrotoxin (seeds of *Anamirta cocculus*, "fishberries"). Picrotoxin blocks presynaptic and postsynaptic inhibition of inhibitory neurons by antagonizing the inhibitory transmitter GABA (γ-aminobutyric acid).
3. Xanthines (caffeine, theophylline, theobromine). These prevent the breakdown of cAMP (adenosine 3′ : 5′-cyclic phosphate). Cyclic AMP acts as second messenger in nerve cells, apparently one effect of which is to alter the active transport system that maintains the sodium/potassium concentration differential.

Depressants

Depressants are agents that decrease the excitability of neurons.

1. Volatile organics (halothane, methylene chloride, carbon tetrachloride, butane, etc.). The exact mechanism by which these chemicals work is unknown; depression is possibly related to partition coefficients and lipophilicity; it is though they stabilize the nerve membrane and decrease ion fluxes (Na^+, K^+, Ca^{++}).
2. Alcohol. This blocks impulse conduction by decreasing sodium and potassium conductance.
3. Barbiturates. Their mechanism remains unknown, but may be related to depressed neuronal metabolism, respiration, and oxygen consumption. These compounds also appear to decrease the release of neurotransmitters at the synapse.

Receptor Antagonists

These agents bind to the postsynaptic receptors without eliciting activity, thereby preventing the actual neurotransmitter from activating the receptor and initiating an impulse.

1. Anticholinergic compounds (atropine, scopolamine, and related belladonna alkaloids). These chemicals competitively bind the receptors

of the cholinergic nerves (i.e., those nerves for which acetylcholine is the neurotransmitter).

2. Antiadrenergic compounds (phenoxybenzamine, phentolamine, tolazoline, propranolol, etc.). These bind to the receptors of adrenergic nerves and prevent the action of the neurotransmitters epinephrine (adrenalin) and norepinephrine.

Anticholinesterase Agents

These agents are specific for the cholinergic nerves for the nervous system. They cause an increased stimulation of these nerves by inhibiting the enzyme acetylcholinesterase, an enzyme that ends the receptor stimulation of acetylcholine by hydrolyzing it into its basic, inactive components (see Figure 7-2).

1. Organophosphate insecticides (parathion, malathion, diazinon, etc.). These insecticides bind to acetylcholinesterase, thereby preventing the breakdown of acetylcholine. Most of these agents produce an irreversible phosphorylation of the enzyme; for this reason exposures may be additive and the effects may last a long time until the new enzyme is synthesized. Treatment consists of administering atropine, which blocks acetylcholine's effects upon the receptor, and pralidoxime (2-PAM), which, if given soon enough, may reactivate the enzyme by removing the phosphorylating group. However, an aging process occurs that limits 2-PAM's effectiveness if it is not used within a short time after the exposure.
2. Carbamate insecticides (carbaryl or sevin, aldicarb, etc.). These are similar in mechanism to the organophosphate insecticides, but the acetylcholinesterase enzyme is carbamylated. The bond formed by this reaction is not as stable, and hydrolysis occurs fairly quickly; thus, the inhibition, while noncompetitive, is relatively brief.
3. Reversible, competitive anticholinesterase inhibitors (physostigmine, adrophonium, neostigmine). These agents act as false substrates and competitively inhibit acetylcholinesterases.

Neuromuscular Blocking Agents

Neuromuscular blocking agents antagonize the junction between the muscle and the nerves controlling it.

1. Curare. This agent is a competitive antagonist of acetylcholine at the postjunctional membrane of the muscle fibers. Curare blocks acetylcholine's transmitter action.
2. Succinylcholine. It causes a persistent depolarization of muscle fiber membrane. The muscle fibers contract or twitch for a brief period as depolarization occurs; paralysis follows as normal repolarization and the return to the normal resting potential are prevented from occurring.

7.3 Anoxia as a Basic Toxicity of the Nervous System

The neurons within the central nervous system are cells with a high metabolic rate but with little capacity for anaerobic metabolism. Consequently, inadequate oxygen flow to the brain causes cell death within minutes and, for some neurons, death may occur before the complete cessation of oxygen or glucose transport. Three types of anoxia (i.e., oxygen deprivation) are generally recognized: asphyxial; ischemic; and cytotoxic.

Asphyxial anoxia is an inadequate oxygen delivery in the presence of adequate blood flow. This may result from a paralysis of or a decrease in respiratory function caused by such compounds as curare, barbiturates, narcotics, etc. Oxygen supply may also be decreased during normal blood flow as a result of interference with the oxygen-carrying capacity of blood by carbon dioxide, carbon monoxide, the methemoglobin-forming nitrites, and methylene chloride. These chemicals are sometimes also referred to as "chemical asphyxiants" because, like asphyxiant gases, they decrease the amount of oxygen carried by the blood. Asphyxiant gases are those inert gases that, by their increased concentration in the air, displace or dilute the oxygen content of the air inhaled to such a point that an inadequate amount of oxygen is delivered to the tissues.

Ischemic anoxia results from a decrease in blood flow while the oxygen content remains adequate. The net result is a decreased oxygen supply to the tissues. Examples of ischemic anoxia are cardiac arrest, hemorrhage, thrombosis, or hypotension.

Cytotoxic anoxia is caused by an interference with cellular metabolism such that, while the blood flow and oxygen content of the blood are normal, the proper utilization of oxygen is blocked. Examples of anoxic agents operating through cytotoxic mechanisms are insulin in excess (hypoglycemia), cyanide, hydrogen sulfide, azide, dinitrophenol, malononitrile, and methionene sulfoxime. (See Chapter 4.)

7.4. Agents That Cause Physical Damage to the Nervous System

Agents Damaging the Myelin Sheath

The myelin sheath covers many neurons within the nervous system and basically serves as an insulator on portions of the nerve and facilitates the speed of impulse conduction along nerve fibers. A number of chemicals produce toxic neuropathies in animals or humans by disrupting or destroying the myelin sheath.

Table 7-1. Demyelinating Neurotoxins.

Acetylmethyl tetramethyl tetralin
Bicyclohexanone oxalyldihydrazone
Chronic cyanide or carbon monoxide
Cyanate
Diphtheria toxin
Ethidium dibromide
Ethylnitrosourea
Hexachlorophene
Isoniazid
Lead
Lysolecithin
Pyrithiamine
Salicylanilides
Tellurium
Thallium
Triethytin

Numerous demyelinating neurotoxins are shown in Table 7-1. The major symptoms noticed when these agents in Table 7-1 affect the brain are dullness, restlessness, muscle tremor, convulsions, loss of memory, epilepsy, and idiocy. The peripheral nervous system symptoms include: neuritis, palsy or muscle weakness, sensory disturbances, and hair loss.

Agents Damaging the Peripheral Motor Nerves

A number of chemicals are toxic to the peripheral nerves. Many of these agents, which include chemicals causing damage to the nerves involved with hearing and sight, are shown in Table 7-2.

The symptoms of these toxic agents include weakness of the lower extremities, abnormal limb sensations, visual and hearing disturbances, irritability, and loss of coordination.

Neurotoxins Capable of Causing Permanent Brain Lesions

Certain chemicals are capable of causing permanent brain damage, and physical and/or personality disorders that are irreversible. Some of the chemicals with this potential are acetylpyridine, DDT, mercury, and manganese. Symptoms include convulsions, personality disorders (e.g., the "madhatter" disorder of mercury), and disorders resembling Parkinson's disease.

Table 7-2. Peripheral Motor Neurotoxins.

Acrylamide
6-Aminonicotinamide
Arsenic
Azide
Bromophenylacetyluria
Carbon disulfide
Chlorodinitrobenzene
Cyanoacetate
Diisopropyl fluorophosphate
Dinitrobenzene
Dinitrotoluene
Disulfiram
Doxorubicin
Ethambutol
Ethylene glycol
Formate
Hexane and 2, 5-hexanedione
Iminodipropronitrile
Iodoform
Methanol
Methyl-*N*-butyl ketone and 2, 5-hexanediol
Methyl mercury
Perhexilene
Phosphorous
Tetraethyl lead
Triorthocresyl phosphate
Vincristine

7.5 Evaluation of Injury to the Nervous System

Assessing damage to the nervous system is possibly the most difficult determination of organ injury facing a physician. The key to evaluating any patient with neurologic complaints is, first, to determine the possible neuro-anatomical sites or distribution of the damage. For example, is the injury or functional deficiency related to a particular portion of the brain?; or is it manifested by motor function deficits?; are only peripheral nerves involved? (etc.). The location of the injury requires an orderly review of nervous function in the various parts of the central and peripheral nervous systems of patients. To determine the anatomical location of damage generally requires a neurologic examination, composed of:

- A careful patient history.
- An evaluation of the patient's mental status.
- An evaluation of the cranial nerve function.
- An evaluation of the motor system and reflex function.
- An evaluation of the sensory system.
- Additional tests, if deemed appropriate.

Patient History

What should be entered in the patient history depends upon what the examiner feels is appropriate to gain a thorough understanding of the problem and to properly diagnose it. Neurologic disorders require a careful analysis of the patient's complaints and symptoms through questions that are based upon an understanding of the symptomology and functional basis of problems associated with specific areas of the nervous system. In this manner the physician can often obtain important information regarding the anatomical origin of the problem and clues about its pathology. Particularly relevant are questions concerning:

- The time of onset of the problem: was it sudden or gradual?
- The duration of the illness: what potential causes are associable with its onset or occurred prior to its onset?
- The course of the illness: is it progressive, unchanged, or improved?
- Is the patient's problem worse distally or proximally?
- Is the problem affected by activity or work?
- Do certain situations seem to bring the problem on?
- Does the patient have other symptoms that are associable with the major problem?

The background history should investigate the patient's health and habits, occupation and occupational surroundings, uncover significant illnesses or injuries, the use of drugs or medications, and obtain some picture of the health of relatives. The information obtained from the history should help to clarify whether or not the nature of the illness is psychotic or psychogenic, a disturbed physiologic mechanism (e.g., migraines), organic (e.g., brain tumor, stroke, or other disease state), or caused by chemical or work-related exposures.

Mental Status

Valuable information may be learned by attempting to determine the patient's mental status (i.e., the patient's memory, language ability, etc.), which might help elucidate the origin of the problem. Notes about the following should be obtained during the examination:

- State of consciousness.
- Appearance, behavior, moods, thought content.
- Intellectual functions.
- Orientation (confusion about time, place, or person is indicative of several disorders).
- Memory.
- Ability to do calculations; ability to concentrate on abstract thoughts.
- Capacity to reason and exercise judgment.
- Language function.
- Cooperation.

The Cranial Nerves

These nerves, which emanate from the brain rather than the spinal column, are usually the first nerves tested for deficits in function. The 12 cranial nerves to be examined are the following (the number of each is denoted in parentheses):

- (1) The olfactory nerve, which carries sensations of taste and smell.
- (2) The optic nerve.
- (3, 4, 6) The oculomotor, trochlear, and abducent nerves, which control the pupil, eyelids, and eye movements.
- (5) The trigeminal nerve, which carries sensations from the face, tongue, and eye, and which innervates muscles of the jaw.
- (7) The facial nerve, which innervates the muscles of the face (expression and facial movement).
- (8) The auditory nerve, which carries impulses of sound from the ear and contains the vestibular nerve, which functions to give us balance or equilibrium.
- (9, 10) The glossopharyngeal and vagus nerves, which are responsible for sensations from the palate and tongue, control muscles in this area, and supply the muscles of the vocal cords.
- (11) The accessory nerve, which provides motor control to muscles of the head and shoulder girdles controlling head movements.
- (12) The hypoglossal nerve, which controls the movements of the tongue.

Motor System and Reflexes

The aspects of the motor system that should be examined are muscle strength, reflexes, and muscle tone and bulk. A survey for involuntary movements or tremors should also be done.

Sensory System

Examination of the sensory system may be divided into an evaluation of primary sensations (e.g., pain, temperature, touch) and cortical sensations (e.g., two-point discrimination, figure writing, object-weight discrimination of hand-held objects).

Additional Tests of the Neurological Examination

Additional information may be obtained by the physician by:

- Observing the patient's stance and gait.
- Testing coordination.
- Clinical measurements of specific parameters of blood and urine.
- Analyzing cerebrospinal fluid.
- Biopsy of the brain or other organs for histopathological examination of injury.

Ancillary Services

Depending upon the findings of the neurological examination, additional information about the nervous disorder may be acquired through the use of fairly sophisticated equipment less often applied to routine neurologic analysis.

X-rays of the chest and skull sometimes in combination with the use of radio-opaque dyes can be used to identify structural defects of various causes or disease states (for example tumors, that may be responsible for the neural damage.).

Computerized axial tomography (CAT scanning) is a very recent technique commonly referred to as "brain scanning." This sophisticated instrument combines X-rays and computer analysis to provide simulated pictures of the various organs and anatomical features. Again, structural and pathological changes may be easily identified.

Electroencephalography (EEG) can be used to measure changes in the electric potentials transmitted by the brain and to localize areas of damage or change. As is probably known by many readers, this instrument records patterns of "brain waves" or the electric potentials of nerve activity as measured by electrodes attached to the scalp. In the hands of an experienced physician much useful information can be added to other findings and may be useful in locating the area of a lesion or injury within the brain.

Electrophysiologic measurements can also be taken from the peripheral nervous system. The electromyograph records the electrical activity of muscle tissue using either surface electrodes or electrodes inserted into the muscle tissue itself. This may be used in the analysis of nervous disorders causing muscle dysfunction and, again, may help localize problems. Similarly, nerve conduction velocity measurements can be used to measure impulse transmission velocities; such measurements can reflect changes in various segments of a nerve. For this measurement, electrodes are attached so that a stimulus can be applied to the nerve while another electrode monitors the time it takes for the impulse to reach the affected muscle. However, these tests cannot be used by themselves and have certain shortcomings. For example, diseases characterized by axonal Wallerian degeneration show less change in conduction velocity measurements than would be expected clinically, while those problems producing segmental demyelination show a conduction velocity that is markedly decreased.

In summary, the examination and determination of neurologic disorders entails a careful and thorough analysis that is somewhat akin to solving a jigsaw puzzle. There is no single, easy, direct test that ensures proper diagnosis. The approach taken is to identify or localize the extent of the problem and the anatomical area from which it emanates. Therefore compiling a careful history and exposure background of the individual is always required.

Summary

The nervous system is probably the most complex and least understood of the organ systems and is responsible to a degree for integrating and coordinating the functions of most of the other organ systems.

- Because chemicals may alter normal neural physiology or biochemistry in a number of adverse ways, it is hard to make generalizations about neurotoxicities and neurotoxins.
- A number of neurotoxins may be better known for other toxicities they produce; for example, certain chlorinated pesticides are usually feared for their liver damage, their persistence, their bioaccumulative capacity, and their carcinogenic activity in animals, yet the fact that their mode of action during acute poisoning is to attack the nervous system is frequently overlooked.

The chapter presented a brief treatment of the basic physiology of nervous activity.

- Familiarity with this can be a basis for understanding the mechanisms of a number of different neurotoxic agents.

- For some other neurotoxins, the mechanisms by which they cause permanent, physical damage to brain cells, thereby producing deficits in certain types of nerve function, may be less clear or detailed. For these toxins, an understanding of the localization of lesion(s) can help in determining what functions are compromised and what symptoms to be alert for.
- It is suggested that the occupational health specialist working with chemicals capable of producing irreversible damage to the nervous system acquire a thorough knowledge of various types of lesions and the resulting functional losses or changes, so that they will be able to aid the occupational physician by being vigilant of injury. Such vigilance is helpful backup to the normal safety procedures employed in the workplace.

References and Suggested Reading

Bennett, M. V. L. (Ed.) 1974. *Synaptic Transmission and Neuronal Interaction*. New York: Raven Press.

Chason, J. L. 1971. "Nervous System and Skeletal Muscle." *In* Anderson, W. A. D. (Ed.) *Pathology*. Vol. 2. 6th Edition. St Louis: C. V. Mosby Co.

Eccles, J. C. 1964. *The Physiology of Synapses*. Berlin: Springer Verlag.

Ecobichon, D. 1982. *Pesticides and Neurological Diseases*. Boca Raton (Florida): CRC Press.

Goodman, L., and Gilman, A. 1975. *The Pharmacological Basis of Therapeutics*. 5th Edition. New York: Macmillan Publishing Co., Inc.

Hodgson, E., and Guthrie, F. E. 1980. *Introduction to Biochemical Toxicology*. New York: Elsevier/North Holland.

Kuffler, S. W., and Nicholls, J. G. 1976. *From Neuron to Brain*. Sunderland (Massachusetts): Sinauer Associates, Inc.

Norton, S. 1980. "Toxic Responses of the Central Nervous System." *In* Doull, J., Klaassen, C. D., and Amdur, M. O. (Eds.) *Casarett and Doull's Toxicology: The Basic Science of Poisons*. New York: Macmillan Publishing Co., Inc.

Roizin, L., Shiraki, H., and Grčević, N. (Eds.) 1977. *Neurotoxicology*. New York: Raven Press.

Spencer, P. S., and Schaumberg, H. H. 1980. *Experimental and Clinical Neurotoxicology*. Baltimore: Williams and Wilkins.

Williams, P. L., and Warwick, R. 1975. *Functional Neuroanatomy in Man*. Philadelphia: W. B. Saunders Co.

Chapter 8

Dermatotoxicity: Toxic Effects in the Skin

Robert L. Rietschel

Introduction

From this chapter the safety and health professional will learn:

- The important role the skin plays in separating our internal environment from our external environment.
- The general composition of the skin.
- The metabolic activity of the skin.
- Types of skin disorders.
- The major clinical tests used by the dermatologist to evaluate occupational skin disease.
- Specific industrial case studies dealing with dermatotoxicity.

Since the skin is the interface between the environment and the body, many kinds of bizarre events occur in relation to it that many people tend to trivialize. People often think that if something is "only a skin problem," it ought to be easy to treat and very simple to understand or explain. Unfortunately, unlike some organs that are more homogeneous, the skin is an extremely diverse organ. For example, the top of the head in no way resembles the bottom of the foot, yet both areas are referred to as skin. One can readily see the difference between the top of the head and the bottom of the foot, but differences that may not be quite as apparent exist between other small areas such as the abdomen and forearm. All of these distinctions account for the differences one will see in the ways skin responds to environmental and internal hazards. At times the two types of responses mimic each other, so that one doesn't know whether the problem is being generated internally or externally.

The skin has more than just toxic responses to substances. In fact, the word "toxic" doesn't apply well to the skin. The terms "irritants," "allergens," or "infectious agents" may be somewhat better than "toxic responses," and this chapter will touch on all of these.

8.1 Skin Composition

Stratum Corneum and Epidermis

I will review the composition of the skin before I discuss the types of responses the skin has. The skin is often thought to be a half inch thick or less. In fact, the epidermis is very thin, a matter of a millimeter or so, and not all of the epidermis has the same effectiveness as a barrier between an individual and the world: the dead outer layer of the skin—the stratum corneum—is the part of the barrier that really does the most protecting. It is the rate-limiting membrane of the skin. For example, if a person is working with a pesticide that happens to get on his skin, the ability of the particular chemicals in that pesticide to penetrate the stratum corneum will determine whether or not they enter the body and produce an adverse reaction. The stratum corneum is the barrier zone.

As an example of the diversity of the skin, the stratum corneum on your palm is much thicker than the stratum corneum in other areas. One would think that the skin of the palm would be a superb barrier, and in some ways it is. However, it is more easily penetrated by some chemicals than is the thinner layer in other areas of the skin. This is because it is fairly porous, though it is very thick. The protective value of its thickness is partly mitigated by its porosity.

If it is not particularly easy for a contaminant to penetrate the skin, it will be cast away because the epidermis entirely replenishes itself at a rate of approximately once every month. It takes about fourteen days to replenish the entire stratum corneum.

Figure 8-1 presents a cross-section of skin. The stratum corneum, located on the surface of the skin, is faintly pigmented. The pigment, made near what we call the basal cell layer of the epidermis, is distributed within the epidermal tissue, and eventually is carried up into the dead stratum corneum. The stratum corneum protects the living tissue in the epidermis and dermis from becoming adversely influenced.

Melanocyte. A melanocyte is a cell in the epidermis that is responsible for skin pigmentation and has various abilities to withstand ultraviolet-induced hazards. Melanocytes tend to have dendritic form; melanosomes fan out

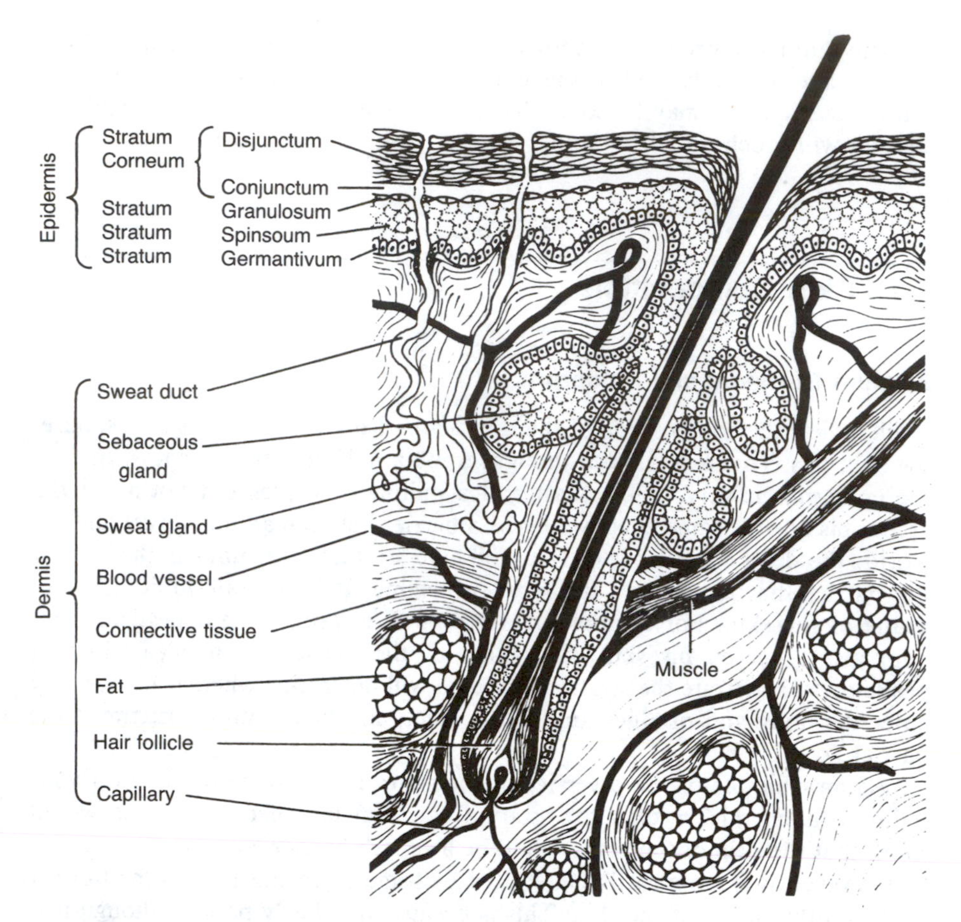

Figure 8-1. Diagram of a cross-section of skin. (Based on Doull, J., *et al.* (Eds.) *Casarett and Doull's Toxicology: The Basic Science of Poisons*. 2nd Ed. New York: Macmillan Publishing Company, 1980.)

through the branches of each cell to serve approximately 36 surrounding epidermal cells, so relative to other epidermal cells, melanocytes are few in number. In a fairly heavily pigmented individual who receives a lot of sun exposure, the skin attempts to compensate for the ultraviolet hazard by increasing the production and distribution of melanosomes to surrounding cells.

Basal Cell Layer. The basal cell layer is where the proliferation of the epidermis begins. Virtually all of it is in the one layer lining the interface between the dermis and epidermis. To protect the skin from ultraviolet injury, melanin is deposited right over the nuclei of the basal cells; this is greatly effective because it prevents the cellular DNA from being ruptured by the melanin-absorbed ultraviolet light. Melanin is one of the body's natural defenses.

Dermis

The blood vessels in the skin do not extend directly into the epidermis. They terminate in the dermis. A toxic material that gets into the skin must therefore penetrate the dermis to be systemically carried to other organs. Once the material passes the stratum corneum, the epidermis poses little resistence to its further penetration into the dermis and dermal blood vessels.

Other organs in the dermis account for various types of reactions. These include hair follicles, sweat glands, blood vessels, nerves, or subcutaneous fat.

Hair Follicles. On hair-bearing skin like the scalp, a unique situation exists because of the many hair follicles in such locations. One might ask why this would make a difference in the way skin responds to contact with toxic substances. In addition to the protective value of hair, follicles serve as diffusion shunts for whatever lands on the skin. If the area has many follicles, a rapid absorption through the follicular epithelium may occur, allowing for greater absorption of a toxic material than would occur elsewhere. Depending on the hairiness of the skin, more or less absorption may occur. The hair may provide a degree of protection against large or particulate toxic substances by impeding their contact with the stratum corneum.

The forearm does not have as dense an area of hair follicles as the scalp does and, consequently, has fewer diffusion shunts. The face is similar to the scalp with regard to diffusion shunts, and chemicals that land on the face will tend to be more rapidly absorbed.

Sweat Glands. The sweat glands are located deep in the dermis near the subcutaneous fat. From the sweat glands, coiled sweat ducts wind upward through the epidermis and ultimately out through the stratum corneum.

One of the consequences of heat stress is a malady called miliaria, a type of prickly heat in which sweat ducts become ruptured. Its symptoms and appearances depend on the location of the ruptured ducts. This type of injury was very common in Vietnam: many soldiers developed a type of heat stress in which the coils deep within the dermis were ruptured. The injury might seem minor but it results in an individual's not being able to evaporate sweat, which means that the body cannot be effectively cooled, a condition presenting the risk of heat stroke. The inability to sweat lasts for two to three weeks after the rupture of the sweat ducts, so the individual is at risk for heat injury during this period of time. After a bout of prickly heat, the Vietnam soldier had to be kept out of active duty for several weeks. The same type of injury can afflict people working around blast furnaces.

Nerves. Within the skin are various cutaneous nerve endings that help to alert us to extremes of heat or cold. Injury to cutaneous nerves after excessive exposure to cold or heat often persists for months to years after the injury and is usually manifested by exaggerated vascular responses to reexposure to heat or cold.

Subcutaneous Fat. The skin undergoes normal anaerobic lipolysis most of the time and tends to use a hexose monophosphate shunt in this process. The skin also manufactures lipids (in addition to carbohydrates), both from the epidermis itself and from the sebaceous glands attached to hair follicles. The lipids consist of fatty acids and esters that have both antibacterial and antifungal properties. Researchers are trying to understand the antibacterial and antimicrobial properties so that they can use them in treating people with infections.

The lipids do have other activities. The surface lipids contain various properties that may combine with substances on the skin and make it easier or more difficult for them to enter the bloodstream, depending on whether it is lipophilic or hydrophilic.

8.2 Metabolic Activity

The skin is a metabolic organ, though it is not as metabolically active as, for example, the liver. Because of its tremendous surface area and the variety of things that land on it, what the skin does with the substances that enter its field is very important. The skin has certain enzymes that will take carcinogens and biotransform them into more potent carcinogens. Although this is not what we like to see happen, it is in fact what does happen with a host of products, particularly petrochemical products. The skin actually induces an enzyme, arylhydrocarbon hydroxylase (AHH), on repetitive exposure to products such as coal tar. The skin also contains several other enzymes that are induced by polycyclic aromatic hydrocarbons in the skin, and these enzymes tend to convert the potential carcinogens into compounds capable of greater carcinogenic activity: in addition to AHH are the enzymes epoxide hydratase (EH) and glutathione *S*-transferase (GST).

Arylhydrocarbon Hydroxylase (AHH)

AHH is a mixed function oxidase (Table 8-1) that is cytochrome-P-450 dependent. The human AHH activity is not uniform and some disease states such as psoriasis result in an altered AHH response. For some reason, people

Table 8-1. Properties of Arylhydrocarbon Hydroxylase.

Mixed function oxidase requiring NADPH + O_2
Present in practically all mammalian tissues
Inducible
Cytochrome-P-450 dependent
Patient with psoriasis cannot induce this enzyme well
Five times more active in epidermis than in dermis

with psoriasis are not able to induce AHH. In tests, those people have been treated actively with substances such as coal tar, and they don't seem to develop skin cancers or any of the adverse responses that dermatologists have come to expect. It is interesting that the thin layer of epidermis that is viable and forming stratum corneum is the area of the skin in which enzymes are formed. It is much more active than the dermis.

Other Metabolic Activity

The skin can perform other activities (Table 8-2). It can even synthesize porphyrins and heme, though usually it does not. It can produce other microsomal enzymes, such as the cytochromes, and it can oxidatively metabolize various chemicals, including heme. The skin is capable of various oxidations of alcohols and aromatic rings. It can reduce carbonyl groups and double-bonded carbons. It can conjugate chemicals by either glucuronidation or sulfation or methylation. We would normally associate all of these activities more with the liver than the skin. However, these processes do occur with chemicals that land on the skin.

Table 8-2. Some of the Metabolic Functions Performed by Human Skin.

1. Synthesis of porphyrins and heme
2. Production of microsomal heme-protein cytochromes P-450 and P-448
3. Oxidative metabolization of heme and other chemicals
4. Oxidations: alcohols, aromatic rings, alicyclic hydroxylation, and deamination
5. Reductions: carbonyl, C=C reduction
6. Conjugations: glucuronidation, sulfation, methylation

8.3 Essential Cutaneous Defense Roles

Direct Influences

The skin attempts to reject all toxic agents, microorganisms, and ultraviolet light. It attempts to retain body fluids, regulate the body thermally to prevent heat and cold injuries, prevent mechanical damage, and protect in some degree low-voltage electrical injury. How does it perform all these activities, and where do they take place (Table 8-3)?

The loss or entry of water is blocked predominantly by the stratum corneum.

Chemical penetration can be limited by a number of factors, including the stratum corneum and the surface lipids. The presence of sweat can be a positive or a negative factor: sweating may facilitate absorption of water-soluble material or it may interfere with and dilute chemicals that are not particularly water soluble. We have already discussed some of the steps the skin takes metabolically to detoxify material that comes in contact with it. The material may evoke an inflammatory response that causes the white blood cells to leave the blood stream and try to engulf—phagocytize—the substance, in order to rid the body of it. The skin may immunologically react with the substance.

Protection from ultraviolet radiation is a function of both the stratum corneum and the epidermis, but predominantly the epidermis. The melanin pigment is distributed to produce that protection.

Table 8-3. Mechanisms of Cutaneous Protection.

Hazard	Manner of defense, and location
Water loss	Stratum corneum
Chemical penetrant	Stratum corneum Surface lipids Eccrine sweat Metabolic detoxification Phagocytosis and immunologic responses
Ultraviolet radiation	Epidermis and stratum corneum Melanin pigmentation
Microorganisms	Epidermal barrier Surface lipids Phagocytosis and immunologic responses
Environmental temperature	Eccrine sweat Dermal vessels Subcutaneous fat

Microorganisms are kept out by the skin's dryness. The wetter the skin, the greater the risk of microbial injury. In general, dryness is preferable for the purpose of avoiding microbial damage to the skin and subsequent infections.

Eccrine sweat is very critical to the ability to regulate body temperature. Some people are born without sweat glands; this is a rare condition called anhydrotic epidermal dysplasia. They are at extreme risk of heat damage because they simply cannot evaporate sweat. However, these people can compensate to some degree by moistening their skin and letting the heat from the body evaporate that water, accomplishing the same function as sweating: dilation of the dermal vasculature and its conduction of heat from the core of the body to the skin helps to evaporate sweat when there is water to evaporate. The subcutaneous fat seems to provide protection from cold injury. The thickness of the collagen and the dermis itself protect against trauma, and common sensory perception also aids in this protection.

Indirect Influences

Not all people in any specific work environment react similarly to the same skin hazards. We have discussed the skin as though it were one heterogeneous organ, and now we are going to discuss the diversity of people (Table 8-4). Factors such as race (Table 8-5) play a role in what kind of response someone is likely to have to an environmental hazard.

Skin Thickness. Black skin tends to be somewhat thicker in the stratum corneum area, so one would assume on that basis alone that black skin provides better protection against environmental hazards, and that seems to hold up in clinical experience.

Table 8-4. Indirect Factors Influencing Cutaneous Defense.

Pigmentation (differs according to race and in different seasons)
Skin thickness
Age and sex
Sweating
Body habitus
Hygiene (personal, environmental)
Medications
Season of the year (an influence in xerosis)
Diseases (atopic dermatitis, for example)
Hardening of the skin

Table 8-5. Indirect Factors Influencing Cutaneous Defense.

	Race	
Factor	Black	Caucasian
Stratum corneum stripping	16	9
Stratum corneum layer counts	21.8	16.7
Buoyant density by sucrose gradient ultracentrifugation	1.18	1.11
Thickness (μm)	6.5	7.2
Density in air (gm/ml)	1.68	1.39
Permeability to water (ml/cm^2 per 24 hr)	0.4	1.4

If you take Scotch tape and repeatedly strip the same area of skin, removing all the dead outer tissue, eventually you will reach an area called the "glistening layer." It takes an average of 16 manipulations of this kind to reach the glistening layer in blacks, but only 9 in caucasians. The number of actual cell layers can be extrapolated from this test and one can see there is some difference. Black skin is actually thinner, not thicker, than white, but it is more compact. It is denser and much less permeable to water and water-soluble materials. Therefore, racial differences account in part for the skin's natural resistance to toxic injury.

Climatic Changes. Another factor that must be considered is climate. Many people develop an itch or dry skin during the winter months. Their skin will crack. This winter-sensitive skin is something that cannot be induced in many people during the summertime. No matter how harsh a soap is used or how much abuse the skin receives during the summer, the effect is very small. During the winter, all kinds of simple things irritate the skin that do not irritate it at other times, and this must be taken into account in assessing the hazards of any setting, such as the workplace, occupied by these individuals.

Trauma. Skin thickness will vary somewhat depending on the experience a particular area of skin has had in the past. The skin will thicken in response to repetitive trauma, and this can give some enhanced protection against both ultraviolet and toxic chemical injury.

Sex. In general, females seem to be a bit more susceptible to irritant dermatitis than do males. It is not clear if this is true in scientific fact but it is something that is rather commonly accepted by practicing dermatologists.

Such a judgment is often found unprovable under controlled scientific conditions.

Sweating. Sweating can have adverse or beneficial effects. If an individual is somewhat obese, any substance on the skin tends to concentrate in the areas where the skin folds over on itself. The material tends to be accentuated in those places and to produce greater amounts of irritation. So the body habitus will be one of the other indirect factors.

Medications. The medications people take will affect skin reactions. This is usually discussed only in terms of ultraviolet light. Many commonly administered medications—such as the antibiotic tetracycline—given to young adults for acne, will tend to render them more photosensitive. Therefore, if they have an outdoor occupation, they may be at greater risk for severe sunburns while taking such medications.

Many antihypertensive medications and some tranquilizers will also increase the photosensitivity of the skin, so that an exposure that previously had not produced any adverse reaction can produce a severe reaction.

Atopic Diathesis. Some people have a tendency to develop eczema or hayfever or asthma. Those three illnesses are lumped together and called atopic diathesis. A person who has one of these illnesses has a somewhat greater likelihood of developing an eczema in flexural areas of the body—about the neck, antecubital fossa, and behind the knees. These areas are at risk for flexural dermatitis under certain circumstances, such as adverse climatic changes. These types of conditions are more likely to set off a sequence of events that may lead to a flaring of atopic dermatitis.

Atopic Dermatitis. Some people are very sensitive to atopic dermatitis, and when it starts to flare, they develop total body erythroderma: they will turn red and scaly from the top of their heads to the bottom of their toes. Many employers do not want to employ individuals with atopic dermatitis to work in areas where they know certain skin hazards are present. Nor do they wish to employ them for jobs in which they will suffer hot or cold environmental temperature stresses. Some atopics have trouble delivering sweat to their skin without triggering itching, and environments in which there is a good deal of heat stress trigger sweating. In a person sensitive to atopic dermatitis, the sweat is ineffectively delivered to the skin and is excreted in the epidermis instead of on the surface. This triggers itching and causes the individual to scratch, which may lead to infections.

Such individuals have repetitive skin difficulties. They may have skin problems in either the summer or winter or during various climatic conditions. Many individuals with this condition can work with little difficulty; however, there is no way to predict how much trouble someone who has had this condition in the past will have in the future. If a person exhibits evidence of active atopic dermatitis, the military services will not accept him, because they cannot be certain where they may want to send the person or in what environment he will be working. They simply avoid the issue by saying that the presence of active atopic dermatitis disqualifies an individual from induction into the military. Some of those people could handle virtually anything, but there is no way to predict who these people are.

Hardening. Finally, there is a phenomenon known as hardening, in which the skin, after repetitive exposure to an irritant or allergen, adapts to the point of no longer being irritated by the material. This area represents one of my own research interests: trying to find out what hardening really is and how we can manipulate it to our benefit. It is not a common phenomenon, and most people who have been repetitively exposed to an irritant get an increased irritation. Hopefully we will eventually know more about the phenomenon of hardening and be able to manipulate it favorably.

8.4 Irritant Contact Dermatitis

Developmental Factors

Within the general range of toxic response, irritant contact dermatitis is the most common injury that the skin encounters. Irritant contact dermatitis can occur in anyone, given an adequate exposure to the toxic material, and unlike allergic dermatitis, it does not require any previous exposure to the material. To develop a true allergy to something, a person must usually have a previous exposure to the chemical and then a subsequent reexposure several weeks, months, or years later. In the case of poison ivy, for example, the first time a person is exposed, nothing happens. Only with the second or third exposure does a reaction occur. It takes a period of time to develop a true allergy, and that is the difference between irritants and allergens.

With irritants, a previous exposure is not required for the development of an inflammatory response. The body requires about two weeks to develop an allergy of the delayed hypersensitivity type commonly called contact dermatitis (for example, poison ivy). With irritants, an adequate exposure will elicit the response from anyone.

Irritants tend to elicit a host of responses, ranging from hives (wheals), which make the skin blotchy red (an erythema), to a bruiselike response known as a purpura, blistering, eczemas or rashes that tend to weep and ooze, erosions in which the skin seems to be scraped away, hyperkeratosis or a simple thickening of the skin (much like a callous), pustules (looking like small

Table 8-6. Factors in the Development of Irritant Contact Dermatitis.

1. Regional variation
   Time required to produce a blister with 50 percent ammonium hydroxide:
   Face, 5.3 min; back, 9.0 min; forearm, 13.0 min.
2. Individual differences
   Inverse relationship to pigmentation:
   Minimal blister time for white persons averages 7–11 min, but some blister in 3 min; others take more than 1 hr.

infections), and skin dryness and roughness. All of these responses could be considered manifestations of an irritant contact dermatitis. Allergic contact dermatitis, on the other hand, in general though not always, is simply an eczematous process. Even water is an irritant for many people. There is no substance that is without risk of being an irritant to someone.

In discussing irritant dermatitis, one can refer either to the number of individuals in a population who are likely to respond to a particular exposure or to how many repeated exposures are required before irritation is manifested. Two factors are therefore involved: the factor of the individual, who may or may not be particularly responsive to irritants, and the factor of the amount of time it takes for any individual to have a particular response. Various types of oils—vegetable oil, esters of fatty acids, ethylene glycols, the higher alcohols, or nonionic surfactants—are the kinds of substances that are more likely to produce irritant dermatitis in a subset of individuals.

Response to irritants varies according to what region of the skin is tested, too (Table 8-6). Ammonium hydroxide applied anywhere on the skin will produce a blistering response within a matter of minutes. On the face, on the average it requires 5 min of exposure to ammonium hydroxide to produce a blister; on the back, 9 min; for the forearm, 13 min. There is variation in the absorption of material over these different body regions. As a result, a toxic substance that lands on facial skin as well as forearm skin might produce a response on the face but not on the forearm because of the greater resistance to penetration manifested by the forearm.

Identifying the Hyper Responder

Individual differences seem to vary inversely with pigmentation (Table 8-6). Generally, we consider black people to have a superior barrier function and to be somewhat more protected, but does that mean the Scotch-Irish are most at risk? It does seem that way. They tend to have somewhat more sensitive skin, especially to ultraviolet injury and to blistering. The blistering phenomenon also varies among different individuals. Some people blister within a matter of minutes, while in others blistering may require hours.

Who is going to be exposed to these irritants and who is least likely to experience difficulty with them? A study done with college students showed that if kerosene is placed on the skin, some individuals do not react after 24 hours even to 100 percent kerosene. However, the majority does react. When kerosene is used as a gauge, about 8 of every 100 people can be exposed to particular irritants and not be bothered by them.

Another study has been conducted to measure the time it takes to generate a hive with DMSO (dimethyl sulfoxide). DMSO in concentrations greater than approximately 80 percent tends to produce a hive at the site where it is applied. It stings, burns, and produces irritation. It is possible to distinguish among people who are hyperactive by the amount of time it takes for these responses to occur. With some people it will take minutes and for others, hours. For some, it almost never occurs. Other similar tests are used.

Some people are referred to as "stingers." If a substance innocuous for most people—sorbic acid, for example—is applied to their skin, they immediately experience a stinging sensation. This group of people seems to be a fairly large subset of the general population, anywhere from 5 percent to 20 percent. As we learn more about these different types of people and their reactions to irritants, it may be possible to minimize their risks of irritation in the workplace.

Table 8-7 lists factors of hyperreactivity.

The Scandinavians have kept good records of the percentage of people experiencing irritant dermatitis in Scandinavian countries. In their experience with contact dermatitis cases in industry, they are finding 75 percent of the cases to be allergic and 17 percent to be irritant. In the United States, those figures tend to be reversed. This may relate in some way to methods of recording in the different countries. For women, allergic reactions tend to predominate over the irritant. If you explore the most common cause of an allergic contact dermatitis among men in industry, it seems that in Scandinavia, and Europe in general, contact with chrome is one of the major sources, followed by contact with various rubber chemicals, and then resins.

Table 8-7. Factors and Components of Irritant Reactions Among Hyperreactive Persons.

1. Females are possibly more susceptible
2. Age, 18–50 yrs: no major variation
   Age, over 65 yrs: hyperreactive
3. Season: more irritancy in winter on exposed skin (compared to summer)
4. Facial stinging and erythema in response to sorbic acid, though
   this is not directly correlated with irritancy

For women, contact with nickel is more common, and the explanation given for this is that women often have their ears pierced, bringing them into contact with nickel and rendering them more likely to develop an allergy to it.

Common Causes of Irritant Contact Dermatitis

General Causes. Table 8-8 indicates the common causes of irritant contact dermatitis in the work environment.

Water is first on the list. Work during which the skin remains wet is most likely to initiate irritancy because water leaches out a fair amount of the natural moisturizing factors that keep the stratum corneum soft and pliable. Up to 25 percent of the moisturizer the skin makes on any given day can be washed out with a simple water extraction.

If soap is used on the skin, the loss will rise to 35 percent to 50 percent, but soap tends not to be as irritating as some organic solvents and detergents. Concentration is a factor, of course.

Alkalis tend to be more of a problem for skin than acids, epoxy resin hardeners are one of the chemicals that can produce both allergic contact

Table 8-8. Common Causes of Irritant Contact Dermatitis.

1. Water
2. Cleansers
   Soap is not bad; organic solvents are more problematic; detergents, depending on concentration
3. Alkalis
   Epoxy resin hardeners, trisodium phosphate, sodium silicate, lime, cement, NaOH, ammonium, soda, soap
4. Acids
   These damage less than alkalis do
5. Oils
   May contain emulsifiers, antioxidants, anticorrosion agents, preservatives, perfumes; organic solvents required for removal
6. Organic solvents
   Aromatic solvents especially irritating
7. Oxidants
   Peroxides, benzoyl peroxide, cyclohexanone
8. Reducing agents
   Thioglycolates
9. Plants
   Orange peel, tulip bulbs, pineapple juice, cucumbers, asparagus
10. Animal substances

dermatitis and, because of their alkalinity, an irritant contact dermatitis, as can some of the other chemicals listed in Table 8-8 that are very common in the workplace.

Acids tend to be less damaging than alkalis inasmuch as the normal pH of skin tends to be in the range of 5.5—slightly on the acid side—with 7 being neutral. Since the skin itself tends to be slightly more acidic than the rest of the body, which has a normal pH of 7.4, substances within the range of 5.5 to 7.4 are well tolerated by the skin.

Oils are a common cause of irritant contact dermatitis among industrial employees; oils produce a host of reactions. They contain various emulsifiers, antioxidants, anticorrosion agents, and preservatives (particularly the water-soluble oils). These additive substances are often the culprits, rather than the oil itself.

Organic solvents are sixth on our list of contact irritants, followed by oxidants such as benzoyl peroxide.

Reducing agents are eighth and plants ninth as contact irritants.

Animal substances may cause irritant dermatitis in poultry workers or in others handling raw meat. One response is a contact urticaria, a hivelike reaction to chicken skin. The chicken skin, if laid for about 15 minutes on any part of the skin that has experienced a rash in the past, induces a hive locally. This is a relatively newly described phenomenon: only since the early 1970s has it been known that certain types of animal protein can cause responses of this type.

Industry-Specific Causes. That animal protein might be a cause of contact irritant dermatitis was first discovered by a Danish dermatologist named Neils Hjorth, who studied Danish sandwich makers. Their work was discovered to be highly structured; they hold the knife in one hand and make the sandwich with the other. Hjorth observed that the sandwich makers develop dermatitis only in the hand that makes the sandwich. He was able to trace this to various food substances on their skin. Before Hjorth's work, no dermatologists would credit an explanation of this kind, maintaining instead that big protein molecules could not penetrate the stratum corneum and, thus, could not cause an irritation. But in fact, proteins do slip through diffusion shunts and can produce very quick reactions in some people. Many poultry workers also work in conditions of moisture and wetness and simply develop irritant dermatitis in skin made raw by the constant dampness. Therefore, animal protein substances and moisture seem to be the main causes of irritant dermatitis among poultry workers. Poultry plant employes do well to establish strict personal hygiene programs and have workers wash their hands with lotions.

Substances capable of causing contact irritant dermatitis differ from one occupation to the next. Some of the common substances producing irritant dermatitis in agricultural workers, for example, are fertilizers, disinfectants, cleansers for milking machines, and petrol and diesel oil. The true allergens for agricultural workers tend to be rubber (both in clothing and in milking equipment), some of the food substances themselves (such as oats, barley, and animal feeds), and ingredients in animal feed (antibiotics, preservatives and additives, and cobalt), veterinary medicinals, cement, plants, pesticides, and wood preservatives.

Anyone interested in knowing the likely causes of contact irritant dermatitis within a specific industry can begin by referring to lists provided in several textbooks, such as Cronin (1980) and Adams (1983). As one compares such lists, one will often find some substance, such as the chromates and nickel, on multiple lists. Rubber and acrylates are other substances that are seen often.

8.5 Pigment Disturbances

Some people develop either an increase or decrease in pigment as a result of exposure to various chemicals (Table 8-9). Increased pigmentation—darkening of the skin—will most commonly occur after contact with coal tar compounds and other petrochemical substances, though it may occasionally occur after handling vegetables and fruits. A few years ago, certain celery pickers were developing a blistering dermatitis, similar to what occasionally happens to people who pick figs and various other fruits and vegetables. A particular organism was found to be growing on the celery; it produced a photosensitizing compound; the photosensitizing compound from the infected celery got on to the workers' skin and, when combined with sunlight, produced a blistering response.

Figs produce a substance called furocoumarin, which is a photosensitizer. As the figs are picked, if sap gets on the skin in the presence of sunlight, either

Table 8-9. Changes in Pigmentation in Response to Various Chemicals.

Disturbance	Cause
Increase	Coal tar compounds Petroleum oils Vegetables Fruits Sunlight Trauma
Decrease	Burns Trauma Chronic dermatitis Monobenzyl ether of hydroquinone Tertiary butyl catechol Tertiary amyl phenol Tertiary butyl phenol

a blistering response or some less intensive reaction may develop as a hyperpigmentation (dark streak). People who drink teas or tonic water outdoors on a sunny day will sometimes squeeze a lime into a glass and unintentionally spray lime juice on their hand at the same time. Lime juice contains various photosensitizing compounds. A subclinical dermatitis will often start and a dark discoloration will develop on the hand, seemingly for no reason. The discoloration is a phototoxic reaction to the photosensitizing chemicals in the lime juice.

At other times chemicals on the skin will decrease pigmentation. The most common industrial chemicals capable of depigmenting skin are the paratertiary butyl phenols and catechols. Their depigmentive effect was determined in experimental models by Gerald Gellin in San Francisco. Depigmentation can begin with a rash that seemingly goes away with no ill effects, but then is followed by a loss of skin pigment. Often the depigmentation comes approximately one month after the chemical affront to the skin.

Erythema ab igne is a skin condition that used to be common in England when people sat in front of fireplaces for extended periods. It is a heat-induced injury in which a mottled pigmentation of the skin develops. The same condition can be produced by the prolonged contact of a hot water bottle on the skin. This condition is of interest because it is what occurs with microwave injury of skin. Microwave technicians who work without adequate protection will get erythema ab igne.

8.6 Ulceration

Ulceration of the skin will occur as a primary phenomenon when individuals are exposed to certain types of compounds—cement and chrome ulcers are fairly notorious. Ulceration can occur on mucous membranes as well as in the skin itself as the result of various acids, burns, and trauma. Ulceration of the skin is a difficult problem for people studying occupational dermatitis, because sometimes they are factitiously induced. However, keeping in mind the chemicals capable of producing ulceration should provide clues as to whether the condition is real or artificially induced. Many times it is very hard for dermatologists to determine if the skin has been damaged by the patient to gain sympathy or compensation or if it has really been injured in a legitimate industrial accident.

8.7 Neoplasms

Environmental hazards can produce neoplastic changes. These changes primarily result from ultraviolet radiation, but they can also result from exposures of other kinds than sunlight—for example, exposures to arc welding light and other sources producing fairly high exposure to carcinogenic ultraviolet light. The same rays that cause skin to tan and burn cause skin cancer, and X rays would also be among the types of exposures one should avoid.

Neoplastic changes resulting from contact with petroleum products are somewhat more problematic. It has become clear as more people have attempted to work with shale oil that some shale oil products are extremely potent cutaneous carcinogens. It has been known for a long time that certain petroleum products painted onto the skin of mice will produce various types of keratosis, keratoacanthomas, squamous cell carcinomas, and other cutaneous tumors. This does not seem to happen often when we treat dermatologic conditions with coal tar, though it can. The more potent these exposures are in animals, the more likely they are to be potent in humans. However, all of these chemicals also happen to be photoactive. When they combine with sunlight, new developments occur that can result in either greater or lesser—though likely, greater—exposures to carcinogenic material. Therefore, an individual with both a carcinogenic exposure from a petroleum product and exposure to sunlight is perhaps at greater risk for neoplastic change of skin.

8.8 Follicular and Acneiform Dermatoses

Other types of occupational skin disease (other than the traditional eczematous irritant and allergic contact dermatitis) are the follicular and acneiform dermatoses. In general, we see plugged sebaceous follicles or blackheads as one of the primary events (Table 8-10). These can lead to the development of nodules and occasionally pus. The most common cause of this condition are insoluble oils and greases, tars, waxes, and chlorinated hydrocarbons. Acne conglobata is a very stubborn form of acne in which individuals tend to get two adjacent blackheads or comedones or, occasionally, triple and quadruple comedones.

Chloracne has been called environmental halogen acne. In general, it manifests itself in the form of many large open comedones. Chloracne is at times accompanied by an eczematous component that makes the picture confusing. In the more typical situation, chloracne is composed of blackheads,

Table 8-10. Folliculitis and Acneiform Dermatoses.

Manifestations	
	Plugged sebaceous follicles, nodules, suppurative lesions
Causes	
	Insoluble oils and greases, tars, waxes, chlorinated hydrocarbons

Table 8-11. Chloracne-Producing Chemicals.

1. Polyhalogenated naphthalenes
2. Polyhalogenated biphenyls
3. Polyhalogenated dibenzofurans, especially tetra-, penta-, hexachlorodibenzofuran, and tetrabromodibenzofuran
4. Contaminants of polychlorophenol compounds
   a. 2, 4, 5-T and pentachlorophenol
   b. 2, 4, 5-Trichlorophenol
   c. 2, 3, 7, 8-Tetrachlorodibenzo-*p*-dioxin (TCDD)
   d. Hexachlorodibenzo-*p*-dioxin
   e. Tetrachlorodibenzofuran
5. Contaminants of 3, 4-dichloroaniline and related herbicides (propanil, methazole, etc.)
6. Other
   a. Dichlobenil (Casoran)—a herbicide
   b. 1, 2, 3, 4-Tetrachlorobenzene

papules, and pustules, and it may be difficult to distinguish chloracne from severe acne vulgaris. What one typically looks for are hundreds and hundreds of comedones. These blackheads develop into inflammatory papules and are extraordinarily persistent.

The halogenated chemicals—for example, the polyhalogenated naphthalenes, biphenyls, dibenzofurans, and dioxins (Table 8-11)—are, in general, the chemicals most commonly believed to produce chloracne. The favored ones are the tetra-, penta-, and hexachlorodibenzofurans and dibenzodioxins. When someone has been exposed to a chloracnogen, he or she may claim a host of other symptoms and difficulties (Table 8-12). I will discuss some of the symptoms in the medical literature that suggest some relationship to chloracnogens. Some examples are cystic swelling and hypersecretion of the meibomian glands around the eyelids. These are oil glands within the skin. Other symptoms are nervousness, dullness, abnormalities of the menstrual cycle, and retarded development of offspring or of the exposed person himself. Among the cutaneous findings would be hyperpigmented changes occurring

Table 8-12. Conditions Associated with Exposure to Chloracnegens.

1. Cystic swelling and hypersecretion of meibomian glands
2. Nervousness, dullness, dysmenorrhea, retarded development
3. "Brown chromodermatosis" on dorsum of hands, fingers, nail beds
4. Abnormal sweating, increased fingernail growth, nonscarring alopecia

predominantly on the fingers and elbows. Some chloracnogens have been alleged to cause a brown chromhidrosis, abnormal sweating, increased fingernail growth, and nonscarring forms of hair loss.

Chloracne can be successfully treated in two ways. One is to put up splash guards and other devices to keep the various materials from coming into contact with skin and to change oil-saturated clothing frequently. These preventive techniques are important, along with one type of medication: tretinoin. Tretinoin is a vitamin A acid derivative that actually unglues the cells making the blackhead. Very few substances can actually unglue skin, but that is one of the properties of this substance. A new drug for the treatment of acne, 13-retinoic acid, is related to this compound. This has recently been introduced and will make it much easier to manage chloracne patients.

8.9 The Effectiveness of Barriers

Barrier Creams

In general, barrier creams seem to be of very limited value. They are not as effective as good industrial hygiene and, in general, will not block allergens. Barrier creams with an oil base are of some use when irritant dermatitis is due to a wet environment. But if allergens are the problem in the work environment, barrier creams will not be effective because infinitesimally small amounts of allergenic materials can provoke responses in people.

Rubber Gloves

Unfortunately, some allergens will go right through rubber gloves. Most notorious among these allergens are the methacrylates. In hospitals where methacrylates are used for orthopedic procedures, it is not uncommon for surgeons to develop hand eczema, even though they have been wearing gloves throughout the entire operation. The same problem occurs in industry.

8.10 Evaluating Occupational Skin Disease: Clinical Allergy Testing

The key clinical test for dermatitis is the patch test. Patch testing seems deceptively simple but is extremely complex. In essence, the materials thought to be causing the dermatitis are applied to the skin, covered with a special apparatus for two days, and then removed; the skin is then examined to see if the dermatitis has been reproduced. This type of work is not, cannot, nor should not ever be performed with materials known to be irritants. Irritant dermatitis is diagnosed on clinical suspicion: one does not put a known irritant on the skin to re-prove that the substance is an irritant. Patch testing should be performed by dermatologists thoroughly familiar with the standard procedures required to obtain meaningful results.

In administering a patch test, the materials to be tested should be properly diluted to nonirritant concentrations. This requires that many tests be performed on nonsensitive subjects to assure that reactions occur only on those individuals specifically allergic to the substance being tested. A small amount of the material is applied to either Finn chambers or Al-test patches and affixed to skin with paper tape for 48 hr. Once removed, the tests are graded for the intensity of reaction after one hr has elapsed (this allows the reaction caused by the tape to subside). The tests are reread one or two days later as well.

Standards kits for conducting proper patch testing with notorious allergens are available from the American Academy of Dermatology, Evanston, Illinois.

Many pitfalls await the novice who would casually attempt to conduct such studies.

8.11 Case Studies

The Case of the Chromium-Caused Eczema

A mill worker complained of an eczematous hand. The condition had persisted for several years and the patient had been seen by another dermatologist who thought the condition was an inherited form of endogenous eczema unrelated to external exposure. Now the patient was beginning to develop additional lesions over other areas of the body, and these were also eczematous.

The patient was given a standard patch test for allergy and had a very weak reaction to potassium dichromate. When a patient is thought to have a chromate allergy, what a dermatologist usually looks for is exposure to cement in its dry form, chrome-tanned leather, yellow paints, and (occasionally) green dyes. This individual was asked if he worked with any of these things. He informed us that he worked with a particular device that had a leather strap over it, which he placed over the back of his hand. No one had ever asked him about that implement before. The problem was indeed easily solved once the patient protected himself by replacing the leather strap in the tool with a plastic strap. Another alternative available to this worker was to wear a glove or to put some other physical barrier between the chromium compound and the skin.

The case was solved with no further difficulty.

The Case of Magnesium Chromate Toxicity

A male complained of eczema on the hand. A biopsy showed a condition called lichen planus. Lichen planus is not normally associated with industrial exposure, except among people who handle color film developers (color film developers, however, are notorious for producing a lichen planus condition similar to contact dermatitis). The case was tricky, because the patient had no contact with color film developers. Incorrect diagnosis would not have led to discovery of the irritating substance and would probably have resulted in continued treatment of lichen planus with topical steroids.

A closer look at some of the red papules on the back of the patient's hand caused me to suspect that something different than lichen planus was at work here. The patient was somewhat suspicious too, because the eczema had begun when he started work with a particular sealant. We then began to think that perhaps the sealant contained a compound also used in color film developers. But this was not so. Of the three chemicals predominant in the sealant—a substance called LP2, magnesium chromate, and toluene—the LP2 was a mercaptan-terminated polysulfide polymer in no way related to any of the compounds used in color film developers, and it did not seem to be responsible for the irritation. Toluene and oil, when tested, were negative. We tested the patient for magnesium chromate, and it gave him a tremendous reaction at a concentration of 0.5 percent. It turns out that the sealant was 30 percent magnesium chromate. Considering that 0.5 percent had set off a reaction, it is no wonder that the patient had difficulty. He limited his exposure to the sealant and his condition cleared up.

Unfortunately, this case was complicated further when we went through our diagnostic evaluation with this individual. He was tested with other standard substances besides the above chemicals. We found that he was allergic to wood alcohol, too. This is one of the fractions of lanolin, which meant that most of the skin creams and barrier creams he might have been given would also be allergenic for him. We were able to identify the allergy correctly and keep him away from the substances with which he had difficulty.

The Case of the Allergy to Nickel

It is easily possible in some cases to identify an exact work-related cause of an individual's dermatologic problem and eliminate it. In other cases, people may claim full disability or general occupational injury but examination by a dermatologist will reveal a problem only partly related to the work setting, or even totally unrelated.

An industrial welding supply house hired a man I had been seeing who was having difficulty working because of chronic hand eczema. He was allergic to nickel. It seemed that in his job at the welding supply house almost everying he worked with was made of metal. The man despaired of being able to work at all.

We went to the supply house to observe him at work. He was required to handle welding tanks which, indeed, were made of metal. To see if the tanks contained biologically available nickel, we tested the tanks with a kit known as the dimethylglyoxine kit. It was found that virtually none of the tanks contained any biologically available nickel.

After extensive testing, we found that nickel was confined to two small areas in the plant where nickel-plated fixtures were used. Once these were identified, it was easy to separate the employee from the potential exposure so that he could continue to work.

Summary

The toxicology of skin involves knowledge of environmental factors that have impact on the worker as well as of the chemicals to which the worker is exposed. Once an irritant substance has come into contact with the skin, the host response will vary, depending on the efficiency of the barrier function, the general health of the worker, the immunologic state of the worker with specific regard to the chemical in question, the inherent toxicity of the substance, and the metabolic pathways of the skin used to detoxify the material.

- The principal toxic responses of the skin are irritant contact dermatitis and allergic contact dermatitis.
- Irritant dermatitis varies greatly based on the integrity of the stratum corneum. This can be affected by age, race, sex, season, previous skin disease, medications, sun exposure patterns, and the ability of the stratum corneum to harden.
- Development of allergic contact dermatitis requires a specific individual and a specific chemical. The constitutional factors important in irritant dermatitis assume minor or inconsequential roles in allergic dermatitis.
- Once acquired, allergic reactions are harder to prevent than irritant reactions and avoidance of further contact with the allergen is more important.

References and Suggested Reading

Adams, R. M. 1983. *Occupational Skin Disease*. New York: Grune and Stratton.

Bickers, D. R. 1980. "The Skin as a Site of Drug and Chemical Metabolism." *In* Drill, V. A., and Lazar, P. (Eds.) *Current Concepts in Cutaneous Toxicity*. New York: Academic Press. pp. 95–126.

Birmingham, D. J. 1971. "Occupational Dermatoses." *In* Fitzpatrick, T. B. (Ed.) *Dermatology in General Medicine*. New York: McGraw-Hill, Inc.

Birmingham, D. J. 1973. "Occupational Dermatoses: Their Recognition, Control, and Prevention." In *The Industrial Environment—Its Evaluation and Control*. Washington, D.C.: U.S. Department of Health, Education, and Welfare, Public Health Service, Center for Disease Control, National Institute for Occupational Safety and Health.

Cronin, E. 1980. *Contact Dermatitis*. New York: Churchill Livingstone. pp. 839–878.

Foussereau, J., Benezra, C., and Maibach, H. 1982. *Occupational Contact Dermatitis*. Philadelphia: W. B. Saunders Company.

Kligman, A. M. 1977. "Cutaneous Toxicology: An Overview from the Underside." *In* Simon, G. A., *et al.* (Eds.) *Current Problems in Dermatology*. New York: S. Karger. Vol. 7. pp. 1–25.

Kligman, A. M. 1980. "Assessment of Mild Irritants in Humans." *In* Drill, V. A., and Lazar, P. (Eds.) *Current Concepts in Cutaneous Toxicity*. New York: Academic Press. pp. 69–94.

Maibach, H. I., and Gellin, G. A. 1982. *Occupational and Industrial Dermatology*. Chicago: Yearbook Medical Publishers.

Chapter 9

Pulmonotoxicity: Toxic Effects in the Lung

G. Michael Duffell

Introduction

This chapter will review aspects of the anatomy and physiology of the lungs and discuss some of the specific diseases that occur as a result of the inhalation of toxic substances. In particular, this chapter will discuss:

- The anatomical makeup of the lungs.
- The defenses the lungs employ to ward off injury.
- The variety of toxic substances that can damage the lungs.
- The synergistic role of cigarette smoking in conjunction with other lung toxicants.
- Ways of identifying, treating, and preventing toxic injuries to the lungs.
- Clinical tests used to evaluate occupational injuries to the lung.

When students are asked which body organ has the greatest exposure to the environment, they almost invariably answer: the skin. The correct answer, however, is the lung.

9.1 Anatomy and Physiology

The respiratory system is an air pump whose basic function is to supply all of the body's cells with a continuous supply of oxygen and to rid the body of the gaseous metabolic byproduct, carbon dioxide. At rest the body of an average-sized person utilizes 250 ml of oxygen per minute and produces approximately

200 ml of carbon dioxide. With muscular activity the body's demands for oxygen markedly increase, as does the production of carbon dioxide. To meet this increased metabolic demand, the heart and lungs must proportionately pump more blood and air so that a constant internal metabolic environment is maintained.

The Bronchial Tubes and the Alveoli

The basic structure of the lungs is fairly simple, consisting of two structures: bronchial tubes, and air sacs or alveoli. The bronchi are a system of branching tubes that serve as air conduits leading fresh air deep into the structure of the lungs and serving as pathways to eliminate air from which oxygen has been removed. The terminal branches of the bronchial tube system empty into approximately 300,000,000 blind air sacs, or alveoli, where active gas exchange occurs between the environment and blood.

The basic function of the lungs is relatively simple. Fresh air is distributed through the bronchial tube system to each of the millions of alveoli. Oxygen molecules diffuse along pressure gradients from the alveoli into capillary blood, where the molecules combine with hemoglobin. Carbon dioxide molecules move along pressure gradients in an opposite direction—that is, from capillary blood into the alveolar gas, which is then exhaled out of the lung.

Although each individual alveolus is quite small, the total surface area of the 300,000,000 alveoli is approximately 70 m^2. This means that the air we breathe is distributed over a very large surface area. In the walls of the alveoli are tiny blood vessels called capillaries. In these capillaries blood is brought into intimate contact with air, and gas exchange occurs. The amount of blood in the pulmonary capillary bed at any given instant is approximately 100 ml. Diffusion of gases occurs over a very large surface area bathed in a very small, thin film of capillary blood.

At rest an average-sized person breathes approximately 8 l of air a minute. This amount increases as muscular activity increases. Therefore, in a single day one breathes an enormous amount of air. This air is conducted through the ever-smaller-branching bronchial tubes, is distributed evenly to the alveoli, and, hence, comes into intimate contact with the very delicate lining layer of the alveoli. Since environmental air is brought into intimate contact with a very large alveolar-surface area, the potential for damage is great if the air contains toxic material.

Air Passages

The air passages are traditionally divided into two segments: the upper air passages, which extend from the nose to the level of the vocal cords; and the lower air passages, which extend from the vocal cords in the neck down to the final terminal branches of the bronchial tube system.

Upper Air Passages. When we breathe air into the respiratory system, it comes first into contact with the nasal passages. The air enters through the nasal openings on either side of the nasal septum and is brought into very intimate contact with small protuberances called turbinates. The air passages are designed in such a way that the little streams of air are made very turbulent. The air then mixes in the nasal passages and is brought into contact with the nasal lining layer, or mucosa. The nasal passages are richly supplied with capillaries, which course just beneath the mucosa. Hence, the air is also brought into close proximity with capillary blood.

The nose serves several important functions. First, it regulates the temperature of the air we breathe. If one breathes air that is −30°C (−22°F), it is brought to body temperature by the time that it reaches the vocal cords and windpipes. In like turn, if one breathes air at 54°C (130°F), it is cooled to body temperature by the time it reaches the level of the vocal cords. Second, the nose provides moisture for the air as it enters and is distributed through the bronchial tube system. Air of very low humidity is completely saturated with water vapor when it reaches the trachea. Nasal regulation of temperature and humidity is essential to normal lung function. Third, the nose cleanses the air that we breathe so that when it enters the bronchial tube system it is free of large-sized bacteria, viruses, particulates, and other potentially injurious sub-

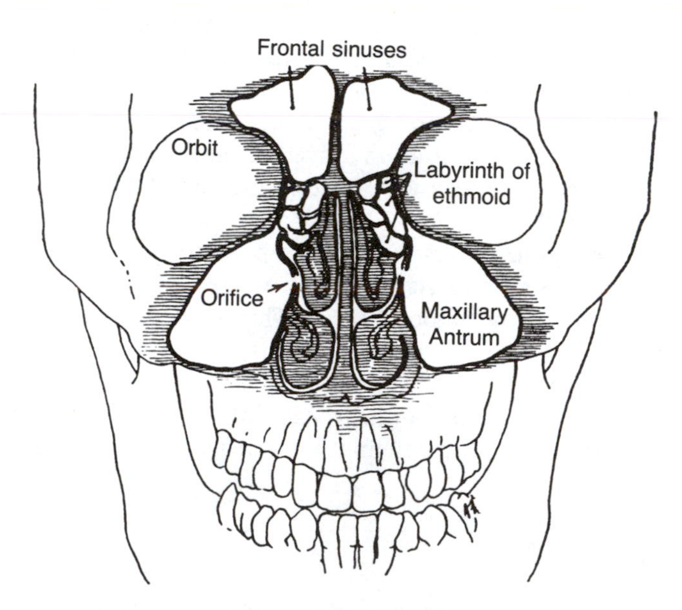

Figure 9-1. Frontal view of the skull, showing frontal, maxillary, and ethmoid sinuses. (Reproduced with permission from Fenn and Rahn, *Handbook of Physiology*. Washington, D.C.: American Physiology Society, 1964.)

stances. The way in which the air is filtered will be discussed later in the chapter.

Sinuses. Figure 9-1 is a frontal view of the skull and Figure 9-2 is a lateral view. In these illustrations four pairs of sinuses can be clearly seen. The sinuses are hollow cavities within the bony skull that are lined with the same delicate lining that lines the nasal passages and tracheobronchial tree. They connect

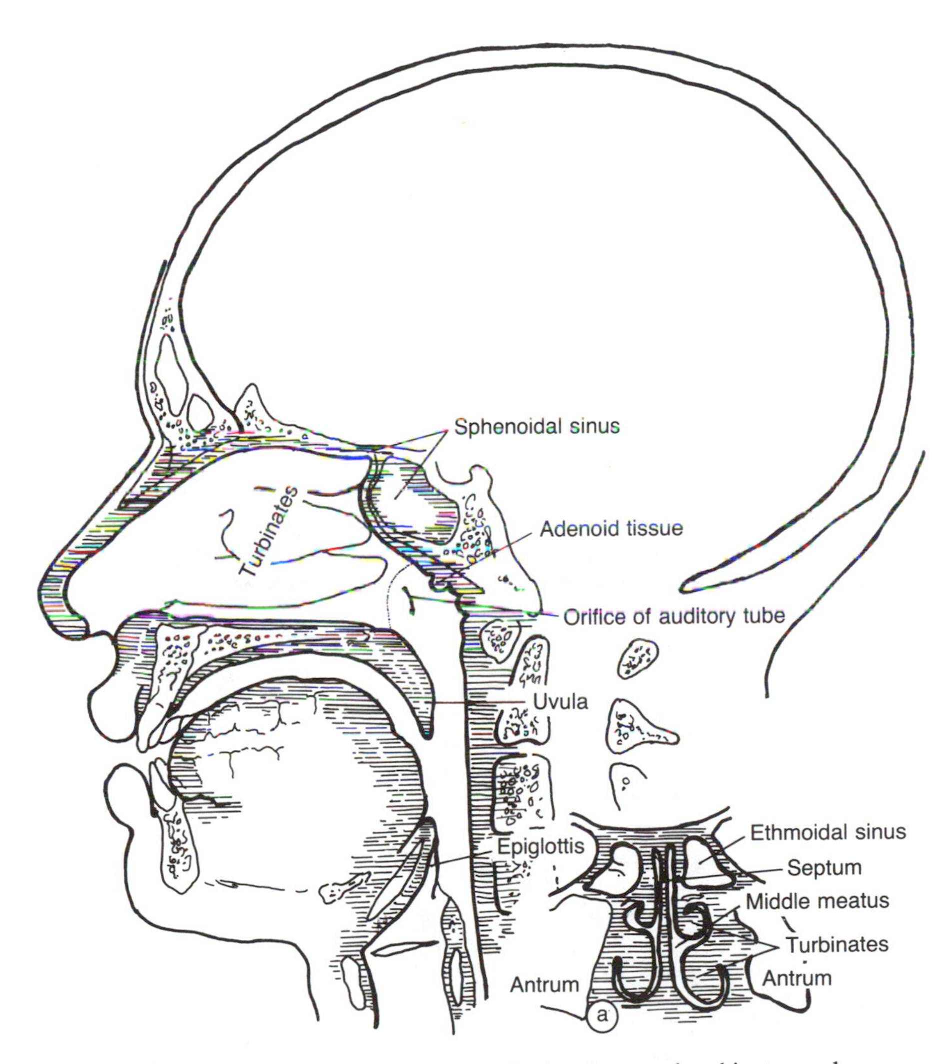

Figure 9-2. Sagittal view of the skull, showing nasal turbinates and sphenoid sinuses. (Reproduced with permission from Fenn and Rahn, 1964.)

with the nasal passages through small openings called ostia. Although the exact function of the sinuses is not known, they do lighten the weight of the skull and, because they open into the nasal passages, they act as resonating chambers, giving some timbre to the voice.

Air breathed through the nose enters the sinus cavities. If the air is contaminated with chemicals, fumes, or other noxious substances, it irritates the delicate nasal lining. It also irritates the delicate lining of the sinus cavities and may produce acute sinusitis. In response to inflammation the mucous glands that compose a portion of the lining of the sinus cavities secrete excess mucus, which ultimately becomes infected. The symptoms of sinusitis can vary from simple postnasal drainage of mucus to full-blown infection with facial pain or severe headache, fever, and tenderness to light pressure over the sinus areas themselves.

Of the four pairs of sinuses, the frontal sinuses are located just above each eye and the maxillary sinuses are just below each eye in the cheekbones. The ethmoid and sphenoid sinuses are located deeper within the nasal passages at the base of the skull. Figure 9-3 shows the flow pattern air follows as it is inhaled. As the air enters the nares it rises and hits the roof of the nasal cavity and then drops back to the posterior nasopharynx. This flow profile produces marked turbulence of air, which in turn mixes the air, bringing it into close

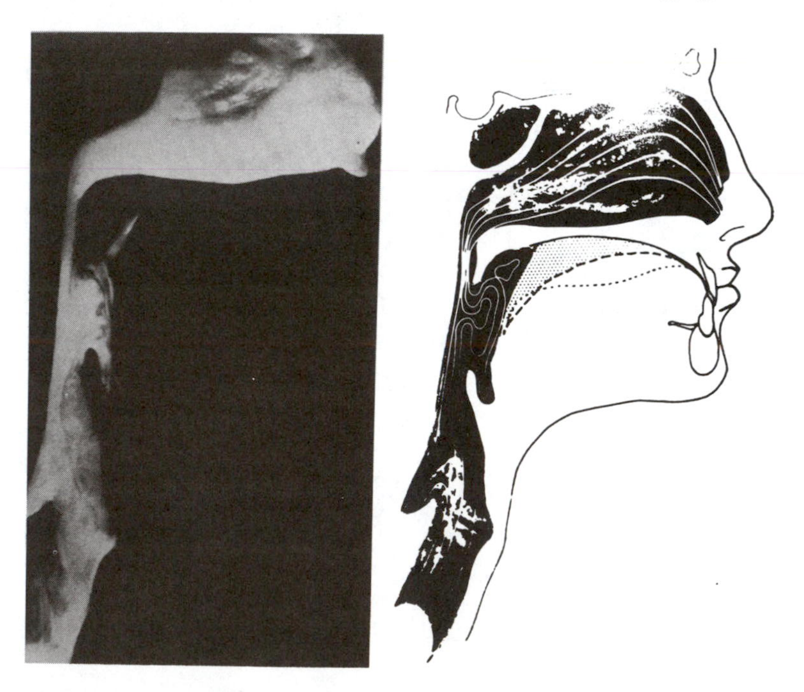

Figure 9-3. Diagram showing air currents (contour lines) in the nasopharynx and the position of the tongue at rest. (From Proetz, *Essays on the Applied Physiology of the Nose*. 2nd Edition. St. Louis: Annals Publishing Company, 1953. Copyright © 1953 by A. W. Proetz.)

contact with the mucosal lining of the nasal passages and the sinus cavities. In this way the air is humidified, and its temperature is regulated to that of the body. It also is cleansed of foreign substances.

Lower Air Passages. Once air has traversed the upper air passages, it moves into the trachea. The trachea traverses the neck and enters the bony thorax, where it then divides into the right and left lung. Each mainstem bronchus in turn divides into two or sometimes three daughter branches or bronchi. This irregular dichotomy continues for approximately 24 to 27 generations of branchings. Each successive branch in an axial pathway from trachea to alveoli becomes progressively smaller. This is shown diagrammati-

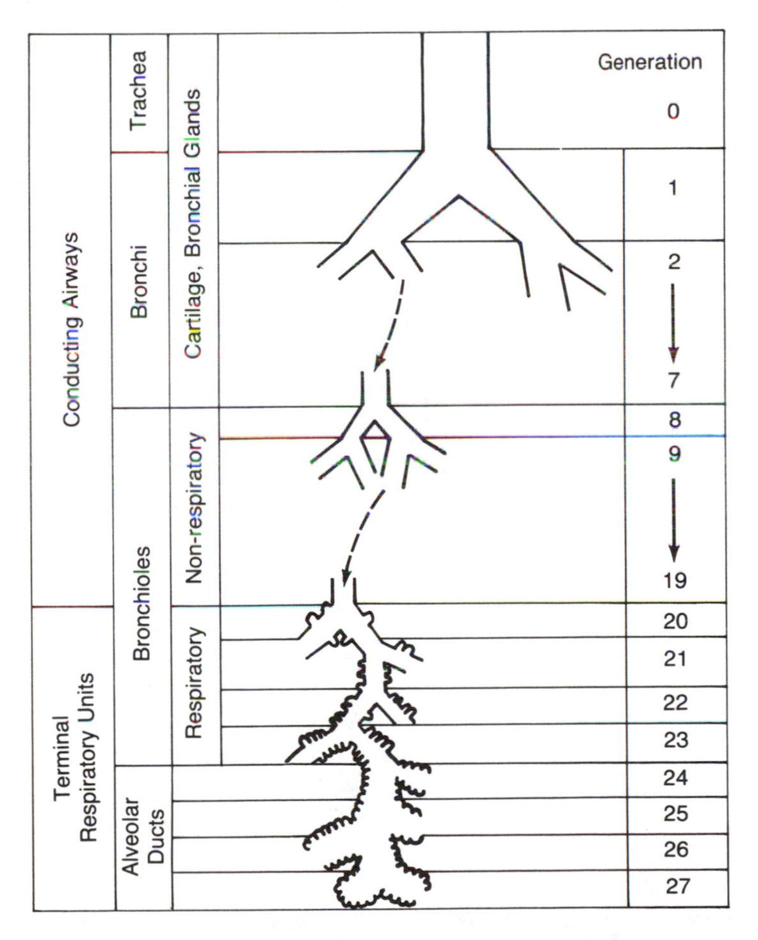

Figure 9-4. Schematic representation of the subdivisions of the conducting airways and terminal respiratory units. (Reproduced with permission from Weibel, *Morphometry of the Human Lung*. New York: Springer-Verlag, 1963.)

cally in Figure 9-4. At about the nineteenth or twentieth division one begins to see outpouchings of alveoli from the lateral walls of the respiratory bronchioles, and a few generations later the respiratory bronchioles become alveolar ducts and then clusters of alveolar sacs. It is at the level of the respiratory bronchiole and beyond to the air sacs that active gas exchange occurs between the environment and the blood.

Figure 9-5 shows that the narrowest portion of the entire air conduit system is in the upper airway at the tracheal level in the neck. The total cross-sectional area here is 2.5 cm^2. As the irregular dichotomous branching occurs, the actual cross-sectional area of the airways increases rapidly and tremendously beyond the first few branchings. By the time air reaches the transitional zone in the bronchial tube system the total cross-sectional area of the airways approaches 130 cm^2. The physiologic significance of this is that most resistance to air flow

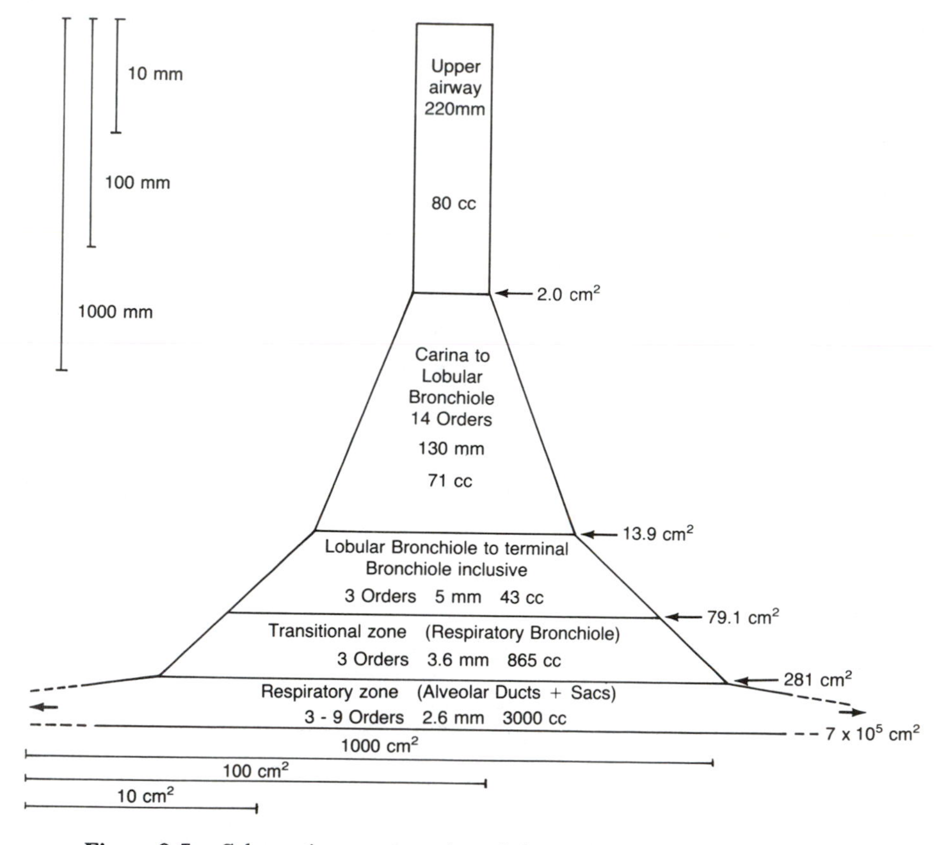

Figure 9-5. Schematic representation of the total cross-sectional area of the airways at successive levels of branching. (From Widdicombe, *Respiratory Physiology*. Woburn (Mass.): Butterworths. 1974. Vol. 2. Figure 1-3.)

occurs in the larger, more central airways, and the least resistance to air flow occurs in the smaller, more peripheral airways. Studies have shown that if one divides the tracheobronchial system into central and peripheral portions at the 2-mm-diameter bronchial level, the central resistance will account for 85 percent of the total resistance to flow, and the peripheral resistance in airways less than 2 mm in diameter will account for some 15 percent of the total resistance to flow.

Diseases such as asthma and bronchitis primarily affect the bronchial tube system. Emphysema, though it is usually classed as an obstructive air flow disease, actually affects the alveoli by destroying the delicate walls separating one air sac from another. This damage is irreversible because the lungs do not regenerate tissue that has been destroyed. Asthma and bronchitis, at least in their early stages, are reversible either by eliminating their causes or with appropriate treatment. Asthma, of course, is generally a very treatable disease and many asthmatics are symptom-free during periods when the condition is not in exacerbation.

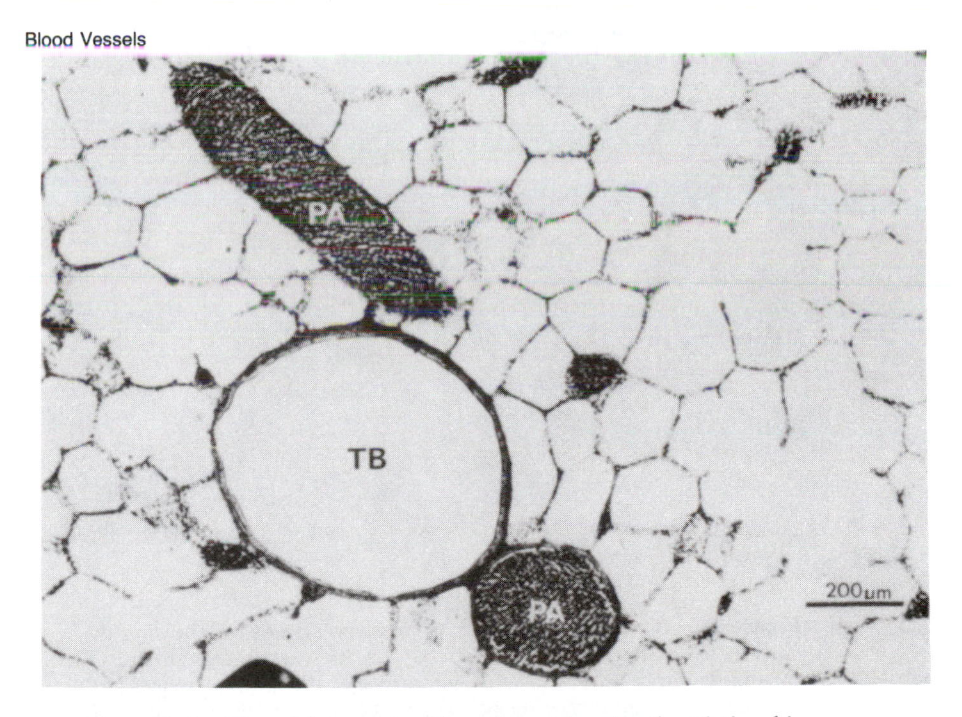

Figure 9-6. Photomicrograph of lung tissue, showing the relationship of a terminal bronchiole (TB) and its accompanying blood vessel, the pulmonary artery (PA), to the alveoli. (Reproduced with permission from Murray, *The Normal Lung. The Basis for Diagnosis and Treatment of Pulmonary Disease*. Philadelphia: W. B. Saunders Co., 1976.)

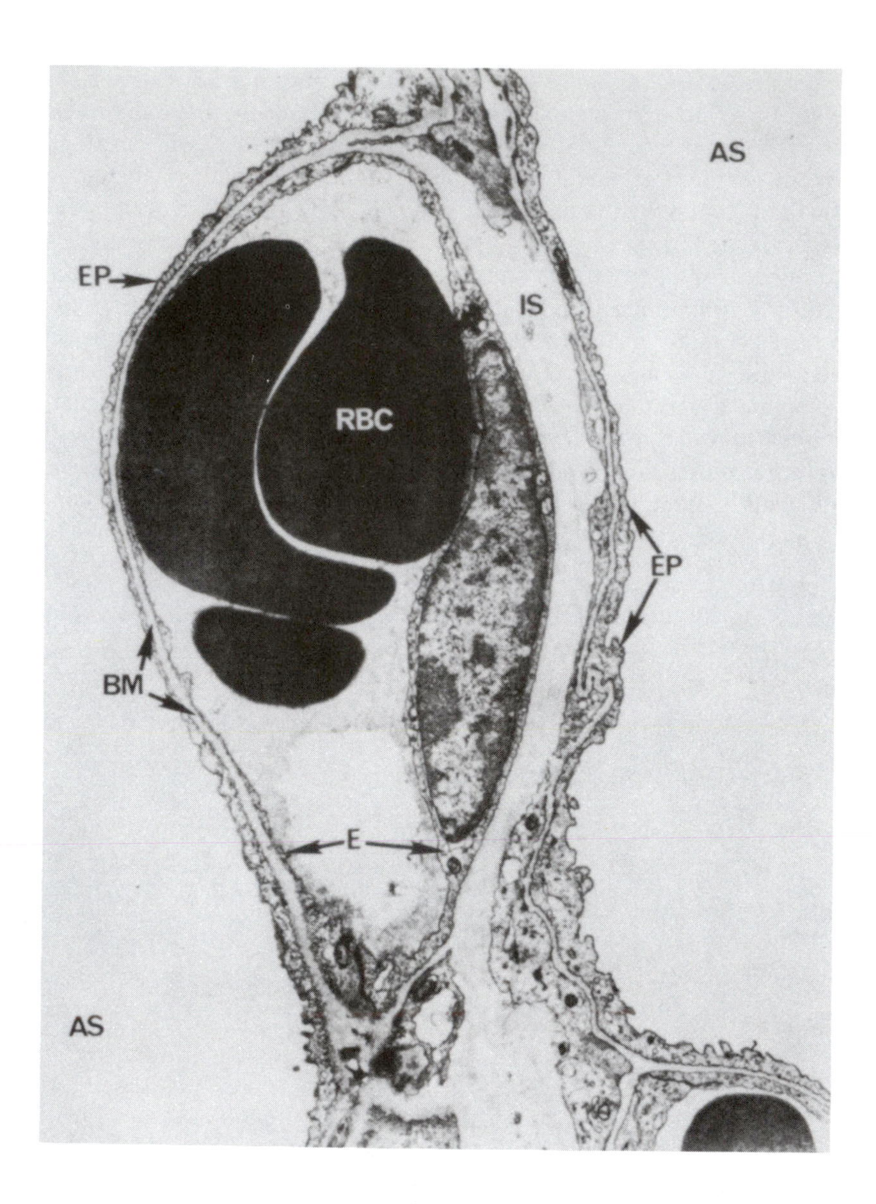

Figure 9-7. Electron micrograph of an alveolar septum, showing the various tissue layers through which oxygen and carbon dioxide must move during the process of diffusion. The surface of the alveolar spaces (AS) is lined by continuous epithelium (EP). The capillary containing red blood cells (RBC) is lined by endothelium (E). Both layers rest on basement membranes (BM) that appear fused over the "thin" portion of the membrane and that are separated by an interstitial space (IS) over the "thick" portion of the membrane. (Reproduced with permission from Murray, 1976.)

Pulmonary Gas Exchange

Figure 9-6 is a microscopic view of lung tissue that has been sectioned transversely. A small blood vessel, a terminal bronchiole, and adjacent alveoli can be clearly seen. The normal alveolar septal walls are quite thin, measuring approximately 2 μm in diameter. The alveolar septum, which separates one adjacent alveolus from its neighbor, is composed of alveolar epithelium, a basement membrane, an interstitial space, a pulmonary capillary endothelial basement membrane, the pulmonary capillary endothelium, blood plasma, and red blood cells. The system is designed in such a way that the alveoli are almost totally bathed in a continuous sheet of capillary blood. As mentioned earlier, if the alveoli were spread uniformly on a surface, they would cover an area of some 70 m^2 and, in turn, would be covered by a 120-ml film of blood. It is this remarkable relationship of a very large alveolar surface area in intimate contact with a very small volume of capillary blood that brings the ambient air into close contact with incoming venous blood.

Figure 9-7 is an electron micrograph of an alveolar septum, and it clearly shows the individual structures noted in the previous paragraph. A red blood

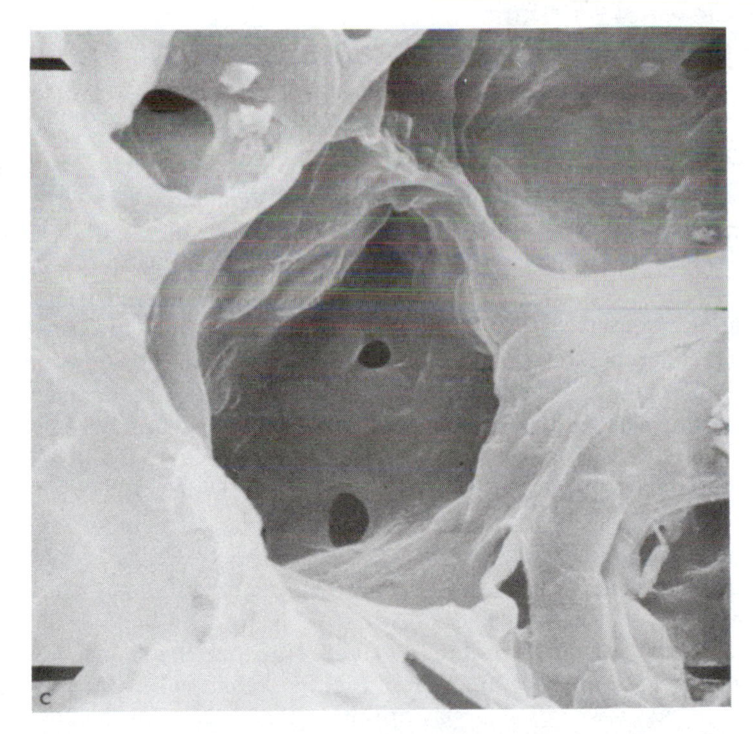

Figure 9-8. Scanning electron micrograph showing interior of an alveolus and its pores of Kohn. (Reproduced with permission from Bates *et al.*, *Respiratory Function in Disease. An Introduction to the Integrated Study of the Lung*. Philadelphia: W. B. Saunders Co., 1971.)

cell in the capillary can be seen in the figure. Oxygen moves from the alveolar space across the alveolar epithelium, the interstitial space, and the capillary endothelium before entering the blood plasma and being taken up by the hemoglobin within the red blood cell.

The alveolar capillary membrane in health is approximately 2 μm in diameter. However, many diseases damage this delicate membrane. The damage is healed by the formation of scar tissue. This scar tissue increases the thickness of the membrane and also alters the elastic properties of the alveolar septum. As a consequence, ambient air is unevenly distributed to the millions of alveoli, and oxygenation becomes abnormal. This is manifested physiologically by a decrease in the tension of oxygen in arterial blood and is manifested clinically by the development of dyspnea, usually in relationship to exertion. Figure 9-8 is a scanning electron micrograph taken as though the observer were standing in a respiratory bronchiole looking out into a cluster of alveoli. The holes that are clearly visible are called pores of Kohn. They provide for some collateral ventilation between adjacent alveoli.

9.2 Pulmonary Defenses

The Mucociliary Mechanism

Function. Figure 9-9 is a diagram of a longitudinal section of a bronchial wall. The mucosa is the portion of the bronchial wall that comes in contact with air in the bronchial lumen. It is composed of pseudostratified columnar epithelial cells and occasional specialized cells, called goblet cells, which produce a special type of thin mucus. Each of the epithelial cells contains several hundred fine hairlike structures called cilia, which project into the lumen of the bronchus. The pseudostratified epithelial cells rest on a thin basement membrane. Beneath this, in the stroma of the bronchial wall, are clusters of mucous glands with coiled ducts, which course upward between the epithelial cells to secrete mucus onto the surface of the cilia. Beneath the mucous gland layer are varying amounts of smooth muscle, which help to regulate the length and cross-sectional area of the bronchus. Beneath the muscle are cartilaginous rings or plates, which help support the bronchial wall. Beneath the cartilage is more stroma, and then the external epithelial cells on the outer portion of the bronchial tube.

The ciliated epithelium is quite specialized in that the thousands of cilia beat rhythmically some several hundred times a second. Their major direction of force is always toward the airway opening—that is, toward the mouth: cilia located in the tracheobronchial tree beat upward toward the mouth and cilia in the nasopharyngeal passages beat downward toward it. The goblet cells and mucous glands secrete their mucus onto the surface of the cilia to form a continuous coating of mucus within the nasopharyngeal and tracheobronchial

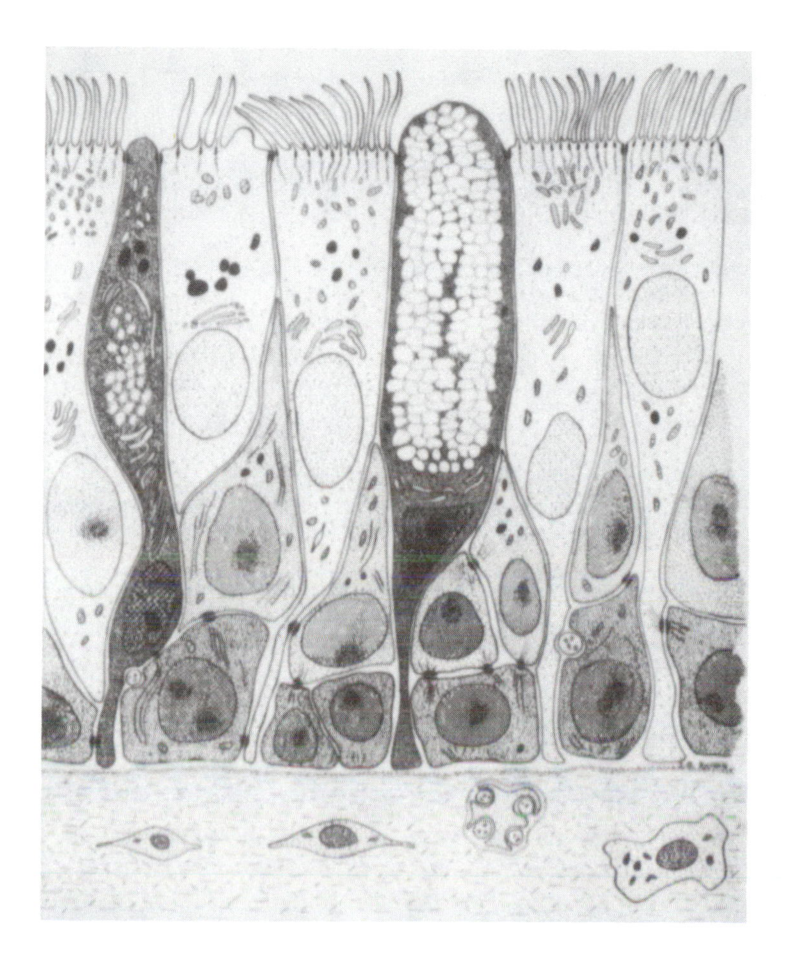

Figure 9-9. Schematic representation of airway mucosa, showing pseudostratified, ciliated epithelial cells and goblet cells. (Reproduced with permission from Rhodin, "Ultrastructure and Function of the Human Tracheal Mucosa." *Amer. Rev. of Resp. Disease, 93*(1966):13.)

systems. Figure 9-10 shows that even the mucous layer is highly specialized. The upper portion consists of very thick, viscid mucus and the lower portion, which comes in contact with the cilia, consists of a thin, aqueous mucus. The lower or sol layer permits the cilia to beat freely and to propel the more viscid upper or gel layer continuously toward the airway opening. As ambient air flows into the nasal passages and down into the tracheobronchial tree, most particulates in the air come into contact with the thick, sticky gel layer and are trapped and moved upward on the continuously flowing river of mucus to the airway opening. There, the mucus and particulate substances are either swallowed or coughed from the body. This mucociliary mechanism is the respiratory system's first line of defense against a potentially hostile environment.

Figure 9-10. Schematic representation of the mucociliary blanket, showing the wavelike motion of the cilia within the sol layer. (Reproduced with permission from Hilding, "Experimental Studies on Some Little-understood Aspects of the Physiology of the Respiratory Tract." *Tr. Amer. Acad. Ophth. and Otol., 1961*: July–August.)

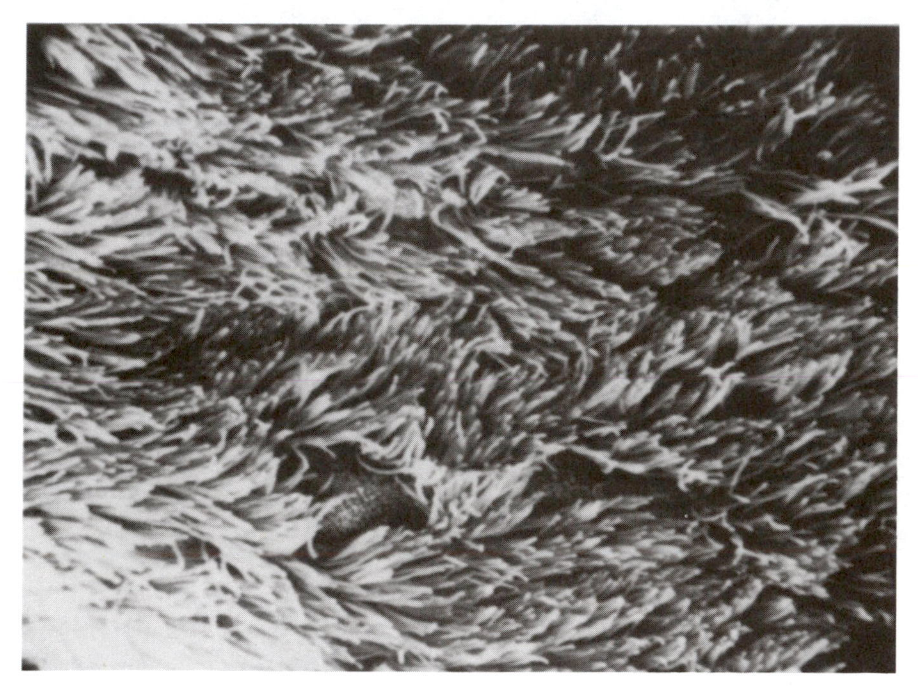

Figure 9-11. Scanning electron micrograph of the luminal surface of a bronchiole, showing the cilia. The mucous layer has been removed. (Reproduced with permission from Ebert and Terracio, "The Bronchiolar Epithelium in Cigarette Smokers." *Amer. Rev. of Resp. Disease, 111*(1975):6.)

Figure 9-11 is a scanning electron micrograph showing the surface of respiratory epithelium. The layer of mucus has been removed so that the cilia can be seen clearly. Several areas that are relatively devoid of cilia are openings of mucous ducts. The bronchial tube system, sinuses, and nasal passages are all covered with this type of ciliated epithelium. This delicate and carefully integrated system maintains clean, sterile air for entry into the alveoli for gas exchange.

Sensitivity to Infection. The mucociliary mechanism, however, is sensitive to a number of substances and diseases, and can be rendered partially or even totally ineffective. One or more of the more-than-2000 chemicals in cigarette smoke can paralyze the cilia over the entire tracheobronchial system and stop the flow of the blanket of mucus. Studies have shown that one puff of cigarette smoke is sufficient to paralyze the cilia for as much as 20 to 40 min. Thus, it is highly probable that the mucociliary mechanism in a smoker who consumes 20 to 40 cigarettes a day is rendered almost continuously ineffective through the paralyzing effect of smoke on cilia. This eliminates one of the lung's major defense mechanisms and is quite probably one of the major factors leading to lower respiratory tract infections and possibly to the eventual development of chronic bronchitis.

The influenza virus also can damage the mucociliary mechanism. During influenza epidemics it is quite common to see severe secondary bacterial infections such as pneumonias or lung abscesses. Many chemical vapors, such as chlorine, ammonia, and oxides of sulfur and nitrogen, also injure the mucociliary mechanism, as do many metal fumes, such as cadmium, nickel, and mercury. High concentrations of oxygen in the inspired air can also interfere with a normal mucociliary function, making individuals receiving the oxygen more susceptible to lower respiratory tract infections.

Alveolar Macrophages

The second major line of defense in the lungs resides in the alveolar macrophages. These are phagocytic cells that have the ability to move about within the alveoli and to phagocytize or engulf foreign particles. When a bacterium, for instance, impinges on the alveolar epithelium, chemotactic substances are released that alert the alveolar macrophages to the presence of the bacterium. These phagocytic cells can ingest or engulf germs, viruses, chemicals, or other foreign substances and, in many cases, destroy them. However, if the macrophages lose the battle, which is the case in many industrial situations, the stage is set for damage to the air sacs.

The macrophages contain powerful proteolytic enzymes, which are used to destroy bacteria, viruses, and other foreign particles. If the substance that is ingested injures and eventually destroys the macrophage, the chemicals within the macrophage are released into the environment of the alveolus and damage the delicate lining layer. The damaged alveolar epithelium is repaired by the formation of scar tissue. If the damage is extensive and results in the formation of a significant amount of scar tissue, lung function is severely altered and

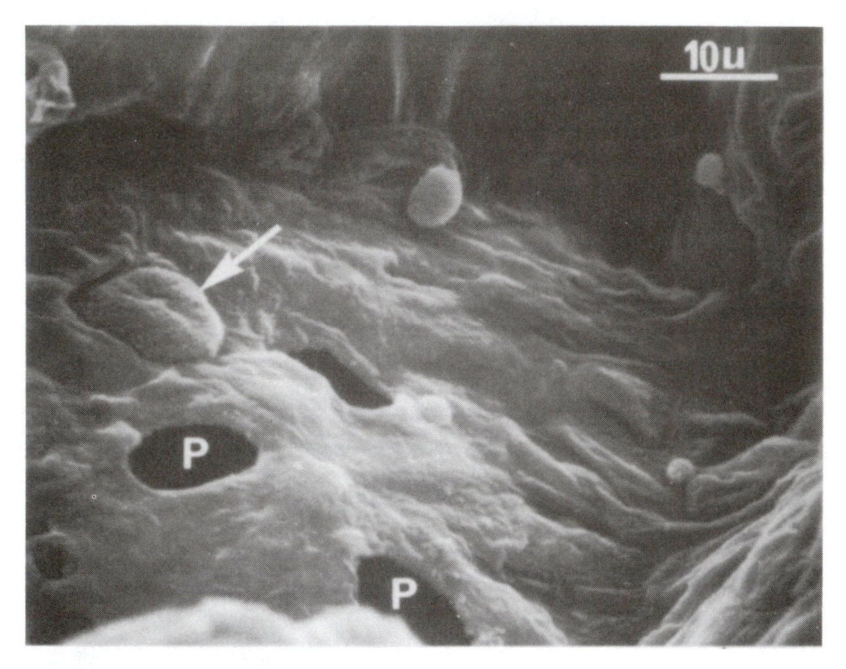

Figure 9-12. Scanning electron micrograph of the interior of an alveolus, showing pores of Kohn (P) and a macrophage (arrow). (Reproduced with permission from Murray, 1976.)

abnormalities of gas exchange occur. Figure 9-12 shows a macrophage in the process of moving across a pore of Kohn from one alveolus to another.

The integrity of the alveoli and their normal function is greatly dependent on the normal function of the alveolar macrophage. Macrophage function, however, can be severely altered by such factors as cigarette smoke; chemical vapors and fumes; high, as well as low, concentrations of oxygen in the inspired air; the influenza virus; corticosteroid drugs; and the ingestion of alcohol.

Lung Clearance Mechanisms

When air containing particles is inhaled, the particles traverse the tracheobronchial tubes and may even reach the alveoli, where they are deposited. A certain number of these particles will be cleared either by the mucociliary mechanism or by inactivation and elimination by alveolar macrophages. This combined process is called lung clearance. The difference between deposition of particles and their clearance is called retention.

As noted earlier, the larger particles impact in the upper airways, the medium-sized particles come in contact with the mucous layer in the smaller airways through the process of sedimentation, and the smallest particles reach the alveoli and settle on the alveolar epithelium through the process of diffusion. There, they may be cleared from the lungs by the cilia.

The structure of a cilium is quite complex. The internal structure is basically composed of nine pairs of tubules, which run the length of the cilium and are arranged around the periphery. These are interconnected by Dynein arms and, in turn, connect to two central or core tubules. The base of the cilium is embedded in the surface of the epithelial cell, again through a complex interconnecting network. The physiology of the cilium is only partially understood at the present time, but recent clinical studies have shown that certain individuals with chronic bronchitis or a disease called bronchiectasis lack one or more of the small interconnecting Dynein arms. This lack of Dynein arms causes the cilia either to become immobile or to beat asynchronously. Individuals with defective cilia are prone to upper respiratory infections, lower respiratory infections, and, in some clinical situations, they become infertile because of immobile sperm. With abnormalities of ciliary function, the blanket of mucus fails to flow properly and the smaller bronchioles tend to become plugged with thick, tenacious mucus; and eventually the bronchioles become secondarily infected.

Individuals with infectious bronchitis usually have thick, tenacious mucus, which tends to collect in the smaller bronchial tubes and interfere with gas exchange. Exposure to fumes, chemical vapors, or other industrial toxins can interfere with normal mucociliary function and can lead to abnormalities of bronchial clearance.

9.3 Inhalation of Pollutants

Deposition of Aerosols in the Lung

An aerosol is a cloud of particles that tend to remain suspended in air for a considerable period of time. In evaluating the dangers of inhalation of aerosols into the respiratory system, one must consider how they are deposited in the lung, and what mechanisms clear the particles from the lung. The difference between the number of particles deposited and the number of particles cleared is the number of particles that are retained within the lung, and which can potentially cause injury to the lung tissues. There are three recognized mechanisms for deposition of aerosols in the lung: impaction, sedimentation, and diffusion.

Impaction. The first, impaction, refers to the tendency of large inspired particles to continue on straight paths through branching tubes. As a result of this tendency, most larger particles impinge on the mucous surface of the nasopharynx as well as at the bifurcation of the larger airways. When these large particles strike the sticky surface of the mucus they become stuck and are eventually carried upward to the airway opening and eliminated from the body. The upper air passages are remarkably efficient in removing large particles from the inspired air, and almost all particles greater than 20 μm in diameter and even 95 percent of the particles 5 μm and larger in diameter are filtered by the nasopharynx.

Sedimentation. The second mechanism for the deposition of aerosols, sedimentation, is due to the gradual settling of particles of medium size, that is 1 to 5 μm. As noted above, the larger particles are removed by impaction, leaving the smaller particles to settle more slowly. The major site of deposition of aerosol particles by sedimentation is in the smaller, more peripheral airways. Because the smaller particles settle slowly toward the periphery of the bronchial tube, they too impinge upon the mucous layer and are eventually swept up and eliminated from the lung. Studies in recent years have shown that most totally asymptomatic smokers display physiologic evidence of abnormalities in the smaller, more peripheral airways. It is quite probable that the earliest changes leading to chronic bronchitis and emphysema occur at this level of the tracheobronchial tree.

Diffusion. The third mechanism of aerosol deposition in the lung is diffusion, which occurs by the random movement of aerosol particles as a result of their continuous bombardment by gas molecules. This has its greatest effect on very small particles less than 0.1 μm in diameter. The deposition of particles of this size occurs primarily at the alveolar level. However, many of these small particles remain suspended in air and are eliminated from the respiratory system on the next expiration.

Atmospheric Pollutants

Table 9-1 lists the most common types of atmospheric pollutants, and Table 9-2 lists the most common sources for these pollutants. Carbon monoxide accounts for almost half of all the pollutants encountered in the atmosphere, and most of this is produced by transportation sources and the burning of gasoline. The remainder of the pollutants is almost equally divided among sulfur oxides, hydrocarbons, oxides of nitrogen, and particulates.

Carbon Monoxide. Carbon monoxide is a colorless, odorless gas that is released from the combustion of almost any material. Most carbon monoxide released into the atmosphere each year comes from the burning of gasoline in automobiles. The major physiological importance of carbon monoxide is its ability to compete with oxygen for binding sites on the hemoglobin molecule

Table 9-1. Current Composition of Polluted Air in the United States.

Pollutant	Percentage
Carbon monoxide	48
Oxides of sulfur	16
Hydrocarbons	15
Oxides of nitrogen	11
Particulates	10

(as discussed in Chapter 4). The exact health consequences of environmental carbon monoxide are unclear. However, there have been a number of reports of carbon monoxide poisoning in traffic policemen during heavy rush hour times of day and no doubt this pollutant adversely affects the health of commuters as well.

Sulfur Oxides. Oxides of sulfur are corrosive and are produced when sulfur-containing fuels, chiefly coal, are burned. If one is exposed to high concentrations of sulfur oxides, the alveolar septa can be damaged and acute pulmonary edema can occur. Exposure over a longer period of time to lower concentrations can result in chronic bronchitis and may also lead to pulmonary emphysema.

Photochemical Oxidants. These substances are produced primarily by the action of sunlight on hydrocarbons and nitrogen oxides in the air. They include ozone, aldehydes, and acrolein. These substances are usually irritating to the delicate mucous membranes of the body and cause inflammation of the eyes and upper respiratory tract. In large concentrations ozone can lead to pulmonary edema.

Particulate Matter. These particles are composed primarily of small carbon particles and dust particles released into the air either through natural mecha-

Table 9-2. Current Sources of Air Pollutants in the United States.

Source	Percentage
Transportation	50
Stationary	22
Industrial	17
Other	11

nisms or human means. Toxic gases become adsorbed to the surface of the particulates, and are breathed into the lung and deposited. It is the adsorbed chemicals, rather than the particulates, that are felt to cause the major damage to the tracheobronchial tree.

Nitrogen Oxides. Oxides of nitrogen are produced primarily from the burning of coal, oil, and gasoline. These gases can cause direct damage to the mucosal membranes of the eyes, upper respiratory tract, or the tracheobronchial tree and alveoli. The site of damage depends on the concentration and the duration of exposure.

Cigarette (Tobacco) Smoke

The health effects of air pollution are small when compared to the health effects of cigarette smoking. Cigarettes are the major cause of chronic bronchitis, pulmonary emphysema, and bronchogenic carcinoma. The nicotine and other substances in cigarette smoke are also a major cause of coronary artery disease as well as other peripheral vascular diseases.

Evidence of Danger. Figure 9-13 is a graph that plots the per capita usage of cigarettes on the ordinate, as well as the superimposed incidence of chronic obstructive lung disease (that is, chronic bronchitis and pulmonary emphysema, and also bronchogenic carcinoma), against years on the abscissa.

In the early 1900s, the per capita consumption of cigarettes was low, relatively constant, and confined primarily to males. In the early 1920s cigarette consumption began to rise, and some 15 to 20 years later the incidence of chronic obstructive lung disease began to rise as well. Some 10 years following this the incidence of bronchogenic carcinoma began to increase.

As can be seen in the graph, the rate of increase in chronic obstructive lung disease and bronchogenic carcinoma has closely paralleled the increasing consumption of cigarettes as the years have progressed. The scientific data that have accumulated over the years linking cigarette smoking in a cause-and-effect relationship to chronic bronchitis, pulmonary emphysema, and bronchogenic carcinoma are massive and totally convincing. There are over 2000 identifiable substances in cigarette smoke. Some of these substances are irritants to the delicate lining of the tracheobronchial tree and respiratory passages; some substances lead through a variety of mechanisms to chronic bronchitis and to pulmonary emphysema; and some substances are carcinogens and lead to bronchogenic carcinoma. In addition to the adverse effects of cigarette smoke on lung function, nicotine and possibly other substances in cigarette smoke also act as potent blood vessel constrictors and play a significant role in the development of arteriosclerosis in the coronary blood vessels, which supply the heart muscle with blood, as well as in the peripheral blood vessels supplying other tissues of the body.

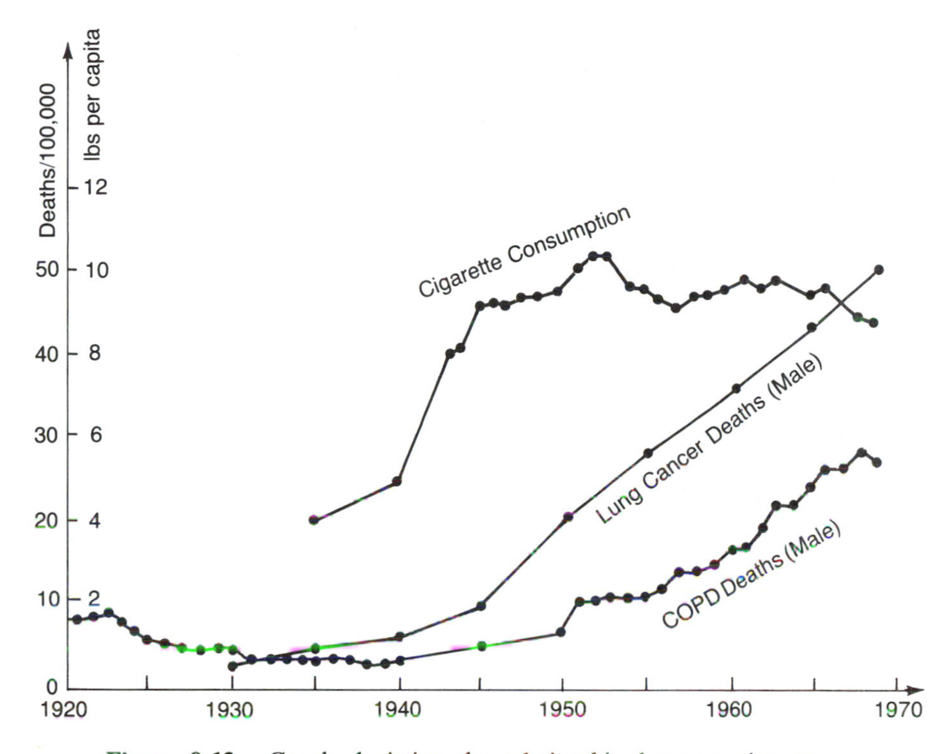

Figure 9-13. Graph depicting the relationship between cigarette consumption and incidence of chronic bronchitis and pulmonary emphysema (COPD plot) and lung cancer. (Based on data from American Thoracic Society, New York.)

Mechanisms of Damage. As we have seen earlier, substances in cigarette smoke directly paralyze cilia function in the respiratory mucosa, leading to partial or total cessation of ciliary motion and to stagnation of the mucus. In addition to this, other substances in cigarette smoke act as irritants; in response to this, the submucosal bronchial mucous glands produce and secrete more mucus in an attempt to wash away the irritant. The combination of the cessation of mucus flow and the increased production of mucus leads to an increase in intraluminal mucus, with partial or even total blockage of smaller bronchi. This of course interferes with gas exchange. Over an extended period of exposure to the irritants in cigarette smoke the submucosal mucous glands eventually hypertrophy or increase in size. This hypertrophy encroaches on the airway lumen and contributes to a decrease in airway caliber and to an increase in resistance to air flow through the bronchi.

Also, as previously mentioned, cigarette smoke contains substances that interfere with the normal function of alveolar macrophages. Thus, chronic cigarette smokers have an increased propensity for lung infections through alterations in normal defense mechanisms. Over a period of years, many of these individuals will develop chronic, irreversible damage to the bronchial tubes, which is called chronic bronchitis, with air-flow limitation.

Recent experimental data have shown that cigarette smokers have an increased number of alveolar macrophages, an increased number of polymorphonuclear leukocytes in the alveoli, and a decrease in the normal amount of proteolytic enzyme inhibitors. Thus, as the macrophages and polymorphonuclear leukocytes age, die, and rupture, they release their proteolytic enzymes into the alveoli. Because of the deficiency of inhibitors of proteolytic enzymes, these strong enzymes are then free to digest the delicate septa separating one alveolus from another. After prolonged exposure to these proteolytic enzymes, the alveolar septa rupture and pulmonary emphysema develops.

Lung Cancer. Bronchogenic carcinoma—that is, lung cancer—is essentially a problem of the twentieth century. The first case of lung cancer was reported in the medical literature in the late 1800s, and by 1900 some 300 cases had been reported on a worldwide basis. Since that time, the incidence of bronchogenic carcinoma has been rising at an alarming rate. In 1983, there were over 130,000 new cases of bronchogenic carcinoma in the United States. Current data suggest that at least 90 percent of this striking rise in lung cancer is directly related to cigarette smoking. Lung cancer has been the leading cancer killer in men for quite a long period of time, and within the next several years it will become the leading cancer killer in women.

For many years lung cancer was essentially a disease of men, and many felt that for unexplained reasons women were immune to its development. The first suggestion that this was not the case was made by Drs. Alton Ochner and Michael DeBakey in the 1950s, when they predicted that cigarettes were a major cause of lung cancer. Based on epidemiologic data available at that time, they pointed out that men had been smoking heavily since the early 1900s and that the incidence of bronchogenic carcinoma had begun to rise sharply some 20 years following this. Using these data, they pointed out that women as a group had begun to smoke heavily in the mid-1940s, and they predicted a sharp rise in bronchogenic carcinoma in women in the mid to late 1970s.

This prediction has proved true. Increasing numbers of women are developing and dying of bronchogenic carcinoma at present. Lung cancer, unfortunately, carries a very high mortality despite the advances of modern medicine. Of every 100 individuals diagnosed as having bronchogenic carcinoma, only five to ten survive beyond five years and are presumably cured.

Tobacco in any form can produce cancer if brought into prolonged contact with any of the body's surfaces. People who dip snuff develop lip cancer, tongue cancer, and mouth cancer. People who smoke cigars or hold pipes in their mouths also have the potential for developing cancer in those same areas.

As previously mentioned, nicotine is a very potent blood vessel constrictor and is an important cause of coronary and peripheral arteriosclerosis. Nicotine is felt to be the major addicting substance in cigarette smoke. When one smokes cigarettes on a regular long-term basis, one becomes physically addicted to nicotine. If the cigarettes are suddenly stopped, the smoker experiences physical withdrawal. The realization that cigarette smokers are physically addicted helps to explain the great difficulty that smokers have in stopping smoking on a permanent basis.

The Effect on Nonsmokers. Many studies have shown that tobacco smoke can have adverse health effects in nonsmokers. Early studies showed that pregnant women who smoke have more miscarriages and give birth to infants with lower birth weights than do nonsmoking women. In addition, the fetuses carried by these women have higher intrauterine heart rates. Other studies have shown that children growing up in homes where one or both parents smoke have more respiratory illnesses than children whose parents don't smoke. There are data suggesting similar adverse health effects in adult nonsmokers within the workplace.

Chemical Toxins

A large number of chemical vapors and metal fumes have been identified as respiratory toxins. To consider each of these in detail is beyond the scope of this chapter. But some attention can be given to the general symptomatology of these types of toxins.

The commonly encountered chemical vapors are chlorine gas, ammonia gas, oxides of sulfur, oxides of nitrogen, and hydrogen sulfide gas. Exposure to these gases causes direct chemical injury to the exposed tissues of the respiratory system. The site of injury depends on the concentration of gas to which the individual is exposed, the duration of exposure, and the solubility of the gas in the body fluids. Exposure can cause symptoms that vary from irritation to the upper air passages, with resulting rhinitis and sinusitis, to tracheobronchitis, with cough, sputum production, secondary infection, and bronchospasm; and, finally, if the chemicals reach the level of the alveolus, alveolar damage will result, with either an alveolitis and subsequent diffuse fibrosis or, if the damage is massive and acute, to sudden pulmonary edema.

A number of metal fumes have been identified as respiratory toxins. These include cadmium, nickel, mercury, and many others. As with chemical vapors, the site of damage within the respiratory system depends on the concentration of the metal fumes, the duration of exposure, the particle size, and the chemical characteristics of the metal itself. The damage can occur in the upper airway, the tracheobronchial tree, or at the alveolar level, and can cause either the sudden onset of symptoms or, insidiously, the development of symptoms over a period of months or years.

9.4 Occupationally Acquired Diseases

Pneumoconiosis

The term pneumoconiosis is used in reference to lung diseases resulting from the inhalation of dust, usually inorganic in nature.

Silicosis. Silicosis is a lung disease seen in industries in which individuals are exposed to quartz rock dust. It is historically a very old disease, having been found in Egyptian mummies. It is seen primarily in individuals who work in rock quarries, slate quarries, in sandblasting, and in other industrial settings where quartz dust is present in respirable size.

Silicosis is caused by the inhalation and retention within the lungs of free silica—that is, of silica dioxide crystals. Those crystals that are not trapped by the mucociliary blanket reach the alveoli and settle onto the epithelial lining. The alveolar macrophages then ingest the crystals in an attempt to cleanse the lungs. The phagocytized silica dioxide crystals produce a weak acid, silicic acid, which damages the macrophages and eventually destroys them. The dead macrophages release chemicals that damage the lungs and lead to fibrosis. In late stages of the disease, the right side of the heart may hypertrophy and the individual may develop congestive heart failure.

Because of the cycle of phagocytosis of silica crystals, injury and rupture of macrophages with release of the crystals, and reingestion by other macrophages, silicosis tends to be a self-perpetuating problem. Once enough silica dioxide crystals have been retained within the lungs, the damage tends to continue despite the fact that the individual may be removed from further exposure. Unfortunately, there is no specific treatment of silicosis. Therapy is entirely confined to amelioration of symptoms. In individuals with moderate to severe silicosis, the prognosis is quite poor.

The major impetus is toward prevention, to keep dust levels as low as possible through effective work practices and engineering controls (discussed in Chapter 20), to insist that individuals who are at risk wear effective respiratory protection (e.g., dust masks), and to establish and maintain effective health programs. The latter should stress the associated health problems of smoking, the potential dangers of exposure to silica crystals, and the importance of wearing a dust mask. The program should also include a medical surveillance, with chest x-ray photos taken on at least an annual basis, and screening pulmonary function testing.

Asbestosis. Asbestos is a filamentous mineral useful to man because of its resistance to heat. Its major use earlier in this century was as an insulating material, but in more recent years it has found widespread use in a number of other industries and is quite ubiquitous at the present time.

At one time diseases related to asbestos exposure were limited to a few occupations, but because of the widespread use of asbestos in recent years, the potential for the development of asbestos-related disease has increased to include a large sample of the population. Those whose occupations require them to use asbestos include shipyard workers, particularly those individuals who are involved in the building and refitting of ships; insulation workers, particularly in years past when asbestos was used almost exclusively; some plumbers; and individuals involved with the relining of brakes and the turning of brake drums. Certain construction jobs also bring individuals into close contact with asbestos, and are therefore potentially hazardous.

As with silica dioxide crystals, the long filamentous crystals of asbestos, if deposited in the alveoli, are ingested by macrophages and, through mechanisms that are not currently clear, lead to the development of diffuse lung damage and the formation of fibrosis. In this respect, asbestosis is very similar to silicosis. In addition, asbestos can also lead to the development of a malignant tumor of the lining of the chest cavity and lung surface called mesothelioma, and, in association with cigarette smoking, can lead to an increased incidence of bronchogenic carcinoma. It is said that individuals who smoke cigarettes and have significant exposure to asbestos have an 80-fold-increased likelihood of developing lung cancer.

As with silicosis, the diagnosis of asbestosis depends on a well-documented history of exposure, an accurate estimate of the degree of exposure, and a characteristic clinical picture. The patterns of impairment seen with asbestos exposure usually take 20 to 40 years to develop. There is evidence that lesser amounts of exposure can lead to the development of mesothelioma. Individuals who develop moderate to severe degrees of asbestosis have a poor prognosis, and those who develop malignant mesothelioma have an exceedingly poor outlook, in that mesothelioma cannot be removed surgically, nor is it amenable to irradiation therapy or to chemotherapy. The outlook for those individuals who develop associated bronchogenic carcinoma is the same as for those who develop lung cancer and who have not been exposed to asbestos—that is, the five-year survival is about 5 percent to 10 percent.

Most individuals living in large cities for a number of years have some asbestos fibers in their lung tissues. This was vividly documented in a study done some years ago in five large cities, in which randomly autopsied lungs were examined for the presence of asbestos fibers. Some 40 percent to 45 percent of the lungs had easily identifiable asbestos crystals, but none of these lungs showed evidence of disease resulting from the asbestos. This same study was repeated in more recent years, and almost every lung examined had identifiable asbestos crystals in the alveoli. The long-term health consequences of this low-dose exposure are unknown at this time. It is highly unlikely that this small degree of exposure could lead to the development of the interstitial fibrotic pattern of impairment, but there is a possibility that such exposure could lead to an increased incidence of mesothelioma and, in smokers, to an increased incidence of bronchogenic carcinoma.

Coal Workers' Pneumoconiosis. Coal miners who work underground for a period of years may develop abnormalities that show up in x-ray photos, as well as mild degrees of pulmonary impairment. This is called simple coal workers' pneumoconiosis, and is diagnosed primarily in individuals who have worked in underground coal mines for a period of eight years or longer. The symptoms are usually those of cough and mild breathlessness, and pulmonary function impairment is usually quite mild.

In approximately 8 percent of miners with coal workers' pneumoconiosis, a more serious degree of impairment occurs, and this is called massive progressive fibrosis. In these individuals the lung nodules become coalescent and form great conglomerate masses of scar tissue, usually around the midzones of the lung or in the apices. These individuals develop progressive shortness of breath and severe impairment of pulmonary function, and are often quite disabled. Many recent studies of coal workers' pneumoconiosis have shown that most of the respiratory symptomatology and impairment is directly related to the smoking of cigarettes rather than exposure to coal dust.

Reactive Airway Diseases

Byssinosis. Byssinosis is an occupational lung disease seen in textile workers exposed to cotton, flax, or hemp. The disease was first noted more than 100 years ago by two Belgian physicians who were struck by what appeared to be an excess of respiratory symptoms in Belgian textile workers. In interviewing a large number of workers, these two physicians identified a symptom complex of chest tightness, wheezing, and shortness of breath in some workers. Later studies corroborated the initial findings and showed byssinosis to be a reactive airway disease occurring in some textile workers on exposure to respirable substances in the air within the textile mill.

The characteristic clinical picture of byssinosis is the development of a sensation of chest tightness; wheezing, which often is audible; and shortness of breath noticeable in susceptible workers several hours after arriving at work on the first day after having been away for two days (a weekend) or longer. The symptoms tend to worsen during the first work day, and lessen when the individual leaves work. These symptoms usually do not recur during the remainder of the week, but on the first day back at work after the next weekend, the symptom complex recurs.

The first-day-back-at-work symptom complex occurs either intermittently or weekly over a period of some six to eight years, and, if the condition is allowed to persist, affected workers then begin to notice the symptoms extending into other days of the work week. The symptoms eventually become continuous, both at work and away from work. In the early years of the disease, when symptoms occur intermittently, the workers have no permanent pulmonary function impairment and, if they are removed from exposure or instructed to wear effective filtration masks, suffer no long-term permanent impairment. However, if the process is not discovered and steps are not taken to prevent airway reactivity, persons with byssinosis develop permanent airway constriction and show a clinical picture that is indistinguishable from chronic bronchitic airway narrowing.

The pathophysiology of byssinosis is felt to be as follows: when the toxic substances in textile-mill air enter the susceptible individual's airways, specialized cells in the mucosal lining of the airways, called mast cells, are stimulated and release chemical mediators—principally histamine—which are potent bronchoconstrictors. Apparently, on initial exposure during the first day back at work after time away, the mast cells release all of their mediators. Later, when the worker is at home, this time spent away from work is not sufficient for the mast cells to regenerate their full complement of chemical mediators. Therefore, on reexposure the next morning, the mast cells do not contain sufficient mediators to cause discernible symptoms, and this situation continues throughout the remainder of the work week.

However, two days or a weekend away from work is sufficient time for the mast cells to regenerate a full complement of mediators, and on first exposure the following Monday the symptom complex and pathogenic sequence recur.

As mentioned above, in the early years of this process the susceptible mill-worker's pulmonary function away from work is normal and the process is totally reversible. This disease can therefore be easily detected by medical surveillance, including medical questionnaires for eliciting the symptom complex of byssinosis, and by screening pulmonary function testing done before an employee's arrival at work on the first day and repeated some six to eight hours later, after exposure to "cotton dust." Persons who are reactors and have byssinosis will show a significant drop in the maximal expiratory flow rates after work exposure.

Currently, federal regulations require textile mills to do preemployment medical evaluations and pulmonary function testing and to perform before-and-after-work-exposure spirometry early in an individual's employment and to repeat the medical questionnaire and spirometry on a yearly basis. Those employees with symptoms suggesting byssinosis, those who show a significant decrement in their pulmonary function after six to eight hours of work exposure, and those whose pulmonary function is below a predetermined value are required to be evaluated by a pulmonary specialist. In addition to this, textile mills are required to meet certain air standards. The institution of these measures has led to a marked decrease in textile-related pulmonary diseases and, it is hoped, will eliminate the problem of byssinosis for future workers.

Many studies have shown that workers who are cotton-dust reactive, who show evidence of byssinosis, and who smoke cigarettes develop much more lung impairment than do those individuals with byssinosis who do not smoke. Thus, many mills have incorporated smoking-cessation clinics as part of their medical surveillance.

Occupational Asthma. Exposure to substances encountered in some occupations can lead to the development of asthma through immunologic pathways. Although the substances producing the asthma vary widely, as do some of the immunologic responses, the net result closely resembles the extrinsic asthma that occurs in atopic individuals.

Individuals working in bakeries and exposed to flour can become sensitized either to the flour dust itself or to organic material in the flour. This has been termed baker's asthma and is in all respects similar to classic allergic asthma.

Exposure to a number of different wood dusts can produce asthma in some individuals. The most commonly encountered wood dust producing bronchospasm in this country is western red cedar, which is widely used in the construction industry today. Other woods that can also lead to the development of asthma are mahogany, teak, and some exotic woods from Africa. Again, the symptomatology is in all respects similar to the usual type of extrinsic atopic asthma. Its treatment is also similar.

In recent years an asthmalike syndrome has been described in grocers who wrap and seal meat or produce in polyvinyl film. This has been termed meat-wrapper's asthma and is thought to be due to sensitization to the chemical fumes released when the polyvinyl chloride film is touched to a hot wire for melting and sealing.

The chemical toluene diisocyanate (TDI) is widely used as an additive in the production of polyurethane varnish, polyurethane foam, and polyurethane paints. Some individuals are quite sensitive to this chemical and, when exposed to it, become sensitized and develop asthma. Once a person has become sensitized, it requires only a very small concentration of TDI molecules to precipitate a severe asthmatic attack. Other isocyanates, such as methylene bisphenyl isocyanate (MDI), can produce similar responses.

An interesting but infrequently encountered cause of industrial asthma is termed oyster-shucker's asthma. This is produced by the inhalation of organic residue secreted onto the exterior of oyster shells by a small sea animal called a sea squirt. Again, in individuals who become sensitized, the symptoms are in all respects similar to classic asthma.

Extrinsic Allergic Alveolitis

A large number of occupational or environmental exposures have been identified that lead to the development of an allergic reaction within the lung, with the major pathophysiology occurring at the peripheral bronchiolar level—that is, in the small terminal branches of the bronchial tube system just prior to their entry into the alveolar sacs.

Inhalation of Actinomyces Spores. The oldest of these diseases has been termed farmer's lung and results from the inhalation of spores of thermophilic *Actinomyces* in moldy hay. When moldy hay is fed to livestock, large numbers of fungus spores are released into the air. If these enter the lungs of individuals capable of becoming sensitized, these individuals develop IgG antibodies, which, on subsequent exposure, react with the antigen in the thermophilic *Actinomyces* to produce an allergic bronchiolitis. The symptoms are cough, sputum production, fever, myalgias, and fatigue. These symptoms are quite similar to pneumonia and, indeed, these individuals are often diagnosed and treated as though they have pneumonia. If the correct diagnosis is not made and if exposure occurs frequently and continuously over a period of years, the lungs can be permanently damaged.

In approximately 10 percent of the individuals who are exposed, a more insidious clinical picture develops. Rather than developing acute febrile symptoms, these individuals develop a slow insidious onset of shortness of breath. In this situation the diagnosis is frequently made quite late in the course of the disease, after a great deal of lung damage has occurred. The treatment in either clinical situation is early diagnosis with removal of the individual from further exposure to thermophilic *Actinomyces* spores.

Inhalation of Other Fungi and Proteins. A similar disease pattern has been described in lumbermen working with sequoia and maple trees, individuals working with malt in the Scotch whiskey industry, cork workers in Portugal (a condition termed suberosis), bird fancier's lung (seen in individuals who raise parakeets, pigeons, chickens, or turkeys and due to the inhalation of and sensitization to proteins in the bird droppings), laboratory workers who work with rats (through sensitization to proteins in rat urine), and—surely one of the most exotic afflictions on record—among archeologists who remove the wrappings from Egyptian mummies (Coptic lung disease).

Many other interesting and rather exotic situations can lead to extrinsic allergic alveolitis. All are due to the inhalation of and sensitization to fungus spores or organic protein molecules. The clinical pictures are similar in all, as are approaches to treatment. Diagnosis demands a high index of suspicion on the part of the physician.

9.5 Evaluation of Lung Disease

Medical surveillance of employees in working environments that could potentially damage the lungs should include a preemployment evaluation and reevaluations at 6-to-12-month intervals.

The preemployment evaluation serves two important functions.

- It identifies individuals with preexisting lung diseases who might be at risk of developing further damage if exposed to industrial toxins.
- It also serves as a baseline reference point to which further evaluations can be compared.

Preemployment evaluation should include a medical questionnaire, a chest x-ray photograph, and simple spirometry (measurement of lung capacity and function).

Simple spirometry includes

- Measurement of forced vital capacity (FVC) in liters.
- Forced expired volume during one second (FEV_1) in liters per second.
- Forced expiratory flow in the middle half of the vital capacity (FEV 25–75 percent) in liters per second.
- The ratio of the observed FEV_1 to FVC $\times$ 100 (FEV_1/FVC percent).

Most of the questionnaires presently used are modifications of one developed by the British Medical Research Council. U. S. textile mills use a questionnaire required by OSHA regulations, and it can be easily modified for other types of industries. It is quite comprehensive, running to about 13 pages in length. It can be found in: *General Industry OSHA Safety and Health Standards (29 CFR 1910).* 1979. Washington, D.C.: U.S. Department of Labor/OSHA Publication No. 2206.

After employment, evaluations should be repeated at least yearly.

- These should include questionnaires and simple spirometry.
- Chest x-ray photography should be repeated if abnormalities are noted on the questionnaire or are evident in spirometry (usually, however, if abnormalities are detected, the employee is referred to a pulmonary specialist who obtains an x-ray photograph as part of his evaluation).

In industrial settings that predispose workers to acute reactive airway disease, such as textile mills and factories using toluene diisocyanate (TDI), pre- and post-work exposure spirometry is helpful in detecting reactors (i.e., individuals whose expiratory flow rates consistently worsen when exposed to the toxins in these environments). This type of testing should be done on all new employees so as to identify reactors among them, and on all employees on an annual basis. Those individuals who are consistent reactors will need to wear effective filtration masks or be transferred to work areas where exposure to toxins will not occur.

Diseases such as byssinosis lend themselves to medical surveillance programs. During the early years of employment, long before permanent lung damage occurs, individuals with byssinosis can be easily identified by positive responses in questionnaires or by decreases in FEV_1 after work exposure. On the other hand, diseases such as silicosis are far advanced by the time symptoms appear, or before spirometry and chest x-ray pictures show abnormality.

When medical surveillance programs identify a possible health problem, the employee should be referred to a pulmonary disease specialist for thorough evaluation.

Summary

The structure of the lungs is beautifully designed to bring large volumes of air into intimate contact with the delicate mucosal lining of the bronchial tubes and alveoli. This contact of air with mucosa facilitates:

- Regulation of temperature and humidification of the air.
- Cleansing of toxic substances.
- The process of oxygen and carbon dioxide transfer with blood.

These mechanisms work well and maintain normal health and function of the lungs provided the ambient air is relatively free of toxic substances. However, if the air is heavily contaminated with tobacco smoke toxins, or industrial and environmental toxins,

- The mucociliary mechanism and alveolar macrophages are overwhelmed and the lungs are damaged.
- If the damage is extensive, gas exchange becomes abnormal and permanent respiratory impairment results.

Most studies of environmental and occupational lung diseases have shown a synergistic or additive effect of cigarette smoking and lung damage.

- Textile workers who smoke and have byssinosis suffer far more lung damage than nonsmoking workers with byssinosis.
- Workers with significant exposure to asbestos fibers who smoke are at an 80-fold-increased risk of developing lung cancer, as compared to nonsmokers.

Three requirements are necessary for the prevention of environmental and occupational lung diseases:

- Ambient air that meets or exceeds occupational health standards.
- An ongoing medical surveillance program.
- A program for smoking cessation.

References and Suggested Reading

Bates, D. V., Macklern, P. T., and Christie, R. V. 1971. *Respiratory Function in Disease. An Introduction to the Integrated Study of the Lung*. Philadelphia: W. B. Saunders Co.

Department of Health, Education, and Welfare (National Institute for Occupational Safety and Health). 1977. *Occupational Diseases. A Guide to Their*

Their Recognition. Revised Edition. DHEW (NIOSH) Publication #77-181. Superintendent of Documents. Washington, D.C.

Dosman, J. A., and Cotton, D. J. (Eds.) 1980. *Occupational Pulmonary Disease. Focus on Grain Dust and Health*. New York: Academic Press, Inc.

Drinker, P., and Hatch, T. 1954. *Industrial Dust: Hygienic Significance, Measurement, and Control*. 2nd Edition. New York: McGraw-Hill Book Co.

Ebert, R. V., and Terracio, M. J. 1975. "The Bronchiolar Epithelium in Cigarette Smokers." *Amer. Rev. of Resp. Disease, 111*:6.

Fenn, W. O., and Rahn, H. 1964. *Handbook of Physiology*. Washington, D.C.: American Physiology Society.

Frazier, C. A. (Ed.) 1980. *Occupational Asthma*. New York: Van Nostrand Reinhold Company.

Hatch, T., and Gross, P. 1964. *Pulmonary Deposition and Retention of Inhaled Aerosols*. New York: Academic Press, Inc.

Hilding, A. C. 1961. "Experimental Studies on Some Little-understood Aspects of the Physiology of the Respiratory Tract." *Tr. Amer. Acad. Ophth. and Otol*. July–August. pp. 475–495.

Holt, P. 1957. *Pneumoconiosis. Industrial Diseases of the Lungs Caused by Dust*. London: E. Arnold.

King, E., and Fletcher, C. 1960. *Industrial Pulmonary Diseases*. Boston: Little, Brown, and Co.

Lanza, A. 1963. *The Pneumoconioses*. New York: Grune and Stratton.

Liebow, A., and Smith, D. 1968. *The Lung*. Baltimore: The Williams and Wilkins Co.

Menzel, D. B., and McClellan, R. D. 1980. "Toxic Responses of the Respiratory System." *In* Doull, J., Klaassen, C. D., and Amdur, M. O. (Eds.) *Casarett and Doull's Toxicology: The Basic Science of Poisons, 2nd Edition.* New York: Macmillan Publishing Co.

Morgan, W. K. C., and Seaton, A. (Eds.) 1975. *Occupational Lung Diseases*. Philadelphia: W. B. Saunders Co.

Muir, D. (Ed.) 1972. *Clinical Aspects of Inhaled Particles*. Philadelphia: F. A. Davis Co.

Murray, J. F. 1976. *The Normal Lung. The Basis for Diagnosis and the Treatment of Pulmonary Disease*. Philadelphia: W. B. Saunders Co.

Parkes, W. R. 1982. *Occupational Lung Disorders*. 2nd Edition. Woburn (Mass.): Butterworths.

Proetz, A. W. 1953. *Essays on the Applied Physiology of the Nose*. 2nd Edition. St. Louis: Annals Publishing Company.

Rhodin, J. A. G. 1966. "Ultrastructure and Function of the Human Tracheal Mucosa." *Amer. Rev. of Resp. Disease, 93*:13.

Steele, R. 1972. "The Pathology of Silicosis." *In* Rogan, J. M. (Ed.) *Medicine in the Mining Industries*. Philadelphia: F. A. Davis Co.

Stokinger, H. E., and Coffin, D. L. 1968. "Biologic Effects of Air Pollutants." *In* Stern, A. C. (Ed.) *Air Pollution*. 2nd Edition. New York: Academic Press.

Tager, I. B., Weiss, S. T., Muñoz, A., Rosener, B., and Speizer, F. E. 1983. "Longitudinal Study of the Effects of Maternal Smoking on Pulmonary Function in Children." *New Engl. J. Med., 309*: 699–703.

Weibel, E. R. 1963. *Morphometry of the Human Lung*. New York: Springer-Verlag.

Weill, H., and Turner-Warwick, M. (Eds.) 1981. *Lung Biology in Health and Disease*. Volume XVII. New York: Marcel Dekker, Inc.

Widdicombe, J. G. (Ed.) 1974. *Respiratory Physiology*. MTP International Review of Science—Physiology Series 1. Vol. 2. Woburn (Mass.): Butterworths.

PART II

Specific Areas of Concern

Chapter 10

The Toxic Effects of Metals

Joel G. Pounds

Introduction

Metals are used extensively in the workplace and employee exposure can result from numerous industrial operations, including welding, grinding, soldering, painting, smelting, and storage battery manufacturing. The objective of this chapter is to present some of the fundamental concepts important to an understanding of the toxicology of metals. Some of these concepts include:

- The chemical and physical properties of metals.
- The classification of metals.
- The absorption, storage, and excretion of metals.
- The mode of action of metal toxicity.
- The role of metals in carcinogenesis.

About 80 elements are classified as metals, of which 50 are of economic and industrial significance. Metals are frequently discussed as though they were a homogenous group of toxicants with a similar range of biological and toxicological properties. In fact, however, metals exhibit a very wide range of physical, chemical, biological, and toxicological properties.

10.1 Classification of Metals

Elements are classified as metals on the basis of physical properties including:

- High reflectivity and metallic luster.
- High electrical conductivity.

- High thermal conductivity.
- Strength and ductility.

From a biological perspective, it is more useful to define a metal as an element that will give up one or more electrons to form a cation in aqueous solution. The chemical and physical properties of the metal in aqueous solution determine the biological and toxicological significance of the metals. These properties include:

- Solubility.
- Oxidation state.

Groups of metals are frequently referred to as heavy, trace, essential, nonessential, or toxic metals. These classifications are somewhat arbitrary and may be confusing. The term "heavy metal" is defined in chemical dictionaries to include metals of specific gravity greater than 4 or 5. The term is more commonly used to denote toxic metals.

The classification of elements into essential metals versus toxic metals is equally confusing, as many essential metals are quite toxic—for example, copper, iron, and cobalt. In addition, some nonmetallic, essential elements such as selenium or arsenic behave much like metals in biological systems and are commonly grouped with metals.

10.2 Metabolism of Metals

In the context of toxicology, metabolism of metals refers to all of the processes by which metals are handled by the body. Absorption, storage, and elimination are the most important processes in the metabolism of inorganic metal salts. Organic metal compounds may be modified biochemically; an example of this is the enzymatic dealkylation of tetraethyl lead to triethyl lead.

Respiratory Absorption

Fumes and dusts are the principal forms of metals to which people become exposed in the workplace. The deposition, distribution, absorption, and retention of inhaled metals in the respiratory tract depends on the physiochemical properties of the inhaled material. A metal may be inhaled as the vapor or aerosol (i.e., fume or dust particulate). Fumes and vapors of cadmium and mercury are readily absorbed from the alveolar space, as are many organic metal compounds such as tetraethyl lead. Large particles of metal aerosol ($> 10\ \mu m$) are trapped by the upper respiratory tract and cleared by mucociliary transport to the pharynx, and then are swallowed. The absorption of these particles is equivalent to oral exposure.

Very small particles (< 1.0 μm) may reach the alveolar or gas exchange portion of the lungs. Water-soluble metal aerosols are then rapidly absorbed from the alveoli into the blood.

Gastrointestinal Absorption

Metals are introduced into the gastrointestinal tract through food and water, and by mucociliary clearance from the respiratory tract. In the workplace, the contaminated hands of workers can introduce metals into the mouth while eating or smoking. Metals are absorbed into the cells lining the intestinal tract by:

- Passive or facilitated diffusion.
- Specific transport processes.
- Pinocytosis.

The gastrointestinal absorption of metals varies widely. Metal salts of lead, tin, and cadmium are poorly absorbed ($\leq$ 10 percent) while the salts of arsenic and thallium are almost completely absorbed (> 90 percent). The absorption of metals depends on many factors, including:

- The solubility of the metal salt in fluids of the intestinal tract.
- The chemical form of the metal (the lipid-soluble methyl mercury is completely absorbed, whereas inorganic mercury is poorly absorbed).
- The presence and composition of other materials in the intestinal tract (these can affect passage of contents through the alimentary tract).
- The competition for absorption sites between similar metals (e.g., zinc and cadmium, or calcium and lead).
- The physiological state of the person who has been exposed, including age (vitamin D enhances the absorption of lead).

Excretion

Kidney. The important route for excretion of most metals is renal elimination. Metals in blood plasma are bound to plasma proteins and amino acids. Metals bound to low-molecular-weight proteins and amino acids are filtered in the glomerulous into the fluid of the renal tubule. Some metals (e.g., zinc, cadmium) are very effectively resorbed by tubular epithelia before they reach the urinary bladder where very little resorption occurs. Tubular resorption of a given metal, and, thus in part, urinary excretion of that metal, depends on the urinary pH, the amount and kind of amino acids and proteins associated with the metal, and the presence or absence of other metals competing for resorption sites. In addition, the presence of a toxic metal or chemical agent that poisons renal tubular epithelia will reduce the resorption and thereby increase the urinary excretion of metals. Chapter 6 describes the toxicological effects of cadmium, mercury, and lead on the kidney.

Gastrointestinal Tract. A second important route for the excretion of metals is the gastrointestinal tract. Metals including cadmium, mercury, and lead can reverse the process and direction of absorption and cross the mucosa and epithelium into the contents of the intestinal tract. In addition, metals bound to the cells lining the intestinal tract are lost, as these cells are shed and excreted.

Enterohepatic Circulation. Absorbed metals may also be excreted into the intestinal tract in bile, pancreatic secretion, or saliva. Metals excreted in bile may be reabsorbed into the blood stream further down the intestinal tract, and returned to the liver for reexcretion in bile (enterohepatic circulation). Biliary excretion can be affected by drugs, chemicals, and preexisting diseases that promote or reduce the secretion of bile by the liver (see Chapter 5).

The large enterohepatic circulation of some metals, specifically methyl mercury, has been used to therapeutic advantage through the oral administration of resins containing thiol groups. Methyl mercury excreted in bile binds tightly to the resin, thus greatly reducing intestinal reabsorption and increasing net gastrointestinal excretion.

Minor Pathways. Minor pathways for the elimination of metals include loss through hair, nails, saliva, perspiration, exhaled air, lactation, and exfoliation of skin. Certain metals such as mercury, zinc, copper, and arsenic will deposit irreversibly in hair. The follicle where the hair starts to grow (within the body) will contain these elements and deposit them on the hair as it grows. Hair grows at a fairly constant rate of about 1 cm per month. If the hair is analyzed segmentally, one may find an indication of what the body's milieu was in relationship to the metal of interest at various points in the past.

10.3 Toxicity of Metals

Acute Toxicity

Acute toxicity is caused by a relatively large dose of a metal over a short period of time. The duration of time from metal exposure to the onset of the clinical signs and symptoms is usually short and the intensity of the effects is often higher than with chronic toxicity. Frequently, the organs and tissues affected are those involved in the absorption or elimination of the metal. This sensitivity is the result of accumulation of high, critical concentrations of the metal at these sites with little opportunity to detoxify, eliminate, or adapt to the metal. The diagnosis of acute metal intoxication is generally easier than the diagnosis of chronic metal intoxication due to the shorter duration between metal exposure and onset of the symptoms. The treatment of acute metal intoxication is designed to:

- Enhance the elimination of the metal through neutralization.
- Prevent irreversible damage to organs and tissues.
- Treat the symptoms of acute toxicity.

Chronic Toxicity

Chronic toxicity is caused by long or repeated exposure to relatively small doses of metals. The time duration from initial exposure to onset of the signs and symptoms of toxicity may be months or even years. The development of symptoms is gradual and the intensity of the symptoms is often, but not always, less severe than with acute exposure. Diagnosis of chronic metal intoxication is more difficult than the diagnosis of acute intoxication because of the longer duration between onset of exposure and the development of symptoms. The diagnosis of chronic metal intoxication commonly includes or requires analysis for the presence of excess metal(s) in blood or urine. Furthermore, organ systems not involved in absorption or elimination of the metal such as the hematopoetic or immune system may be affected by chronic exposure to metals.

Many metals can elicit two different symptoms, one for acute and one for chronic toxicity. For example, the symptoms of acute inorganic mercury intoxication include nausea, headache, diarrhea, abdominal pain, and metallic taste. On the other hand, symptoms of chronic inorganic mercury intoxication include ataxia, dysarthria (imperfect articulation in speech), dysphagia (difficulty in swallowing), impaired vision, hearing, taste and smell, and loss of coordination in arms and legs. The treatment of chronic metal intoxication is designed to:

- Decrease body and tissue stores of the metal.
- Prevent irreversible injury to organs and tissues.
- Treat the symptoms of chronic toxicity.
- Identify the source of metal and prevent continued exposure.

Table 10-1. Target-Organ Toxicity of Selected Metals and Metalloids.

Metal	Renal system	Nervous system	Liver	Gastrointestinal tract	Respiratory system	Hematopoietic system	Bone	Endocrine system	Skin	Cardiovascular system
Aluminum		+			+					
Arsenic		+	+	+	+	+		+		
Beryllium					+				+	
Bismuth	+		+						+	
Cadmium	+	+		+	+		+			+
Chromium	+	+	+		+				+	
Cobalt		+		+	+			+	+	+
Copper				+		+				
Iron		+	+	+	+	+		+		
Lead	+	+		+		+			+	
Manganese		+			+					
Mercury	+	+		+	+					
Nickel		+			+				+	
Selenium	+			+					+	
Strontium										+
Thallium	+	+	+	+	+			+		
Tin (Organic)		+		+						
Zinc				+		+	+			

It is also important to note that acute and chronic toxicity are not absolute and clinically distinct syndromes, but rather represent the extremes of a continium of effects.

10.4 Mechanisms of Metal Toxicity

As is the case with other toxicants, there is often little correlation between the sensitivity of an organ or tissue to the toxic effects of a metal and the concentration of the metal in that tissue. For example, 95 percent of the body burden of lead in adults is found in calcified tissue (bone and teeth); however, toxicity is manifested primarily in the nervous system, renal system, and hematopoietic system. Some tissues can sequester toxic metals in more or less biologically inactive forms. Kidney tubular cells and hepatocytes of some species can store large amounts of lead and bismuth in metal-protein complexes that are relatively nontoxic.

Table 10-1 provides an overview of the target-organ toxicity for selected metals and metalloids. The sensitivity of the metabolic system(s) to disruption, and the metabolic demand or requirements of an organ or tissue, are equally important in determining the toxic effects of a metal. Thus, metals that impair DNA synthesis and cell replication may affect the rapidly proliferating tissue of intestinal mucosa and bone marrow to produce intestinal ulceration and anemia.

The mechanisms by which metals exert toxic effects can be grouped into the following categories:

1. Enzyme inhibition. Most toxic metals have a high affinity for functionally essential amino acid side chains such as sulfhydryl, histidyl, or carboxyl groups, and can react directly with proteins to alter enzymatic or structural function. Examples of enzyme inhibition include the effect of mercury and lead on Na^+-K^+ ATPase.
2. Indirect effects. Many metals may bind to cofactors, vitamins, and substrates, thereby altering the availability of these cell constituents for biological function.
3. Substitution for essential metals. Since particular metals play essential roles in protein structure, enzyme catalysis, osmotic balance, and transport processes, these biological processes are sensitive to alteration by toxic metals that are chemically similar to the essential metal.
4. Metal imbalance. An excess of a given metal through dietary, occupational, or environmental exposure may lead to depletion or repletion of

an essential metal at numerous biological levels: at molecular, cellular, tissue or organ, and systemic levels of organization. For example, copper deficiency can be produced by excessive, but otherwise not toxic, exposure to zinc. Cadmium intoxication produces necrosis in the intestine, which is prevented by sufficient zinc. Lead intoxication alters tissue levels of many essential elements, including iron, zinc, copper, and calcium.

10.5 Selected Metal Toxicants

Lead

Lead is a trace element with many industrial applications. The largest uses of lead are in storage batteries, gasoline additives, cable sheathing, pigments, alloys, and the electronics industry. The U.S. annual consumption of lead is about 2 million metric tons.

Sources of human exposure to lead include both environmental and occupational routes. The most important sources of occupational lead exposure include:

- Lead mining and smelting where air concentrations may exceed 1.0 mg/m^3.
- Cutting and welding lead-painted structures.
- Manufacture of lead storage batteries.
- Production of lead-based paints.

Inorganic lead salts are poorly absorbed from the gastrointestinal tract. About 10 percent of ingested lead is absorbed by adults; however, up to 50 percent may be absorbed by children. Dietary insufficiencies of calcium or iron enhance lead absorption and toxicity.

Following absorption, lead is transported to all organs and tissues of the body by the blood. Approximately 95 percent of the lead in blood is associated with the erythrocytes and the remainder with plasma proteins. Lead accumulates in bone throughout life. Nearly 90 percent of the human total body burden of lead is found in bone with most of the remaining 10 percent in the kidneys and liver. The biological half-life of lead in bone is 10 to 20 years, while the half-life of lead in soft tissues is several months.

Repeated exposure to lead or lead salts results in a diverse spectrum of clinical signs and symptoms. The organ systems most affected are the gastrointestinal, hematopoetic, and nervous and neuromuscular systems. The renal and cardiovascular systems may also be affected.

The most frequent signs and symptoms of lead poisoning (plumbism) are:

- Muscle weakness.
- Lassitude.
- Insomnia.
- Anorexia, weight loss.
- Facial pallor.
- Anemia.
- Paralysis of extensor muscles of the wrist (wrist drop).
- Colic.
- Constipation.
- Lead line on gingiva.
- Headache.
- Loss of memory.
- Irritability.

A thorough laboratory evaluation is necessary to confirm a preliminary diagnosis of lead intoxication, as the signs and symptoms of intoxication are nonspecific and may be confused with other disease states.

A variety of special tests may be used to identify lead poisoning.

- Blood lead: The concentration of lead in blood is the best single indication of recent excessive lead exposure. Blood lead values are a crude estimate of lead exposure and often do not adequately predict the severity or symptoms of lead intoxication.
- Heme metabolism: Lead inhibits delta-amino-levulinic acid dehydratase, an enzyme involved in the synthesis of porphyrins and heme. The inhibition of this enzyme results in an accumulation of the substrate aminolevulinic acid (ALA), or other components of the heme biosynthetic pathway, in blood and urine which can be measured biochemically as a measure of lead toxicity.
- Nerve conduction velocity: Lead decreases the velocity at which a nerve impulse is conducted along an axon. This velocity can be easily measured in peripheral nerves of the arm.
- Ca EDTA mobilization test: The Ca EDTA mobilization test is used to estimate the body burden of lead. In this test, a single dose of the chelating agent Ca EDTA is administered intravenously, and the urine is collected for 24 hours. If the total amount of lead excreted in the urine exceeds 500 to 600 μg, excess lead exposure is indicated.

Acute lead intoxication is treated by the intravenous administration of Ca EDTA (Versenate), 1 to 4 g per day. Ca EDTA is toxic and treatment should not exceed five days. The patient should be monitored for renal function and irregularities of cardiac rhythm during the course of therapy.

Cadmium

Cadmium is a soft, ductile metal which is obtained as a by-product from the smelting of lead and zinc ores. The principal use of cadmium is as a constituent in alloys and in the electroplating industry. Other uses of cadmium include paint and pottery pigments, corrosion-resistant coating of nails, screws, etc., in process engraving, in cadmium-nickel batteries, and as fungicides.

It is clear from animal studies and clinical or industrial studies that cadmium is a potent metal toxicant. Nearly all cadmium salts are hazardous. As is the case with many metals, human exposure may result from both occupational and environmental sources. The general population is mainly exposed to cadmium through food, although tobacco smoking may provide a significant additional source of cadmium. The more important sources of occupational cadmium exposure include:

- Smelting of ores.
- Production and handling of cadmium powders.
- Welding or remelting of cadmium coated steel.
- Use of solders containing cadmium.

Cadmium salts are poorly absorbed from the gastrointestinal tract. Only about 5 percent of ingested cadmium is absorbed by adults, although absorption in children is probably higher. Inhaled cadmium is absorbed more efficiently (10 to 50 percent) depending on the particle size and the solubility of the cadmium particle.

Absorbed cadmium is bound to plasma proteins and transported to the liver from which it is slowly released and finally accumulated in the kidney. The elimination of cadmium from the body is very slow, so cadmium accumulates over the lifetime. The biological half-life is about 20 years. Cadmium accumulates in the kidney and reaches a maximum, "normal" concentration of 10 to 50 μg/g at about 50 years of age. Renal tubular damage occurs when the cadmium concentration reaches or exceeds 200 μg/g wet weight in the kidney cortex.

Cadmium produces its toxic effects through two principal biochemical mechanisms:

- Displacing or replacing zinc from the many (over 200) enzymes requiring zinc as a catalytic or structural component.
- Reacting with $-$SH groups required for enzyme structure and function.

Some aspects of experimental cadmium toxicity in animals can be reduced or ameliorated by supplementing the animals' diet with excess zinc.

Acute exposure to cadmium fumes or dusts causes respiratory damage and may result in death. Within a few hours of postexposure, the symptoms include:

- Irritation of upper respiratory tract.
- Cough.
- Metallic taste in the mouth.
- Chest pain.

A delayed reaction may develop within 1 to 7 days following acute exposure to include the following signs and symptoms:

- Severe dyspnea.
- Cough.
- Hemoptysis.
- Wheezing.
- Anorexia, nausea.
- Pneumonitis.
- Abdominal pain, diarrhea.
- Pulmonary interstitial edema.

The long term sequela of acute or chronic cadmium exposure includes:

- Testicular atrophy.
- Emphysema.
- Renal tubular necrosis.
- Anemia.
- Cardiovascular effects.
- Teratogenic effects.
- Liver damage.
- Osteomalacia.

The diagnosis of cadmium intoxication requires both an estimation of cadmium levels and measures of cadmium toxicity. The most common requirements for diagnosis are:

- History of exposure.
- Increased urinary cadmium (or blood cadmium).
- Reduced pulmonary function.
- Impaired renal tubular function (proteinuria).

The most effective treatment for cadmium poisoning is prevention. Following diagnosis, most therapy is symptomatic and includes:

- Prevention by adequate ventilation and use of personal protective equipment (e.g., respirators).
- Oxygen, decongestants, bronchodilators.
- Prophylactic administration of antibiotics.
- British antilewisite (BAL) and dimercaprol chelation therapy is contraindicated due to increased renal toxicity of the cadmium complex. EDTA is not effective for decreasing tissue stores of cadmium.

10.6 Carcinogenicity and Mutagenicity of Metals

The field of metal carcinogenesis is a relatively new and small part of the study of chemical carcinogenesis. However, both epidemiological and experimental studies confirm the belief that certain metals cause mutations and tumors. Nickel in the form of nickel carbonyl produces cancer of the nose and lungs following inhalation, as does chronic exposure to chromium dust. Exposure to high levels of lead causes renal tumors in rodents, but most epidemiological studies do not identify lead as a human carcinogen.

The metal carcinogens listed in Table 10-2 cause genetic damage in short-term tests for mutagenicity in bacteria, yeast, or mammalian cell-culture assays for mutagenicity. The significance of these observations is unclear since several metals generally recognized as not carcinogenic (e.g., aluminum, mercury) also produce genetic damage in these tests.

Table 10-2. Nonradioactive Metals Causing Tumors in Man or Experimental Animals.

Metal	Route of exposure or administration	Type of tumor
Beryllium	Inhalation	Pulmonary carcinoma
Cadmium	sc, im Intratesticular	Sarcoma Leydigioma Teratoma
Chromium	sc, im, ip Intraosseous Intrapleural Intrabronchial	Sarcoma Squamous cell carcinoma Adenocarcinoma
Cobalt	sc, im Intraosseous	Sarcoma
Copper	Intratesticular	Teratoma
Iron	sc, im	Sarcoma
Lead	sc Dietary	Renal cell carcinoma Renal adenoma Lymphoma
Nickel	sc, im Inhalation	Anaplastic carcinoma Adenocarcinoma Squamous cell carcinoma Sarcoma
Selenium	Dietary	Hepatoma Sarcoma Thyroid adenoma
Titanium	im	Fibrosarcoma Hepatoma Lymphoma
Zinc	Intratesticular	Leydigioma Seminoma Chorionepithelioma Teratoma

Note: iv = intravenous; sc = subcutaneous; im = intramuscular; ip = intraperitoneal.

Summary

This chapter has discussed the fundamental concepts of metal toxicity.

- Metals are widely used in industry and many are ubiquitously present in the environment as well as in man.
- In the workplace, metals are primarily absorbed into the body by either inhalation or ingestion.
- Although metals are excreted in a number of ways, renal elimination is the most important for the majority of metals. Consequently, the blood and urine are commonly used as specimens for biological monitoring.
- Metals can exert their adverse effects by a number of mechanisms, including enzyme inhibition, by binding to other substrates and altering their biological functions, and by substituting for or altering the level of essential metals within the body.
- The adverse biological effects of metals can vary greatly and may range from renal and nervous system damage to mutations and tumors.

It should be remembered that individual metals may have unique toxicological properties; the References and Suggested Reading list at the end of this chapter provides sources of information for specific metals.

References and Suggested Reading

Berman, E. 1980. *Toxic Metals and Their Analysis*. Philadelphia: Heyden and Sons, Inc.

Browning, E. 1966. *Toxicity of Industrial Metals*. 2nd Edition London: Butterworths.

Friberg, L. (Ed.) 1976–1978. *Toxicology of Metals*. Three Volumes. Research Triangle Park (North Carolina): Health Effects Research Laboratory, Office of Research Development, U.S. Environmental Protection Agency.

Friberg, L., Nordberg, G. F., and Vouk, V. B. 1979. *Handbook on the Toxicity of Metals*. New York: Elsevier/North Holland.

Friberg, L. T., and Vostal, J. J. 1972. *Mercury in the Environment*. Cleveland: CRC Press.

Hepple, P. (Ed.) 1972. *Lead in the Environment*. Essex: Applied Science Publishers.

Lederer, W. H., and Fensterheim, R. J. (Eds.) 1983. *Arsenic: Industrial, Biomedical, Environmental Perspectives*. New York: Van Nostrand Reinhold Company.

Lee, D. H. K. (Ed.) 1972. *Metallic Contaminants and Human Health*. New York: Academic Press.

Luckney, T. D., and Venugopal, B. 1977. *Metal Toxicity in Mammals, Volume 1. Physiologic and Chemical Basis for Metal Toxicity*. New York: Plenum Press.

McMichael, A. J., and Johnson, H. M. 1982. "Long-Term Mortality Profile of Heavily Exposed Lead Smelter Workers." *J. Occup. Med.*, *24*:375–378.

Nordberg, G. F. 1976. *Effects and Dose-Response Relationships of Toxic Metals*. New York: Elsevier.

Plunkett, E. R. 1976. *Handbook of Industrial Toxicology*. New York: Chemical Publishing Co., Inc.

Polson, C. J., Green, M. A., and Lee, M. R. 1983. *Clinical Toxicology*. Philadelphia: J. B. Lippincott Co.

Sittig, M. 1976. *Toxic Metals: Pollution Control and Worker Protection*. Park Ridge (New Jersey): Noyes Data Corporation.

Stopford, W. 1979. "Industrial Exposure to Mercury." In Nriagu, J. D. (Ed.) *Biogeochemistry of Mercury*. Amsterdam: Elsevier/North Holland.

Venugopal, B., and Luckney, T. D. 1977. *Metal Toxicity in Mammals, Volume 2. Chemical Toxicity of Metals and Metalloids*. New York: Plenum Press.

Waldron, H. A., and Stofen, D. 1974. *Sub-Clinical Lead Poisoning*. London: Academic Press.

Chapter 11

The Toxic Effects of Pesticides

Woodhall Stopford

Introduction

This chapter will discuss:

- Commonly used organic pesticides.
- The acute effects of pesticides.
- The persistent effects of pesticides, both after excessive acute exposures and with low-level chronic exposures.
- Case studies of occupational exposures to pesticides.

The term pesticide describes a chemical capable of being used to control pests to humans, agricultural crops, commercial operations, and households. These compounds are selectively sought or formulated for specific toxic properties. The regulatory requirements (Federal Insecticide, Fungicide, and Rodenticide Act) for registration of pesticides have promoted research that has resulted in an understanding of their toxicologic properties and environmental fates. From a toxicologic standpoint, probably the most thoroughly understood chemicals are pharmacologic compounds and pesticides.

Modern pest control strategy attempts to use chemicals that are selective toward the specific target pest and nonpersistent in the environment. Meeting these goals has not been entirely successful. For example, nonpersistent insecticides (carbamates and organophosphates) are presently in use, but technology has not yet perfected truly selective compounds.

Since many pesticides are toxic to humans, they are occupational hazards in the workplace. For example, in California, statistics show that agricultural workers have significantly higher rates of occupational disease compared to total workplace disease rates. Much of this increased risk is probably due to pesticide use. Additionally, workers who formulate and manufacture pesticides are also at risk of pesticide exposures. Clearly, pesticide toxicology is an appropriate topic for occupational safety and health specialists.

11.1 Organophosphate Pesticides

The development of organophosphates has had an important impact upon the chemical control of insects. They are effective against a wide range of insects and are not environmentally persistent, compared to halogenated hydrocarbon insecticides. Organophosphates were first produced in Germany just prior to World War II and have been manufactured as chemical warfare agents.

The main drawback is that some organophosphates are extremely toxic to mammals and therefore are also hazardous to pesticide handlers and applicators. More instances of acute poisoning have occurred with these compounds than with any other pesticide group. They are used as crop surface sprays, plant systemics, animal systemics, aerosols, baits, and fumigants. Table 11-1 lists and illustrates commonly used organophosphates.

Mode of Action

The basis for the acute toxic effects of organophosphate pesticides is the binding to and inactivation of acetylcholinesterase. Acetylcholinesterase (AChE) is found at synapses (see Chapter 7) within the central and autonomic nervous systems and at the nerve endings in striated muscles (neuromuscular end plates).

In the normal sequence of neuronal transmission several reversible steps take place at the synapse. First the neurotransmitter, acetylcholine, is released from the presynapse and binds to the protein receptor at the postsynapse. This leads sequentially to a conformational change, to the opening of ionic channels, and to a depolarization of the postsynaptic membrane. Two events combine to terminate the depolarization: as acetylcholine is released by the receptor, it diffuses away and its further action is reduced or prevented through hydrolysis by AChE. If AChE is inhibited, hydrolysis is prevented and the effects of acetylcholine are greatly prolonged; as a result its activity terminates only after acetylcholine diffuses out of the synaptic cleft. In the absence of hydrolysis, acetylcholine remains in the cleft for relatively long periods and therefore is repeatedly bound to receptors; in this way one acetylcholine molecule can sequentially open many channels leading to continued nerve excitation (i.e., rigid paralysis).

Table 11-1. Examples of Common Organophosphate Pesticides.

1. **TEPP (tetraethyl pyrophosphate)**

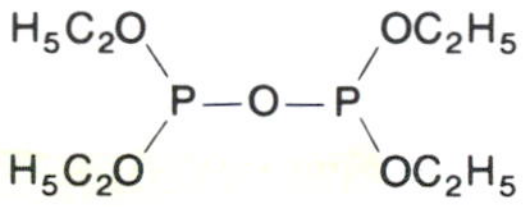

$$(H_5C_2O)_2P{-}O{-}P(OC_2H_5)_2$$

2. **Parathion**

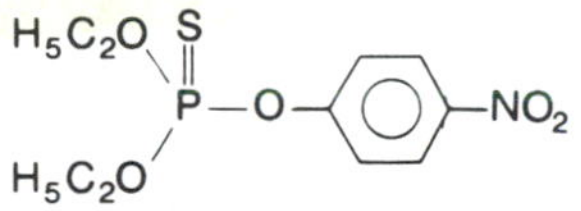

$$(H_5C_2O)_2P(=S){-}O{-}C_6H_4{-}NO_2$$

3. **Dichlorvos**

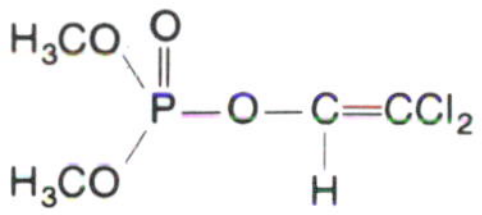

$$(H_3CO)_2P(=O){-}O{-}C(H){=}CCl_2$$

4. **Diazinon**

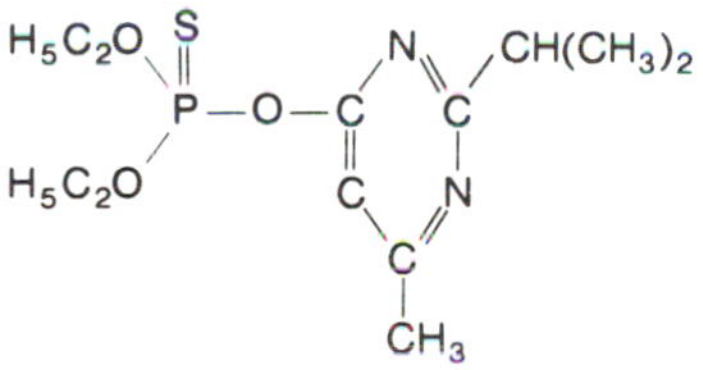

$$(H_5C_2O)_2P(=S){-}O{-}C_4HN_2(CH_3)(CH(CH_3)_2)$$

5. **Dimethoate**

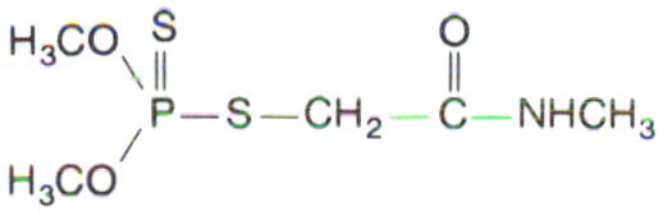

$$(H_3CO)_2P(=S){-}S{-}CH_2{-}C(=O){-}NHCH_3$$

6. **Malathion**

$$(H_3CO)_2P(=S){-}S{-}CH(COOC_2H_5)CH_2COOC_2H_5$$

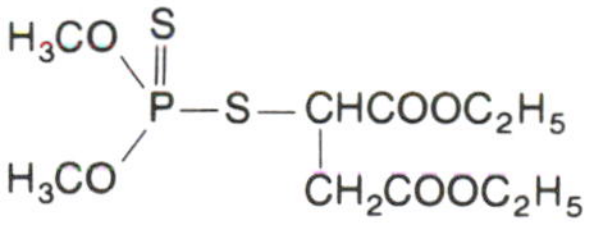

Absorption and Metabolism

Organophosphate pesticides are rapidly absorbed through the skin, lungs, and gastrointestinal tract. For organophosphate pesticides that do not require metabolic activation, initial toxic effects are seen at the sites of contact and include excessive sweating and fasciculations (for skin contact), chest tightness and excess bronchial secretions (for inhalation), and nausea and vomiting (after oral ingestion).

Only a few organophosphate pesticides, such as paraoxon, will inhibit acetylcholinesterase *in vitro.* Most organophosphate pesticides used today contain a thiono (=S) moiety that must be metabolized to the corresponding (=O) analog for activation. Such metabolism occurs in the mixed-function oxidase system found primarily in the liver. Organophosphate pesticides are rapidly metabolized and excreted. However, they readily cross the blood–brain barrier and can be metabolically activated in brain tissue, a mechanism potentially important for explaining their effects on the central nervous system. When an organophosphate pesticide binds with acetylcholinesterase (Figure 11-1), the inactivation will persist until hydrolysis of the phosphorylated cholinesterase occurs. In some cases, such as with trichlorfon, the hydrolysis can occur in a matter of hours. In other cases, such as with paraoxon, a dealkylation and stabilization of the phosphorylated enzyme ("aging") occurs such that hydrolysis can no longer take place and the enzyme is irreversibly

(a) Acetylcholine metabolism

Acetylcholinesterase (AChE)

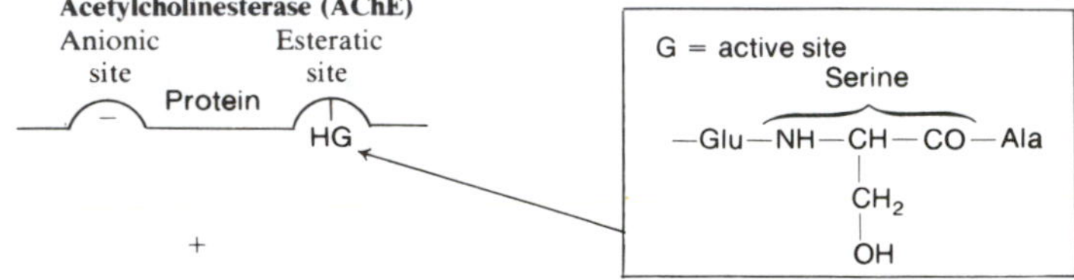

Acetylcholine

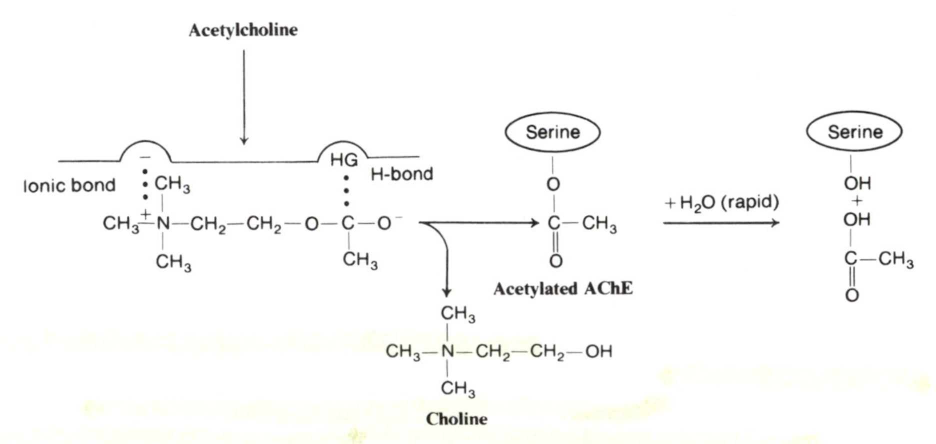

Figure 11-1. The scheme of hydrolysis of acetylcholine by acetylcholinesterase (part a) and reactions of the acetylcholinesterase-binding insecticide paraoxon (part b).

inhibited. In such cases return of acetylcholinesterase activity parallels the time required to resynthesize this enzyme.

Acute Intoxication

Three groups of acute symptoms result from excessive exposure to organophosphate pesticides. One group of symptoms results from the inhibition of function of neuromuscular junctions: muscular twitching from excessive contraction of muscles, extreme weakness, and paralysis may result (these are termed nicotinic effects). The primary muscles of concern with regard to acute exposure are the muscles of respiration: paralysis of the diaphragm and chest muscles can lead to depression of respiration.

(b) Organophosphate (parathion) metabolism

S
OC_2H_5
O_2N—O—P
OC_2H_5

Parathion

Biotransformation

OC_2H_5
O_2N—O—P
O OC_2H_5

Paraoxon

O_2N—OH

***p*-Nitrophenol**

Serine
O OC_2H_5
P
O OC_2H_5

Phosphorylated AChE (most stable)

Aging

Serine
O OH
P
O OC_2H_5

+ H_2O (slowest)

Serine
OC_2H_5
OH + HO—P
O OC_2H_5

Figure 11-1. (*continued*)

If the transmissions in the autonomic nervous system are inhibited, a number of other problems can result from effects at muscarinic receptors. Secretions can increase in the respiratory system, filling up the bronchioles and bronchi with fluid and, in effect, drowning the affected individual. Intensely painful spasms can occur in the gut or bladder. The smooth muscle in the respiratory tract can go into spasm, causing constriction of the airways and an asthmatic response. One of the effects in some exposures is miosis, a marked constriction of the pupils.

A number of symptoms of exposure to organophosphate pesticides result from affectation of the central nervous system: tremor, confusion, slurred speech, disequilibrium such that the person tends to bump into things and has difficulty coordinating hand movements, and, in cases of extreme exposure, convulsions.

Chronic Effects

Neuropathy. A majority of organophosphate pesticides can cause some form of a neuropathy. In one experiment in which chickens were treated with atropine to prevent acute intoxication, subacute dosing was done with 30 different organophosphate pesticides, and 22 of these resulted in paralysis. This paralysis reversed itself usually within one month after treatment was stopped. Similar effects have been seen in man: in one study in which 200 separate cases of poisoning with organophosphate pesticides were followed, 52 of these individuals developed a paralysis. This paralysis was delayed in onset up to 85 hr after the poisoning occurred and, for the most part, reversed after another 72 hr. Teased nerve preparations of affected individuals disclosed axon damage.

A persistent neuropathy has been associated with exposure to a limited number of organophosphate pesticides and has a delayed onset after acute or subacute exposure. This type of neuropathy has been called a central–peripheral distal axonopathy because of its involvement of both the central and peripheral nervous systems (Table 11-2).

Nerve biopsies performed on human subjects or laboratory animals that have developed this type of neuropathy demonstrate that there is damage primarily to nerve axons. In such axons, the neurofibrils, which are usually parallel, increase in number and become entangled in bunches. This tangle of neurofibrils appears to delay the transport of nutrients along the nerve. With severe involvement, the nerve distal to this tangle deteriorates.

When a person with neuropathy is clinically examined, it is generally observed that even though sensory changes can be found, motor effects predominate. These people have paralysis mainly in the lower legs, with wasting of muscles, loss of strength, and difficulty with movement. Although these symptoms are largely due to motor nerve damage, the loss of some sensation and the inability to locate one's joints in space can effectively limit such a person's ability to walk without visual clues.

Central nervous system effects of neurotoxic organophosphate pesticides include damage to the spinal cord and constriction of peripheral visual fields. With time the affectation of the peripheral nervous system can be reversed. However, the effects in the spinal cord are persistent and a condition nearly identical to amyotrophic lateral sclerosis (ALS) can dominate the clinical picture.

Low-level exposures to neurotoxic organophosphate pesticides can be significant. Experiments with chickens have shown that the cumulative low-level chronic dose of a pesticide capable of causing a specific level of paralysis is less than the amount required to cause a similar paralysis with just one dose. The cumulative effect persists even during periods of time when no exposure occurs.

Myopathy. Acute or subacute exposures to organophosphate pesticides have been associated with muscle damage. Rats treated acutely or chronically with paraoxon develop necrosis of skeletal muscles. This damage is prevented by nerve section or treatment and appears to be related to persistent high levels of acetylcholine. Similar effects have been seen in humans intoxicated with Diazinon, parathion, and mipafox (an experimental organophosphate pesticide). Such effects show themselves acutely in muscle tenderness, changes in surface electromyography (EMG) and elevated muscle enzymes such as CPK (creatine phosphokinase).

Psychiatric Changes. Subacute or chronic exposure to organophosphate pesticides has been associated with adverse psychiatric effects. Gershon and

Table 11-2. Central–Peripheral Distal Axonopathy: Organophosphate Pesticides.

Dose
Cumulative low-level dose (chronic)
Less-than-acute high level for same effect
Distribution of organophosphate pesticides in the nervous system
Long fibers affected first
Sensory and motor nerves
Central and peripheral nervous system
Distal locations affected earlier and more severely than proximal locations
Morphology
Neurofibrillar tangles proximal to the nodes of Ranvier
Late demyelination
Clinical findings in poisoned individuals
Motor symptoms predominate
Wasting and loss of strength distally
Ataxia
Permanent spasticity
Psychiatric disturbances, memory impairment, and optic nerve involvement with severe poisoning

Shaw (1981), in a study of ten greenhouse workers with recurrent episodes of acute organophosphate poisoning after exposures to parathion and malathion, noted that eight of these ten workers developed either acute psychosis or severe depressions within two months of an acute episode of intoxication. In each case the affected individual had difficulty with memory. Adverse psychiatric states persisted for a minimum of six months but, in most cases, cleared within one year. The authors also noted similar cases identified among farm workers and scientists working with organophosphate pesticides. Persistent effects in those individuals they studied were depression, headaches, impaired memory, decreased concentration, irritability, nightmares, confused states, speech difficulties, sleepwalking, instability, spatial disorientation, tremor, ataxia, and emotional instability.

A controlled study has been done comparing individuals with high organophosphate pesticide exposure and individuals with low organophosphate pesticide exposure. A series of psychological tests showed that the individuals with higher exposure had poorer high-level adaptive functioning and showed changed electroencephalogram data compatible with dysfunction of the frontal lobe of the cerebrum.

Biological Monitoring

In the past the primary way of monitoring individuals with organophosphate exposures was by measuring red blood cell cholinesterase (which corresponds with acetylcholinesterase levels at the neuromuscular junction) or serum cholinesterase (which appears not to relate to nerve function). Of the two, the red blood cell cholinesterase values appear to be least influenced by factors not directly related to organophosphate exposure. Serum cholinesterase is synthesized in the liver, and serum cholinesterase values in a person with liver disease may not accurately correlate with acute or chronic neurotoxicity. However, the systems of persons with chronic exposure to organophosphate pesticides appear to make an adaptation such that significant depressions of red blood cell cholinesterase are produced without signs of intoxication; hence, even this measurement needs to be used cautiously.

A more sensitive means of monitoring workers exposed to organophosphate pesticides is that of surface electromyography (EMG). In workers exposed to organophosphate pesticides, such studies show dysfunction of the neuromuscular junction similar to that seen after acute high-level exposure. These changes have been documented in agricultural pesticide applicators and in formulators of dimethylvinylphosphate pesticides. In this latter group, EMGs showed that increasing neuromuscular dysfunction was occurring as the work week progressed.

Treatment

The classical treatment for organophosphate pesticide poisoning is a combination of atropine and praladoxime. Atropine is useful in counteracting autonomic nervous system effects. To be effective it has to be used in high doses. To counteract its toxicity and to increase its effectiveness it is used in combination with metaraminol. Praladoxime acts by hydrolyzing phosphorylated acetylcholinesterase and is effective in counteracting the paralytic effects of organophosphate pesticides if it is administered prior to the "aging" of the phosphorylated enzyme. Praladoxime has been used on a chronic basis in treatment of individuals with constricted visual fields.

Structure-Toxicity Relationships

Consider the following general organophosphate structure:

$$(R'O)(R''O)P(=O(S))-O(S)-X$$

R′ and R″ denote groups that affect acute toxicity and stability; longer alkyl chains reduce toxicity and increase stability. The change from P═S to P═O is a metabolic activation step accomplished in mammals by mixed-function oxidases in liver microsomes. X is a side group that primarily affects solubility, stability, selectivity, and potency. Detoxification of organophosphates is achieved through hydrolytic dealkylation and hydrolysis of side chains:

$$(RO)_2P(=O)-OX \quad \textbf{General structure}$$

$$(RO)_2P(=O)-OH \longleftarrow (RO)_2P(=O)-OX \longrightarrow (HO)(RO)P(=O)-OX$$

Nontoxic forms

Malathion produces its effect in the following manner. The pesticide is selectively toxic to insects; the basis is differential metabolism:

$$\underset{\textbf{Malathion}}{(H_3CO)_2P(=S)-S-CH(COOC_2H_5)CH_2COOC_2H_5} \xrightarrow{\text{Metabolism in mammals}} \underset{\textbf{Nontoxic form}}{(H_3CO)_2P(=O)-S-CH(COOH)CH_2COOH}$$

$$\downarrow \text{Metabolism in insects}$$

$$(H_3CO)_2P(=O)-S-CH(COOC_2H_5)CH_2COOC_2H_5$$

The activation reaction (P═S → P═O) predominates in insects while the deesterification reaction is dominant in mammals.

Case Study: Organophosphate Pesticide Poisoning

A case report will convey a good deal about what actually occurs when a person is exposed to an organophosphate pesticide (Table 11-3). A 28-yr-old male was establishing a garage in an old warehouse so as to maintain trucks and cars. In the warehouse was a large amount of white dust, which he noticed would turn his skin red and cause him to grow short of breath. He took an air hose and blew all the dust out of the building to clean the area completely so he could begin work. At that time, he developed blisters and was unsteady on his feet; his hands shook to the degree that he could not disconnect the air hose. His vision became blurred; he wheezed and coughed up a copious amount of sputum, was dizzy, and had a headache.

He was seen in a local emergency room, told that there was nothing wrong with him, and sent home. Unsatisfied, he went to a state agency that inspected his warehouse and found that the white dust contained significant amounts of an organophosphate pesticide, imidan, and phenyl mercury acetate. He then underwent further tests; the levels of mercury in his blood were extremely high. However, he was not evaluated for possible acute effects of organophosphate poisoning.

When seen three months after his exposure, he had memory difficulty and his visual fields had constricted; he showed muscular weakness and a fairly remarkable sensory peripheral neuropathy. He was fatigued and remarked that he no longer had a sexual drive. His muscle enzyme levels were elevated, and

Table 11-3. Organophosphate Case Report.

Patient: 28-yr-old male

Exposure
- Cleaning out warehouse used to bag captan and imidan.

Acute effects
- Second-degree burns, ataxia, blurred vision, tremor, wheezing, copious sputum, dizziness, headache

Chronic effects
- Loss of memory, constricted visual fields, muscle weakness, peripheral neuropathy, fatigue, and loss of sex drive.

Exam
- Tremor
- Generalized weakness
- Loss of memory
- Constricted visual fields
- Reduced dark adaptation
- Increased creatine phosphokinase (CPK, muscle enzyme)
- Reduced acetylcholinesterase levels
- Transient increased motor endplate antibody
- Decreased endplate conduction
- Renal effects

electrical studies of his muscles showed abnormalities at the neuromuscular end plate. Cholinesterase values, even at this late date, were depressed.

The only mercury-related effect that could be documented was found in the kidneys. He had renal insufficiency and proximal tubular damage. His kidney function gradually improved with time, with reversal of the proximal tubular defect.

His muscular function gradually improved over a period of four years, and now he has difficulty only toward the end of the day; he still has muscle-electrical abnormalities. Most of the time he is not weak, but if he uses his muscles a great deal, he becomes weak, as someone with myasthenia gravis would. His visual impairment has lessened remarkably, and his peripheral vision is improved, but he still has difficulty with dark adaptation (night vision).

11.2 Carbamate Pesticides

Carbamate pesticides are similar to organophosphate pesticides in that they act by inactivating acetylcholinesterase. The carbamylated enzyme, however, is rapidly hydrolyzed and reactivated. Acute symptoms after carbamate pesticide exposure include lightheadedness, nausea and vomiting, increased sweating, blurred vision, increased salivation, weakness and muscular fasciculations,

Table 11-4. Examples of Common Carbamate Insecticides.

1. Propoxur (Baygon)

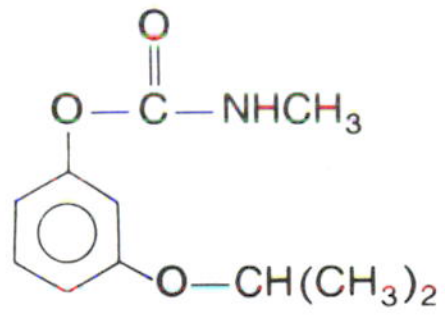

   O—C(=O)—$NHCH_3$; O—$CH(CH_3)_2$

2. Carbaryl (Sevin)

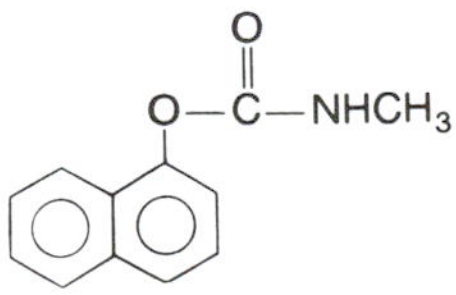

   O—C(=O)—$NHCH_3$

3. Aldicarb (Temik)

   H_3C—S—C(CH_3)(CH_3)—CH=N—O—C(=O)—$NHCH_3$

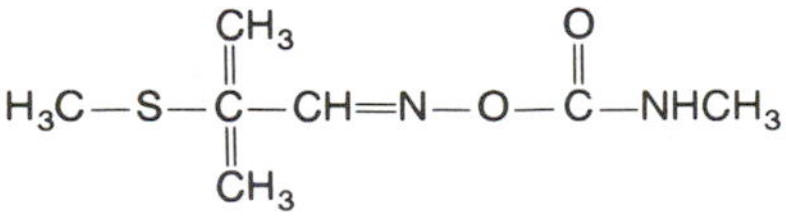

miosis, and, in severe cases, convulsions. The only effective treatment for carbamate poisoning is treatment with atropine. With poisoning, acetylcholinesterase levels are depressed; however, these levels return to normal and symptoms resolve usually within two hours. Examples of carbamates are provided in Table 11-4.

One carbamate pesticide, carbaryl, has been associated with birth defects when administered to dogs. Dogs, however, appear to be unique in that dogs are unable to metabolize carbaryl to 1-naphthol (a major metabolic pathway in most other species, including man). In other experimental animals, the dose required to produce teratogenic effects is near the material toxic dose. As a result, it is believed that dogs have an increased sensitivity to the teratogenic action of carbaryl.

Mode of Action

The mode of action of carbamate pesticides is the same as that of the organophosphates except that the acetylcholinesterase enzyme is carbamylated (Figure 11-2). One noteworthy difference is that carbamate poisoning is more rapidly reversible (i.e., decarbamylation of the inhibited enzyme more readily occurs). For this reason, praladoxime is not administered in poison treatment. With some carbamates praladoxime is contraindicated because it increases the chlorinesterase inhibition. Bioactivation, required for the effectiveness of phosphothiionate insecticides, is not required for carbamate pesticides.

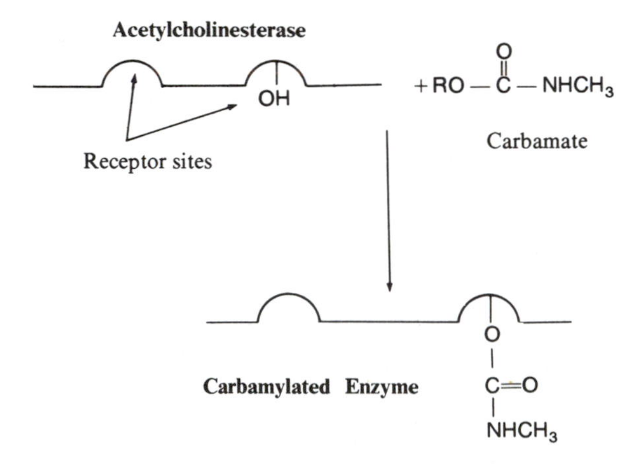

Figure 11-2. Basic mode of action of carbamate pesticides.

11.3 Organochlorine Pesticides

As a class the organochlorine pesticides are considered to be less acutely toxic, but to have greater potential for chronic toxicity, than the organophosphate or carbamate pesticides. Organochlorine pesticides are the oldest class of synthetic insecticides and, as a group, these compounds have the following characteristics.

- They are very stable in the environment and bioaccumulate in ascending trophic levels in food chains. These features mean that environmental problems are associated with their use.
- In general they are weakly water soluble and all have high lipid solubility. In exposed mammals, organochlorines accumulate in fatty tissues.
- They exhibit a wide range of acute toxicity values and produce chronic effects. Several are weak-to-moderate carcinogens. As a group they are not as acutely toxic as the organophosphates.
- Organochlorine pesticides are most efficiently absorbed by ingestion and, in general, they act to stimulate or depress the central nervous system.

Although signs and symptoms of toxicity vary with specific chemicals, an overview of the effects of organochlorine pesticides is provided in Table 11-5.

Chloroethane Derivatives

DDT (dichlorodiphenyltrichloroethane) was the first synthetic insecticide to find widespread use. Its insecticidal properties were discovered in 1939 and it was widely used during the 1940s and 1950s. In spite of its reputation as a dangerous pesticide, DDT is one of the safest compounds in terms of docu-

Table 11-5. Effects of Organochlorine Pesticides.

Effects of Organochlorine Pesticides
Acute
Adverse effects on the central nervous system
Irritability
Dizziness
Tremor
Convulsions
Headaches
Chronic
Reduced sperm count
Effects in the central nervous system
Opsoclonus
Tremor
Loss of memory
Personality changes

mented effects on human health. Despite the fact that it was applied directly to humans (in lice control) there is no documentation of fatal human poisoning. During the 1960s DDT was banned because of its adverse environmental impact (biomagnification, or an increased concentration at higher levels in the food chain) and its potential for accumulation in humans.

Excessive acute exposure to DDT results primarily in neurotoxic effects, and these include paresthesia of the face, foreboding, irritability, dizziness and poor equilibrium, tremor, and convulsions. DDT has a direct estrogenic effect and subacute dosing can result in a decrease in testicular size. DDT inhibits ATPase in the axonal membranes with a resultant alteration in membrane permeability. DDT is lipophilic, and body fat acts as a sink. In humans who have been chronically dosed with low levels of DDT, no adverse clinical effects have been noted. However, during episodes of starvation, fat stores of DDT are released and acute toxic effects can result.

A derivative of DDT, methoxychlor, is now increasingly used. It has a low order of toxicity to mammals and, compared to DDT, it has relatively low persistence. Methoxychlor is rapidly metabolized by *O*-demethylation and subsequent conjugation and excretion. This pathway is catalyzed by microsomal enzymes in mammals. Additionally, enzymes in soil organisms and other biota are able to metabolize methoxychlor. As a result of its rapid metabolism and its low storage in fat (only 0.01 to 0.1 times its chronic intake), methoxychlor presents far fewer problems with regard to environmental persistence and biomagnification.

Chlorinated Cyclodiene Pesticides

This group of pesticides includes chlordane, heptachlor, aldrin, dieldrin, endrin, and endosulfan. Chlorinated cyclodienes have similar acute toxic effects: they tend to produce convulsions even prior to other signs of toxicity. Other signs of intoxication include headaches, nausea and vomiting, dizziness, and mild chronic jerking of muscular groups. The neurotoxic effect seems to be secondary to release of betaine esters from brain mitochondria. Such betaine esters, when injected intracranially, rapidly produce convulsions.

In chronic doses in animal studies, dieldrin, heptachlor, and chlordane (which contains approximately 10 percent heptachlor) have all produced liver cancers in mice. Aldrin, dieldrin, and endrin cause birth defects (webbed feet and cleft palate) when given to pregnant mice and hamsters. Aldrin and heptachlor are fetotoxic, and dosing of experimental animals with these two pesticides has resulted in increased fetal and perinatal mortality.

Kepone

Kepone (also called chlordecone) is a pesticide used primarily for the control of fire ants, and was initially produced by the Allied Chemical Company.

Later, its production was contracted to a small company that worked out of a renovated garage in Hopewell, Virginia. The controls for preventing excessive worker exposure were minimal, in terms of both respiratory and environmental controls. Within a year of the inauguration of the Hopewell facility, a work-related illness was identified by an internist in the area, who noted opsoclonus during a routine examination. The physician recognized this as a sign of organochlorine pesticide exposure. An occupational history and blood measurements documented excessive exposure to Kepone as the most likely cause of the finding. Other workers in the facility were also examined, and most were found to have various problems (Table 11-6) and elevated Kepone levels.

Studies performed on Kepone metabolism in the workers demonstrated that Kepone accumulates in the fat and in the liver. It was readily excreted in the bile, where it was immediately reabsorbed by the gastrointestinal tract, redeposited in the liver, reexcreted, and so on, via a circular pathway of excretion and reabsorption (called the enterohepatic circulation). Cholestyramine, an agent that binds with bile salts, was used to interrupt the enterohepatic circulation of Kepone, and this resulted in increased excretion of Kepone in the feces. Stores of Kepone decreased and the clinical situation among the workers improved. This case is discussed in further detail in Chapter 19.

Table 11-6. Effects of Exposure to Kepone.

In the central nervous system
Tremor
Opsoclonus
Ataxia
Hallucinations
Irritability
Cerebral edema
Endocrinal (estrogenic)
Anovulation
Spermatogenic arrest
Low sperm motility
Carcinogenic: hepatocellular carcinoma
Teratogenesis
Nerve and muscle damage
Metabolic
Accumulates in liver and brain
Crosses placenta
Is enterohepatic-circulatory: only 15 percent of Kepone in bile is found in stool.

11.4 Herbicides

Herbicides are formulated so as to be toxic to plant biochemical systems that are absent in mammals. Consequently, they are generally considered to be weakly toxic to man. However, there can be major exceptions to this generality and fatal poisonings of humans have occurred.

Chlorophenoxy Compounds

This group of chemicals is used to kill broadleaf weeds in agriculture and as vegetation eradicants in right-of-way maintenance. This group includes

- 2,4-Dichlorophenoxyacetic acid (known as 2,4-D).
- 2,4,5-Trichlorophenoxyacetic acid (known as 2,4,5-T).

These compounds exert their toxicity by interfering in the hormonal systems of plants; treated plants exhibit uncontrolled growth symptoms. Pure preparations of phenoxyacetic acid herbicides are weakly toxic to mammals through a poorly understood mode of action. In other animals death occurs from ventricular fibrillation; early symptoms involve muscular impairment, ataxia, paralysis, and weakness. The compound 2,4,5-T contains 2,3,7,8-tetrachlorodibenzo-*p*-dioxin (TCDD, or dioxin) as a trace impurity, and this impurity is among the most potent toxins known. The following data are representative of this potency:

Animal	LD_{50}
Monkey (oral)	70 μg/animal
Guinea pig (oral)	2 ppb
Mouse (oral)	284 ppb
Rabbit (oral)	115 ppb
Rabbit (dermal)	275 ppb

The mode of action of dioxin is unknown. Treated animals succumb after a lag period of several days to weeks. Various organs are affected depending upon the species of test animal studied. There is speculation that dioxin inhibits cell division and that organs and tissues slowly degenerate.

In the workplace chloracne is the conspicuous symptom of dioxin exposure. Dioxin is also an extremely potent teratogen and animal carcinogen. Teratogenic effects have been observed even for 1/400 of the maternal LD_{50} dose in rats. Dioxins are persistent in the environment and can be contaminants whenever chlorophenols are used in chemical manufacture. Synthesis reactions using 2,4,5-trichlorophenol are particularly significant since they can produce the 2,3,7,8-isomer. Seventy-five chlorinated dioxin isomers are possible; most have not been toxicologically characterized. Chapter 19 presents a dioxin case study and describes the toxicology in more detail.

Dinitrophenols

These compounds are used in weed control. Symptoms of acute human poisoning include nausea, sensation of heat, rapid breathing, sweating, rapid heart rate, and coma. The effects are rapid and death or recovery generally occurs within one to two days following exposure. These compounds uncouple oxidative phosphorylation in the mitochondrial cytochrome system and cause the symptoms of increased metabolic activity. Chronic effects include fatigue, anxiety, sweating, thirst, and weight loss. Treatment for acute exposure includes providing ice baths and oxygen. Atropine is contraindicated so it is vital not to misdiagnose dinitrophenol poisoning as an organophosphate poisoning.

Paraquat

Paraquat is a popular weed killer that has resulted in several fatal poisonings. It causes lung, liver, and kidney damage. Lung fibrosis can occur even when the exposure is not respiratory. The lethal dose in test animals is less than 50 ppm; the mode of action is presently unknown.

Summary

This chapter has discussed the toxicology of the most common groups of organic pesticides:

- Organophosphate pesticides.
- Carbamate pesticides.
- Organochlorine pesticides.
- Herbicides.

From the discussion, the following can be concluded:

- The organochlorine pesticides are the most persistent in the environment and pose the greatest risk of chronic exposure for man.
- Of the pesticides discussed, organophosphate pesticides have the greatest acute toxicity.

Since pesticides are manufactured to be poisons, adequate protection should be taken to avoid exposures. This includes:

- The proper use of personal protective equipment (e.g., gloves and respirators) by employees.

- Proper storage and handling procedures should be adhered to.
- Appropriate engineering controls (e.g., local exhaust ventilation and closed chemical processes) should be implemented wherever feasible.

References and Suggested Reading

Aldridge, W. N., Barnes, J. M., and Johnson, M. K. 1969. "Studies on Delayed Neurotoxicity Produced by Some Organophosphorus Compounds." *Ann. N.Y. Acad. Sci., 160*: 314–322.

Cohn, W. J., Boylan, J. J., Blanke, R. V., Fariss, M. W., Howell, J. R., and Guzelran, P. S. 1978. "Treatment of Chlordecone (Kepone) Toxicity with Cholestyramine." *New Eng. J. Med., 298*: 243–248.

Doull, J. 1976. "The Treatment of Insecticide Poisoning." *In* Wilkinson, C. F. (Ed.) *Insecticide Biochemistry and Physiology*. New York: Plenum Press.

Ecobichon, D. J. 1982. *Pesticides and Neurological Diseases*. Boca Raton: CRC Press.

Eto, M. 1974. *Organophosphorus Pesticides: Organic and Biological Chemistry*. Boca Raton: CRC Press.

Frear, D. E. H. (Ed.) 1961. *Pesticide Index*. State College (Pennsylvania): College Science Publishers.

Gershon, S., and Shaw, F. H. 1981. "Psychiatric Sequelae of Chronic Exposure to Organophosphorus Insecticides." *Lancet, 1*: 1371–1374.

Hayes, W. J., Jr. 1975. *Toxicology of Pesticides*. Baltimore: Waverly Press, Inc.

Heath, D. F. 1961. *Organophosphorus Poisons; Anticholinesterases and Related Compounds*. London: Pergamon Press.

Jager, K. W., Roberts, D. V., and Wilson, A. 1970. "Neuromuscular Function in Pesticide Workers." *Br. J. Ind. Med., 27*: 273–278.

Melnikov, N. N. 1971. *Chemistry of Pesticides*. New York: Springer-Verlag.

Murphy, S. D. 1980. "Pesticides." *In* Doull, J., Klaassen, C. D., and Amdur, M. O. (Eds.) *Casarett and Doull's Toxicology: The Basic Science of Poisons*. New York: Macmillan Publishing Company, Inc.

Namba, T., Nolte, C. T., Jackrel, J., and Grob, D. 1971. "Poisoning Due to Organophosphate Pesticides: Acute and Chronic Manifestations." *Am. J. Med., 50*: 475–492.

O'Brien, R. D. 1967. *Insecticides, Action, and Metabolism*. New York: Academic Press.

Straub, W. E. 1977. "Pesticides." *In* Key, M. M., Henschel, A. F., Butler, J., Ligo, R. N., Tabershaw, I. R., and Ede, L. (Eds.) *Occupational Diseases: A*

Guide to Their Recognition. Washington, D.C.: USDHEW, PHS, CDC, National Institute for Occupational Safety and Health.

Taylor, J. R., Selhorst, J. B., Houff, S. A., and Martinez, A. J. 1978. "Chlordecone Intoxication in Man. I. Clinical Observations." *Neurology, 28*: 626–635.

Weil, C. S., Woodside, M. D., Carpenter, C. P., and Smyth, H. F., Jr. 1972. "Current Status of Tests of Carbaryl for Reproductive and Teratogenic Effects." *Toxicology and Applied Pharmacology, 21*: 390–404.

Chapter 12

The Toxic Effects of Organic Solvents

Robert C. James

Introduction

The group of chemical compounds that can be classified as organic solvents is so large that a detailed discussion of each chemical is impossible. Instead, the approach taken in this chapter will be to discuss:

- General toxicities of the major classes of organic solvents.
- How functional groups of a chemical's structure alter its toxicologic effects.
- The toxicology of many specific chemicals commonly used as industrial solvents.
- The toxicology of many organic chemicals not traditionally considered to be solvents.
- Exposure hazards resulting from the immediate handling of various organic chemicals.

12.1 General Principles

The first common toxic effect to consider for any organic chemical is its depression of central nervous system (CNS) activity.

Depression of Central Nervous System Activity

A CNS depressant acts like a general anesthetic, inhibiting activity in the brain and spinal cord, lowering the functional capacity of the exposed person and making them less sensitive to stimuli until eventually the affected individual is rendered unconscious or comatose. A general feature of gases used as general anesthetics in surgery today is their highly lipophilic character. As discussed in Chapter 3, a lipophilic chemical has a high affinity for lipids and conversely has a low affinity for water. Therefore, these compounds accumulate in the lipid or fat portions of the body and tissues. By mechanisms that still are unclear, as these chemicals accumulate in the lipid membranes of nerve cells, they disrupt the normal excitability of these tissues and suppress the conduction of normal nerve impulses. Since most organic chemicals with few or no functional groups (which usually serve to increase water solubility) are highly lipophilic, they all possess varying degrees of CNS-depressant activity.

In general, CNS-depressant activity of an organic compound increases with the length of its carbon chain. Increased toxicity is particularly noticeable when larger arrangements of carbon molecules are added to the functional groups in small organic compounds, because the increase in size decreases water solubility and increases lipophilicity; thus the volume of distribution of the chemical—the amount bound by tissues—is increased. As a practical consideration for hazardous exposures, however, this is only of concern for chemicals up to about five carbons in size. As size increases larger than a five-carbon molecule for any of the functional classes (i.e., amines, alcohols, ethers, etc.), the vapor pressure, which is generally small, is further diminished. Thus, the likelihood that chemicals larger in size than five carbons will achieve hazardous exposure levels is relatively small.

Usually, the CNS-depressant properties of an organic molecule are greatly enhanced by halogenation and, to a lesser extent, by alcoholic functional groups. For example, while methane and ethane have no real anesthetic properties and asphyxiate only at high concentrations, both methanol and ethanol are potent CNS depressants. Likewise, while methylene chloride (i.e., dichloromethane) has appreciable anesthetic properties, chloroform is much stronger than methylene chloride, and carbon tetrachloride is more potent still. Lastly, unsaturated compounds—organic chemicals in which hydrogens have been removed to create one or more double bonds between the carbon atoms—are more potent CNS-depressant chemicals than their saturated analogs.

Irritation of Membranes and Tissues

The second common toxic response of consideration for organic chemicals is the irritation of membranes and tissues.

All organic chemicals have some irritant properties. Because cell membranes within the body are largely a protein–lipid matrix, organic solvents are ideal for extracting the fat or lipid portion out of the membrane. This defatting of the skin causes irritation and cell damage and may seriously injure the skin, lungs, or eyes if it is extensive. In general, any functional group added to an

Table 12-1. Approximate Ranking of the CNS-depressant and Irritant Activity of Organic Solvents.

CNS-depressant activity:
Alkanes < Alkenes < Alcohols < Organic acids < Esters < Ethers < Halogenated compounds

Irritant activity:
Alkanes < Alcohols < Aldehydes or ketones < Organic acids < Amines

organic molecule increases the irritant properties of the chemical. Amines and organic acids add corrosive properties to the molecule, while alcohol, aldehyde, and ketone groups increase the damage caused to cell membranes because, at high concentrations, they precipitate and denature proteins.

Likewise, unsaturated compounds are stronger irritants than their saturated analogs. Again, as the size of the molecule increases, the importance of the irritant properties of the functional group decreases and the solvent or defatting action of the hydrocarbon portion becomes more important. Table 12-1 gives approximate rank orderings for the functional groups an organic molecule may have and indicates how these functional groups affect the chemical's CNS-depressant and irritant properties. These approximate rankings are based upon comparisons with unsubstituted chemical analogs and, therefore, quickly lose applicability to larger and more complex multisubstituted compounds.

Other Toxic Properties of Organic Solvents

CNS-depressant and irritant properties are common to chemicals usually referred to as "organics," "hydrocarbons," or "solvents." These two properties are the focus of this chapter because they are consistently observed to some degree in all of the chemical classes discussed. However, these classes of chemicals do produce several other acute toxicities upon which some generalizations can be made. After systemic absorption, other acute toxicities are hepatotoxicity, nephrotoxicity, and cardiac arrhythmias induced by a sensitization of the heart to catecholamines (i.e., adrenaline). While these are not often observed in occupational settings, animal studies and accidental human poisonings have shown that these and occasionally other organ toxicities may be produced by acute exposures to organic solvents. This fact may be obvious for certain classes such as halogenated hydrocarbons, particularly in chronic exposure situations, but it is emphasized here so that attention will be given to these and other organs when overexposure occurs. These chemicals are commonly found in household products and are often ingested by small children (e.g., approximately 25 percent of all pediatric poisoning cases are hydrocarbon ingestions). With oral ingestion, aspiration into the lungs and the resultant chemical-pneumonitis become major considerations in treating such cases.

Lastly, the symptomology of any particular hydrocarbon poisoning, be it from an inhalation or oral exposure, often provides few clues to the specific solvent in question. Acute overexposure to organic solvents usually produces a "chemical malaise" that may encompass the entire range of possible subjective complaints. Therefore, treatment is usually symptomatic with regards to the systemic toxicities that develop and attempts should be made to limit further systemic absorption.

To illustrate the problems faced by the health specialist in attempting to diagnose the causative solvent in an unknown exposure situation, the following chemicals and the symptoms reported with exposure to them are provided:

- Methanol (wood alcohol): Euphoria, conjunctivitis, headache, dizziness, nausea, vomiting, abdominal cramps, sweating, weakness, bronchitis, narcosis, delirium, coma, decreased visual acuity.
- Benzene: Euphoria, excitement, headache, vertigo, nausea, vomiting, dizziness, irritability, narcosis, coma.
- Carbon Tetrachloride: Conjunctivitis, headache, dizziness, nausea, vomiting, abdominal cramps, nervousness, narcosis, coma.

12.2 The Toxic Properties of Aliphatic Organic Solvents

The Saturated Aliphatic Solvents

$$C_nH_{2n+2}$$

Alkane homologs

The aliphatics or alkanes are the least potently toxic class of solvents when acute toxicities are considered. The vapors of these solvents are only mildly irritating to the mucous membranes at the high concentrations required to produce their weak anesthetic properties. The first four chemicals in this series are gases with little toxicity and their hazardous nature is limited almost entirely to their flammability and explosivity potential. In fact, the first two members, methane and ethane, are so toxicologically inert that they belong to that group of nontoxic gases known as asphyxiants whose systemic toxicity is related to the hypoxic conditions they produce at high concentrations.

The liquid members of this chemical class do have some CNS-depressant and irritant properties, but this is primarily a concern of the lighter, more volatile fluid compounds in this series (i.e., pentane, hexane, heptane, octane, and nonane). The liquid paraffins, beginning with decane, are fat solvents and primary irritants capable of dermal irritation and dermatitis upon repeated or prolonged contact.

The symptoms of acute poisoning are similar to those previously described for other organic solvents (i.e., nausea, vomiting, cough, pulmonary irritation, vertigo or dizziness, slow and shallow respiration, narcosis, coma, convulsions, and death), with the severity of the symptoms dependent upon the magnitude and duration of exposure. Accidental ingestion of large amounts (about 1 ml/kg body weight) may produce systemic toxicity. If less than 1 ml/kg is ingested, a cathartic, used in conjunction with activated charcoal (to limit absorption), is usually the therapeutic approach taken. In both situations, aspiration of the solvent into the lungs is probably the primary health concern. Low viscosity hydrocarbons are of the most concern because their low surface tension allows them to spread over a large surface area, thereby producing extensive damage to the lungs after exposure to only small quantities. A chest x-ray may be taken to monitor any pulmonary complications which might develop, and the ultimate severity of this adverse effect can be determined within the first 24 hours. These chemicals may also sensitize the heart to epinephrine (adrenaline), but this is usually not a practical consideration since only a narrow dose separates this effect and fatal narcosis.

Chronic exposure to some of the aliphatics, in particular hexane and heptane, has been found to produce a polyneuropathy in humans and animals, characterized by a lowered nerve conduction velocity and a "dying back" or degenerative change in distal nerve axons. Symptoms include muscle pain and spasms, muscular weakness, and parathesias. Metabolites have been implicated as the causative agents with 2,5-hexandione and 2,6-heptandione as the respective toxic metabolites of hexane and heptane. As these metabolites represent oxidations, first to an alcohol and then to the respective diketone, structurally similar alcohols and ketones will also produce this neuropathy.

The Unsaturated Aliphatic Solvents

$$C_nH_{2n}$$

Olefin (alkene) homologs

The alkenes, which are unsaturated alkanes, are commonly referred to as olefins, and are toxicologically very similar to the alkanes. The double bond(s) enhances the irritant and CNS-depressant properties, but this change may not be sufficient to be of great practical value. For example, ethylene is a more potent anesthetic than its corresponding alkane, ethane, a simple asphyxiant. However, since a concentration of 60 percent ethylene is required to induce anesthesia, the potential for hypoxia and the explosive hazard are obvious and major drawbacks to its clinical use. Moreover, such a concentration in an industrial setting would sufficiently dilute the oxygen present so that asphyxiation (as is the case with ethane) would be the major toxicity rather than narcosis and respiratory arrest. Of perhaps greater toxicological interest is the indication that the unsaturated double bond in the hexene and heptene series apparently abolishes the neurotoxic effects observed with chronic hexane or heptane exposure, a change that could be related to an alteration in the metabolism of the olefinic compounds.

12.3 The Toxic Properties of Alicyclic Solvents

H_2C—CH_2—H_2C (three-membered ring): **Cyclopropane**

H_2C—CH_2 / H_2C—CH_2 (four-membered ring): **Cyclobutane**

The alicyclic hydrocarbons are merely alkanes to which both ends of an otherwise straight chain carbon compound have been bonded to form a cyclic or ring structured carbon compound. Toxicologically, they resemble their open chain relatives, the alkanes, and are generally anesthetics or CNS-depressant. Industrial experience seems to indicate that minimal or no chronic toxicities are produced after prolonged exposures to these compounds. The smaller alicyclics, like cyclopropane (cyclane), have been used as a surgical anesthetic. The larger compounds, like cyclohexane, are not as useful because the margin of safety between narcosis and a lethal concentration is very small. The irritant qualities of cycloolefins tend to be greater than that of their unsaturated counterparts.

12.4 The Toxic Properties of Aromatic Hydrocarbon Solvents

General Introduction and Toxicology. The chemical class of organic solvents commonly referred to as "aromatics" are chemicals composed primarily of one or more benzene rings. The smallest and simplest member of this chemical class is, of course, benzene itself followed by the alkylbenzenes and then the aryl- and alicyclic-substituted benzenes. Diphenyl and polyphenyl compounds are also included in this class, as well as the polynuclear aromatics (PNAs) which are also referred to as polycyclic aromatic hydrocarbons (PAHs). Benzene and its alkyl derivatives are industrially very important organic compounds in the U.S. with well over 1.5 billion gallons of benzene being produced or imported each year. Even larger quantities of several of the alkylbenzenes are produced. Benzene and the alkylbenzenes are used as solvents or as reactants for chemical synthesis in the ink and dye, oil, paint, plastics, rubber, adhesives, chemical, and drug industries as well as being major components of gasoline. Most gasolines contain at least one percent benzene (sometimes higher), and alkylbenzenes may be added to unleaded brands to levels reaching 30–35 percent.

The aromatics are far more irritating than the aliphatics previously discussed. These compounds are primary irritants causing dermatitis and a severe defatting of the skin that may result in tissue injury or chemical burns if dermal contact is repeated or prolonged. Both conjunctivitis and corneal burns have been reported when benzene or its alkyl derivatives have been splashed into the eyes, and naphthalene has been reported to cause cataracts in animals. If the aromatics are aspirated into the lungs after being ingested, they are capable of causing a severe pulmonary edema, chemical-pneumonitis and even hemorrhage. Inhalation of high concentrations can result in bronchial irritation, cough, hoarseness, and pulmonary edema. Once they have been absorbed and reach the systemic circulation, they are considerably more toxic than aliphatics and alicyclics of comparable carbon number. In addition, even though CNS depression is a major acute systemic effect of this class, it differs considerably from that produced by the aliphatics. The aliphatic-induced coma (e.g., kerosene or gasoline) is characterized by an inhibition of the deep tendon reflexes. In contrast, benzene tends to produce a coma characterized by motor restlessness, tremors and hyperactive reflexes and is sometimes preceded by convulsions.

Benzene

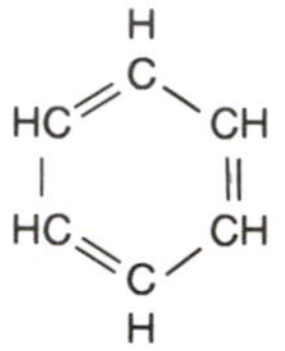

```
        H
        C
  HC⁄⁄     ⁀CH
  |         ||
  HC        CH
        C
        H
```

Benzene

Benzene is a colorless liquid with a pleasant, balsamic, "aromatic" odor. The term benzene should not be confused with benzine which refers to a low-boiling petroleum fraction composed primarily of aliphatics. Benzene is toxic by all routes of administration. In animals the LC_{50} begins at 10,000 ppm; in humans 20,000 ppm has been reported fatal in 5–10 minutes. At concentrations of 7,500 ppm in man, severe acute symptoms are induced in approximately 30–60 minutes, but even brief exposure levels in excess of 3,000 ppm are very irritating to the eyes and respiratory tract and 250 ppm will produce vertigo, drowsiness, headache, and nausea. Ingested benzene is far more systemically toxic than the aliphatics. The fatal human dose has usually been reported to be around 0.2 ml/kg or about 10–15 ml for an adult, but amounts as low as 2–5 ml have also been reported fatal. While its CNS effects tend to overshadow other systemic toxicities in acute situations, cardiac sensitization and arrhythmias may occur, particularly when the intoxication is severe. The pathological findings observed in acutely poisoned victims are respiratory inflammation, edema and hemorrhage of the lungs, renal congestion, and cerebrel edema.

Benzene is an irritating liquid causing erythema, vesiculation, and a dry, scaly dermatitis of the skin. If splashed into the eyes, it produces a transient corneal injury. Prolonged dermal contact with benzene (or toluene and other alkylbenzenes) may result in lesions resembling first or second degree burns, and skin sensitization has been reported. As mentioned previously, inhalation of high concentrations may produce bronchial irritation or even pulmonary edema.

Benzene differs from most organic solvents in that it is a myelotoxin. In most cases the hematological findings are variable, but effects are usually noted in red cell count, hemoglobin, platelet counts, and leukocyte counts. The most commonly reported effect is a fall in the white cell count and benzene was actually used between 1910 and 1920 to decrease the number of circulating leukocytes in leukemia patients.

Three separate stages or degrees of severity can be identified in the benzene-induced change in blood tissues. The first sign of toxicity may be blood clotting defects, as well as a decrease of all blood components (a mild pancytopenia or aplastic anemia). If quickly and properly diagnosed, this stage is readily reversible. With continued exposure, the bone marrow first becomes hyperplastic and a stimulation of leukocyte formation may be the earliest clinical observation. Hypoplasia will follow with continued benzene exposure, and if severe, produces anemia and hemorrhage. Thus, even though chronic benzene exposure is probably better known for its link to leukemia, aplastic anemia is a more likely chronic health problem.

Patients suffering from chronic benzene poisoning usually have a red cell count that is only 50 percent of normal. Metabolites of benzene have been implicated as the causative toxic agents and leukopenia and anemia have been induced in animals with chronic hydroquinone and pyrocatechol administration. However, the benzene syndrome has not been observed in humans exposed to phenol, hydroquinone, or catechol. Lastly, it should be noted that some chromosomal changes have been observed in the lymphocytes of workers chronically exposed to benzene.

Substituted Compounds

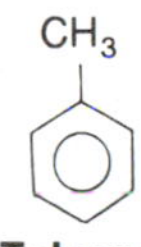

Toluene

Toluene (methylbenzene) is a more powerful CNS depressant than benzene. Exposures in men for periods of 8 hr at 200 ppm produce fatigue, weakness, confusion, and paresthesia of the skin. The fatigue lasts for several hours. At 400 ppm, mental confusion becomes a symptom. At 600 ppm, extreme fatigue, confusion, exhilaration, nausea, headache, and dizziness result in 3 hr. In contrast to benzene, no definitive evidence links toluene exposure to permanent blood disorders.

The acute toxicity of the xylenes (xylol and dimethylbenzene), $C_6H_4(CH_3)_2$, is greater than that of toluene but the symptoms are similar. Inhaling naphthalene ($C_{10}H_8$) vapors causes headache, confusion, nausea, and profuse perspiration. Severe exposures may cause optic neuritis and hematuria.

Cataracts have been produced experimentally in rabbits and one case has been reported in humans. Naphthalene is also an irritant and hypersensitivity has been reported.

Polycyclic Aromatic Hydrocarbon (PAH) Compounds

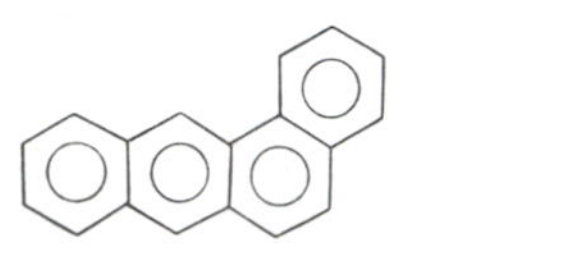

Benz[*a*]anthracene

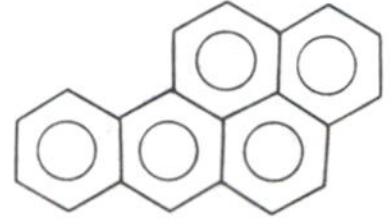

Benzo[*a*]pyrene

General Toxicology. The most widespread mode of contact with the polycyclic aromatic hydrocarbons (also called polynuclear aromatic hydrocarbons) is inhalation of hydrocarbon particulates formed during combustion of tobacco and by exposure to air contaminated with industrial combustion products (vehicular exhaust and other effluents).

PAHs are nonpolar, lipid soluble compounds that may be absorbed via the skin, lungs, or digestive tract. Once absorbed, these compounds are distributed to all tissues, but become particularly concentrated in those organs with a high lipid content. They are metabolized by a subpopulation of cytochrome P-450, which they also induce, commonly referred to as aryl hydrocarbon hydroxylase (AHH) or cytochrome P-448. Since PAHs are composed of aromatic rings with little or nothing else to metabolize, hydroxylation is the first goal in the body's attempt to metabolize PAHs to more water soluble forms that can be readily excreted. In attempting to accomplish this objective, arene oxides are commonly generated during their metabolism and these represent reactive, potentially toxic and carcinogenic metabolites. Recent evidence indicates, however, that the simple or initial epoxide metabolites are not the ultimate carcinogens. Instead diol-epoxide intermediates now appear to be the proximal carcinogens because these secondary metabolites have been shown to be more potently mutagenic and carcinogenic, and because they will form DNA adducts which are more resistant to DNA-repair processes.

While this evidence suggests that our continual exposure to PAHs would inevitably lead to carcinogenesis, other routes of metabolism have been identified that act as protective mechanisms by detoxifying reactive PAH metabolites. Similarly, natural or added constituents of foods such as flavenoids, selenium, vitamins A, C, and E, phenolic antioxidants, and additives like BHT and BHA all exert protective effects. Thus, the interactions that may ultimately lead to or prevent carcinogenesis after chronic exposure to the PAHs need to be better elucidated.

PAHs can be acutely toxic, but generally only in very high doses, making acute systemic toxicity observable in some animal tests, but not a likely occurrence in the industrial setting. At these high, acute doses, PAHs are toxic to many tissues and degenerative changes may ultimately be observed in the kidney and liver, but the thymus and spleen are particularly sensitive to the acute effects of PAHs. This probably results from the fact that these compounds are widely distributed throughout the body and are capable of inhibiting the cell's normal mitotic cycle. It is not surprising then that rapidly proliferating cells are the ones most often affected or that toxicity in the hematopoietic and lymphatic systems is a common observation. In general, it appears that the carcinogenic potency of these compounds correlates well with those potencies measured for other toxicities. For example, the noncarcinogenic acenaphthene given in doses as high as 2,000 mg/kg produces only minor changes in the liver or kidney and is relatively nontoxic when compared to the hematoxicity produced by 100 mg/kg of dimethylbenzanthrene.

The teratogenic/embryotoxic effects of PAHs have only been documented for a few of the more potent, carcinogenic PAH compounds, but it seems likely that this toxicity will also correspond to the carcinogenic potency of the congeners in this chemicals class. Mutagenic activity has been more clearly demonstrated for the PAHs in a number of test systems, particularly the Ames assay. Likewise, the carcinogenicity of some of the PAH compounds has been convincingly demonstrated in both lung and skin tissue by numerous animal experiments. In fact, results of skin-painting experiments using these compounds may be credited with the development of the two-stage theory of carcinogenesis (i.e., initiation–promotion), an important addition to current theories and understanding of chemical-induced carcinogenesis. Unfortunately, this animal evidence concerning the carcinogenicity of these compounds was preceded by a long history of epidemiologic evidence (see Table 12-2).

Table 12-2. History of the Epidemiologic Basis for Concern about Human PAH Exposure.

Researcher	Year	Workers exposed	Cancer type
Pott	1775	Chimney sweeps	Scrotum
Volkmann	1875	Tar/paraffin workers	Scrotum
Rehn	1895	Synthetic dye workers	Bladder
Luehe	1907	Carbon workers	Skin
Kennaway	1947	Asphalt and coal gas workers	Lung
Doll	1952	Coal gasification workers	Lung
Sexton	1960	Coal liquefaction workers	Bladder
Soy/Redmond/ Maxemdar/Lloyd	1968–1975	Coke producers or workers	Lung

12.5 The Toxic Properties of Alcohols

Comparative Toxicity of Alcohols

R — OH

R = hydrocarbon group

Alcohol compounds

Alcohols are far more powerful CNS depressants than their aliphatic hydrocarbon analogs. In general, tertiary alcohols are more potent than secondary alcohols, which, in turn, are more potent than primary alcohols. The alcohols are slight irritants. They are stronger irritants than organic chemicals without functional groups, but far less irritating than amines, aldehydes, or ketones. The irritant properties of the alcohol group decrease with increasing total molecular size, because the alcohol group contributes less to the chemical characteristics of a moleculear compound as the size of the molecule increases. General systemic toxicity increases with molecular weight as water solubility is diminished and lipophilicity is increased. Alcohols and glycols rarely represent serious hazards in the workplace, because their vapors are usually below irritant levels which prevents their CNS manifestations as well.

Methanol

CH_3OH

Methanol

Although methanol is only industrially significant because of its unique toxicity in the eye, it has received considerable attention from the medical community over the years due to misuse and accidental human consumption. Consequently, the interested reader can find considerable information on methanol in the medical literature. To illustrate this point, the following paragraphs on methanol are summarized from the Goodman and Gilman text on pharmacology, *The Pharmacological Basis of Therapeutics* (Fifth edition).

Methanol (also called methyl alcohol and wood alcohol) is the simplest of the alcohols. It is widely employed industrially as a solvent. It is also used as an adulterant to "denature," and thereby make unfit to drink, the ethyl alcohol used for cleaning, paint removal, and other applications.

Methanol is purely of toxicologic interest and has no clinical or recreational value. Poisoning results from its ingestion as a substitute for, or as an adulterant of, ethyl alcohol. For example, methanol caused 6 percent of all blindness in the U. S. Armed Forces during World War II.

Industrial exposure can cause serious or fatal poisoning. Poisoning by methanol results from a combination of the following:

1. A minor factor of CNS depression, similar to that produced by ethyl alcohol.
2. A major factor of acidosis owing to the production of formic and other organic acids.

3. A specific toxicity of the oxidation products of methanol (probably formaldehyde) for the retinal cells.

Methanol is less inebriating than ethanol; indeed, inebriation is not a prominent symptom of methanol intoxication, unless a very large amount is consumed or ethanol is also ingested. An asymptomatic latent period of 8 to 36 hr may precede the onset of symptoms. If ethanol is simultaneously imbibed in sufficient amount, methanol poisoning may be considerably delayed, or, on occasion, even averted. In such cases ethanol intoxication is prominent, and methanol ingestion may not be suspected.

Symptoms and signs of methanol poisoning consist of headache, vertigo, vomiting, severe upper abdominal pain, back pain, dyspnea, motor restlessness, cold or clammy extremities, blurring vision, hyperemia of the optic disc, and, occasionally, diarrhea. The visual disturbance can proceed to blindness, and the pupils then do not react to light. Restlessness and delirium may be marked. Blood pressure is usually unaffected. The pulse slows in severely ill patients, and bradycardia constitutes a grave prognostic sign. Coma can develop with amazing rapidity in relatively asymptomatic subjects. In moribund patients the respiration is slow, shallow, gasping, and "fish mouth" in type. Death may be sudden, or it may occur only after many hours of coma. Death occurs with inspiratory apnea and convulsions.

The oxidation of methanol, like that of ethanol, proceeds independently of its concentration in the blood. The rate of oxidation, however, is only one-seventh that of ethanol, so that complete oxidation and excretion of methyl alcohol usually requires several days. Oxidation occurs mainly in the liver and kidney. It is still generally agreed that in man alcohol dehydrogenase is involved in the first step of oxidation. The fact that this same enzyme is responsible for the oxidation of ethyl alcohol presumably explains the finding *in vitro* that ethanol considerably depresses the rate of oxidation of methanol. The common biochemical pathway of oxidation of both alcohols also accounts for the clinical observation that simultaneous administration of ethanol may ameliorate the toxic sequelae of methanol poisoning. The products of oxidation of methanol are much more toxic than methanol itself, and, therefore, the degree of poisoning is minimized if the rate of oxidation of methanol is reduced as much as possible. As little as 15 ml of methanol has caused blindness and 70 to 100 ml may be fatal. Besides blindness, other neurologic damage may follow methanol poisoning, possibly leading to permanent motor dysfunction.

Ethanol

CH_3CH_2OH

Ethanol

Ethanol (also called ethyl alcohol) acts as a local irritant, injuring cells by precipitating and dehydrating cell protoplasm. The irritant action of ethanol is mild to moderate in strength.

Alcohol affects the CNS more markedly than any other system. The apparent initial stimulation results from the activity of various parts of the brain that have been freed of inhibition through the depression of their control mechanisms. Ethanol raises the pain threshold 35–40 percent at moderate doses.

Ethanol produces a vasodilation of cutaneous blood vessels. Because of this, its use is contraindicated when the exposed person is suffering from hypothermia or exposure to cold. Ethanol causes a cardiovascular depression, which is of CNS origin. It directly damages tissue at high doses, producing skeletal myopathy and cardiomyopathy. Because ethanol increases gastric secretion at high concentrations, it causes an erosive gastritis, which can increase the severity of ulcers. In the liver it promotes accumulation of fat, and chronic use leads to cirrhosis, which may convert to liver cancer or become fatal in itself. It promotes urine flow by inhibiting the release of antidiuretic hormone from the pituitary and causes the release of hormones from the adrenal glands (steroids and adrenaline). Lastly, it has direct depressant action on the bone marrow and leads to a depression of leukocyte levels in inflamed areas, explaining the poor resistance of alcoholics to infection.

Other Alcohols

High concentrations of propanols (propyl alcohol, C_3H_7OH) cause intoxication and CNS depression. They are bactericidal. Isopropyl alcohol ($(CH_3)_2CHOH$) is less toxic than propanol ($CH_3(CH_2)_2OH$), but both substances are more toxic than ethanol. In humans, an exposure of 3–5 min at 400 ppm of isopropyl alcohol causes mild irritation of eyes, nose, and throat. *n*-Butanol (C_4H_9OH) is potentially more toxic than lower homologs, but it is less volatile. Its symptoms include irritation of eyes, nose, and throat; vertigo; headache; drowsiness; contact dermatitis; and corneal inflammation. Yet no systemic effects have been noted for exposure under 100 ppm. Skin irritation is common from allyl alcohol; absorption through the skin leads to deep pain and can cause severe burns of the eye. It is not particularly anesthetic, but it is highly irritating. Symptoms include lacrimation, photophobia, and blurring vision. Allyl alcohol is metabolized by the liver to allyl aldehyde, which is a potent hepatotoxin.

Glycols

$$R-\underset{\displaystyle OH}{\underset{|}{C}}-\underset{\displaystyle OH}{\underset{|}{C}}-R$$

Glycol compounds

In general, the larger alkyl-chain glycols have less acute oral toxicity than the monohydroxy alcohols, and are not significantly irritating to eyes or skin, and their vapor pressures are so low that toxic concentrations are not usual at room temperature. Ethylene glycol is the compound of primary concern when considering the glycol family. A single oral dose of 100 ml is lethal for most people. It is metabolized to oxalate, a dicarboxylic acid that is toxic to the kidneys and may cause complete renal failure. Again, ethanol can be used as a competitive inhibitor of ethylene glycol by blocking its metabolism by aldehyde dehydrogenase—the same method by which ethanol inhibits methanol poisoning.

12.6 The Toxic Properties of Aldehydes

Comparative Toxicity of Aldehydes

$$\begin{matrix} H \diagdown & \\ & C{=}O \\ R \diagup & \end{matrix}$$

Aldehyde compounds

General Toxicology. Aldehydes cause primary irritation of the skin, eyes, and mucosa of the respiratory tract. The effect is common to nearly all the aldehydes, but it is most characteristic and important in those with lower molecular weights and those with unsaturated aliphatic chains. While aldehydes can produce narcosis, this effect is usually prevented because the irritation accompanying exposure serves as a warning against overexposure. Some aldehydes, such as fluoroacetaldehyde, are metabolically converted to the corresponding fluorinated acids, giving them an extraordinarily high degree of systemic toxicity. These fluorinated acids are very potent poisons that inhibit normal cellular metabolism and are therefore toxic to all cells. The irritant properties of the dialdehydes have not been intensively studied, but in some instances concentrated solutions can be severe irritants to the skin or the eyes. With some exceptions, acetals and the aromatic aldehydes have a much lower degree of primary irritant action.

One toxicity reported for aldehydes (not common to most other organic solvents) is sensitization. Formaldehyde is the major causative agent among the aldehydes with respect to this problem and sensitization reactions have been reported in persons exposed to formaldehyde merely by wearing "permanent press" fabrics containing melamine-formaldehyde resins. Because their irritant effects limit inhalation exposure, the industrial use of aldehydes is relatively free of problems associated with systemic or organ toxicities of a serious nature. However, if tolerable levels are exceeded, animal studies suggest that damage to the respiratory tract is possible.

The unsubstituted aldehydes like acrolein and ketene are particularly toxic. The double bond in close association with the aldehyde functional group makes these compounds far more reactive, and therefore far more toxic than their unsubstituted analogs. For example, ketene and acrolein are found to be about 100 times more potent when measuring their acutely lethal air concentration than are either acetaldehyde or proprionaldehyde. Reflecting this increase in potency is the fact that their damage to the respiratory system is also more severe, resembling the deep lung damage of phosgene. If absorbed, the systemic toxicity of the unsaturated aldehydes is also more severe.

Lastly, certain aldehydes (e.g., acrolein, formaldehyde and acetaldehyde) are clearly mutagenic in a number of test systems, raising concerns for the carcinogenic potential of these compounds.

Formaldehyde

$$\mathrm{H_2C{=}O}$$

Formaldehyde

Formaldehyde is the first member of the aldehyde family of organic solvents. It is the most important aldehyde in both commerce and the environment, and some 7–8 billion pounds are produced in the U.S. annually. Because of its instability in the pure form, it is generally marketed in an aqueous solution ranging from 37 to 50 percent formaldehyde (known as formalin). Formaldehyde is also sold in two other forms, the cyclic trimer, trioxymethylene, and the low molecular weight homopolymer, paraformaldehyde. It is used primarily in the plastics and resins industries, in the synthesis of chemical intermediates, and to a lesser extent, in sealants, cosmetics, disinfectants, foot care creams, mouthwashes, embalming fluids, corrosion inhibitors, film hardeners, wood preservatives, and biocides.

Formaldehyde is a fairly strong irritant and its local actions tend to predominate the adverse effects observed with excessive exposure rather than the systemic effects that might otherwise be possible. The following list, adapted from the 1981 National Academy of Sciences review of aldehydes, roughly describes the concentration-effect relationship for formaldehyde exposures and its irritative effects.

Health effects reported	Approximate formaldehyde concentrations (ppm)
None	0–0.5
Odor threshold	0.05–1.50
Eye sensation/irritation	0.05–2.0
Upper airway irritation	0.10–25
Lower airway/pulmonary effects	5–30
Pulmonary edema/inflammation	50–100
Death	100

Formaldehyde produces dermal sensitization reactions in approximately 4 percent of the persons patch-tested, making it the tenth most common cause of dermatitis. Studies have demonstrated that the repeated application of high, irritating solutions of formaldehyde induces sensitization in about 8 percent of the male subjects tested, but with lower concentrations (about 2 percent) the incidence is only about 5 percent or less. While a number of reports have documented that dermal contact to formaldehyde may sensitize the individual, only a few reports have suggested that inhaling formaldehyde might also lead to a sensitization of the individual. The reports that have put forth this suggestion tend to be anecdotal in nature and without a definitive diagnosis with which to support this contention. Thus, it appears that the primary irritant effects of formaldehyde are likely the actual problem.

The fatal oral dose of formaldehyde is estimated to be some 60–90 ml of formalin. Depending upon the dose, ingestion may cause headaches, corrosion of the G.I. tract, edema of the lungs, fatty degeneration of the liver, renal tubular necrosis, unconsciousness, and vascular collapse.

Formaldehyde has been tested in animals and found not to cause reproductive toxicities or teratogenicity. However, concerns have been raised with the recent findings of its mutagenic activity in certain test systems and with its corresponding carcinogenicity in rodents. While the exact mechanism of its mutagenic actions remains to be resolved, a recurrent finding in several test systems was that formaldehyde produces crosslinks within DNA which are frequently recognized and repaired by the DNA repair enzymes present in those cells. Similarly, the rodent carcinogenicity tests reveal a steep dose-response relationship which strongly suggests that a definite threshold exists for this toxicity. Consistent with this suggestion are the following:

- Formaldehyde is a common metabolite of normal cellular metabolic processes and serves as a cofactor in the synthesis of several essential biochemical substances. (In fact, tissue concentrations of formaldehyde are several ppm.)
- The mechanism for its carcinogenicity appears to be a recurrent tissue injury and hyperplasia caused by an unreasonably high, irritating and necrotizing exposure.
- No epidemiologic studies to date have linked formaldehyde exposure to any form of cancer in humans, nor have they indicated that persons chronically exposed to formaldehyde are at any increased risk.

Acetaldehyde

$$H_3C{-}CH{=}O$$

Acetaldehyde

Acetaldehyde is the next smallest aldehyde and is a common industrial chemical. It is a relatively innocuous chemical systemically and subchronic and chronic tests in animals have not revealed major toxic effects. Acetaldehyde is a metabolite of ethanol metabolism and is believed to cause the "hangover" associated with ethanol. It is less reactive than formaldehyde and therefore, is generally less irritating and toxic than formaldehyde. It has not been found to be carcinogenic in animal tests, but its carcinogenic potential has probably not been adequately examined by appropriate long-term tests. Acetaldehyde is embryotoxic and teratogenic in animal tests and is probably the proximal toxicant inducing these effects when ethanol has been tested.

At concentrations of 200 ppm, it is capable of eye irritation marked by conjunctivitis and redness, and at 25–50 ppm, some individuals may experience its irritant effects. The odor threshold is 0.07 ppm and affords a considerable margin of safety if used as a warning property (as the TLV is 100 ppm). While it is a minimal health hazard industrially, it does represent an explosion hazard.

Acrolein

Acrolein is the unsaturated analog of propionaldehyde and the presence of the double bond greatly enhances its toxicity. Acrolein is toxic by all routes of adminstration and is capable of severe eye and pulmonary irritation. Since contact with the skin may produce necrosis, its direct contact with the eyes must be carefully avoided. Even though it is a reactive chemical, no carcinogenicity has been observed in tests thus far. The TLV for acrolein is 0.1 ppm and is not far below that level considered to be moderately irritating, 0.25 ppm.

Chloral and Chloral Hydrate

```
        Cl    H
        |     |
  Cl —  C  —  C — OH
        |     |
        Cl    OH
```

Chloral hydrate

In general, halogenation greatly increases absorption of aldehydes and thus their oral and dermal toxicities. Chloral ("Mickey Finn") is the 3,3,3-trichloro derivative of acetaldehyde. Halogenation increases lipid solubility and usually also enhances the chemical's activity on the CNS. Thus chloral is more of a depressant of this system than is acetaldehyde. The formula of chloral is Cl_3CCHO. Because chloral is an unstable, disagreeable oil that does not lend itself well to pharmaceutic formulations, it was introduced into medicine in the form of its hydrate, which is formed by adding one molecule of water to the carbonyl group. The formula of chloral hydrate is $Cl_3CCH(OH)_2$. Chloral hydrate is quite irritating to the skin and mucous membranes, which accounts for its gastrointestinal side effects, which are particularly likely to occur.

12.7 The Toxic Properties of Ketones

Comparative Toxicity of Ketones

$$R'(R)C{=}O$$

Ketone compounds

General Toxicology. The lack of reports in the literature of serious injury indicates that these substances do not present serious hazards to health, probably because they have fairly effective warning properties. Ketones are CNS depressants, but the vapors at concentrations high enough to cause serious sedation are irritating to the eyes and respiratory passages and thus are avoided. Lower concentrations, however, may be easily inhaled and accumulated to levels that impair judgment, and death owing to overdose results from respiratory failure. In general, the toxic properties increase with increasing molecular weight, and the unsaturated compounds are more toxic than the saturated compounds.

Acetone

$$(H_3C)_2C{=}O$$

Acetone

It has been reported that several daily ingestions of acetone in doses of 15–20 g produced no ill effects other than drowsiness. Skin irritation occurs only after repeated prolonged contact. Persons unaccustomed to acetone may experience eye irritation at 500 ppm, while workers used to daily exposure tolerate up to 2500 ppm. At 9300 ppm, irritation of the throat and lungs occurs. The odor threshold rises after immediate contact in acclimated people. Studies of exposed employees with average exposure concentrations of 2000 ppm revealed no serious injury; the human toxicity of acetone is very low.

Other Ketones

Methylisopropenylketone is more toxic than most of the saturated aliphatic ketones. Eye contact causes severe and possibly permanent damage. It causes moderate skin irritation and may result in a burn if the exposure is long enough.

Acetophenone (phenylethylketone) was used in the past as an anesthetic, but it is used today in perfumes because it has a persistent odor not unlike orange blossoms or jasmine. It is a strong skin irritant and may even burn, but it is not very toxic as a depressant. Eye contact causes marked irritation and transient corneal burns.

Methyl-*n*-butyl ketone is quite neurotoxic and is metabolized to 2,5-hexandione, the neurotoxic metabolite of hexane. Therefore, it induces a polyneuropathy like that described previously for hexane.

12.8 The Toxic Properties of Carboxylic Acids

$$R-C(=O)-OH$$

Carboxylic acid

General Toxicology. These compounds can be classified primarily as irritants. As with aldehydes and ketones, the irritant properties tend to present the most concern and mask the CNS-depressant properties. The acidity, and therefore irritancy, decreases with increasing size. Halogenation of carboxylic acids increases the strength of the acid formed and makes a stronger irritant. Likewise, dicarboxylic acids and unsaturated carboyxlic acids are also more corrosive. Alpha-carbon substitution by a hydroxyl group or halogen greatly enhances acidity. For example, acetic acid (CH_3COOH) is moderately irritating, but the unsaturated acids acrylic acid ($CH_2{=}CHCOOH$) and crotonic acid ($CH_3CH{=}CHCOOH$) or the halogenated trichloroacetic acid (CCl_3COOH) produce severe burns and tissue damage.

12.9 The Toxic Properties of Esters

$$R-C(=O)-OR'$$

Ester compounds

General Toxicology. Esters are more potent anesthetics than alcohols, aldehydes, or ketones, but are weaker than ethers or halogenated hydrocarbons. Esters are broken down in the bloodstream by plasma esterases to their constituent carboxylic acids and alcohols.

As irritants, the lower-carbon esters are more potent than the alcohols and are known to cause lacrimation. Halogen substitution increases the irritant effects, and unsaturated double bonds in the side chain may increase the toxicity tenfold. Unsaturated esters have increased irritancy, and some may act to cause CNS stimulation rather than CNS depression. Additional functional groups other than halogens or double bonds between carbons in the alkyl chain tend to reduce the vapor pressure and systemic toxicity. Therefore, these esters are usually of less concern as irritants to the skin and eyes.

Phosphate esters, used largely as plasticizers, may produce damage to the CNS, are very irritating, and act as either convulsants or depressants. Because of the wide variety of esters and the dramatic differences additional functional groups may make to their toxicity, the exact consequences of each compound should be found in the literature, and few generalizations other than the ones indicated here should be made about these compounds.

12.10 The Toxic Properties of Ethers

$$R—O—R'$$

Ether compounds

General Toxicology. Ethers are very effective anesthetics, and this is a property that increases with the size of the molecule. The utility of this property is limited by the irritant effects of ethers and the fact that they are easily oxidized to peroxides, which are quite explosive.

Ethyl Ether

```
     H   H       H   H
     |   |       |   |
  H—C—C—O—C—C—H
     |   |       |   |
     H   H       H   H
```

Ethyl ether

Diethyl ether is rapidly absorbed through the lungs and rapidly excreted through the lungs. Widely used as an anesthetic at one time, it is slightly irritating to the skin and contact with the eyes should be avoided. It produces anesthesia in humans in a concentration range of 3.6 percent–6.5 percent in air, but respiratory arrest occurs at 7 percent–10 percent. Ether can produce profound muscular relaxation by means of corticospinal and neuromuscular blockade. However, nausea and vomiting are common side effects.

Isopropyl Ether

```
        CH3     CH3
         |       |
   H3C — C — O — C — CH3
         |       |
         H       H
```

Isopropyl ether

Isopropyl ether is more toxic than ethyl ether but causes irritation at much lower concentrations than those required to produce anesthetic effects. In man, 500 ppm for 15 minutes caused no irritation, but the odor was noticeably unpleasant at 300 ppm. At 800 ppm, irritation of the eyes and nose is noticeable.

Other Ethers

Divinyl ether is more potent than ethyl ether. The unsaturated ethers are, in general, more toxic than the saturated ethers, produce anesthesia faster, and possibly cause liver damage. Halogenated ethers can cause very severe irritation to the skin, eyes, and lungs. For example, the vapors of chloromethyl ethers are painful at 100 ppm. The chlorinated ethers may also be potent alkylating agents, and compounds such as bis(dichloromethyl) ether are known carcinogens. Aromatic ethers, on the other hand, are less volatile, less irritating, and less toxic than the alkyl ethers.

12.11 The Toxic Properties of Phenols

General Toxicology. Aromatic alcohols (phenols), like alkyl alcohols, have the ability to denature and precipitate proteins. Because of this property, phenol is bacteriostatic at a concentration of 0.2 percent and bactericidal at 1.0 percent. Phenolic compounds have some local anesthetic properties and, in general, are CNS depressants. These compounds can be quite corrosive—the irritation they produce is so strong that severe burns may result from direct contact.

Dihydroxy compounds act like phenols but are largely local irritants. Trihydroxy compounds have an additional adverse effect, which is to reduce the oxygen content of blood, and chlorinated phenols are strong irritants but also produce muscle tremors, muscle weakness, and, in overdoses, convulsions, coma, and death. Chlorinated phenols also have an increased oral toxicity because of their direct inhibition of cellular respiration.

Phenol

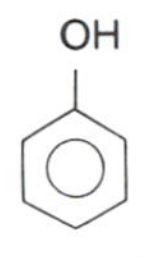

Phenol

Phenol, with its ability to complex with and denature proteins, can be cytotoxic to all cells and tissues if sufficient cellular levels are reached. Since it is easily and rapidly absorbed and because it forms a rather loose complex with protein, phenol may quickly penetrate the skin and underlying tissue causing deep burns and considerable tissue necrosis. This penetrability and its non-specific toxicity give it a high handling hazard and all routes of exposure must be limited or prevented. When splashed on the skin, it produces redness and irritation with dermal changes or injury ranging from eczema or discoloration and inflammation to necrosis or gangrene, depending on the exposure.

If accidentally ingested, the extensive local necrosis it produces in the mucous membranes of the throat, esophagus, and stomach causes severe pain, vomiting, and a tissue corrosion that may rapidly lead to shock and death. If inhaled at sufficient concentrations, a serious case of chemical pneumonitis may be induced. Like many other organic solvents, it may cause tissue damage and necrosis in the liver and kidneys when absorbed systemically. However, it is far more potent in this respect than most organics, and phenol is capable of producing degenerative and necrotic changes in the heart and urinary tract as well. Blood pressure may also fall as a result of phenol's CNS-depressant properties on vasomotor control, and because it exerts direct, toxic actions on the myocardium and smaller blood vessels. The predominant CNS effect of phenol is that of CNS depression. If acute poisoning occurs, death is usually the result of respiratory depression. However, a brief period of CNS stimulation and convulsions may be initially observed in the poisoned person, producing the same end result.

As mentioned previously, phenol is quite hazardous and toxic by all routes of exposure. Various estimated lethal oral doses have been reported, some suggesting that a fatal amount of phenol is as little as 1–2 grams. Regarding its inhalation hazard, it might be noted that on a ppm basis, the TLV of phenol is approximately that of cyanide (i.e., 5 ppm compared to 4.7 ppm). Its dermal hazard becomes quite evident from reports involving persons on whom phenol was accidentally splashed. In one case, a man accidentally sprayed about his thighs died within 10 minutes of the exposure (illustrating its rapid dermal absorption), even though he had attempted to remove it with water.

Other Phenolic Compounds

Substituted phenolic compounds like catechol (o-dihydroxybenzene), resorcinol (m-dihydroxybenzene), hydroquinone (p-dihydroxybenzene), and cresol (methylphenol) have toxicologic properties that are very similar to phenol. These substitutions tend to increase the toxicities of these compounds and all may be considered to be more toxic than phenol. Catechol may induce methemoglobinemia in addition to those toxicities already described for phenol, and hydroquinone is a more reactive compound than phenol and shows a significant increase in toxicity.

12.12 The Toxic Properties of Halogenated Alkanes

General Toxicology. The halogenated aliphatic compounds are excellent organic solvents of low flammability. They have excellent anesthetic properties, and several chemicals are the systemic anesthetic agents of choice in surgery today. Other toxic properties include sensitization of the heart to adrenaline, which can lead to cardiac arrhythmias and cardiac arrest, and the potential for liver and kidney damage. Systemic toxicity increases with increasing molecular size. The capability for CNS depression and liver/kidney injury increase with the degree of chlorination, and unsaturated compounds are usually even more potently toxic for these properties. Aromatic ring substitution greatly decreases the systemic responses. Halogenated compounds can be strong irritants. Brominated compounds are more toxic systemically and locally than chlorinated compounds, while fluorine replacement of the chlorine decreases the toxicity further. A drawback to the widespread use of halogenated alkanes is that many of these compounds may induce liver cancer in rodent bioassays, and the compounds are environmentally persistent.

Halogenated Methane Compounds

Methyl chloride is used as a refrigerant, aerosol propellant, solvent, and chemical intermediate. Chronic and subacute exposures predominantly affect the CNS, producing the following symptoms: ataxia, staggering gait, weakness, tremors, vertigo, speech difficulties, blurred vision, nausea, abdominal pain, and diarrhea. Acute toxicity generally mimics inebriation by alcohol. It is a weak to moderate irritant.

Methyl bromide is a strong irritant and this property may warn victims before they become heavily sedated. Serious skin burns and lung irritation occur from contact with its vapors. Human exposure can produce severe nervous system effects, and recovery from these may be slow or incomplete.

Methylene chloride is probably the least toxic of the four chlorinated methanes and causes a "drunken" state only at high vapor concentrations. It is only mildly irritating to the skin. Eye contact will be painful but not likely to cause serious injury. Adaptation to the odor occurs with repeated exposure and decreases the victim's ability to detect exposure.

A primary acute effect of chloroform is CNS depression when absorbed through the lung, the gastrointestinal tract, and skin. But acute exposure may also result in liver or kidney damage and cardiac sensitization. In man, 200–300 ppm appears to be the odor threshold; 1000 ppm causes fatigue and headache; 1500 ppm, dizziness and salivation within minutes; 4000 ppm, fainting and nausea; and 14,000–16,000 ppm, almost immediate narcosis.

Although carbon tetrachloride is an effective CNS depressant, the damage to the liver and/or kidneys presents the primary concern with this compound. Exposure may cause centrilobular necrosis in the liver and with chronic injury lead to cirrhosis similar to that seen after chronic ingestion of alcohol.

Vinyl Chloride

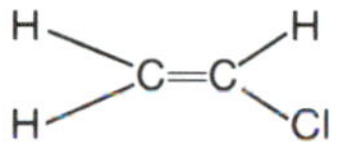

H H
C=C
H Cl

Vinyl chloride

Vinyl chloride is a skin irritant, and contact with the liquid may cause frostbite upon evaporation. The eyes may be immediately and severely irritated. Vinyl chloride depresses the central nervous system, causing symptoms that resemble mild alcoholic intoxication. Lightheadedness, some nausea, and dulling of visual and auditory responses may develop in acute exposures. Severe vinyl chloride exposure has been reported to result in death. Chronic exposure of workers who have had to enter reactor vessels and clean them by hand may have caused development of a triad complex combining arthro-osteolysis, Raynaud's phenomenon, and sclerodermatous skin changes. Chronic exposure may also damage the liver and liver cancer (angiosarcoma) is now a well-established risk for chronic exposures to vinyl chloride at the old TLV of several hundred ppm. Increased rates of cancer of the lung, lymphatic, and nervous systems have also been reported. Experimental evidence links vinyl chloride to tumor induction in a variety of organs, including liver, lung, brain, and kidney, and to nonmalignant alterations, such as fibrosis and connective tissue deterioration; the oncogenic and toxicologic effects of vinyl chloride appear to be multisystemic.

The mutagenic, carcinogenic, and reproductive hazard of vinyl chloride will be further discussed in Chapters 14, 15, and 16.

12.13 The Toxic Properties of Amines

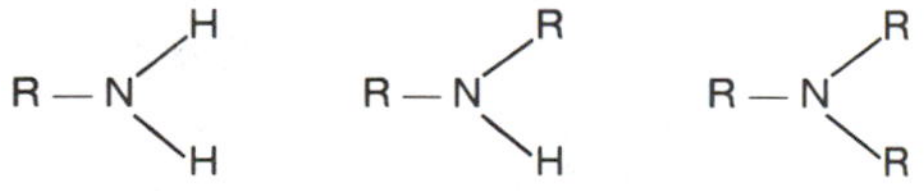

Primary	Secondary	Tertiary
R—N(H)(H)	R—N(R)(H)	R—N(R)(R)

Amine compounds

General Toxicology. Amine-substituted chemicals are among the most toxic solvent or organic chemicals easily and commonly produced. As stated in the introduction, they are strong irritants and therefore probably represent a greater handling hazard than any of the other chemical classes discussed in this chapter. Their strong irritant properties stem from the fact that the amine portion of the molecule is a very corrosive functional constituent. The skin has some resistance to changes in pH and can withstand chemical attack in the pH range of 1 to 10 for short periods without significant damage. However, if the base strength of the chemical is much above 10 or the acid has a pH lower than 1, significant skin injury may occur very quickly after initial contact with the chemical. As the following listing shows, the base strength of most of the simple amines is in the immediately injurious range of pH greater than 10.

Amine compound	Dose strength (pK)
Methylamine	10.6
Dimethylamine	10.6
Trimethylamine	10.7
Ethylamine	10.8
Diethylamine	11.0
Triethylamine	10.7
Propylamine	10.6
Butylamine	10.6
Allylamine	9.5
Cyclohexylamine	10.5

As this listing shows, the size of the chemical and the degree of substitution (i.e., primary, secondary, or tertiary) have little effect upon the corrosiveness of the amine group itself. Thus, while the irritant nature of the other functional groups (e.g., alcohols, ethers, carboxylic acids, etc.) is decreased as the size of the organic portion of the chemical increases, the irritation of the amines is not affected.

Another feature of amines that adds to their acute toxicity and handling hazards is that they are easily and well absorbed by all routes. Thus, they represent a high dermal hazard not only because of the injury they produce to the skin, but also because they tend to penetrate the skin so easily that the dose lethal or acutely toxic when absorbed via the skin tends to be equivalent to the ingested dose capable of producing the same toxicity. Therefore, contact with the skin should always be avoided. Here is an example of this problem:

Amine	Oral LD_{50}	Skin LD_{50}	Dermal effects
Methylamine	0.02 mg/kg	0.04 mg/kg	Necrosis
Ethylamine	0.4 mg/kg	0.4 mg/kg	Necrosis
Propylamine	0.4 mg/kg	0.4 mg/kg	Necrosis
Butylamine	0.5 mg/kg	0.5 mg/kg	Necrosis
Hexylamine	0.7 mg/kg	0.4 mg/kg	Slight necrosis

(Note that LD_{50} is the amount lethal to 50 percent of the animals tested. Necrosis means cell or tissue death.)

Because of their tissue-penetrative and tissue-corrosive characteristics, amines are toxic to all tissues in which they are absorbed in measurable amounts and adversely affect a number of organs. What organs will be most affected probably depends to a large extent upon the distribution of the chemical within the body. Some of the systemic effects observed for lethal exposures are edema and hemorrhage of the lungs, necrosis of the liver, necrosis and nephritis in the kidneys, and muscular degeneration of the heart.

Two other common, characteristic toxicities observed in amine compounds are methemoglobin formation in the red blood cells, discussed in Chapter 4, and sensitization to the chemical itself. Sensitization to amines probably happens because the amine group is fairly reactive. The chemicals may bind to cellular proteins to form haptens, or molecules that go unrecognized by the body's immune defenses. The body produces antibodies against these haptens, and upon repeated chemical exposure a severe allergic reaction ensues. During this allergic antibody-hapten response, the body releases histamine. Histamine then induces an arterial vasoconstriction, capillary dilation, an overall fall in blood pressure, an itching of the nerves, and a bronchoconstriction in the respiratory system. These effects explain many of the effects seen in a severe allergic response: labored, difficult breathing; fainting or possibly anaphylactic shock; and a reddening or irritationlike response where contact with the skin has occurred. Some of the amine compounds are potent sensitizing chemicals, and, again, exposure to the skin should be avoided.

A somewhat uncommon toxicity possessed by the alkyl amines is their ability to simulate the actions of epinephrine (adrenaline) within the body. Epinephrine is an important neurohormonal transmitter within the body. Chemically, it consists of a propyl amine side chain attached to a catechol ring group. Thus it is easy to understand why some of the alkyl amines may be able to mimic some of epinephrine's physiologic responses. Following is a list of some of the general conclusions from reports concerned with alkyl amine-induced sympathomimetic activity within the body:

- Activity increases with the size of the alkyl chain, up to six carbons.
- For alkyl amines larger than six carbons, a slowing of heart rate and a dilation of blood vessels is seen more often than the epinephrinelike response of increasing heart rate and blood vessel constriction.
- A branching of the alkyl carbon chain decreases the activity of the chemical.
- Pressor activity (constriction of blood vessels) follows primary amines, secondary amines, tertiary amines.
- Repeated exposure causes cardiac depression and vasodilation.
- Convulsions may cause mortality at high, acute exposures.

Lastly, human exposure to amine compounds should be avoided where possible, because many of these chemicals are carcinogens. In particular, the aromatic or diphenylamine compounds like benzidine, 2-naphthylamine, or 4-aminodiphenyl, which induce bladder tumors and are responsible for the so-called "aniline tumors" seen in dye-industry workers, should be avoided. Many other amine compounds are carcinogenic as well, and although most of these are aromatic amine chemicals, the alkyl amine group of nitrosamines are also carcinogenic. The nitrosamines, liver carcinogens, represent an interesting human risk, as any alkyl amine (including those generated during the digestion of food) absorbed orally may be converted to nitrosamines by the acid and nitrite found in the gut. However, animal studies have failed to demonstrate that this is a significant problem in humans.

In summarizing the general toxicology of amine compounds, it needs to be repeated that these are very toxic chemicals presenting extreme dermal and handling hazards. The chemicals are hazardous in many ways, and specific problems with any of the amines should be individually reviewed wherever they are used.

Putting this caution aside for a moment, for the amines it may generally be said that

- Irritation increases up to six carbons in size, then decreases with the loss in volatility and exposure probability.
- For the smaller alkyl amines, the irritant toxicity is essentially equivalent regardless of degree of substitution (i.e., $1° = 2° = 3°$).
- Unsaturated amines have greater systemic and dermal toxicity.
- The irritation produced by these chemicals is not usually affected by other functional groups.
- The salts of amine chemicals are usually weaker irritants.
- Additional functional groups may increase the sensitization potential of

the chemical.

- Sensitization is greater in aromatic amines.
- Aromatic rings do not decrease the various toxicities.
- Aromatic rings add methemoglobinemia and cancer hazards to the danger already posed by the chemical.

12.14 The Toxic Properties of Amides

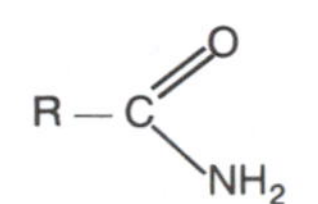

Amide compounds

General Toxicology. Above formamide the simple amides are solids, thereby reducing the exposure hazards. Amides are generally not hazardous by the usual routes of exposure and are rapidly hydrolyzed to the corresponding organic acid. Formamide can be absorbed through the skin without any real irritant effects, though some sensitization occurs in rare instances. The unsaturated amides are very toxic and have pronounced CNS, liver, and kidney toxicities. Alkyl substitutions increase skin absorption and therefore compound the toxicity, while aromatic ring substitutions minimize toxicity and may add analgesic and antipyretic properties to the amide, as in the case of acetanilide.

12.15 The Toxic Properties of Aliphatic Nitro Compounds

General Toxicology. These compounds have little industrial use and in general they are oily liquids of low solubility and volatility. Toxicologically, they can be classified as moderate irritants, because their anesthetic symptoms are mild. Unsaturated compounds may be absorbed via the skin to a significant extent, while the saturated chemicals are not. Halogenation produces definite skin irritation, some systemic absorption and, therefore, toxicity. Aromatic nitro compounds are a class of chemicals, several members of which may produce methemoglobin formation and/or sensitization to the compound.

Nitro derivatives of benzene and toluene also have prominent toxic effects other than sensitization, CNS depression, or methemoglobinemia. The trinitrotoluene or dinitrobenzene compounds are well absorbed by all routes of exposure. These compounds can uncouple oxidative phosphorylation, and liver injury is often seen along with the toxicities previously mentioned. Other problems that have been observed are dermatitis, anemia, heart irregularities, and peripheral neuritis. Persons deficient in glucose 6-phosphate dehydrogenase are sensitive to hemolytic anemia. Some of these compounds cause bladder tumors as well. Again, the specific toxicities of any individual member of this class of compounds should be reviewed individually if it is to be used.

12.16 The Toxic Properties of Nitriles (Alkyl Cyanides)

$$RC{\equiv}N$$

Nitrile compounds

General Toxicology. The nitriles are organic cyanide compounds. They are nonpolar and are readily absorbed by all routes. Because some of these compounds dissociate to produce cyanide, the adverse effects they produce are comparable to those of cyanide poisoning. However, many of these compounds do not readily release cyanide once absorbed and their toxicity cannot simply be characterized as that of cyanide itself (see Chapter 4). Therefore, few generalizations can be made concerning these compounds, and the toxicity of each compound should be reviewed individually from the literature. The unsaturated nitriles have comparable systemic toxicities, but are, of course, more irritating.

Summary

The groups of chemicals discussed in this chapter cover a very broad range. Most of them have organic solvent properties, but many of them are not considered industrial solvents in the traditional sense of being commonly used for that purpose. Because a detailed toxicologic discussion of each of these chemicals would be impractical, this chapter has summarized the chemical classes from the standpoint of the two toxicities most common to the organic nature of these chemicals:

- CNS depression.
- Irritation.

This material should provide practical insights into

- The reversible acute toxicities of organic compounds.
- The immediate handling of the exposure hazards these chemicals represent.

Additional information has summarized the chronic toxicities to be expected from certain chemicals within each chemical class. This information should encourage the reader to pursue a more specific and therefore definitive source of information for individual chemicals; the reader now should have a basic perspective on the toxicities of many organic compounds.

References and Suggested Reading

Browning, E. 1965. *Toxicity and Metabolism of Industrial Solvents*. New York: Elsevier.

Clayton, D., and Clayton, F. E. 1979. *Patty's Industrial Toxicology*. Volume II. 3rd Edition. New York: John Wiley & Sons.

Cornish, H. H. 1980. "Solvents and Vapors." *In* Doull, J., Klaassen, C., and Amdur, M. (Eds.) *Casarett and Doull's Toxicology: The Basic Science of Poisons*. New York: Macmillan Publishing Co.

Finkel, A. J. 1983. *Hamilton and Hardy's Industrial Toxicology*. Boston: John Wright PSG, Inc.

Formaldehyde and Other Aldehydes. 1981. Published by the National Research Council, National Academy of Sciences, Washington, D.C.

Goodman, L., and Gilman, A. 1975. *The Pharmacological Basis of Therapeutics*. 5th Edition. New York: Macmillan Publishing Company.

Gosselin, R. E., Hodge, H. C., Smith, R. P., and Gleason, M. 1976. *Clinical Toxicology of Commercial Products: Acute Poisoning*. Baltimore: The Williams and Wilkins Co.

Harger, R., and Forney, R. 1967. "Aliphatic Alcohols." *In* Stolman, A. (Ed.) *Progress in Chemical Toxicology*. Vol. 3. New York: Academic Press, Inc.

Hayes, W. J., Jr. (Ed.) 1975. *Essays in Toxicology*. Vol. 6. New York: Academic Press, Inc.

Proctor, N. H., and Hughes, J. P. (Eds.) 1978. *Chemical Hazards of the Workplace*. Philadelphia: J. B. Lippincott Co.

Sax, N. I. 1984. *Dangerous Properties of Industrial Materials*. 7th Edition. New York: Van Nostrand Reinhold Company.

Smyth, H. F., Jr. 1952. "Physiological Aspects of Glycols and Related Compounds." *In* Curme, G. O., Jr., and Johnston, F. (Eds.) *Glycols*. New York: Reinhold Publishing Co.

Toxic and Hazardous Industrial Chemicals Safety Manual for Handling and Disposal, with Toxicity and Hazard Data. 1979. Published by the International Technical Information Institute. Tokyo, Japan.

Truhaut, R. 1971. "Benzene." *In* Jenks, W. (Director-General). *Occupational Health and Safety*, Vol. 1. Geneva: International Labour Office.

von Oettingen, W. F. 1964. *The Halogenated Hydrocarbons of Industrial Toxicological Importance*. New York: Elsevier.

Chapter 13

Occupational Epidemiology

Woodhall Stopford

Introduction

Occupational epidemiology is the study of the distribution of a disease or physiologic condition among a working population and the factors that influence this distribution. This discipline attempts to identify relationships of diseases to occupational exposures when these diseases occur infrequently or, as is more important in many occupational diseases of current interest, when their onset is delayed, as it is in various forms of occupationally associated cancer. The purpose of this chapter is to explore occupational epidemiology with particular emphasis on the distribution of diseases or physiologic conditions associated with toxic substances. This chapter includes a discussion of the following:

- Exposure to a causative agent and its manifestations.
- Measurement of disease frequency.
- Comparison of rates of disease.
- Comparability of populations.
- Bias.
- Control of confounding factors.
- Standardization of data.
- Causality.
- Epidemiologic surveillance.

13.1 Exposure and Manifestations

Exposure

Two definitions are important in epidemiologic studies: that of exposure, and that of manifestations. Exposure to a causative agent can be defined in a number of ways. It can be defined in terms of a worker's average exposure, based either on industrial hygiene samples taken in a designated place (i.e. area samples) or on individual monitoring. It can be based on biological measurements that correlate with exposure, such as the urine azide iodide test for workers exposed to carbon disulfide or the test for phenols in the urine for workers exposed to benzene.

Exposure can also be defined more broadly by a knowledge of a person's job and correlation of job descriptions with various potential types and levels of exposure. This latter methodology has been used in several studies.

In an industry where workers are mobile between various jobs, an exposure can be defined through examining each job description for the various job titles in the industry and then establishing an exposure record by correlating job title with known or hypothesized exposures. This process yields person-years of exposure to each substance encountered in a particular job. This information can then be used to establish a picture of a worker's exposure over his entire life or for any particular segment of his life and can be used to provide the exposure base for an epidemiologic study.

Manifestations

The other parameter one needs to define is the outcome or manifestation of the disease. For example, the outcome may be death, data being obtained from records of sickness and benefit programs, from death certificates, or from payments by the Social Security Administration. The outcome could be morbidity, which can be defined by hospitalizations for specific diagnoses (obtainable through sickness or benefit plans) or through periodic screening programs, annual examinations, or specific screening programs for potential exposure.

One can define the outcome as a change in some physiologic variable. Examples of such manifestations include changes of enzyme levels or a change in the level of prophyrin in the blood as is caused by lead exposure, developments that would not normally be correlated with morbidity or disease.

13.2 Measurement of Disease Frequency

The Incidence Rate and the Prevalence Rate

The primary means of determining whether or not a population with a defined exposure or outcome is of epidemiologic concern is to compare the population of interest with a suitable control population. The control population might be one without the exposure or, if one is interested in the incidence of disease, without the disease. Before one can compare populations, however, one must precisely define what is going to be compared. The basic factor that is going to be compared is a rate or a ratio: this is an incidence or prevalence rate.

Incidence rates represent the appearance of a disease in the population over a period of time. For example, a five-year study may show that a specific population develops lung cancer at an incidence rate of 15 cases per 100,000 members of the population per year.

A prevalence rate eliminates the time factor. For example, the prevalence rate of lung cancer in a population is the ratio of the number of persons in a defined population who have lung cancer at any instant to the number in the population base.

The incidence rate is a measure of the risk of getting the disease. The prevalence rate does not really measure risk because a number of factors can affect this rate. For example, if the incidence of a disease increases, the prevalence of that disease will also increase. If the duration of the disease increases, perhaps because better medical care keeps victims alive longer and so decreases mortality rates, more people among the population will have the disease at any specific time, and the prevalence rate will increase. If no cure is known for the disease and it becomes prolonged, the prevalence rate will also increase.

Attack Rate

Another measure is the attack rate, which combines the data of the incidence and prevalence rates. The attack rate includes the units of prevalance—that is, the number of cases of disease or the outcome per unit of population—but it also includes a time factor. Suppose one is monitoring a work force that is divided into fifteen departments. In order to determine the attack rate per year for lung cancer, the population is analyzed at the end of each year by examining the sickness records for the preceding year. Although this measure involves the prevalence of lung cancer per unit population—the number of people in each department—a time factor is also involved. In analyzing an entire epidemic of specific duration, whether it lasts for five days or five years, one can determine an attack rate.

13.3 Rate Comparison

Relative Risk

Once rates of disease or outcome have been defined within the control population and the population of interest, the rates can be compared. Rates of disease can be compared in several ways.

Relative risk is a comparison of two risks: an example would be the rate of disease in the exposed population as compared to the rate of disease in the

Table 13-1. Examples of Rate Comparison in Occupational Epidemiology.

Rate comparison:

A standard 2 × 2 table is used,

	D	$\overline{D}$	
E	a	b	
$\overline{E}$	c	d	
			N

where: E is the exposed population
$\overline{E}$ is the nonexposed population
D is the number with disease
$\overline{D}$ is the number without disease
N is the total population.

Relative risk:

$$\frac{a/(a+b)}{c/(c+d)}$$

which is the ratio of the rate of disease in an exposed group to the rate in an unexposed group.

Attributable risk:

$$\frac{a}{a+b} - \frac{c}{c+d}$$

used for absolute comparison of rates.

Odds ratio:

$$\frac{ad}{bc}$$

used for approximation of the relative risk in case-control studies.

unexposed population. In Table 13-1 the entire unexposed population—both those with and without the disease—is $c + d$, and the entire exposed population is $a + b$. Those who have been exposed and have the disease fall into block a, so the incidence of disease in the exposed population is $a/(a + b)$. Similarly, the incidence among the nonexposed population is $c/(c + d)$. The relative risk of disease, given exposure, is then

$$\frac{a/(a+b)}{c/(c+d)}.$$

A relative risk of 1.0 would be interpreted as meaning there was no excessive risk of a disease in an exposed population as compared to the risk in an unexposed population.

The Odds Ratio

The odds ratio was developed primarily for case-control studies, and it is an approximation of relative risk, as seen in Table 13-1. It is the cross-product ratio of the standard 2 × 2 table.

As indices, the relative risk and the odds ratio do not provide the data necessary for broad policy changes. However, suppose a firm wishes to know whether or not for health reasons it should eliminate all smokers from its work force. This is a major policy decision. Studies of comparable groups may show a lung cancer mortality of 1/100,000 and 2/100,000 for nonsmokers and smokers respectively, which might be a statistically significant relative risk; but the difference between the rates is not sufficient to support a policy decision that eliminates jobs. The firm would do better to ascertain an attributable risk.

Attributable Risk

To understand the relative impact of a problem, one should consider the criterion of attributable risk: it is the actual rate of disease among exposed persons minus the rate of disease among nonexposed persons. For instance, in a given industrial population with radon daughter exposure, suppose the relative risk of death from lung cancer is approximately the same for smokers and nonsmokers. The management of a factory wishes to learn if smoking is increasing the risk of lung cancer among workers and discovers an excessive mortality rate from lung cancer of 1/100 for radon-exposed nonsmokers, and 10/100 for radon-exposed smokers. The comparison of the risk of death from lung cancer attributable to radon daughter exposure in the two populations provides information that can be used as the basis for policy decisions concerning employment of smokers in areas where they will be exposed to radon.

13.4 Comparability of Populations

Study Design

There are many types of studies that can be used to examine two populations and develop a comparison between them. One that is not available for industry is the experimental study: a study in which, at a specific time, the exposed population, the controlled population, and exposure are assigned before the experiment commences. In industry, employees may or may not already be part of an exposed population. Studies available for comparing populations that do or do not possess a certain characteristic include longitudinal studies and cross-sectional studies. Within the longitudinal studies are two types of cohort studies. A *cohort study* defines a specific population, based on whether or not there is exposure, often by time of entry into the work force, and follows it over time. When a current population is defined and followed into the future, the study is called a *prospective cohort study*. When the population is defined as employees admitted to the work force at some time in the past, and that population is followed to the present, the study is called a *retrospective cohort study*.

In a prospective cohort study, the study begins at the time when the exposed population is defined, and the incidence of disease or death is observed after that time. In a retrospective cohort study, the population is defined after the disease has become apparent, and the analysis is made retrospectively by defining exposure or nonexposure at some time in the past.

In a case-control study, the population is selected after disease has become apparent. A population with a disease or an outcome of interest is compared with a population lacking that characteristic, which is then examined for the presence or absence of exposure.

A cross-sectional study measures both exposure and the occurrence of disease at a specific time. A prevalence rate is used to describe the results of the cross-sectional study. A cross-sectional study can define whether or not a problem exists, but cannot provide any measure of risk.

13.5 Bias

Bias is probably the most important concept in epidemiologic studies. Bias is most likely to be a problem when one tries to compare two essentially noncomparable populations, or two populations that differ in terms of age, sex, smoking history, life style, or other significant factors. To obtain a meaningful measurement of comparison, one needs to eliminate bias that would make the comparison yield the end result one is looking for regardless of whether the result is warranted.

Selection Bias and Observational Bias

Selection bias consists of designing a study based on exposure-versus-nonexposure (or disease-versus-nondisease) and failing to separate the opposite populations adequately. In a retrospective study where there are only historical measures of exposure, individuals can easily be misclassified as to whether they fall into an exposed or nonexposed population.

Observational bias occurs when data are either collected or analyzed in a nonequivalent manner, thereby distorting some of the data. To prevent observational bias when analyzing the data, the analyst should take care to remain unknowledgeable about which group any individual subject belongs to;

a study in which an analyst does not know the backgrounds of the subjects is called a blind study. If the study is not blind, the analyst may unconsciously judge people in one group differently from people in another, especially if there is real interest in obtaining a specific result from the study.

Confounding Bias

The most important type of bias in epidemiologic studies is called confounding bias. This type of bias occurs when an unstudied factor capable of causing the disease of interest is present in both the exposed and nonexposed populations.

For example, suppose a study is designed to explore the relationship between exposure to asbestos and lung cancer, and the study fails to take smoking into account. Very little basis will exist for meaningful comparison of the asbestos-exposed and the asbestos-nonexposed populations because of the profound influence of smoking on lung cancer rates in either group.

13.6 Control of Confounding Bias

Confounding bias can be controlled in several ways. The easiest way is to collect data and then to analyze them by stratification. In stratification one group (smoking or nonsmoking, say) is analyzed in the absence of the other.

Matched Case-Control Studies

One can also control confounding through matched case-control studies. For example, the hypothetical study of lung cancer mentioned earlier could be designed so as to include only nonsmokers or only smokers. The confounding factor of smoking would then be eliminated from the study. Provisions for controlling a confounding characteristic should be made early in a study. Confounding variables can be identified by comparing outcome ratios in subgroups of the exposed population with or without the potential confounding variable. In a comparison of asbestos-exposed smokers versus nonsmokers, is there more lung cancer in a smoking population? If there is, then a control group needs to be designed for this factor. Similarly, is there more lung cancer by ethnicity, by age, by sex, or by past exposure? If no differences exist, then there is no need to control for the potentially confounding variable, and the study can be simplified.

However, excessive controlling for confounding in a study can also cause problems if one of the factors controlled for is in the chain of causality between exposure and disease. For instance, in a study looking at the relationship between salt intake and strokes, if one controls for hypertension between the exposed and control populations, the comparison will be invalid—hypertension is in the causal chain between excessive salt intake and the occurrence of cerebrovascular accidents.

Indirect Standardization

One method to control for confounding is by standardization of rates or ratios during the data analysis phase of a study. With indirect standardization, the study population is stratified for the confounding variable of interest (e.g., age). A standard or nonexposed population is then used to determine strata-specific rates for the disease of interest. A standardized risk ratio (SRR) can then be determined for the disease of interest as follows:

$$\text{SRR} = \frac{\Sigma\ \text{Observed Cases}}{\Sigma\ \text{Reported Cases}}.$$

The number of expected cases for each strata is calculated by multiplying the strata specific rate for the standard population by the total number of individuals in that strata.

Standardized Mortality Rate

A standardized mortality ratio (SMR) is a SRR used commonly in occupational medicine where the mortality rate for some occupational group is compared to that of the general population. In this type of analysis, age and time are usually controlled for by determining the number of person-years of exposure in each age and time strata of the study population. The strata-specific rate for the general population (usually expressed as death rate per 1000) is multiplied by the person-years of exposure in each strata to derive the expected number of deaths. The SMR is then expressed as follows:

$$\text{SMR} = \frac{\Sigma\ \text{Observed Deaths}}{\Sigma\ \text{Expected Deaths}},$$

the ratio being multiplied by 100 by convention.

Proportionate Mortality Rate

Proportionate mortality ratios (PMRs) can be used to control for confounding in a study of a population where data to calculate person-years of exposure is not available. In this analysis, the expected number of deaths from a specific cause (e.g., cancer) are determined by stratifying overall deaths in the study population by age and time and multiplying each strata by the age and time-specific rate for the standard population. The PMR is then expressed as follows:

$$\text{PMR} = \frac{\Sigma\ \text{Observed Deaths}}{\Sigma\ \text{Expected Deaths}} \times 100.$$

For rare causes of death, the PMR is a reasonable approximation of the SMR.

13.7 Causality

The most basic purpose of epidemiological studies is to identify cause-and-effect relationships. Studies often attempt to determine the strength of a potential causal relationship. How strong is the causal association between exposure and disease or death? A number of factors are involved in determining the extent of this association.

One factor is, how many studies support the association. If ten studies are conducted and only one associates lung cancer and exposure to asbestos in nonsmokers, the causal relationship is not very strong. However, if a majority of studies demonstrate the relationship between asbestos and lung cancer in smokers, that will be a convincing relationship and the causal association would be strengthened.

Specificity

A study may be undertaken to determine whether or not a problem is associated with a specific exposure. For example, is the occurrence of mesothelioma, a type of lung cancer, specifically associated with one particular exposure more than any other? As it happens, mesothelioma and exposure to asbestos are fairly specifically associated. The association between angiosarcoma of the liver and exposure to vinyl chloride or arsenic is also relatively specific. With a high specificity between exposure and outcome one will find that measures of relative risk, such as SMR, will be high.

Dose-Response Relationship

Another criterion that would strengthen the causal relationship is the dose-response relationship, discussed in Chapter 2. Many studies classify various exposure groups as low, intermediate, or high. In these studies, employee populations are categorized in ascending levels of exposure, and each group is evaluated according to the incidence of disease. If the incidence of the disease increases as the exposure increases, and these results are reproducible, a strong causal relationship can be argued.

Coherence

Coherence is another significant criterion: is there some biological basis for what is seen in a relationship? For instance, can the causal relationship between exposure and outcome hypothesized after an epidemiologic study be duplicated under experimental conditions? Does the investigator find an excessive occurrence of leukemia when exposing rats to benzene? The fact that this finding cannot be reproduced detracts from the causal connection between exposure to benzene and the occurrence of leukemia.

Temporal Relationship

Temporal relationships are important. Does the exposure or cause precede the effect? In cross-sectional studies that do not measure whether exposure or outcome comes first, there can be no measure of risk and, therefore, no assessment of causality. Selection bias, as can be seen in retrospective cohort studies, can distort apparent temporal relationships.

Statistical Significance

A final element in establishing causality is statistical significance. A test of the statistical significance of a measure of relative risk is a test of the stability of this measure: is the study population large enough to limit possible values of relative risk to a range that can occur by chance alone? A test of statistical significance does not strengthen the causal relationship: it does not reflect the magnitude of relative risk, and measurement of relative risk is important in establishing a causal relationship.

13.8 Epidemiologic Surveillance

In a number of industries little is known about the potential risk to workers from exposures within those industries. Epidemiologic methods can be used to follow an industrial population prospectively in order to identify potentially hazardous exposures.

Exposure Definition

In designing such an epidemiologic surveillance system, one can use an exposure-based system or a disease-based system. Exposure-based systems are most often found in industry. With industrial populations there essentially have to be three measurements. One is exposure. The simplest and least costly method of defining exposure is from work or employment records—for example, how much time did a certain person spend in a particular department? The exposure can also be defined according to specific measurements. Area measurements or personal measurements can be made in the department being examined. Also, exposure can be defined by using biological monitoring, such as the urine azide iodide test for carbon disulfide exposure.

Outcome Measurement

The second measurement needed is the disease state. This can be acquired from insurance data. Who is becoming ill, what are the accident rates by department, and who is cashing an insurance policy death benefit? One can also measure the incidence of ill health by means of a baseline examination and periodic repetitions of the examination that can be compared to the baseline. This technique has become so sophisticated that some companies actually repeat the baseline examination three times before they allow a person to become exposed to a dangerous substance. This gives a more reliable data base for comparison. Often, however, it is better to utilize group norms for data and to compare people who are or are not normal, than to compare people to their baseline values.

Confounding

The third measurement that has to be made in designing a surveillance system is an assessment of confounding. This is primarily done at the time of the preemployment examination. At that time it is possible to identify a worker's ethnicity, age, sex, history of past exposure, and such social habits as drinking or smoking. This information can be updated periodically, perhaps every five or ten years—it is not the type of data that must be collected annually.

Data Analysis

The analysis of these data is relatively simple. Surveillance studies are primarily concerned with trends. The primary measurements used to establish trends are attack rates (the incidence of a disease per year in a specific department) or age-standardized incidence rates (since employment began, as opposed to annually). If a trend begins to emerge—for example, a department that uses benzene may begin to see cases of leukemia—the health and safety specialist might want to do more complex studies, such as a cohort study or a case-control study. Such a study would better define the risk of an outcome for a particular exposure.

Registry

One can also conduct a surveillance program by using registries. For example, there is a beryllium registry. This is an exposure-specific registry. There are also outcome-specific registries such as for hepatomas, glaucoma, and brain cancer. One can use a registry to try to correlate the disease or the exposure in the registry with some exposure or effect. For the working epidemiologist, however, data obtained directly from an industry are likely to be more meaningful than information obtained by working backwards from a registry.

Summary

Epidemiologic techniques are necessary to define relationships between exposure and effect where the adverse outcome is rare or occurs after a prolonged latency. This chapter has examined:

- Means of comparing outcomes in populations that have or have not been exposed to a condition of interest.
- The importance of bias in study design and data analysis.
- The design of an epidemiologic surveillance program.

There are numerous potential problems in drawing conclusions from epidemiologic information that has been improperly gathered or analyzed. However, these problems can be minimized with careful attention to study parameters, including:

- Sample size.
- Sample representation.
- Appropriateness of classification.
- Test validity.

A properly designed study can be an invaluable tool in the identification and control of physiologic impairments related to toxic substances encountered in the workplace environment.

References and Suggested Reading

Beyer, W. H. (Ed.) 1981. *CRC Handbook of Tables for Probability and Statistics*. 2nd Edition. Boca Raton (Florida): CRC Press.

Chiazze, L., Lundin, F. E., and Watkins, D. (Eds.) 1983. *Methods and Issues in Occupational and Environmental Epidemiology*. Ann Arbor (Michigan): Ann Arbor Science Publishers.

Friedman, G. D. 1974. *Primer of Epidemiology*. New York: McGraw-Hill.

Hill, A. B. 1971. *Principles of Medical Statistics*. New York: Oxford University Press.

Kleinbaum, D., Kupper, L., and Morgenstern, H. 1982. *Epidemiologic Research: Principles and Quantitative Methods*. Belmont (California): Lifetime Learning Publications.

Leidel, N. A., Busch, B. A., and Lynch, J. R. 1977. *Occupational Exposure Sampling Strategy Manual*. Washington, D.C.: National Institute for Occupational Safety and Health, U.S. Department of Health, Education, and Welfare. NIOSH Publication 77-173.

Levy, P., and Lemeshow, S. 1982. *Sampling for Health Professionals*. Belmont (California): Lifetime Learning Publications.

Lilienfeld, A. M., and Lilienfeld, D. E. 1980. *Foundations of Epidemiology*. 2nd Edition. New York: Oxford University Press.

MacMahon, B., and Pugh, T. F. 1970. *Epidemiology: Principles and Methods*. Boston: Little, Brown and Co.

Mausner, J. S., and Bahn, A. K. 1974. *Epidemiology, An Introductory Text*. Philadelphia: W. B. Saunders Co.

Monson, R. R. 1981. *Occupational Epidemiology*. Boca Raton (Florida): CRC Press.

Chapter 14

Mutagenesis

Christopher M. Teaf

Introduction

Modification of the human genetic material by chemical agents or physical agents (e.g., radiation) represents one of the most serious potential consequences of exposure to toxicants in the occupational environment. The discipline of genetic toxicology combines study of the physically or chemically induced changes in the hereditary process with prediction and prevention of these effects.

This chapter will discuss:

- Possible types of genetic alteration.
- Methods for determination of genetic change.
- The significance of test results from animals and humans in identification of potential mutagens.
- The proposed relationship between mutagenesis and carcinogenesis.

14.1 Induction of Mutations and Their Possible Consequences

A mutation may be defined as a transmissible change in the genetic material of an organism. The actual heritable change in the genetic constitution of an individual may be referred to as a genotypic change because the genetic material has been altered. While all mutational changes result in alterations of the genetic material in the parent cells, not all are immediately expressed in daughter cells as functional (phenotypic) changes. This is discussed in greater detail in the next section.

Consequences of Mutagenic Events

Radiation represents the best-studied example of a dose-dependent mutagen, and many of the characteristics of radiation-induced mutation are common to chemically-induced mutation. This is particularly true for molecules known as radicals, which are formed in both radiation and chemical toxic events. Radicals contain unpaired electrons and are therefore strongly electrophilic, a feature that is well correlated with both mutagenic and carcinogenic potency. Such reactive molecules are probably responsible for the alterations of nucleic acid sequences observed in genotoxic processes.

Several undesirable functional conditions or disease states have been linked to heritable changes. In the case of some cancers, a change in the genotype of a cell results in a change in phenotype that is grossly defined by rapid cellular division and a reversion of the cell to a less specialized type (dedifferentiation). The subsequent generations eventually form a growing tumor mass within the affected tissue. This simplified sequence is known as the *somatic cell mutation theory of cancer*. While not all chemically-induced cancers can be explained by this hypothesis, the general applicability of somatic cell mutation theory is supported by the facts that:

- Many demonstrated chemical mutagens have been found to be carcinogenic in animals.
- Chemical complexes in which carcinogens are bound to DNA have been isolated from cells treated with carcinogens.
- Heritable defects in DNA-repair capability such as occur in the sunlight-induced disease *xeroderma pigmentosum* predispose the affected individual to cancer development.
- Tumor cells can be initiated by chemical carcinogens but remain dormant for many cell generations—a finding consistent with a permanent change in DNA.
- Most cancers display chromosomal abnormalities.
- All cancers display altered gene expression (i.e., a phenotypic change).

Thus, for a significant proportion of human cancers it is thought that the genotypic cell change is initiated as a mutational event.

While mutational changes in somatic cells are of grave concern because such consequences as cancer can terminate the life of the organism in which they are induced, mutational changes in germ cells (sperm or ovum) may have even more serious consequences because of the potential for affecting subsequent human generations. If a *lethal* and dominant mutation occurs in a germinal cell the result is a nonviable offspring. On the other hand, a dominant but *viable* mutation can be transmitted to the next generation, and it need only be present in single form (heterozygous) to be expressed in the phenotype of the individual. Once again, if the phenotypic change is severely debilitating or

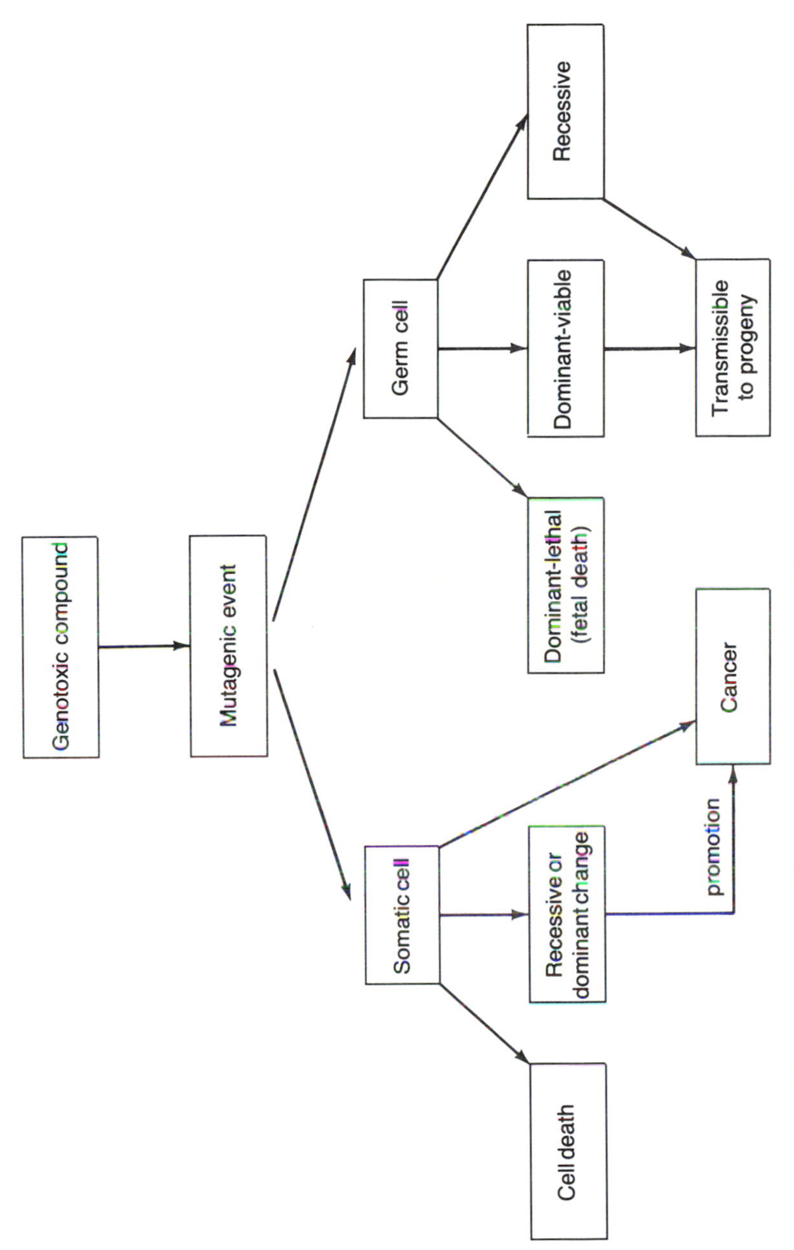

Figure 14-1. Possible consequences of a mutagenic event in somatic and germinal cells.

undesirable, there is little chance of it becoming established in the human gene pool. In contrast, individuals heterozygous for recessive genes represent unaffected carriers that are essentially impossible to detect; hence, recessive mutations pose the greatest risk and are of the greatest concern. These mutations can cause effects ranging from minor to lethal whenever two heterozygous carriers produce an offspring that is homozygous (genes present in both copies) for the recessive trait. Therefore, recessive mutations in germ cells may, because they go undetected, have the greatest potential impact on the human gene pool. Figure 14-1 shows the potential consequence of various mutagenic events.

Workplace Mutagens versus Spontaneous Mutations and Naturally Occurring Mutagens

When considering the effects of chemicals upon humans, it must be recognized that both physical and chemical mutagens occur naturally in the environment. Radiation is a ubiquitous feature of our daily lives, sunlight being the obvious example. Incomplete combustion produces mutagens such as benzo[*a*]pyrene, and some mutagens occur naturally in the diet, or may be formed during normal cooking or food processing. In addition, drinking water and swimming pool water have been shown to contain potential mutagens formed during chlorination procedures. Thus, the mutational events that continue to contribute to the human evolutionary process should be viewed as a combination of the normal background incidence of spontaneous mutations occurring during cellular division coupled with exposure to naturally occurring chemical and physical mutagens.

Mutagenic chemicals in the workplace or introduced into the environment via industrial operations represent another potential contribution to the genetic burden. It has been estimated that 60,000 synthetic chemicals are currently in use and that this number is increasing at the rate of 1000–2000 per year (ICPEMC, 1983). Only a very small fraction of these have been confirmed as human carcinogens and no compound has been unequivocally shown to be mutagenic in humans. However, animal and bacterial tests have demonstrated the mutagenic potential of many occupational and environmental compounds, and prudence dictates that we err on the side of caution in limiting or eliminating human exposure to these compounds.

14.2 Genetic Fundamentals

Transcription and Translation

The deoxyribonucleic acid (DNA) molecule is the structural unit on which heredity and genetics are based. Subunits of the DNA molecule are grouped into genes that contain information necessary to produce a cellular product.

This product is nearly always a polypeptide or protein, which may have a structural, enzymatic, or regulatory function in the organism. Figure 14-2 illustrates how the sequence of messages on the DNA molecule is transcribed into the RNA (ribonucleic acid) molecule and ultimately translated into the polypeptide or protein. The sequence of base pairs in the DNA molecule specifies the appropriate complementary ("mirror-image") sequence in formation of the messenger RNA (mRNA). Transfer RNAs (tRNA), each specific to a single amino acid, are matched to the appropriate segment of the mRNA. When the amino acids are released from the tRNAs and are linked in a continuous string, the polypeptide (protein) chain is formed.

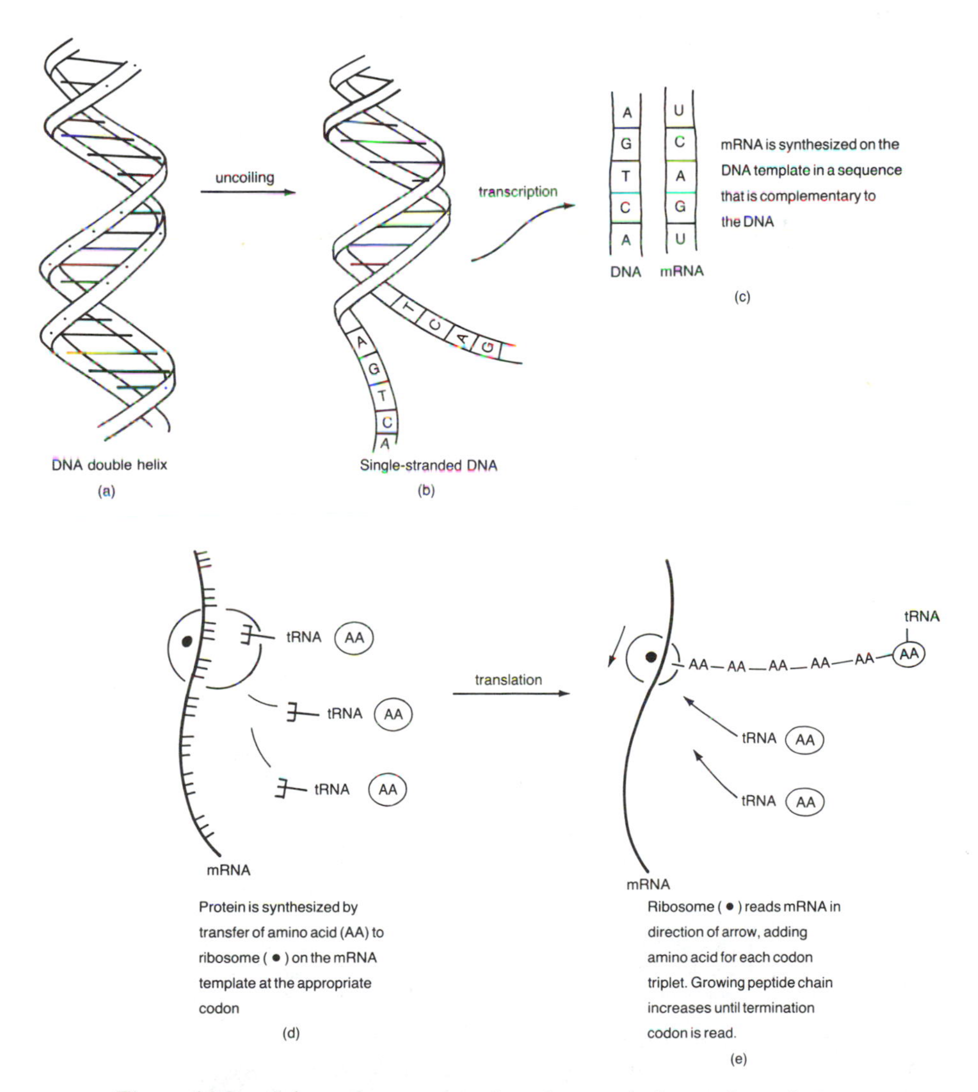

Figure 14-2. Schematic representation of transcription and translation.

Recognition of the mRNA regions by the tRNA–amino acid complex is accomplished by three-base (triplet) codons (mRNA) and complementary anticodons (tRNA). The critical features of this coding system are that it is simultaneously "unambiguous" and "degenerate." In other words, no triplet codon may call for more than a single specific tRNA–amino acid complex (unambiguous), but several triplets may call for the same tRNA–amino acid (degenerate). This results from the fact that four nucleotides form DNA and RNA (DNA is made up of adenine, cytosine, guanine, and thymine, abbreviated as A, C, G, and T, and RNA is made up of A, C, G, and uracil, abbreviated as U), and the nucleotides in RNA may be combined in triplet form in 64 different ways ($4 \times 4 \times 4$). The 20 amino acids and several initiator/terminator codes represent less than half of the available possibilities, leaving well over 30 codons unaccounted for. The biological significance of this degeneracy may lie in the fact that such a degeneracy will effectively minimize the influence of minor mutations (single deletions or additions) because codons differing only in minor aspects may still code for the same protein. The significance of having an unambiguous code is clear; the formation of proteins must be reproducible and exact. Table 14-1 depicts the amino acids coded for by the various triplet codons of DNA, as well as the initiation and termination signal triplets.

Table 14-1. Correspondence of the Genetic Code with the Appropriate Amino Acids (Unambiguity and Degeneracy Should Be Noted).

First position in triplet	Second position in triplet				Third position in triplet
	U	C	A	G	
A	Isoleucine	Threonine	Asparagine	Serine	U
	Isoleucine	Threonine	Asparagine	Serine	C
	Isoleucine	Threonine	Lysine	Arginine	A
	*Methionine	Threonine	Lysine	Arginine	G
C	Leucine	Proline	Histidine	Arginine	U
	Leucine	Proline	Histidine	Arginine	C
	Leucine	Proline	Glutamate	Arginine	A
	Leucine	Proline	Glutamate	Arginine	G
G	Valine	Alanine	Aspartate	Glycine	U
	Valine	Alanine	Aspartate	Glycine	C
	Valine	Alanine	Glutamate	Glycine	A
	Valine	Alanine	Glutamate	Glycine	G
U	Phenylalanine	Serine	Tyrosine	Cysteine	U
	Phenylalanine	Serine	Tyrosine	Cysteine	C
	Leucine	Serine	STOP	STOP	A
	Leucine	Serine	STOP	Tryptophan	G

*The sequence AUG, in addition to coding for methionine, is part of the initiator sequence that starts the translation process by which mRNA is formed from the DNA template.

The process of mutagenesis results from an alteration in the DNA sequence; if it is not too radical a rearrangement, this change is normally faithfully transmitted through the mRNA to protein synthesis, which results in a gene product that cannot perform the normal function for which it was designed. Such changes may be correlated with carcinogenesis, fetal death, fetal malformation, or biochemical dysfunction depending upon which cell type has been affected.

The initiation and termination of the transcription of DNA is regulated by a separate set of regulatory genes. Most regulatory genes respond to chemical cues, so that only those genes needed at a given time are expressed. The remaining genes are in an inactive state. The processes of gene activation and inactivation are believed to be critical to cellular differentiation, and interruption of these processes may result in expression of abnormal conditions such as tumors.

Chromosome Structure and Function

DNA in mammalian species including man is packaged together with specialized proteins (predominantly histones) into units termed chromosomes, which are found in the nucleus of the cell. The proteins are thought to "cover" certain segments of the DNA and act as inhibitors of expression for some regions. Chromosomes may be present singly (haploid), as in germ cells (sperm or ovum), or in pairs (diploid), as in somatic cells or in fertilized ova. In haploid cells, all functional genes present in the cell can be expressed. In diploid cells, one allele may be dominant over the other and in this case only the dominant gene of each functional pair is expressed. The unexpressed allele is called recessive, and recessive genes are only expressed when *both* copies of the recessive type are present. Some cell types in mammals are found in forms other than diploid. Functionally normal liver cells, for example, are occasionally found to be tetraploid (two chromosome pairs instead of one pair).

Features and terminology important to the study of chromosomes include:

- Karyotype: the array of chromosomes, typically taken at the point in the cell cycle known as metaphase, which is unique to a species and often forms the basis for cellular taxonomy; may be used to detect damage from physical or chemical agents.
- Centromere: the primary constriction, which represents the site of attachment of the spindle fiber during cell division; useful in identifying specific chromosomes, as its location is relatively consistent.
- Nucleolar organizing region: the secondary constriction, which represents the site of synthesis of RNA, which is used in ribosomes for protein synthesis.
- Satellite: the segment terminal to the nucleolar organizing region; useful in specific chromosome identification.

- Heterochromatin: the tight-coiling region of DNA, with low content of functional DNA; considered relatively inactive.
- Euchromatin: the loose-coiling region of DNA, with high content of functional DNA; considered the primary site of activity.

Mitosis

The actual process by which a cell divides into two daughter cells is called mitosis. The first stage of mitosis is called *prophase*, during which the spindle is formed and the chromatin material (DNA and protein) of the nucleus becomes shortened into well-defined chromosomes. During *metaphase*, the centriole pairs are pushed far apart by the growing spindle, and the chromosomes are pulled tightly by the attached microtubules to the very center of the cell, lining up in the equatorial plane of the mitotic spindle. With still further growth of the spindle, the chromatids in each pair of chromosomes are broken apart, a stage of mitosis called *anaphase*. All 46 pairs (in humans) of chromatids are thus separated, forming pairs of daughter chromosomes that are pulled toward one mitotic spindle or the other. In *telophase*, the mitotic spindle grows longer, completing the separation of daughter chromosomes. A new nuclear membrane is formed and shortly thereafter the cell constricts at the midpoint between the two nuclei, forming two new cells.

14.3 Alteration of DNA and Chromosome Structure

Basic Types of Genetic Alteration

A transmissible change in the linear sequence of DNA can result from any one of three basic events: infidelity in DNA replication, point mutation, or chromosomal aberration (Figure 14-3). The first—infidelity or inexact copying of a DNA strand during normal cellular replication—may result from inaccurate initiation of replication, failure of the transcription enzymes to accurately "read" the DNA, or interruption of the transcription process by agents that interpose (intercalate) themselves within the DNA molecule or between the DNA and an enzyme.

Point mutations may be subdivided into base-pair changes and frameshift mutations (Figure 14-4). The former result from transition or transversion of DNA base pairs so that the number of bases is unchanged but the sequence is altered. Because the genetic code is "degenerate" this may or may not result in an altered product after transcription and translation. A frameshift mutation, however, results from insertion or deletion of one or more bases from the

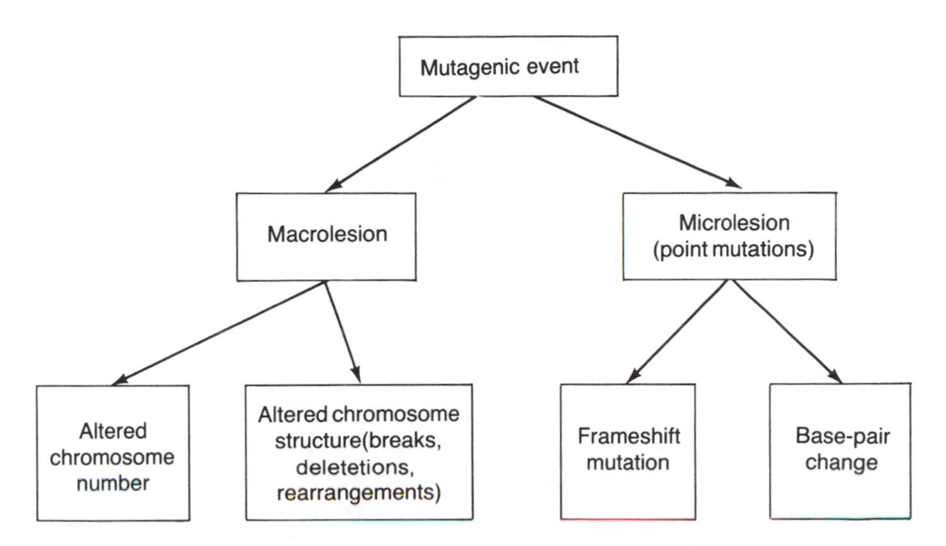

Figure 14-3. Types of mutagenic changes. (Adapted from Brusick, *Principles of General Toxicology*. New York: Plenum Press. 1980.)

linear sequence of the DNA. This causes the transcription process to be displaced by the corresponding number of bases and virtually assures an altered genetic product.

Chromosomal aberrations may be present as chromatid gaps or breaks, symmetrical exchange (exchange of corresponding segments between arms of a chromosome), or asymmetrical interchange between chromosomes. Point mutations can result in altered products of gene expression, but chromosomal aberration or alteration in chromosome numbers passed on through germ cells can have disastrous consequences, including embryonic death, teratogenesis, retarded development, behavioral disorders, and infertility. Some naturally occurring abnormalities of human chromosomal structure or number are shown in Table 14-2. The frequency of these events may be increased by mutagenic agents.

It is extremely difficult to detect those alterations in mammalian DNA caused by insertions or deletions of one or a few bases, except in rare instances where the specific protein product is known and its formation can be monitored. It is somewhat easier in bacterial or prokaryotic systems and this has led to the use of bacterial or *in vitro* screening assays to detect potential mutagens.

Genotoxicity Tests in Higher Organisms

Tests for genotoxicity in higher organisms may be placed into one of three basic categories. These are gene mutation tests, chromosomal aberration tests, and DNA damage tests. For the purpose of this discussion, the principles of

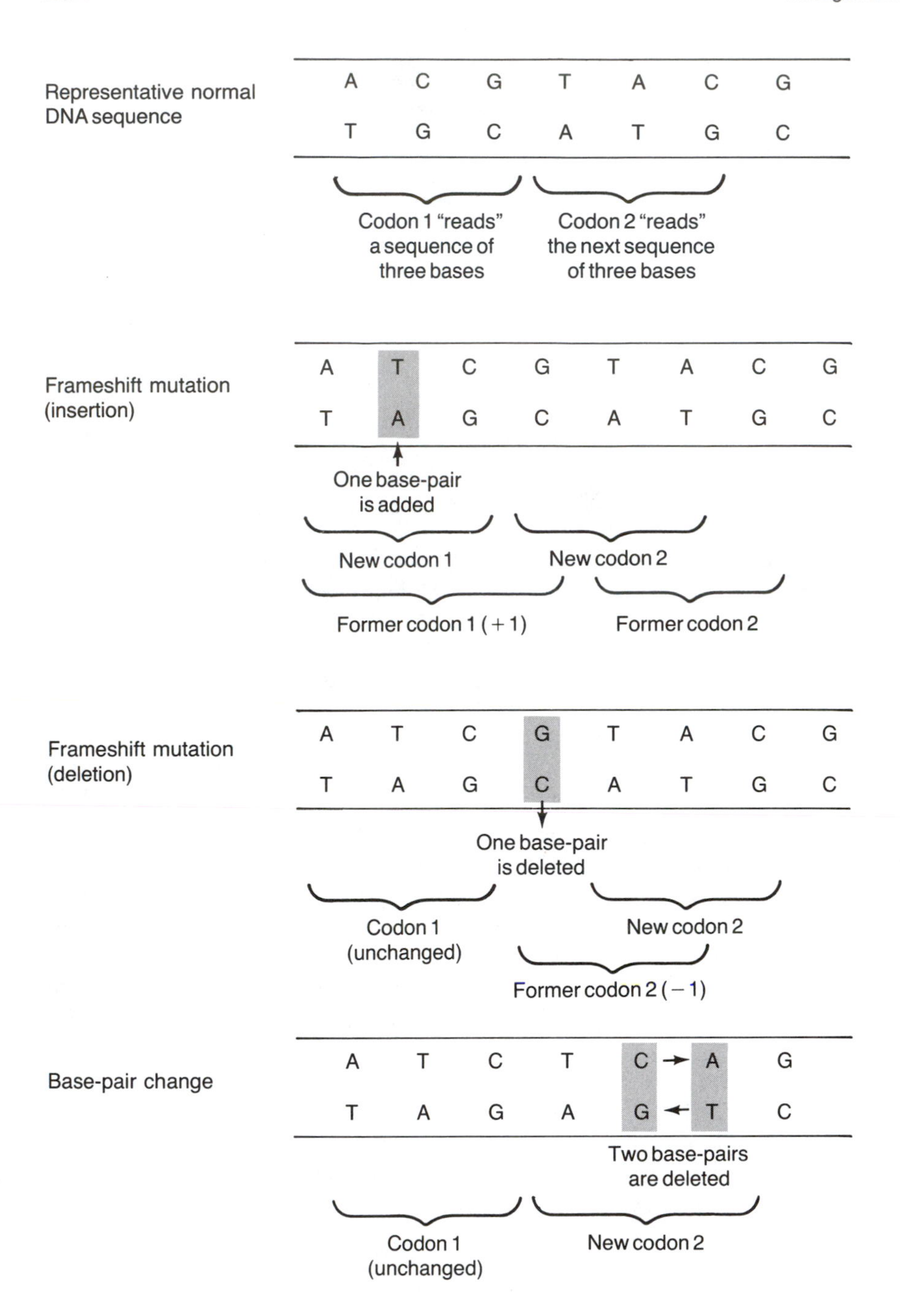

Figure 14-4. Schematic representation of point mutations (frameshift and base-pair changes).

each test category will be reviewed and specific tests will be discussed by broad phylogenetic classifications.

Gene Mutation Tests. Gene mutation tests measure those alterations of genetic material limited to the gene unit, and which are transmissible to progeny unless repaired. Brusick (1980) referes to gene mutations as microlesions because the actual genetic lesion is not microscopically visible. Microlesions are classified as either base-pair substitution mutations or frameshift mutations. These two categories of gene mutations are induced by different mechanisms and, many times, by distinctly different classes of chemical mutagens. Yet both types of gene mutations are virtually always monitored by measuring some phenotypic change in the test organism. As described earlier, the base-pair changes induced by point mutations (Figure 14-3) alter RNA codon sequences, which in turn change the amino acid sequence of the peptide chain being formed, which may result in an alteration of some measurable cellular function. The phenotypic changes that can be monitored by this type of test include auxotrophic changes (i.e., acquired dependence upon a formerly endogenously synthesized substance), altered proteins, color differences, and lethality.

Table 14-2. Examples of Human Genetic Disorders.

Chromosome Abnormalities
Cri-du-chat syndrome (partial deletion of chromosome 5)
Down's syndrome (triplication of chromosome 21)
Klinefelter's syndrome (XXY sex chromosome constitution; 47 chromosomes)
Turner's syndrome (X0 sex chromosome constitution; 45 chromosomes)
Dominant Mutations
Chondrodystrophy
Hepatic porphyria
Huntington's chorea
Retinoblastoma
Recessive Mutations
Albinism
Cystic fibrosis
Diabetes mellitus
Fanconi's syndrome
Hemophilia
Xeroderma pigmentosum
Complex Inherited Traits
Anencephaly
Club foot
Spina bifida
Other congenital defects

Chromosomal Aberration Tests. The second classification of mutagenicity tests involves monitoring changes in the chromosomes. Because these genetic lesions may be visualized by microscopy, they are referred to as macrolesions. One type of macrolesion is caused by an incomplete separation of replicated chromosomes during cell division. This is characterized by the abnormal chromosome numbers that result in the daughter cells and may be recognized as a change in the number of haploid chromosome sets (ploidy changes) or in the gain or loss of single chromosomes (aneuploidy). A second type of macrolesion caused by damage to chromosome structure (clastogenic effects) is categorized by the abnormal chromosome morphology that results.

Two theories are currently available to explain the mechanism of chromosome aberration. One is the classical breakage-first hypothesis. This theory assumes that the initial lesion is a break in the chromosomal backbone that is indicative of a broken DNA strand. Several possibilities exist following such an event: (1) the ends may repair normally and rejoin to form a normal chromosome, (2) the ends may not be repaired, resulting in a permanent break or, (3) they may be misrepaired or join with another chromosome to cause a translocation of genetic material.

A second theory is the chromatid exchange hypothesis. If the exchange occurs with a chromatid from another chromosome, an "exchange figure" results. This theory assumes that the *initial* lesion is not a break and that the lesion can be either repaired directly or may interact with another lesion by a process called exchange initiation. Most chromosomal abnormalities result in cell lethality and, if induced in germ cells, generally produce dominant lethal effects that cannot be transmitted to the next generation. The classical method for determining chromosomal aberrations is the direct visual analysis of chromosomes in cells frozen at metaphase of their division cycle. Thus, metaphase-spread analysis evaluates both structural and numerical chromosome anomalies directly.

Dominant lethal assays can be performed in any organism where early embryonic death can be monitored. The male is treated with the test agent, then mated with one or more untreated females for a period consistent with the entire spermatogenic cycle. An increase in early fetal death observed as deciduomata (resorptions) or dead fetuses, is generally considered to be due to a dominant lethal mutation in the paternal germ cells caused by a chemically induced chromosomal aberration.

The fruit fly (*Drosophila melanogaster*) has received wide use in the sex-linked recessive lethal test. The endpoint phenotypic change monitored in this test is the lethality of males in the F_2 generation. Brusick has gone to the extent of labelling *Drosophila* an "honorary mammalian model" by virtue of its widespread application and correlation with positive mutagens in mammalian testing. *Drosophila melanogaster* has also been utilized to monitor two types of chromosomal aberration endpoints through phenotypic markers: loss and nondisjunction of X or Y sex chromosomes and heritable translocations.

The monitoring of translocations has the advantage of a very low background rate, facilitating comparisons between controls and treated groups. Dominant lethal assays are also done with insects and can theoretically be done in any organism where early embryonic death can be monitored. The male is treated with the test agent, then mated with the female. If early fetal deaths occur these are demonstrative of a dominant-lethal mutation in the germ cells of the treated male.

Chemicals inducing changes in chromosomal number or structure also may be identified by the micronucleus test, an assay that assesses genotoxicity by observing micronucleated cells. It is a relatively simple assay because the number of cells with micronuclei are easily identified microscopically. At anaphase, in dividing cells that possess chromatid breaks or exchanges, chromatid and chromosome fragments may lag behind when the chromosome elements move toward the spindle poles. After telophase, the undamaged chromosomes give rise to regular daughter nuclei. The lagging elements are included in the daughter cells, too, but a considerable proportion are included in secondary nuclei, which are typically much smaller than the principal nucleus and are therefore called micronuclei. Increased numbers of micronuclei represent increased chromosome breakage. Similar events can occur if interference with the spindle apparatus occurs, but the appearance of micronuclei produced in this manner is different, and they are usually larger than typical micronuclei.

Tests for Primary DNA Damage. The third category of genotoxicity tests are those designed to measure DNA damage. Two types that have been applied to germ cells are the sister chromatid exchange test and the unscheduled DNA synthesis test. In sister chromatid exchange (SCE), there is no change in the chromosome morphology; instead, the chromatid exchanges represent an interchange of DNA between replication products of apparently the same sections on opposite chromosome arms. The process is presumed to monitor damage caused by DNA breakage, and then to repair the DNA through exchange of DNA with the sister chromatid. It has been suggested that when DNA is damaged by a mutagen, an enzymatic system repairs the chromatid and causes an increase of SCE. A controversy has developed in this area because recent evidence correlates SCE more closely with cell lethality than with the mutagenic potency of a compound. At this point, all that can be definitely said is that although many mutagens are correlated with identifiable SCE, many are not.

Another assay for DNA damage, unscheduled DNA synthesis monitoring, is based on the fact that cells not undergoing replication (scheduled DNA synthesis) should not exhibit significant DNA synthesis. Thus, incorporation of radiolabelled tracer molecules into the DNA of these cells should be minimal. However, if a chemical mutagen damages the DNA, the DNA repair system may be activated, causing unscheduled DNA synthesis. If such is the case, radiolabelled tracers will be incorporated into the DNA; these can be monitored by autoradiography or by direct measurement of radioactivity in the repaired DNA.

14.4 *In Vitro* Nonmammalian Mutagenicity Tests

Because results from bacterial or prokaryotic assays often establish priorities for other testing approaches, it is of interest to briefly describe the assays currently used to screen for mutagenic capacity, particularly those done in industrial settings.

Rapid cell division and the relative ease with which large quantities of data can be generated (approximately 10^8 bacteria per test plate) have made bacterial tests the most widely utilized routine means of testing for mutagenicity. These systems are the quickest and most inexpensive procedures. However, bacteria are evolutionarily far removed from the human model. They lack true nuclei as well as the enzymatic pathways by which most promutagens are activated to form mutagenic compounds. The DNA lacks the protein coats that are seen in eukaryotes. Nevertheless, bacterial systems have great utility as a preliminary screen for potential mutagens.

In addition to bacteria, fungi have been used in genotoxicity assays. The *Saccharomyces* and *Schizosaccharomyces* yeasts, as well as the molds *Neurospora* and *Aspergillus*, have been utilized in forward mutation tests similar in design to the *Salmonella* histidine revertant assays that will be described in the next section.

Widely Used Bacterial Test Systems

The most widely utilized bacterial test system for the monitoring of gene mutations and the most widely utilized short-term mutagenicity test of any type is the *Salmonella typhimurium* microsome test developed by Dr. Bruce Ames and co-workers and commonly called the Ames assay. The phenotypic marker utilized for the detection of gene mutations in all the Ames *Salmonella* strains is the ability of the bacteria to synthesize histidine, an amino acid essential for bacterial division. The tester strains of bacteria have mutations rendering them unable to synthesize histidine and, thus, they must depend on histidine included in the culture medium in order to be able to multiply. Bacteria are taken directly from a prepared culture and incorporated with a trace of histidine into a soft agar overlay on a dish containing minimal growth factors. The bacteria undergo several divisions, which are necessary for the expression of mutagenicity, and after the available histidine has been used up a fine bacterial lawn is formed. Bacteria that have backmutated in their histidine operon sites (thus, reverted to the ability to synthesize histidine) will keep on dividing to form discrete colonies. The nonmutated bacteria will die. A chemical that is a positive mutagen will demonstrate a statistically significant dose-related increase "revertants" (colonies formed) when compared to the spontaneous revertants in control plates.

Five Ames *Salmonella typhimurium* tester strains are recommended for routing mutagenicity testing: TA1535, TA1537, TA1538, TA98, and TA100. The TA1535 tester strain detects base-pair substitution mutations. The TA1538 tester strain detects frameshift mutagens that cause base-pair deletions. The TA1537 tester strain detects frameshift mutagens that cause base-pair additions. The TA100 and TA98 strains are sensitive to effects caused by certain compounds, such as nitrofurans, which were not detectable with the previous three strains.

The lack of oxidative metabolism to transform promutagens (those mutagens requiring bioactivation to the active form) is overcome in these bacterial assays by two means. First, a suspension of rat liver homogenate containing appropriate enzymes may be added to the bacterial incubation. The liver preparation is centrifuged at $9000 \times g$ for 20 min at 4°C and the resultant supernatant (S-9) is added to the culture medium. In a slightly more complex procedure, called the host-mediated assay, the bacterial tester strains are injected into the body cavity of a test animal such as the mouse. This host is treated with the suspected mutagen and after a selected period the bacteria are harvested and assayed for mutation (revertants) as described earlier. Other bacterial species used in mutagenicity screens include *Escherichia coli* and *Bacillus subtilis*.

Assays that measure DNA repair in bacterial systems have also been developed. These tests are based on the premise that a strain deficient in DNA repair enzymes will be more susceptible to mutagenic activity than a similar strain that possesses repair enzymes and can correct the mutagenic damage. A "spot" test consists of placing the chemical to be tested in a well or on a paper disc on top of the agar in a petri dish. The test chemical will diffuse from the central source, causing a declining concentration gradient as the distance from the source increases. A strain deficient in repair enzymes will exhibit a greater diameter of bacterial kill than the repair-sufficient strain tested with a mutagen. In a "suspension" test a given number of bacteria are preincubated with and without the test compound. The bacteria are then plated and the colonies counted. The repair-deficient strain will demonstrate a greater percentage kill than the sister DNA-repair-sufficient strain. A liver S-9 activation system can also be incorporated in bacterial DNA repair tests. The most widely used bacterial DNA repair test utilizes the $polA^+$ and $polA^-$ strains of *E. coli*. The $polA^-$ strain is deficient in DNA polymerase I, whereas the $polA^+$ strain is sufficient in this enzyme.

14.5 *In Vivo* Mammalian Mutagenicity Tests

Testing of chemicals for mutagenicity *in vivo* in mammalian systems is the most relevant method for learning about mutagenicity in humans. Mammals such as the rat or mouse offer insights into human physiology, metabolism, and reproduction that cannot be duplicated in other tests. Furthermore, the route of administration of a chemical to a test animal can be selected to parallel normal human environmental or occupational conditions of oral, dermal, or inhalation exposures.

Human epidemiologic findings may also be compared with the results of tests done in animals. While the monitoring of human exposures and their effects does not constitute planned, controlled mutagenicity testing, human epidemiology offers the opportunity to monitor and test for correlations suggested by other mutagenicity tests. Thus, these studies are the only opportunity for direct human modeling of a chemical's mutagenic potential. It is worth noting that despite extensive investigation, to date no chemical substances have been positively identified as human mutagens.

One perceived disadvantage of *in vivo* mammalian test systems is the time they require and their cost. A larger commitment of physical resources and personnel is required than is required with *in vitro* testing. Human epidemiology studies are further complicated by the fact that not all of the environmental variables can be controlled. Frequently the duration and extent of exposure to single or multiple compounds can only be estimated. Nevertheless, progress is being made to lessen the cost and decrease the time required for *in vivo* mammalian testing. Also, new data handling, statistical techniques, and increased cooperation from industry have increased the reliability of human epidemiology studies. More regular sampling of workplace exposures has helped to improve the quality and accessibility of human data.

Germ Cell Mutation Tests

A basic test used to detect specific gene mutations induced in germ cells of mammals is the mouse specific locus assay. This test involves treatment of wild-type mice, either male or female, with a test compound before mating them to a strain homozygous for a number of recessive genes that are expressed visibly in phenotype. If no mutational events occur, then all offspring will be of the wild type. If a mutation has occurred at one of the loci in the treated mice, then the recessive phenotype will be visibly detectable in the offspring. The mouse specific locus test is of special significance in human modeling because it is the only standardized assay that directly measures heritable germ cell gene mutations in the mammal. A major drawback of the mouse specific locus test is that extensive physical plant facilities are required to execute this assay: it has been estimated that one scientist and three technicians could execute 10 single-dose mouse specific locus tests in one year, provided there are facilities for 5000 cages.

New and promising test procedures have been described for detecting germ cell mutations by using alterations in selected enzyme activity as the phenotypic endpoint. A large group of somatic cell enzymes can be monitored for changes in activity and kinetics in the F_1 generation. These changes will reflect changes in the parental genome.

A similar biochemical approach has been proposed for identifying germ cell mutations in humans through the monitoring of placental cord blood samples. The activity of several erythrocyte enzymes, such as glutathione reductase, can be monitored because the enzyme proteins are the products of a single locus and because heterozygosity of a mutant allele for the chosen enzymes will result in abnormal levels of enzyme activity. Likewise, it has been proposed that gene mutations be directly monitored in mammalian germ cells by searching for phenotypic variants with biochemical markers such as lactic acid dehydrogenase-X (LDH-X), an isozyme of lactic acid dehydrogenase found only in testes and sperm. The test is based on the fact that a monospecific antibody for rabbit LDH-X reacts with rat but not mouse LDH-X in sperm. The rat sperm fluoresce as a result of the reaction but the mouse sperm do not flouresce unless a phenotypic variant is present. If adapted to man, this test has potential use as a noninvasive screening test of germ cell mutations in males.

It has been proposed that the induction of behavioral effects in offspring of male rats exposed to a mutagenic agent represents a genotoxic endpoint. For example, studies have demonstrated that the mutagen cyclophosphamide induces genotoxic behavioral effects in the progeny of male rats and that these effects correspond to genetic damage caused in the spermatozoa following meiosis.

Mammalian germ cells can be monitored for chromosomal aberrations, and normally the testes are used as the cell source. Mammalian male germ cells are protected by a biological barrier comparable in function to that which retards the penetration of chemicals to the brain. The blood–testes barrier represents a complex system composed of membranes surrounding the seminiferous tubules and the several layers of spermatogenic cells organized within the tubules. This barrier restricts the permeability of large-molecular-weight compounds to the developing male germ cell. An advantage of *in vivo* mammalian germ cell mutagenicity testing is that the protective contribution of this barrier is automatically taken into account. Conventional procedures for harvesting mammalian male germ cell tissue for metaphase-spread analysis involve mincing or teasing the seminiferous tubules to liberate meiotic germ cells in suspension. This homogenate is centrifuged, the centrifuged pellet is discarded, and the suspended cells are collected and analyzed. However, it was found that the tissue fragments discarded during this conventional procedure contained more spermatogonial cells and meiotic metaphases than the suspension.

Thus, the method has been refined by using tissue fragments and adding collagenase to dissociate them. After collagenase treatment, the tissue fragments are gently homogenized and centrifuged, and the pellet containing meiotic cells is resuspended and prepared for microscopic analysis.

Dominant Lethal Assays

As described earlier, mammals are commonly used in dominant lethal assays. The male animals (usually mice or rats) are treated with the suspected mutagen before being mated with one or more females. Each week these females are removed and a new group of females is introduced to the treated male, this process being repeated for a period of six to ten weeks. The females are sacrificed before parturition, and early fetal deaths are counted in the uterine horns. This test has become standardized and a large number of compounds have been screened in mouse studies. As with most *in vivo* mammalian assays, costs and commitment of resources can be extensive. However, the applicability of the data is typically quite good. The dominant lethal test in rodents is of significance for human modeling because it gives an indication of heritable chromosomal damage in a mammal. Even though the endpoint of early fetal death may seem of minor significance when considering only its effects on the human gene pool, it does provide a signal that viable heritable types of chromosomal damage and gene mutations may also be produced.

Heritable Translocation Assay

Results of dominant-lethal assays frequently correlate well with another test used for determining clastogenic effects in mammalian germ cells: the heritable translocation assay (HTA). Translocation represents complete transfer of material between two chromosomes. In HTA procedures, male mice are mated to untreated females after treatment with the test compound, and the pregnant females are allowed to deliver. Male offspring are subsequently mated to groups of females. If translocations are produced through genotoxic action, then the affected first generation male progeny will be partially or completely sterile; this can be noticed in the litter size produced from those females to which they were mated.

Micronucleus Tests

Application of the micronucleus test to mammalian germ cells has recently been reported. This test procedure is analogous to the bone marrow micronucleus test but it involves the sampling of early spermatids from the seminiferous tubules of male rats. The number of micronuclei are quantified by using fluorescent stain and counting the micronucleated spermatids.

Sperm Head Morphology Assay

Some relatively new tests have been developed that evaluate the ability of a test chemical to induce abnormal sperm morphology when compared to controls. It has been proposed that an increase in abnormal sperm morphology is evidence of genotoxicity because there seems to be an association between abnormal sperm morphology and chromosome aberrations. However, recent investigations have reported that induction of morphologically aberrant sperm can be caused by nongenotoxic actions, such as dietary restriction. In addition, some known mutagens, including 1,2-dibromo-3-chloropropane (a pesticide with mutagenic, carcinogenic, and gonadotoxic properties), were reportedly unable to induce production of sperm-head abnormalities in mice, when tested. It should be noted that sperm abnormalities are fairly frequent in humans and may occur at rates of 40–45 percent. Thus, more verification is needed before strong conclusions can be drawn about the mammalian sperm-head morphology assay.

Occupational Monitoring

Recently, nondisjunction of the Y chromosome in sperm has been monitored in human males. This was done by a technique for quinacrine-staining the sperm, which causes the Y sex chromosome to fluoresce. A sperm with one fluorescent spot indicates a normal complement of one Y chromosome. However, two fluorescent spots indicate two Y chromosomes and can be used as a monitor for nondisjunction of the Y chromosome. Workmen exposed to 1,2-dibromo-3-chloropropane (DBCP), a chemical known for the sterility it causes in exposed males, have been shown to have a threefold increase in double-Y-fluorescing bodies in their sperm when compared to nonexposed individuals. However, the fluorescent body test for Y chromosome nondisjunction is in need of further validation.

Tests for Primary DNA Damage

A historical test thought to monitor primary DNA damage in mammalian germ cells *in vivo* involves the monitoring of sister chromatid exchange. The observation of sister chromatid exchanges through differential staining involves exposing the cells to bromodeoxyuridine for two rounds of replication, so that the chromosomes consist of one chromatid substituted on both arms with 5-bromodeoxyuridine and the other substituted only on a single arm. Differential staining between sister chromatids is due to the differences in bromodeoxyuridine incorporation in the sister chromatids.

Unscheduled DNA repair has been induced by chemical mutagens in mammalian male germ cells from the spermatogonial to mid-spermatid stages of development. However, male germ cells have lost DNA repair capability when they have advanced to the late spermatid and mature spermatozoa stages; unscheduled DNA synthesis cannot then be induced by chemical mutagens. The genotoxic agents methyl methanesulfonate, ethyl methanesulfonate, cyclophosphamide, and Mitomen have been shown to induce unscheduled DNA repair *in vivo* in male mouse germ cells.

14.6 *In Vitro* Mammalian Mutagenicity Tests

Test systems have been developed that use mammalian cells in culture (*in vitro*) to detect chemical mutagens. Disadvantages in comparison with *in vivo* mammalian tests are that *in vitro* tests lack organ–system interactions, require a route for administration of the agent that cannot be varied, and lack the normal distributional and metabolic factors present in the whole animal. The obvious advantages are that costs are decreased and that experiments are more easily replicated, which facilitates verification of results. Cases where human cells have been cultured successfully (e.g., lymphocytes) provide the only viable *in vitro* experiments on the human organism.

Endpoints of *In Vitro* Testing

Several endpoints can be used during testing of potential mutagens *in vitro*. One of the most common involves the monitoring of mutations in specific well-characterized gene loci, such as those coding for hypoxanthineguanine phosphoribosyl transferase (HGPRT), thymidine kinase (TK), or ouabain resistance (OVA^r). Mutagenic modification in the segments of the DNA coding for these proteins (enzymes) results in an increased sensitivity of the cell, which can often be evaluated by the cell's heightened susceptibility to other agents (e.g., bromodeoxyuridine or 8-azaguanine).

As described in the section on *in vivo* mammalian testing, evaluations of sister chromatid exchange, DNA repair activity, and chromosomal aberration through interpretation of metaphase spreads may be applied to *in vitro* testing of mutagens. An additional procedure that has been correlated with chemical mutagenicity is examination for cell culture transformation: following treatment with mutagens, some cells in culture lose their normal, characteristic arrangement of monolayered attachment and begin to pile up in a disorganized fashion. A major drawback in looking for this feature is that considerable expertise is necessary to interpret the results accurately, and that the criteria for evaluation are somewhat more subjective than for other mutagenicity assays.

Occupational Monitoring

Cytogenetic analysis (chromosome evaluation) has been a standard industrial technique for monitoring human genetic damage. However, several limitations are inherent in the conventional use of human lymphocytes as indicators of exposure to genotoxic chemicals or radiation:

- Individual and population variability in normal levels of chromosomal aberrations may mask small changes in the frequency of mutations. To overcome this obstacle, specific defects that occur with low frequency in normal individuals may also be tested for, but typically several thousand cells must be scored per individual to achieve sufficient sample size.
- Evaluation of chromosomal aberrations is subject to substantial variation between laboratories. Therefore, replicate readings should be obtained; this substantially increases the effort required when thousands of samples are involved.
- Since chromosomal aberrations are considered indicators of relatively gross damage, the techniques may miss many more subtle effects of mutagens.

Evaluation of sister chromatid exchange (SCE) may be potentially valuable in answering some of these difficulties. For example, SCEs are elevated in patients undergoing chemotherapy, which is not unexpected, as many of the chemotherapeutic agents are powerful mutagens. These elevations tend to be dose-related, which supports the usefulness of the technique as a potential screening device. It must be emphasized that SCE may not be a damaging lesion in itself, but may prove a useful marker for other detrimental effects on the DNA induced by the agents in question. This caveat is underscored by the observation that SCE is poorly correlated with radiation exposure and exposure to other agents that break DNA. Agents that alkylate the DNA (bind tightly to the molecule) show a better correlation with mutagenic potential and may be a more sensitive indicator for the monitoring of chromosomal aberrations.

14.7 The Occupational Significance of Mutagens

Areas of Concern: Gene Pool and Oncogenesis

The potential significance of occupationally acquired mutagenesis can be divided into two areas. The first is concern for the protection of the human

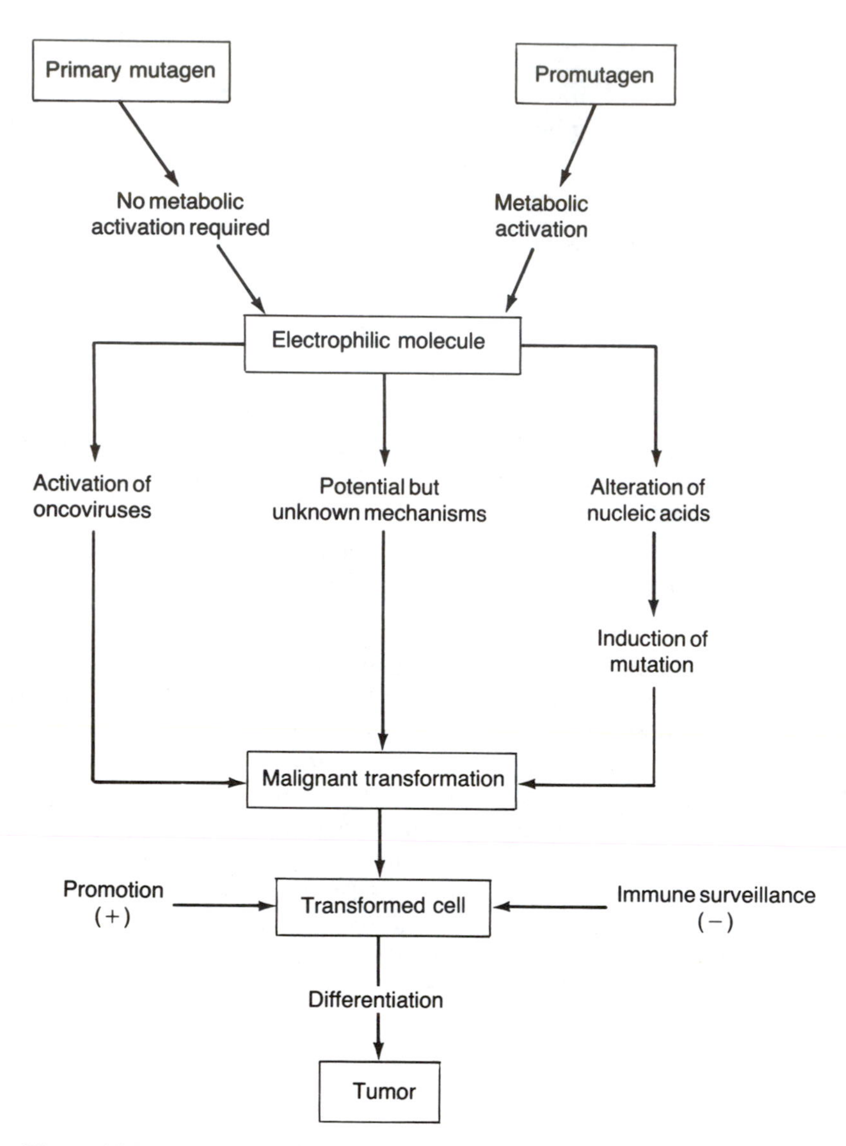

Figure 14-5. Proposed relationships between mutagenicity and carcinogenicity. (Adapted from Fishbein, *Potential Industrial Carcinogens and Mutagens*. Amsterdam: Elsevier Scientific Publishing Co. 1979.)

gene pool. This factor may represent the most significant reason for genetic testing, but it is often underemphasized by nongeneticists involved with safety evaluation because of the inability to demonstrate induced mutation in humans to date. The second area is that of oncogenesis (Figure 14-5). The intimate relationship between the tumorigenic and genotoxic properties of many chemicals makes genetic testing a potentially powerful screening technique for establishing priorities for future testing of chemicals of unknown cancer-causing potential. This factor has been one of the primary driving forces behind the rapid expansion of genetic toxicology as a discipline. Once again, however, the paucity of proved human carcinogens compared with the number of demonstrated animal carcinogens suggests weaknesses in the process of extrapolating from animal studies to human exposure in the workplace.

At the heart of the present legal and regulatory approach toward environmental and occupational mutagens is the possibility that they are causing human genetic damage. Two important assumptions underlie this central concept:

- Environmental mutagens may cause aneuploidy, chromosome breaks, gene-locus mutations, or other genetic damage in humans.
- Environmental (and occupational) mutagens that can be controlled by regulatory effort represent a significant component of total human exposures.

Much of the interest in potential environmental and occupational mutagens is related to the prevalent opinion that many cancers are initiated by a mutagenic event. This premise is supported by the strong correlation between some specific occupational chemical exposures and cancer incidence in man. One good example is the relationship between liver cancer (angiosarcoma) and exposure to vinyl chloride in the plastics industry. Another is the respiratory tract cancers caused by the exposure to bis(chloromethyl) ether (Weisburger and Williams, 1980).

A Multidisciplinary Approach

Although each of the mutagenicity tests described in this chapter has individual strengths, likewise each is weak in some facets of detection. Clearly, therefore, the accurate and efficient testing of chemicals and protection from potential occupational mutagens require a multidisciplinary approach integrating toxicology, clinical chemistry, microbiology, pathology, epidemiology, industrial hygiene, and occupational medicine. Testing intact animals has the advantage of increasing the reliability of any extrapolations that must be made from the data; however, cost often limits the application of *in vivo* mammalian assays except when it is expected that they will verify lower tier assays.

There are three areas where the results of mutagenicity testing of a given substance may be applied:

- Extrapolation of test results in order to make a quantitative evaluation of the hazard of exposures for humans.
- Prioritization of human hazards posed by specific compounds.
- Institution of remedial procedures that should be undertaken to minimize the human hazard.

One of the most difficult areas of analysis is the correct application of mutagenicity tests to arrive at a quantifiable human hazard from exposure to a given substance. There is frequently good correlation between the mutagenicity and carcinogenicity of a substance in animal tests (Table 14-3). However, this may be misleading because models of carcinogenicity determination are often characterized by chronic procedures utilizing very high doses in nonprimate species. These may bear little resemblance to such aspects of exposure in the human model as magnitude and route of exposure, metabolic patterns, and environment (which are qualitative factors), and exposure dose (which is quantitative).

As is the case with many areas of toxicology, one may choose between *in vivo* and *in vitro* test systems, each with their attendant advantages and disadvantages. The testing of chemicals in experimental animals has all the advantages of any intact *in vivo* system; that is, it has all of the biochemical and physiological requirements to make anthropomorphization more reliable. However, *in vivo* mutagenicity testing may require an investment of many thousands of dollars and a long period of time. These disadvantages often

Table 14-3. Comparative Mutagenicity of Various Compounds.

Compound	Established human carcinogen	Bacteria	Yeast	*Drosophila*	Mammalian cells	Human cells
Epichlorohydrin	N	N	O	Y	Y	Y
Ethyleneimine	N	O	Y	Y	Y	Y
Trimethyl phosphate	O	Y	O	Y	Y	Y
Tris	N	Y	O	Y	Y	Y
Ethylene dibromide	N	Y	Y	Y	Y	N
Vinyl chloride	Y	Y	Y	Y	Y	Y
Chloroprene	Y	Y	O	O	O	Y
Urethane	N	Y	Y	Y	Y	O

N = no; Y = yes; O = not tested.

force the tester to use a less expensive, well-established short-term bioassay, such as the Ames *Salmonella* bacterial assay, to determine the mutagenicity of a chemical, and then extrapolate these results into the *in vivo* mutagenicity model.

Areas for Future Progress

Many mutagenicity assays have been proposed, each with a unique attribute and measurable biochemical or visible endpoint. These tests are being incorporated into routine safety assessment programs in all regulatory agencies. Furthermore, these tests have been proposed as part of regulatory decision-making policy by the Occupational Safety and Health Administration (OSHA) for the classification of chemical carcinogens in the workplace, and by the Environmental Protection Agency (EPA) for the regulation of pesticides and for regulating the disposal of toxic wastes. These short-term mutagenicity tests actually serve two purposes. Not only do they assist in the assessment of a chemical's potential for cancer induction but they also assess the potential for inducing germ cell mutations in humans as well. Some of the organizations involved in the development of guidelines for germ cell mutagenicity tests are the International Commission for Protection Against Environmental Mutagens and Carcinogens (ICPEMC), the World Health Organization, and the Commission for European Communities. In the past, most estimates of genotoxic risks were more qualitative than quantitative, and the emphasis has rested on somatic effects (e.g., those leading to cancers) rather than on germinal cells (sperm and ovum). On the basis of evidence in animals demonstrating germinal cell effects, it is imperative to develop human screening methods capable of detecting such effects. Therein lies one of the premier challenges to genetic toxicology and occupational medicine.

The uncertainties of accurate extrapolation of mutagenicity test data to a human hazard model have supported the philosophy that if uncertainty is to occur in extrapolation it should favor the side of safety. This concept is particularly important in the consideration of whether or not threshold characteristics may exist. In the case of carcinogens, discussed further in the next chapter, good evidence supports the view that genotoxic (DNA-damaging) carcinogens may be distinct from epigenetic carcinogens (those that induce or potentiate cancer by means other than direct DNA interaction). For the purposes of this discussion, mutagens are assumed to exert nonthreshold effects. That is, even as one approaches zero dose, there is still a calculable risk of DNA effects.

The concern for the potential mutagenic hazards in the workplace from exposure to chemicals should include routine tests of nonpregnant females and males as well as the more traditional monitoring of pregnant and lactating women. For example, vinyl chloride, mentioned earlier in relation to its suggested role in angiosarcoma of the liver, has been correlated with an increased incidence of nervous system malformations in infants fathered by exposed workers (Barlow and Sullivan, 1982). It has also been demonstrated to cause elevations in chromosomal aberration in the occupationally exposed. 1,2-dibromo-3-chloropropane (DBCP), a pesticide linked to sterility in exposed male workers, causes increases in indices of mutagenic capacity in humans and animals (Barlow and Sullivan, 1982).

Monitoring of male populations may prove particularly important in that the spermatogenic cycle is continuous in adults and therefore poses continuous opportunities for genetic damage to be expressed as damaged chromosomes. Since the female carries the full lifetime complement of ova at birth, susceptibility to propagation of genetic alteration during cell division is reduced except in those periods of division following conception. By the same token, the cessation of exposure in the male should allow for recovery from a mutagenic event in premeiotic spermatocytes, providing that spermatogonia are not affected. If chromosome damage occurs in sperm or ovum, then fetal death frequently occurs. Greater than 50 percent of spontaneous abortions in humans show chromosomal defects.

Once mutagenic potential is established for a compound, the risks posed by exposure under expected conditions must be assessed. As discussed, complications may be encountered in situations where mutagenic effects are due to "multihit" phenomena and therefore reflect threshold-type responses. A more complete discussion on risk assessment is presented in Chapter 17.

Summary

Modification of genetic material by mutagenic agents poses a serious environmental and occupational threat. Chemical or physical mutagens may induce cancer or lead to germ cell alteration.

- The mutagens that lead to cancer alter the DNA of somatic cells so as to cause modifications in gene expression, which results in tumorigenesis.
- Germ cell (sperm, ovum) mutagens may exert their effects through decreased fertility, birth defects, spontaneous abortion, or through changes that may not become evident for several subsequent generations (such hidden mutagenic effects remain essentially undetectable except when expressed as a gross malformation).

Many screening tests have been developed to reveal the mutagenic potential of chemical agents.

- These assays use bacteria, insects, mammals, and various cells in culture.
- Although *in vitro* tests are less expensive and less complex, *in vivo* mammalian tests give results that can be extrapolated to human circumstances more realistically; but *in vivo* studies are expensive and labor-intensive.

Persons whose occupations expose them to potential mutagens may undergo chemically-induced changes at a greater rate than the general population does.

- Epidemiology seeks to identify groups with increased susceptibility to chemical mutagens, or increased incidence of exposure, in order to limit exposures.
- No single method currently stands out as the most comprehensive and thorough screen for identifying mutagenic agents; often, a multidisciplinary approach employing several tests is best suited to the accurate identification of industrial mutagens.

Once mutagenic potential has been demonstrated for a compound, typically an analysis must be made of the risks posed to exposed individuals. Such a determination is essential in the qualitative evaluation of the occupational hazard of mutagens.

References and Suggested Reading

Barlow, S. M., and Sullivan, F. M. 1982. *Reproductive Hazards of Industrial Chemicals*. New York: Academic Press.

Berg, K. (Ed.) 1979. *Genetic Damage in Man Caused by Environmental Agents*. New York: Academic Press.

Brusick, D. 1980. *Principles of Genetic Toxicology*. New York: Plenum Press.

Calabrese, E. J. 1978. *Pollutants and High Risk Groups*. New York: Wiley Interscience.

Cohen, B. H., Lilienfeld, A. M., and Huang, P. C. (Eds.) 1978. *Genetic Issues in Public Health and Medicine*. Springfield (Ill.): Charles C. Thomas Co.

Fishbein, L. 1979. *Potential Industrial Carcinogens and Mutagens*. Amsterdam: Elsevier Scientific Publishing Co.

Hollaender, A. (Ed.) 1971–1984. *Chemical Mutagens: Principles and Methods for Their Detection*, Vol. 1–8. New York: Plenum Press.

International Commission for Protection Against Environmental Mutagens and Carcinogens (ICPEMC). 1983. "Regulatory Approaches to the Control of Environmental Mutagens and Carcinogens." *Mutation Research, 114:* 179.

Miller, E. C., Miller, J. A., Hirono, I., Sugimura, P., and Takayama, S. (Eds.) 1979. *Naturally Occurring Carcinogens, Mutagens, and Modulators of Carcinogenesis*. Baltimore: University Park Press.

Rom, W. N. (Ed.) 1983. *Environmental and Occupational Medicine*. Boston: Little, Brown and Co.

Shaw, C. R. (Ed.) 1981. *Prevention of Occupational Cancer*. Boca Raton (Florida): CRC Press.

Sorsa, M., and Vainio, H. (Eds.) 1982. *Mutagens in Our Environment*. New York: Alan R. Liss Inc.

Weisberger, J. H., and Williams, G. M. 1980. "Chemical Carcinogenesis." *In* Doull, J., Klaassen, C. D., and Amdur, M. O. (Eds.) *Casarett and Doull's Toxicology: The Basic Science of Poisons*. New York: Macmillan Publishing Co.

Chapter 15

Carcinogenesis

Robert C. James and Christopher M. Teaf

Introduction

This chapter will discuss:

- Definitions for the commonly used terminology in the field of carcinogenesis.
- The differences between benign and malignant tumors.
- The properties of carcinogenic chemicals.
- The mechanistic differences in cancer causation.
- Methods for testing chemicals for carcinogenic activity.
- The risks associated with occupational and environmental carcinogens.

Cancer remains a dreaded disease that is responsible for the deaths of about 440,000 Americans per year or about one person every seventy-two seconds. Despite the gains made in cancer research it is the second most common cause of death and ranks only behind heart disease. Cancer has been responsible for one of five deaths in recent years and the ever increasing average age of our population is expected to raise this figure to one of three in the near future. Cancer is of great concern to modern society for many reasons:

- About 900,000 people are diagnosed each year as having cancer.
- Only 38 percent of the patients who develop cancer are alive five years after the diagnosis (if heart attacks and old age are taken into consideration then this number increases to 46 percent).
- Cancer is an insidious disease; it may remain dormant for years after the causal event.
- Cancer may be caused in some cases by small or almost incidental exposures to carcinogens whose other effects at the time may seem to be

minor. Such exposures may go unnoticed or continue for a long period of time.

- Cancer treatments often depend upon differential destruction of only certain cells within the body. The toxicity of drugs used in therapy may put the patient at risk or cause illness during treatment.
- The annual cost of cancer is approximately $9 billion even though this figure excludes such factors as the loss of worker productivity.

In addition to the above problems, associated with the disease itself, fears, doubts, and misconceptions have been created in people's minds such that any discussion of the subject or the possible causal agents is often hard to control or deal with in a calm and reasonable manner. Thus, it is very important to understand the biological and theoretical basis of chemical-induced cancer, as it represents possibly the most dreaded toxic event and probably the hardest for which to provide reassuring safety precautions.

15.1 Definitions and Descriptions of Cancerous Tissue

As we will be discussing a complex biological process, it may be helpful to provide some definitions of the pathological changes in tissues relevant to the subject of cancer.

Unchecked Growth and Lack of Contact Inhibition

Cancer is a process in which cells undergo some change that renders them abnormal—i.e., a phenotypic change or a change in the expression of their genetic code—and they embark on a phase of uncontrolled growth that causes them to spread. In normal tissues, the cells making up the tissue are limited to specific sizes and numbers such that organs generally have a specific size or proportion within the body. Even after some form of injury, cells or connective tissue in the injured area are replaced in a very specific and controlled manner. The danger posed by cancer cells or cancerous tissue is that cancer cells are capable of uncontrolled growth. They lack contact inhibition; that is, they fail to recognize the normal limits and boundaries of their specific tissue. Therefore, cancerous tissue may invade and destroy normal tissues.

Cancer can be divided into a series of recognizable stages; as it progresses from one stage to the next it becomes less controlled, increasingly destructive, and more invasive. For this reason, it is important to detect cancer as early as possible, in the hopes of catching it at a stage in which it is localized and easier to treat or remove. It is estimated that approximately 150,000 persons per year die who might have been saved by earlier diagnoses and prompt treatments of their specific cancers.

Neoplasms

Cancer tissues can be categorized under the more general classification of neoplasms. "Neo" means new, and "plasia" means growth; thus any neoplasm simply refers to the fact that such tissue is new growth. But the implication of the term neoplasm is that it refers to an area of abnormal tissue that is growing more rapidly than normal.

The following growth-related terms are used to describe tissue changes or are associated with neoplasia:

- Growth or changes leading to tissues or organs smaller than normal.
  1. Aplasia: a developmental defect in which there is complete or almost complete failure of a tissue or organ to develop.
  2. Hypoplasia: failure of an organ to achieve its full size.
  3. Atrophy: a decrease in tissue or organ size that may result from a decrease in cell size, cell number, or both.
- Growth or changes leading to tissues or organs larger than normal.
  1. Hypertrophy: increased organ or tissue size caused by an increase in individual cell size.
  2. Hyperplasia: increased organ or tissue size caused by an increase in cell number. Both hypertrophy and hyperplasia may occur in the same tissue.
- Aberrant growth or changes in tissues or organs.
  1. Metaplasia: a reversible change in which one differentiated cell (i.e., specific cell type) is replaced by another cell type.
  2. Dysplasia: a reversible change in cells, which may include an altered size, shape, or organizational relationship. This change is usually observed in epithelial tissues and, in this case, is equivalent to a form of hyperplasia but is not necessarily a neoplastic response.
  3. Anaplasia: a marked regressive change from a specific cell type (differentiated) towards replacement by cells that are more embryonic in nature. Thus, tissue is formed that is less organized and functional than the normal tissue. This is usually an irreversible change and, as such, is a good criterion for malignant or cancerous tissue.
  4. Neoplasia: relatively autonomous new growth containing some or all of the above features. The neoplasm is composed of abnormal cells the growth of which is more rapid than that of other tissues and is not coordinated with the growth of other tissues. The term "tumor," which literally means "swelling," is generally used synonymously with neoplasm.
  5. Metastasis: the process by which tumor cells invade and destroy other tissues. This occurs because many malignant cells are capable

of detaching from the tumor site, migrating via blood vessels or lymph channels, and then lodging and growing in a new location.

Benign Versus Malignant Tumors

Although there may be a range in the characteristics of tumors, histopathologically they are all classified as one or the other of two basic types: benign, or malignant. Medically, this distinction has important consequences, because the malignancy of a tumor is what defines human cancer as a disease state.

The basic difference between benign and malignant tumors is that malignant tumors undergo metastasis. That is, a portion of the tumor may break away, enter the vascular system, and lodge in a distant site, forming a tumor in a new area, whereas a benign tumor is noninvasive and tends to remain more circumscribed in the tissue that is formed. The invasive nature of the malignant tumor presents a greater likelihood that the host tissue or tissues will be consumed by the tumor and that this will ultimately lead to death. The benign tumor does not grow beyond its boundaries but may inflict damage by localized obstruction, compression, or interference with metabolism. An additional difference between the two tumor types is that benign tumors also represent a lesser health risk: once a benign tumor is removed, so too is its harmful influence. A malignant tumor, if removed, may still leave a health risk behind in the form of metastases in other areas of the body. In short, malignant tumors may be much harder to treat.

Since benign and malignant are important distinctions for tumors, the following classification further distinguishes these two tumor types.

Benign	Malignant
Encapsulated	Nonencapsulated
Noninvasive	Invasive
Highly differentiated	Poorly differentiated
Rare cell division	Cell division common
Slow growth	Rapid growth
Little or no anaplasia	Anaplastic to varying degrees
No metastases	Metastases

Tumor Nomenclature

Neoplasms can be formed in any tissue, and a variety of types of benign and malignant tumors may occur throughout the body. In general these neoplasms are classified or named according to their microscopic structures and original cells or tissues from which the tumor cells are derived. The suffix "-oma" is commonly used to denote *benign* neoplasms; "carcinoma" is used to denote *malignant* tumors of epithelial cells; "*sarcoma*" is used to denote malignant tumors of mesodermal tissue origin, such as of muscle or bone:

Tissue	Benign Tumor	Malignant Tumor
Epidermis	Epidermal papilloma	Epidermal carcinoma
Biliary tract	Cholangioma	Cholangiocarcinoma
Adrenal cortex	Adrenocortical adenoma	Adrenocortical carcinoma
Stomach	Gastric polyp	Gastric carcinoma
Liver	Hepatoma	Hepatocellular carcinoma
Fibrous tissue	Fibroma	Fibrosarcoma
Cartilage	Chondroma	Chondrosarcoma
Fat	Lipoma	Liposarcoma
Muscle	Rhabdomyoma	Rhabdomyosarcoma
Bone	Osteoma	Osteosarcoma
Kidney	Renal Adenoma	Renal adenocarcinoma
Lung	Pulmonary adenoma	Pulmonary adenocarcinoma

15.2 Definitions of Carcinogenic Chemicals

Although the term "carcinogen" literally means any cancer-producing substance (i.e., physical agent or chemical compound) and should be properly used only to refer to substances causing carcinomas or sarcomas (i.e., malignancies), in toxicology this term is frequently used in a much broader sense. That is, our concern for chemicals that produce tumors or neoplasms is so great that operationally we do not usually distinguish between those chemicals that increase the incidence of benign tumors from those producing malignant tumors.

An operational definition of a carcinogen is "any chemical capable of producing an increased number of tumors in some test species." On the basis of this definition, four characteristic results of animal testing have become accepted as evidence of a chemical's tumorigenicity (tumor-causing capacity):

- A statistically significant increase in the incidence (number) of tumors in the exposed group compared to the incidence observed in the control (unexposed) groups.

- Tumors occur much earlier in the exposed group than in the control group; at earlier times there are significant differences in tumor incidence.
- Tumor types develop in the exposed group that are not observed in the control group.
- In the exposed group, an increased number of tissues develop tumors.

Chemicals may produce cancer by several different mechanisms and may produce markedly different effects under differing test conditions. These differences may have a profound impact upon the relevance of the mathematical extrapolations, to human exposure, of the risks that are often derived from this animal test data. Therefore, it is not surprising that new definitions of "carcinogens" are evolving rapidly. The following definitions should help clarify subsequent discussions about chemical-induced carcinogenesis:

- A carcinogen, as defined in this chapter, is a chemical capable of inducing tumors in animals or man.
- Carcinogenesis is the origin or production of cancer; operationally speaking, this includes any tumor, either benign or malignant.
- A procarcinogen is a chemical that requires metabolism to another chemical form before the carcinogenic action can be expressed (e.g., polyaromatic hydrocarbons).
- Direct or primary carcinogens are agents reactive or toxic enough that they act directly to cause cancer in the parent, unmetabolized form (e.g., alkylating agents, radiation).
- A cocarcinogen is a chemical that increases the carcinogenic activity of another carcinogen when coadministered with it. While not carcinogenic itself, the agent may act to increase absorption, increase bioactivation, or inhibit detoxification of the carcinogen it is administered with.
- An initiator is a chemical that induces the carcinogenic process in a cell or tissue. That is, it initiates that change in the cell that irreversibly converts it to a cancerous or precancerous state. This is considered to be a mutational change that ultimately alters the phenotypic expression of the cell. Although the change may be irreversible, it often remains "latent" or unexpressed until an added stimulus (e.g., a promoter) causes the cell to grow and form a tumor.
- A promoter is a chemical that can increase the incidence of response to a carcinogen *previously* administered to the test species, or shorten the latency period for the carcinogenic response. That is, it "promotes" or stimulates the growth of the carcinogenic response induced or initiated by another chemical.

15.3 Cancer Causation and Classifications of Carcinogens

Somatic Cell Mutation Theory

The essential feature of a tumor cell that differentiates it from a normal cell is a heritable change (transmissable from one generation to the next) that frees the tumor cell from responding to the signals that keep each cell type within certain tissue-specific bounds. Since a mutation is defined as a heritable change in DNA that is transmitted to progeny, it has been proposed that the etiology of cancer is in many cases, cellular mutation, which changes the genetic expression (phenotype) of the cell; the cell then no longer behaves or responds normally.

In cancer, phenotypic change is grossly characterized by rapid cellular division, dedifferentiation of the cell, and abnormal growth patterns. Daughter cells and subsequent generations eventually form a growing mass within the affected tissue, which is labelled a tumor. This sequence of events is generally referred to as the somatic cell mutation theory of cancer etiology. Although not all chemical-induced cancers can be explained by this hypothesis, many can, and the somatic cell mutation theory is supported by several lines of evidence (see Chapter 14, on mutagenesis). The importance of this theory is that it defined the initial approach taken by the scientific community to quickly identify possible carcinogens. That is, a large battery of tests were developed to identify mutagenic chemicals; these then became likely candidates for carcinogenesis testing.

Initiation and Promotion

It is now well established that cancer causation is a multifaceted process and that chemical-induced carcinogenesis may evolve in a number of different ways. In general, chemical-induced carcinogenesis is thought to involve at least two distinct stages. The first is an initiation stage leading to a permanent genetic alteration (mutation) in the cell. The cell is then capable of a phenotypic change in which it loses many characteristics of normal cells in the same tissue. The second state is thought to be a promotional stage in which physiological and biochemical changes facilitate the growth and expression of the initiated cell (see Figure 15-1). By separating cancer induction into two distinct stages we are able to explain two troubling characteristics of chemical-induced carcinogenesis: the sometimes dormant nature of the cancer process itself, and a vast and growing literature concerned with the many different mechanisms by which chemicals may increase the incidence of tumors.

Stage 1: Initiation

C

C*

A chemical (C) enters a cell where it is metabolized to a carcinogenic agent (C*). The carcinogen attacks and irreversibly alters the genetic code of the cell.

The mutation becomes "fixed" — i.e., not repaired — and subsequent divisions yield a phenotypically different cell.

Stage 2: Promotion

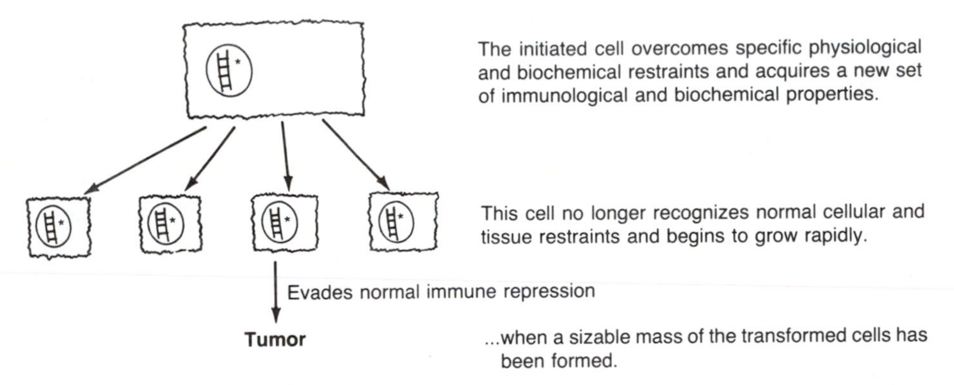

Figure 15-1. Hypothetical stages of cancer induction.

Genotoxic and Epigenetic Carcinogens

Once the diverse nature of carcinogens was recognized it was only natural that schemes to classify carcinogens according to various properties would be proposed and developed. Probably the most comprehensive and versatile classification is that of Weisburger and Williams (1981). This scheme divides chemical carcinogens into two general classes based upon whether or not the chemical acts in a genotoxic fashion. Accordingly, the two main groups of carcinogenic chemicals are:

- Genotoxic carcinogens.
- Epigenetic carcinogens.

Genotoxic carcinogens are those chemicals thought to act by directly altering DNA or genetic expression (i.e., they are initiators). This group of carcinogens includes those chemicals that produce phenotypic change by way of mutational and clastogenic changes or by changes in the fidelity of DNA replication. Epigenetic carcinogens are those chemicals whose mechanism does not involve direct interaction with genetic material. The general goal of this classification is to separate carcinogens into two general groups representing qualitatively different hazards.

In very general terms, in this chapter we shall interpret the Weisburger and Williams scheme to suggest that carcinogens of the epigenetic group are more likely to promote cancer, or to increase cellular division such that tumorigenesis is likely to occur in previously initiated cells or when the normal background incidence of spontaneous mutation increases as cell turnover increases.

Table 15-1. Classification of Carcinogenic Chemicals.

Type	Possible or probable mechanism	Examples
Group 1: Genotoxic carcinogens		
1. Direct-acting or procarcinogenic	An electrophile, the toxicant, alters the genetic code via mutagenic or clastogenic processes	Bis(chloromethyl) ether, nitrosamines, benzanthracene epoxides, dimethyl sulfate, nitrosoureas.
2. Inorganic carcinogenic	Alters the fidelity of DNA replication	Cadmium, chromium, nickel
Group 2: Epigenetic carcinogens		
1. Solid state	Mechanism unknown, but physical form of the chemical is crucial	Asbestos, metal foils, plastics
2. Hormonal	Disrupts cellular dedifferentiation, promotes cellular growth	Estrogens, androgens, thyroid hormone
3. Immunosuppressant	Depression of the immune system allows the proliferation of initiated cells or tumors	Azathioprine
4. Cocarcinogenic	Modifies the response of genotoxic carcinogens when coadministered	Ethanol, solvents, catechol
5. Promoter	Enhances cell growth, promotes response of initiator or genotoxic carcinogen	Phorbol esters, catechol, ethanol
6. Recurrent cytotoxicity or cellular injury	Increases incidence of spontaneous mutations, promotes cell turnover	Burns, repeated freezing, hepatotoxins like tetrachloroethylene

Source: Weisberger and Williams, 1981.

Thus, Table 15-1 represents an adaptation of previous classifications by Weisburger and Williams and also includes recurrent cytotoxicity or cellular injury as an epigenetic mechanism.

15.4 The Importance of Mechanistic Differences in Cancer Causation and the Impact of This on Hazard Evaluation and Risk Estimates

The One-Hit Hypothesis

When considering any toxic event, the mechanism of action is an important consideration in hazard evaluation and risk estimation, but for carcinogens it is a particularly important aspect that must not be overlooked. Genotoxic or initiating carcinogens, because they change the genetic material of a cell, pose a clear and qualitatively greater hazard to humans than do epigenetic carcinogens. Genotoxic carcinogens are occasionally effective after a single exposure, may act in a cumulative manner such that each exposure adds to the risk created by previous exposures, and may act together in an additive fashion with other genotoxic carcinogens to affect the same organ.

The greater qualitative human hazard posed by this class of carcinogens may best be simplified to two basic considerations. First, the mutagenic or genotoxic agents that are capable of permanently altering DNA are also capable of initiating other serious health hazards caused by genotoxic changes (e.g., teratogenicity, certain disease states, etc.). That is, cancer in single or multiple tissues may not be the only risk of impaired health faced by the person who is exposed. Unfortunately, because we do not know which additional adverse effects are possible and therefore do not monitor the dose-response relationships of these other toxic effects, their potential is not considered during the risk assessment. Second, inasmuch as genotoxic agents are capable of permanently altering that most basic cellular material from which all other cellular processes, cellular structure, and cellular products emanate—namely, DNA—there is an increased likelihood that the carcinogenic insult will persist in the organism or its offspring. This persistence is in contrast to a direct cellular change that may, for example, inhibit synthesis of a cellular protein. While direct cellular damage may alter the phenotypic expression of the cell, this change may be corrected or compensated for as the influence of the chemical is removed from the body by metabolism, excretion, or as the cell divides; the potential for "reversal" exists because new or replacement proteins can be synthesized from the intact genetic material. Thus, if a tumor is found to have developed, then it may be surmised that high doses or long exposures to a genotoxic carcinogen have succeeded in sustaining the change and have persisted long enough to alter the genetic material.

It is this line of reasoning that led to the so called "one-hit" hypothesis, which describes a worst-case but theoretically possible situation in which a single unrepaired event involving a critical portion of the genome leads to a malignancy. The possibility that a cancer cell could be initiated by a single genotoxic event and later develop to a malignant tumor led in turn to the current basis for extrapolating carcinogenic risk—that is, the risk that cancer will occur has *some* possibility at any and all exposures. In other words, no threshold exposure is designated because each exposure carries with it some mathematically derived risk, regardless of how small it might be, and therefore, mathematically the risk of a response (cancer) is not zero at any dose. If this basis for estimating carcinogenic risk is used, then "safe exposure" is defined in terms of a reasonable or acceptable degree of risk (for example, 1 in 1,000,000) rather than zero risk.

Promoting Carcinogens

In comparison to genotoxic or initiating carcinogens, epigenetic or promoting carcinogens represent a lesser qualitative hazard. For this type of carcinogen the tumorigenic potential is often significant only at high and/or sustained exposure levels such that a prolonged physiologic or biochemical abnormality (e.g., hormonal imbalance, immunosuppression, recurrent injury and cellular regeneration, or loss of cellular communication) acts as "growth stimulus" or "promotional pressure" that stimulates cell growth and division of latent, already initiated cells. This is probably how hormones such as estrogens and androgens promote cancer, and yet our normal bodily levels of these hormones seem to be largely without an adverse effect. Consequently, the risk from exposure to these agents may be more quantitative than qualitative in nature, and an accurate assessment of risk must take into consideration the mechanism of the critical process and a dose-response curve for it. Thus, for the epigenetic carcinogens it is theoretically possible to establish a threshold dose, or a dose that does not result in the biochemical changes that progress to tumorigenesis. While genotoxic or initiating carcinogens may also trigger events that approach thresholds, owing to DNA repair or other mechanisms that limit the critical event, genotoxic events are essentially impossible to detect in humans in advance. Since thresholds for these genotoxic carcinogens are not easily detected and verified, the possibilty that they may also have thresholds currently tends to be overlooked.

Risk Assessment

Although responses to chemicals with carcinogenic activity have been modeled according to the traditional assumption that there is no threshold and some risk is attendant to each dose, it is clear that some chemicals, while carcinogenic, display this toxicity only above some identifiable exposure level. Some scientists have criticized the "one-hit theory" by invoking arguments based on the normally dynamic nature of cells (homeostasis). The homeostatic basis suggests that constant synthesis and replacement of important cellular macromolecules imposes a lower limit on the dose-response relationship existing between cells and the adverse effects produced by chemicals. Another proposal advanced by critics of the "one-hit theory" is that thermodynamic kinetics and quantum mechanics can be applied to describe carcinogenic processes in cells and that these considerations of chemical-cell interactions establish the basis for thresholds of carcinogenicity. At present the simplest and most direct way to examine alternative arguments is to list evidence in support of and in opposition to the idea of existence of thresholds for carcinogens. Truhaut (1977) lists the following arguments (paraphrased here for brevity):

Arguments Opposing the Existence of Thresholds for Carcinogens.

1. The self-replicating nature of the cancer cell, the origin of which may be the result of a single change in DNA.
2. Evidence that small doses of some carcinogens are additive and that the incidence of cancer in some animals is related to the total dose given regardless of dose size and dosage regimen.
3. Evidence from initiation/promotion experiments indicating a permanent heritable cell transformation.
4. The fact that cancer can occur after a single dose of some chemicals, long after the chemical has been eliminated from the organism.
5. Evidence for mutations in cancer cells based upon their aberrant gene expression. Essentially all cancer cells demonstrate phenotypic change.

Arguments Supporting the Existence of Thresholds for Carcinogens.

1. The time-to-tumor concept proposes that lifespan can exert a threshold if the latency period is longer than the species in question typically lives.
2. As the dose becomes smaller the chance of a carcinogen reacting with the "right" portion of the "right" cell becomes an infinitesimal probability.
3. The existence of DNA repair and other protective mechanisms in the cell establishes the need for a sufficient amount of carcinogen to be present to produce heritable damage persisting in spite of the repair.
4. Immunologic reactions work to keep cancer cells in check and prevent or delay tumor formation.

5. Epidemiologic evidence and empirical evidence in animals suggest that carcinogens (e.g., cigarette smoke) have thresholds.

Since such conflicting evidence can be gathered by studying chemical carcinogens, it is clear that important differences exist among carcinogens, and that these alter the human hazards posed by the carcinogens, both qualitatively and quantitatively. That this is apparent is receiving widespread scientific acceptance, but as long as problems remain in determining the mechanism leading to cancer or in establishing the existence of thresholds for cancer induction by chemicals, it is not likely that regulatory agencies will agree to view carcinogenicity in the workplace consistently. Nevertheless, the technical problems do not diminish the current importance of distinctions between carcinogens or the need to separate carcinogens by category. Table 15-2 lists

Table 15-2. Characteristic Effects of Chemical Carcinogens upon Tissue.

Effect	Genotoxic or initiating carcinogens	Epigenetic or promoting carcinogens
1. Chemical produces increases in spontaneous tumors at multiple sites.	+	−
2. Chemical increases malignancy of tumors.	+	−
3. Chemical produces a tumor capable of transplantation.	+	±
4. Chemical produces a tumor that does not regress once chemical is removed.	+	±
5. Chemical must be mutagenic.	+	−
6. Chemical is effective at doses that generally produce no observable cellular toxicity.	+	−
7. Chemical may cause the effect after a single dose.	±	−
8. Chemical should increase the effect of other carcinogens acting on the target organ.	+	±
9. Progression of neoplastic differentiation does not require the continued presence of the chemical.	+	−

+ = Generally.
± = Sometimes.
− = Seldom or never.

some characteristic effects of carcinogens that should assist one in identifying the category to which a particular chemical is likely to belong.

Based upon the preceding remarks, it is clear that mathematical calculations of risk assuming no threshold and a probability of some risk at all exposures should *not* be fitted to epigenetic carcinogens. However, current regulatory practice does not weigh mechanistic distinctions nor use a mathematical approach that considers a threshold when estimating the human carcinogenic risk based on animal data, although these ideas have been discussed within regulatory agencies. Therefore, the likelihood of predicting safe doses from animal data—e.g., doses in which threshold processes are not overcome—and then verifying the safe dose in humans via medical surveillance remains a very remote possibility for genotoxic carcinogens, and would be difficult even for agents acting by an epigenetic mechanism (unless there was scientific agreement concerning the mechanism of carcinogenicity).

15.5 Testing Chemicals for Carcinogenic Activity

Guidelines for Conducting Chronic Bioassays in Small Rodents

The National Toxicology Program (NTP) is responsible in the United States for the testing of chemicals for carcinogenic activity, a responsibility formerly held by the National Cancer Institute. The commonly recommended requirements for a thorough assessment of carcinogenic potential in experimental animals generally include testing with:

- Two species of rodents, and both sexes of each species.
- Adequate controls.
- Sufficient numbers of test animals to provide a statistical basis to defend a carcinogenic effect.
- Treatment and observation extending to most of the lifetime of the animals at a dose range that includes a dosage likely to yield maximum expression of carcinogenic potential.
- Detailed pathologic examination at the termination of the treatment period, and a statistical evaluation of results.

Positive results obtained in at least one species are considered sufficient evidence for carcinogenic potential. Positive results in more limited tests (e.g., when the observation period is shortened to be less than the animal's lifetime), if the tests employ experimentally adequate procedures, are also accepted as evidence of carcinogenicity. On the other hand, negative results, regardless of test duration, are not considered to be definitive evidence that a chemical *lacks* carcinogenic activity, unless there is sufficient, additional evidence to suggest that the chemical is truly not likely to be a carcinogen.

Commonly Used Laboratory Animals

The animals used most often for carcinogenesis bioassays are mice, rats, and hamsters. These animals are most often used because (1) their lifespans are short; (2) they are easier and cheaper to test in large numbers than are larger animals; (3) the availability of inbred strains—i.e., genetically homogeneous test populations—provides consistent and known "background" cancer rates and susceptibility to carcinogens at specific organ sites (historical colonies). Adequately designed and performed studies in other mammalian species also provide useful information on carcinogenicity, but mice and rats are generally used.

For human risk evaluation, any data from bioassays using nonmammalian species are considered to provide suggestive evidence only if positive. Negative studies permit no conclusion. This stems from the fact that many differences may exist between species and these represent potential confounders to any nonhuman test results when the results are used in extrapolating the quantitative risk humans face through exposure to a carcinogen. Therefore, the greater the phylogenetic difference is between test organisms the greater the potential is for significant differences in response to the chemical, and the less certain we can be with our predictions.

Even though rodents are the most frequently used test animals, certain problems are inherent in their use. For example, certain tumor types have a high background frequency, reaching from 50 to 100 percent in some strains. Examples of these rodent strains and their tumor types are: strain A mice for lung adenomas, strain AKR mice for lymphomas, strain C3H/HeN male mice for liver cell tumors, C3H female mice for mammary tumors, and females of several rat strains for mammary fibroadenomas.

The carcinogenic activity of chemicals can be clearly demonstrated in each of the above strains by detecting substantial decreases in the period of latency for a particular tumor type, by significant increases in incidence or multiplicity of these tumor types, and by the induction of tumors of other histologic types in the same or other organs. However, caution must be used when evaluating the significance of a higher incidence of these tumors in the treated group versus the concurrent controls when the incidence in the treated animals falls within a range commonly seen in historical controls for this strain.

False Negative and False Positive Results

Both false negative and false positive results are a potential problem in carcinogen bioassays. Ideally, the number of animals required to provide adequate negative evidence should be great enough that even a false negative test (a test failing to detect existent carcinogenicity) will not allow an excessive risk to go unnoticed. The likelihood that such a risk will not be detected during the evaluation of bioassay data is dependent upon two factors (excluding species differences in response): the number of animals tested and the extent to which the test dose exceeds the usual level of human exposure.

The probability that a test will generate a false negative result is also affected by the background tumor rate in the control animals. As the background incidence of tumorigenesis increases, so does the number of animals required to detect any actual percent increase in the rate of tumor development. This means that it may be difficult to detect small increases for tumor types with large spontaneous rates in test groups of only 50 to 100 animals.

Thus to increase the safety of extrapolation the number of animals tested may need to be increased if the number of humans to be exposed to the chemical is expected to be large or if a small margin of safety exists between the animal dose tested and the expected human exposure. In general, however, resource limitations are such that only 50 animals of each sex are tested at each dose for both species; this limits the total number of animals tested to about 400 animals, plus 200 animals to serve as controls.

Test Doses

"Testing should be done at doses and under experimental conditions likely to yield maximum tumor incidence." This recommendation of a Food and Drug Administration advisory committee summarizes the issue of test doses. Bioassays done with only a few dozen or even a few hundred animals have a relatively low sensitivity for detecting weak carcinogenic effects compared to the millions of people with varying degrees of sensitivity who may be exposed to the substances under evaluation. Therefore, the largest possible dose of the carcinogen should be used.

Although a test animal cannot be strictly viewed as the "best surrogate" for a large number of people without oversimplification, the role of the animal test is to provide maximum detectability of carcinogenic effects. Thus, the larger the dose used in the test the more likely carcinogenic activity, if inherent, will be detected. The greater the ratio of the animal test dosage to the probable human dosage, the greater margin of safety provided by any negative result in a carcinogenesis bioassay.

It is generally recommended that at least two doses be tested. All carcinogens should elicit a positive dose-response relationship, but the maximal tumor incidence in the test may not occur at the highest dose tested if some form of competing toxicity occurs at this dose. Thus, the highest test dose that can be effectively used in any carcinogenesis bioassay is limited by the conditions of absorption, by the daily amount the animal can tolerate during a lifetime of administration without developing serious and unwanted side effects, and by the effect the chemical may have on the animal's nutrition when it constitutes a large proportion of the diet.

Maximum Tolerated Dose (MTD)

From the preceding discussion then it is clear that it is important to estimate the highest dose level that can be tolerated by the test animals during lifetime administration. This dose is called the maximum tolerated dose (MTD). The MTD is operationally defined for testing purposes as the highest dose that can be administered to the test animals for their lifetime but which will not produce (1) clinical signs of toxicity or pathologic lesions other than those related to a neoplastic response; (2) an alteration of the normal longevity of the animals from toxic effects other than carcinogenesis; and (3) more than a relatively small percent inhibition of normal weight gain (i.e., 10 percent or less). The MTD is determined on the basis of prechronic tests and other relevant information. A lower dose is also used; it is generally 1/2 to 1/4 the estimated MTD (EMTD).

In summary, to maximize the detection limits of sensitivity for a test and to improve the reliability of the test even though a relatively small number of animals are used (it is not economically feasible to greatly increase the number of animals tested), the maximally tolerated dose is given for the lifetime of the rodent species tested. Even though this creates certain problems, which will be discussed subsequently, these test conditions maximize the opportunity for expression of a chemical's carcinogenicity and currently provide the best basis for identifying and removing environmental carcinogenic hazards.

Short-Term Cancer Bioassays

In addition to the chronic carcinogenesis bioassay, there are a number of tests that can be performed in less time; these are usually only several months in duration. These include but are not necessarily limited to:

- Testing for skin tumor induction by painting chemicals on the shaved backs of mice.
- Testing for lung tumors in strain A mice.
- Testing for breast tumors in female mice.
- Testing for the induction of altered foci in rat livers.

These tests have the advantage of requiring less time and are therefore less costly in terms of animal maintenance. They can also provide valuable information. For example, skin tumor experiments with polynuclear aromatic hydrocarbons and phorbol esters have helped establish that cancer is a two-stage phenomenon requiring both initiation and promotion. While these tests provide suggestive evidence that a chemical may be carcinogenic, by themselves they are far less powerful evidence than the chronic bioassay.

Some of the potential problems that may be inherent in the various short-term bioassays include the following: (1) the animals being tested have such high background tumor rates that mere promotion of a background rate yields false positive conclusions; (2) the tests may be too limited and only specific tumor types or routes of administration are tested; (3) the tumors observed in the test may be benign and of questionable relevance to the malignant transformation we are concerned for in humans; and (4) the test endpoint is not tumorigenesis but is instead preneoplastic change that may or may not develop into tumors (again, the relevance for extrapolation to humans is suspect). Thus, these tests are probably best used as adjunctive evidence to mutagenicity tests and to further suggest that chronic bioassays should be conducted. In addition, they may be thought of as prelimary tests to identify chemicals that are potential epigenetic carcinogens, which have yielded negative data during mutagenicity screening.

Testing Tiers: An Approach for Identifying and Evaluating Carcinogenicity

The testing of chemicals for carcinogenesis is a costly and time-consuming enterprise. A chronic bioassay currently costs in the neighborhood of half a million dollars and probably takes at least three years for the testing and analysis of the data. It should be obvious that not every chemical in the workplace can be tested in the near future, as some 50,000 chemicals are estimated to be under present manufacture or use and many new chemicals are added each year. Thus, it is imperative that some strategy be utilized to identify the most likely carcinogenic candidates and to test them first. An additional impetus for developing a testing strategy, as discussed in previous sections, is to be able to determine any important mechanistic differences that exist between carcinogens. If it is accepted that genotoxic carcinogens, in general, represent qualitatively a greater hazard to humans than do epigenetic carcinogens, then it is clear that there should be some mechanism for discerning which type of carcinogen a chemical is likely to be.

A number of tiered testing schemes have been proposed. Their common feature is that candidate chemicals are first tested in short-term tests that are predictive of their carcinogenic potential. An example of such a scheme is given in Figure 15-2. This and similar plans are useful because they help identify by structure and mutagenicity tests those chemicals likely to be potent genotoxic or initiating carcinogens.

Use of such a scheme might help federal agencies rank chemicals and choose those that should be tested in phases III and IV, as these later test phases will be far more costly and consume more time. However, hard decisions are not completely eliminated by this or any other systematic screening approach. For example, polychlorinated biphenyls or 2, 3, 7, 8-tetrachlorodibenzo-*p*-dioxin (TCDD) are carcinogenic in mice and rats, and the

Phase I: Structure activity relationships

Use structure/activity information of chemicals by class to identify or predict which chemicals are likely to be mutagenic/carcinogenic.

↓

Phase II: Mutagenicity testing

Identify mutagens, which are clear hazards and likely genotoxic carcinogens.

- Bacterial mutagenesis tests (Ames assay)
- DNA repair tests
- Chromosomal analysis/micronucleus test (clastogenic potential)
- Mammalian mutagenesis
- Cell transformation (*in vitro* carcinogenicity)

↓

Phase III: Short-term carcinogenesis bioassays

Obtain some idea of *in vivo* potency of mutagenic compounds; test probable epigenetic carcinogens.

- Skin tumor induction in mice (test for initiation/promotion)
- Pulmonary tumor test in mice
- Mammary tumor test in rats
- Rat liver foci test
- Rat liver initiation/promotion test

↓

Phase IV: Chronic rodent carcinogenesis bioassay

Final test: generate dose-response data for risk extrapolations, get an idea of species/strain/tissue specificity of a compound, and obtain relative ranking of potency for comparison.

Figure 15-2. Testing scheme for carcinogen evaluation. (Adapted from Weisburger and Williams, 1981).

latter compound is an extremely potent carcinogen. Yet neither chemical is a potent mutagen. Thus, by relying solely on tests in phases I and II as means of selecting chemicals for more serious testing, the carcinogenicity of a chemical like TCDD may be missed or testing may be unnecessarily delayed. This only serves to emphasize what is probably obvious to the reader: there are no straight forward, easy tests for identifying epigenetic carcinogens. Therefore, testing schemes may actually be misleading in screening for carcinogens or ranking chemical-testing priorities if they do not include tests for promotion and other epigenetic events. Regardless, testing schemes are helpful when they provide a basis for separating chemicals into genotoxic/epigenetic categories, which is an important mechanistic distinction.

15.6 Evaluating Carcinogenicity Data: The Confounders and Problems

Even after testing schemes have helped identify potential mechanisms, or chronic bioassays have clearly identified carcinogenic potential, several problems with current testing methodologies may undermine the applicability of conclusions when an attempt is made to extrapolate the data to humans.

The following paragraphs summarize several confounders and problems. The intent of this section is not to erode the reader's confidence in current testing methods but, rather, to give perspective to the problems associated with evaluating data, as well as to point out some of the limitations to conclusions drawn from chronic test data.

The Doses of Chemicals Tested Are Too High and Are Not Predictive of the Human Situation

We are constantly stuck on the horns of this dilemma. It is only logical to try to maximize the sensitivity of any carcinogenesis test in the hopes of eliminating false negatives, to ensure statistical significance for small changes, and in the case of the chronic bioassay, to help set some manageable limit to the number of animals to be tested. At present, there is no solution to this problem; there is no method to alter the testing of any chemical such that lower doses can be used without first learning what occurs at maximally tolerated doses. For if lower doses give negative results, objections can always be raised about the adequacy of the test.

Regardless of the reasons supporting the using of maximally tolerated doses, a number of problems inherent in this approach undermine the use of data eventually generated. The biological arguments against the applicability of results generated by administration of high doses are:

- Large doses may alter metabolism and form reactive, toxic metabolites that are not present at lower doses; or these metabolites may be generated in such quantity that normally adequate protective mechanisms of the body are overwhelmed.
- Large doses may lead to chronic tissue irritation or injury, conditions that may lead to an increase in cancer that cannot occur at nontoxic doses.
- High doses may alter immune responses, disrupt nutrition, cause stress and/or hormonal imbalances, etc., all of which may contribute to (i.e., promote) the background rate of tumorigenesis—but again, only at overtly toxic doses.
- Large doses may alter or overwhelm DNA repair, change the fidelity of replication, induce enzymatic imbalances in the cell, etc., but again represent changes occurring largely or only at unreasonably high or overtly toxic doses.

On review, the preceding problems essentially reduce to a single concept: at high chemical doses biochemical events may be produced that induce tumorigenesis, but these events may not even occur at lower doses. In other words, a threshold dose may exist below which the critical event does not occur. Thus, a number of chemicals positive for cancer in animals at high doses may never be capable of producing cancer in humans at the lower doses representative of actual environmental or occupational situations. Yet, at present this distinction is rarely made or is not attempted.

The Route of Administration in the Test Differs from the Route of Human Exposure

While this is a relatively minor objection, the route of administration can certainly affect absorption, body distribution, metabolism, and elimination of chemicals (see Chapter 3). Thus, the animal test conditions might produce vastly different results if the actual route and amount of human exposure were mimicked.

Which Animal Species Represents the Best Human Model for the Test Compound?

The use of mice and rats is generally a compromise aimed at decreased costs. While primates or dogs might better represent humans, they cannot routinely be used for many reasons. In general, use of rodents can be criticized because rodents typically have a faster rate of metabolism than humans. Thus, at high doses, the metabolic pattern and percentage of compound metabolized may be far different than in humans. If the active carcinogen is a metabolite, then the animal surrogate may be "hypersensitive" to the chemical (may produce more metabolite) and may provide results that are not relevant to the human situation. This is a particularly pertinent concern when the chemical tested is positive in only one species. However, the problem of false negatives also applies in that selection of an insensitive species may yield a conclusion of noncarcinogenicity whereas further testing would uncover the actual tumorigenic activity. Since the selection of animal models is always a debatable issue, an open attitude of critical evaluation of data demonstrating species differences should probably be maintained when extrapolating risk to humans by means of bioassay data.

Some Test Animals Are Too Sensitive to Certain Carcinogenic Responses

The use of data in which the observed tumor in the animal is one with a high background (normal) incidence raises several questions. For example, are the mechanisms of cancer initiation or promotion the same for this chemical in humans? Is the potency of the chemical ranked erroneously high because of the observation of tumors not initiated but only promoted by the compound?

For example, liver tumors are probably the most common type identified by chronic bioassays to date and the background incidence of these tumors in rodents is usually high. Yet it is a relatively rare cancer in humans. In addition, a number of chemicals, particularly hepatotoxins, are positive for liver tumors only in rodents, a finding in keeping with an epigenetic or recurrent injury mechanism for selective liver-toxicity at high chemical doses. Along this line, several recent reviews have discussed the relevance of results in which the only positive finding is liver tumorigenesis in rodents.

Detection of Benign Tumors in Test Animals Has Little or No Value in Predicting the Carcinogenic Potential of a Chemical

As discussed earlier, human cancer is clinically determined by malignancy. Yet, an increase of animal tumors, either benign or malignant, is frequently used as evidence of a chemical's carcinogenic potential. In fact, for animal cancer bioassays certain definitions have been adopted for liver tumors that differ significantly from the description of human liver cancer. The proposed classification scheme of the National Cancer Institute (NCI) is extensive and specific in defining neoplastic lesions. It was concluded by the NCI that benign hepatic cell tumors (i.e., tumors without the potential for malignant behavior) could not be consistently diagnosed correctly. Therefore, the NCI considers a term such as "adenoma," which refers to a benign tumor, to be imprecise and does not recommend it to describe liver tumors.

Contrary to human clinical tradition, then, detection of vascular invasion or metastasis is not considered by the NCI to be essential for the diagnosis of hepatocellular carcinoma. Instead, the use of cytological criteria describing changes that may or may not have the potential to become the disease process we call cancer, including many responses that will never progress to cancer, has been accepted at the expense of the more definitive criteria of malignant or invasive neoplasms. Thus, it seems that the criteria for establishing liver cancer varies in a critical way between animal and human studies. Again, the relevance of using such data can be questioned, but the alternative, which is to ignore such data and thereby make carcinogenesis tests more restrictive and therefore less sensitive to potential species differences, is likewise unpalatable.

Overview of the Problems

The preceding should indicate that there are certain problems associated with the evaluation of carcinogenicity data or its extrapolation, which do not necessarily invalidate such tests but do raise questions regarding the relevance of their application to all human situations. In other words, while we continue our attempts to identify chemicals with carcinogenic potential, such efforts do not necessarily provide definitive evidence that a clear human hazard or risk exists. Thus, just as we are concerned with identifying potentially hazardous chemicals and eliminating them from situations that lead to human exposure, we should also be concerned about eliminating potentially useful and beneficial chemicals from commerce when animal data does not reflect the nature of the human response to the chemicals.

15.7 Occupational Carcinogens

Intensive toxicological research and careful scrutiny of industrial health records have produced numerous lists of chemicals with carcinogenic potential. The criteria by which these lists were compiled vary and often inclusion or exclusion is dependent upon different evaluative criteria applied to the same data by different reviewers. However, even though there is considerable disagreement on the strength or weakness of data supporting carcinogenic potential for specific compounds, some consistency has developed over the years. That is, there have emerged classes of compounds, or occasionally only specific members of a class, which have repeatedly demonstrated mutagenic or tumorigenic capacity in a variety of test systems. In addition, some very solid statistical correlations have been drawn from these chemicals between occupational exposure and the rate of specific tumors in exposed populations.

The recent reviews by IARC (1982), Fishbein (1979), and Stokinger (1981) are useful in evaluating large amounts of data generated recently on occupational compounds with carcinogenic potential. The information from these recent reviews on occupational carcinogenics has been adapted in Table 15-3, which lists classes of compounds, with representative examples, that have been widely recognized as posing an occupational cancer hazard to man. The positive result of animal studies and an evaluation of positive association in human epidemiology are listed.

Structural Activity Relationships

Figure 15-3 illustrates the general structural formulae of compounds representative of the carcinogenic groupings included in Table 15-3. It is important to note that while structural similarity may be suggestive of related carcinogenic potency, this is not universally true. For example, the closely related compounds ethylene chlorohydrin and epichlorohydrin differ only slightly in structure but the latter is a much more potent carcinogen in animal species than the former compound.

Other lists of carcinogenic chemicals have been developed, and these vary according to the purpose of the list, extent of data considered, etc.

Table 15-3. Selected Compounds of Occupational Importance that Show Carcinogenic Potential for Humans.

Category and compound	Demonstrated mutagen	Animal carcinogen	Suspect human carcinogen	Probable human carcinogen
I. Metals and physical agents				
Arsenic	L	N	Y	Y
Asbestos	N	Y	Y	Y
Chromium (hexavalent)	Y	Y	Y	Y
Nickel	N	Y	Y	N
Radiation	Y	Y	Y	Y
II. Alkylating agents				
Azaridine (ethyleneimine)	Y	Y	Y	Y
Bis(chloromethyl) ether	L	Y	Y	Y
Dimethyl sulfate	Y	Y	Y	N
Epichlorohydrin	Y	Y	Y	N
Ethylene dibromide	Y	Y	Y	N
Formaldehyde (gas)	Y	Y	Y	N
III. Hydrocarbons				
Aromatic				
Benzene	L	L	Y	Y
Benzo[a]pyrene	Y	Y	Y	N
Soots, tars	Y	Y	Y	Y
Aliphatic				
Vinyl chloride	Y	Y	Y	Y
IV. Hydrozines, carbamates				
Hydrazine	Y	Y	Y	N
Urethane	Y	Y	Y	N
V. Aromatic amines				
Benzidine	Y	Y	Y	Y
2-Naphthylamine	Y	Y	Y	Y
o-Toluidine	Y	Y	Y	N
VI. Unsaturated nitrites				
Acrylonitrile	Y	Y	Y	Y

Y: Data indicate carcinogenic potential.
N: Data inadequate to conclusively indicate carcinogenic potential.
L: Limited data indicate carcinogenic potential.
Sources: Fishbein (1979); IARC (1982); and Stokinger (1981).

The number of chemicals tested for carcinogenicity has been estimated at about 7000 chemicals, with some 1000 or more of these positive in one type of carcinogenicity test or another. Many of these positive compounds are merely structural analogs tested, for example, in the skin tumor bioassay for the purpose of determining structure/activity relationships. Therefore, no definitive statement can be made about the percentage of positive carcinogens to be expected for each 100 compounds tested. Although it is not the purpose of this

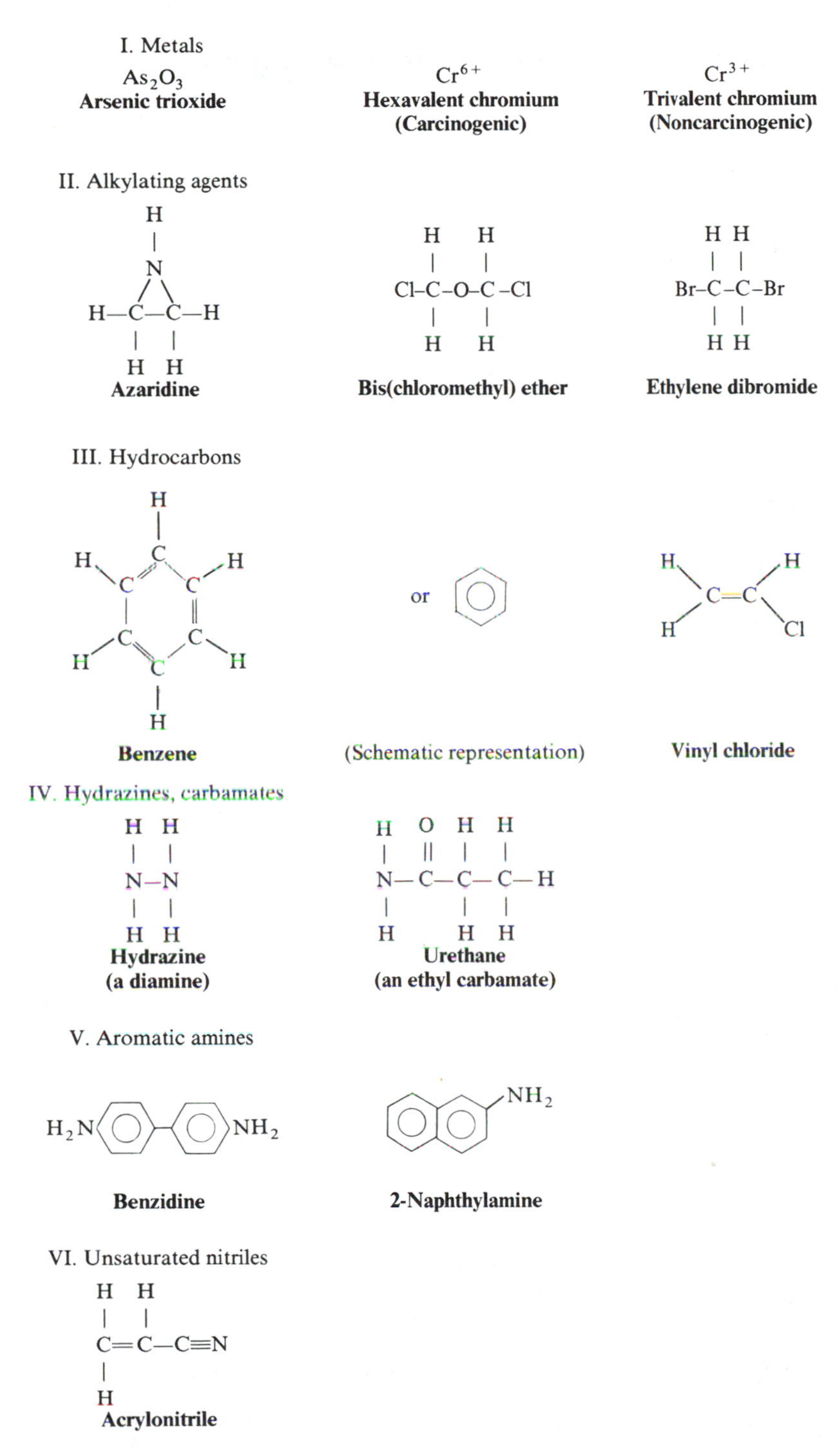

Figure 15-3. Selected structural formulae for some classes of carcinogenic compounds.

Table 15-4. Carcinogens Listed in the Third Annual Report of the National Toxicology Program.

A. Known human carcinogens

4-Aminobiphenyl
Arsenic and certain arsenic compounds
Asbestos
Auramine manufacture
Benzene
Bis(chloromethyl) ether
Chlorambucil
Chlornaphazine
Chromium and certain chromium compounds
Coke oven emissions
Cyclophosphamide
Diethylstilbestrol
Hematite
Isopropyl alcohol manufacture
(strong acid process)
Melphalan
Mustard gas
2-Naphthylamine
Nickel refining
Soots, tars, mineral oils
Thorium dioxide
Vinyl chloride

B. Suspected or reasonably anticipated to be human carcinogens

2-Acetaminofluorene
Acrylonitrile
Aflatoxins
2-Aminoanthraquinone
1-Amino-2-methylanthraquinone
Amitrole
o-Anisidine and *o*-anisidine hydrochloride
Aramite
Benz[*a*]anthracene
Benzo[*b*]fluoranthrene
Benzo[*a*]pyrene
Beryllium and certain beryllium compounds
Carbon tetrachloride
Chloroform
p-Cresidine
Cupferron
Cycasin
2, 4-Diaminoanisole sulfate
2, 4-Diaminotoluene
Dibenz[*a*, *h*]acridine
Dibenz[*a*, *j*]acridine
Dibenz[*a*, *h*]anthracene
7*H*-Dibenzo[*c*, *g*]carbazole
Dibenzo[*a*, *h*]pyrene
Dibenzo[*a*, *i*]pyrene
1, 2-Dibromo-3-chloropropane
1, 2-Dibromoethane
3, 3′-Dichlorobenzidine
1, 2-Dichloroethane
Diepoxybutane
Di(2-ethylhexyl)phthalate
3, 3′-Dimethoxybenzidine
4-Dimethylaminoazobenzene
3, 3′-Dimethylbenzidine
Dimethylcarbamoyl chloride
Dimethyl sulfate
1, 4-Dioxane
Direct Black 38
Direct Blue 6
Ethylene thiourea
Formaldehyde
Hexachlorobenzene
Hydrazine and hydrazine sulfate
Hydrazobenzene
Indeno[1, 2, 3-*cd*]pyrene
Iron dextran complex
Kepone (chlordecone)
Lead acetate and lead phosphate
Lindane and other hexachlorocyclohexane
isomers
4, 4′-Methylenebis(2-chloroaniline) (MOCA)
4, 4′-Methylenebis(*N*, *N*-dimethyl)benzenamine
Michler's ketone
Mirex
Nickel and certain nickel compounds
Nitrilotriacetic acid
5-Nitro-*o*-anisidine
Nitrofen
Tris(1-aziridinyl)phosphine oxide
N-Nitrosodi-*n*-butylamine
N-Nitrosodiethanolamine
N-Nitrosodimethylamine
p-Nitrosodiphenylamine
N-Nitrosodi-*n*-propylamine
N-Nitroso-*N*-ethylurea
N-Nitroso-*N*-methylurea
N-Nitrosomethylvinylamine
N-Nitrosomorpholine
N-Nitrosonornicotine
N-Nitrosopiperidine
N-Nitrosopyrrolidine
N-Nitrososarcosine
Oxymetholone
Phenacetin
Phenazopyridine and phenazopyridine hydrochloride
Phenytoin and sodium salt of phenytoin
Polybrominated biphenyls
Polychlorinated biphenyls
Procarbazine and procarbazine hydrochloride
β-Propiolactone
Reserpine
Saccharin
Safrole
Selenium sulfide
Streptozotocin
Sulfallate
2, 3, 7, 8-Tetrachlorodibenzo-*p*-dioxin
(TCDD)
Thioacetamide
Thiourea
o-Toluidine and *o*-toluidine
hydrochloride
Toxaphene
2, 4, 5-Trichlorophenol
Tris(1-aziridinyl)phosphine oxide
Tris(2, 3-dibromopropyl)phosphate
Urethane

Source: National Toxicology Program, *Third Annual Report on Carcinogens*. Washington, D.C.: U.S. Department of Health and Human Services, 1982.

chapter to present an exhaustive list of the chemicals that have been tested for carcinogenicity, several additional lists are provided.

Occupational Carcinogens Listed by the National Toxicology Program

Table 15-4 is a list contained in the Third Annual Report on Carcinogens, which was prepared in 1982 by the National Toxicology Program of the U.S. Public Health Service and will be updated at intervals in the future. The table lists 117 chemicals. Section A includes a few technological processes along with 19 substances considered by the weight of evidence to be "known human carcinogens." The other 98 chemicals, in section B, are compounds "which may reasonably be anticipated to be carcinogens." These are chemicals for

Table 15-5. The IARC Listing of Chemicals and Industrial Processes Associated with Carcinogenicity.

A. Industrial exposures and chemicals causally associated with cancer (i.e., by sufficient human evidence)

1. Industrial exposures
   - Auramine manufacture
   - Boot and shoe manufacture and repair (certain occupations)
   - Furniture manufacture
   - Isopropyl alcohol manufacture (strong acid process)
   - Nickel refining
   - Rubber industry (certain occupations)
   - Underground hematite mining (with exposure to radon)

2. Chemicals
   - 4-Aminobiphenyl
   - Analgesic mixtures containing phenacetin
   - Arsenic and arsenic compounds
   - Asbestos
   - Azathioprine
   - Benzene
   - Benzidine
   - *N*, *N*-Bis(2-chloroethyl)-2-naphthylamine (chlornaphazine)
   - Bis(chloromethyl) ether and technical-grade chloromethyl methyl ether
   - 1, 4-Butanediol dimethanesulfonate (myleran)
   - Certain combined chemotherapy for lymphomas (including MOPP)
   - Chlorambucil
   - Chromium and certain chromium compounds
   - Conjugated estrogens
   - Cyclophosphamide
   - Diethylstilbestrol
   - Melphalan
   - Methoxsalen with ultraviolet A therapy (PUVA)
   - Mustard gas
   - 2-Naphthylamine
   - Soots, tars, and oils
   - Treosulphan
   - Vinyl chloride

B. Chemicals and technologies probably carcinogenic to humans (i.e., limited human evidence exists for such an association)

- Acrylonitrile
- Aflatoxins
- Benzo[*a*]pyrene
- Beryllium and beryllium compounds
- Combined oral contraceptives
- Diethyl sulfate
- Dimethyl sulfate
- Manufacture of magenta
- Nickel and certain nickel compounds
- Nitrogen mustard
- Oxymetholone
- Phenacetin
- Procarbazine
- *o*-Toluidine

Table 15-5. (Continued)

C. Chemicals possibly carcinogenic to humans (i.e., there are inadequate human data but sufficient animal data with possible mutagenic evidence, or generally limited human and animal data)

- Actinomycin D
- Adriamycin
- Amitrole
- Auramine (technical grade)
- Benzotrichloride
- Bis(chloroethyl)nitrosourea (BCNU)
- Cadmium and cadmium compounds
- Carbon tetrachloride
- Chloramphenicol
- 1-(2-Chloroethyl)-3-cyclohexyl-1-nitrosourea (CCNU)
- Chloroform
- Chlorophenols (occupational exposure to)
- Cisplatin
- Dacarbazine
- Dichlorodiphenyltrichloro-ethane (DDT)
- 3, 3′-Dichlorobenzidine
- Dienestrol
- 3, 3′-Dimethoxybenzidine (*o*-Dianisidine)
- Dimethylcarbamoyl chloride
- 1, 4-Dioxane
- Direct Black 38 (technical grade)
- Direct Blue 6 (technical grade)
- Direct Brown 95 (technical grade)
- Epichlorohydrin
- Estradiol-17β
- Estrone
- Ethinylestradiol
- Ethylene dibromide
- Ethylene oxide
- Ethylene thiourea
- Formaldehyde (gas)
- Hydrazine
- Mestranol
- Metronidazole
- Norethisterone
- Phenazopyridine
- Phenytoin
- Phenoxyacetic acid herbicides (occupational exposure to)
- Polychlorinated biphenyls
- Progesterone
- Propylthiouracil
- Sequential oral contraceptives
- 2, 3, 7, 8-Tetrachlorodibenzo-*p*-dioxin (TCDD)
- 2,4,5-Trichlorophenol
- Tris(aziridinyl)-*p*-benzoquinone (Triaziquone)
- Tris(1-aziridinyl)phosphine oxide (Thiotepa)
- Uracil mustard

D. Industrial chemicals not classified as to carcinogenic potential to humans (i.e., inadequate human evidence and limited-to-inadequate animal or mutagenicity data exist)

- Aldrin
- Aniline
- Chlordane
- Chlorinated toluenes (benzyl, benzoyl, or benzal chlorides)
- Chloroprene
- (2, 4-Dichlorophenoxy)acetic acid (2, 4-D) and esters (occupational exposure only)
- Dichlorobenzenes
- Dieldrin
- Heptachlor
- Hexachlorocyclohexane
- Leather, lumber, pulp, and paper manufacture (certain exposures)
- Lead and lead compounds
- Magenta
- Methylene chloride
- 1-Naphthylamine
- Pentachlorophenol
- Phenyl-2-naphthylamine
- Styrene and styrene oxide
- 2, 4, 5-Trichlorophenoxy)acetic acid (2, 4, 5-T) and esters (occupational exposures only)
- Tetrachloroethylene
- Trichloroethylene
- 2, 4, 5-Trichlorophenol
- Vinylidine chloride

Table 15-6. Occupational Carcinogens Listed by the ACGIH.

Category I. Human carcinogens (substances associated with industrial processes, recognized to be carcinogenic or to have carcinogenic potential in humans).

Acrylonitrile	Chromium(VI) (certain water-insoluble compounds)
4-Aminobiphenyl (*p*-xenylamine)	Coal tar pitch volatiles
Asbestos	β-Naphthylamine
Benzidine	Nickel sulfide roasting, fumes, and dust
Bis(chloromethyl) ether	4-Nitrobiphenyl
Chromite ore processing (chromate)	Vinyl chloride

Category II. Industrial substances suspected of having carcinogenic potential in humans.

Acrylonitrile	Ethylene oxide
Amitrol	Formaldehyde
Antimony trioxide production	Hexachlorobutadiene
Arsenic trioxide production	Hexamethyl phosphoramide
Benzene	Hydrazine
Benzo[*a*]pyrene	4, 4′-Methylenebis(2-chloroaniline)
Beryllium	Methylhydrazine
1, 3-Butadiene	Methyl iodide
Cadmium oxide production	2-Nitropropane
Carbon tetrachloride	*N*-Nitrosodimethylamine
Chloroform	*N*-Phenyl-β-naphthylamine
Chloromethyl methyl ether	Phenylhydrazine
Chromates of lead and zinc, as Cr	Propane sultone
Chrysene	β-Propiolactone
3, 3′-Dichlorobenzidine	Propyleneimine
Dimethylcarbamyl chloride	*o*-Tolidine
1, 1′-Dimethylhydrazine	*o*-Toluidine
Dimethyl sulfate	Vinyl bromide
Ethylene dibromide	Vinyl cyclohexene dioxide

Note: The ACGIH has compiled additional categories for a "Proposed Classification of Experimental Animal Carcinogens."

which there is limited but suggestive evidence in humans, or for which there is animal evidence to suggest a potential human hazard.

The IARC Listing of Carcinogens

Table 15-5 lists those chemicals tabulated by the IARC, a program of the World Health Organization, which publishes critical reviews of carcinogenicity data on groups of chemicals in its series of monographs entitled *IARC Monographs on the Evaluation of the Carcinogenic Risk of Chemicals to Humans*. Table 15-5 represents a compilation taken from volumes 1–29 of the IARC monographs. In Table 15-5 the chemicals are separated into various categories depending upon the strength of the human data, animal data, and mutagenicity test results. Assignment of a chemical to a category was done by the IARC working group. For more information, the reader is referred to IARC (1983) in the reference list at the end of the chapter.

The ACGIH and OSHA Listings of Occupational Carcinogens

Table 15-6 lists those chemicals or substances considered by the American Conference of Governmental Industrial Hygienists (ACGIH) to be either human carcinogens or compounds suspected of having human carcinogenic potential.

Workplace carcinogens are regulated by the Occupational Safety and Health Administration (OSHA) and a list of these chemicals is provided in Table 15-7. As one can readily observe, this listing is much briefer than the other lists. This difference is due in part to the mechanism required to establish these regulations; in many cases the process falls far behind the current

Table 15-7. The OSHA Listing of Suspected Workplace Carcinogens.

Acrylonitrile	Ethyleneimine
2-Acetylaminofluorene	Inorganic arsenic
4-Aminobiphenyl	Methyl chloromethyl ether
Asbestos	α-Naphthylamine
Benzidine	β-Naphthylamine
Bis(chloromethyl) ether	4-Nitrobiphenyl
Coke oven emissions	*N*-Nitrosodimethylamine
1, 2-Dibromo-3-chloropropane	β-Propiolactone
3, 3′-Dichlorobenzidine	Vinyl chloride
4-Dimethylaminoazobenzene	

toxicological data. As a result, OSHA's allowable airborne levels for some chemicals may be much higher than what the scientific data would suggest.

An example of this is benzene. In the late 1970s concern for the carcinogenicity of benzene led to the lowering of the permissible exposure limit. However, various special interest groups were able to defeat this new standard. Many other chemicals are still regulated under the ACGIH levels that were initially adopted by OSHA in 1970.

15.8 Cancer and Our Environment: Factors That Alter Our Risks to Occupational Hazards

Over the past years, a number of publications have discussed the role that environment plays in cancer causation. Depending upon the reviewer and the definition of environment, estimates of this role vary widely. While predicted or accepted estimates may vary for each environmental factor for each review of the subject, there is common agreement that lifestyle is a major contributing factor in cancer causation.

The importance of this fact is twofold. First, it is important to view cancer as a phenomenon intimately involved with normal biological processes and therefore affected by those factors that may influence biologic functions—e.g., diet. Second, environmental factors often are an overwhelming influence in epidemiologic studies of occupational hazards. That is, they may mask or exacerbate weak carcinogenic responses and therefore are difficult confounders to any study that is not normalized in a manner that removes these influences from any association being studied.

Table 15-8 lists various factors and their contribution to the human "cancer load." From this table it is clear that smoking and diet, two factors that many

Table 15-8. Estimated Causes of Cancer by Percent Contribution to Total Cancer Load.

Cancer cause	Estimated percent of total cancer incidence
Occupation	1–5
Leukemia, lymphoma, etc.	10–15
Tobacco smoking	23
Diet	
Contaminants (mycotoxins, etc.)	5
High fat, low fiber, etc.	44
Tobacco/Alcohol	5
Iatrogenic	1

people underrate, are probably the most important factors regarding cancer risk. The following summarizes the evidence for the contribution of various environmental factors. It is hoped that the reader will gain an appreciation of why certain persons run increased occupational risks, and why epidemiologic studies may provide equivocal evidence for weak carcinogenicity for chemicals except when the agent is potent or produces rare forms of cancer.

Genetic Makeup of Individuals

The genetic makeup of an individual affects predisposition to cancer from industrial chemicals and environmental factors alike. There are a number of cancers that are linked to heredity alone and are not dependent upon environmental exposures. Likewise, genetic makeup may determine the extent and route of metabolism of chemicals, thereby affecting the extent to which toxic metabolites are generated during detoxification of chemicals. Other possible or probable genetic differences influence immune surveillance of precancerous states, cellular responses to injury and repair, and the capacity for DNA repair or the fidelity with which genetic material is replicated. Regardless of the biochemical mechanism, it is clear that heritable differences among individuals affect susceptibility to cancer.

Smoking

The American Cancer Society estimates that cigarette smoking is responsible for 83 percent of lung cancer cases in men and 43 percent in women or more than 75 percent of the lung cancer in all persons. Smoking affects cancer at other sites as well, and the cancer death rate for male smokers is double that of nonsmokers. In addition, smoking is a major cause of heart disease and has been linked to chronic bronchitis, emphysema, and gastric ulcers. There may even be passive smoking hazards to nonsmokers who breathe the smoke of others.

Industrial workers are especially susceptible to lung disease for which smoking greatly increases the risk. For example, workers who have combined asbestos and cigarette exposure have a cancer mortality risk that is 60 times that of persons not exposed to either substance. Likewise, it has been suggested that the association of cigarette smoking and inhalation of fumes from rubber or fluorocarbon polymers, chlorine, and carbon or cotton dust may increase the risk of cancer. Research has identified more than 40 carcinogenic substances emitted in cigarette smoke. Many of these compounds are initiating (genotoxic) carcinogens capable of inducing cancer by themselves at sufficient doses. In apparent contradiction to this, studies of persons who have quit smoking suggest that the effects of smoking on lung cancer may be more of a promotional phenomena. This aspect has been suggested by studies that indicate that as the duration of abstinence increases, a person's risk of cancer becomes lower until it eventually approaches the risk faced by a nonsmoker. If cigarettes acted largely as initiating or genotoxic carcinogens, the cumulative exposure would then be proportional to the risk, and cessation of cigarette exposure would not lessen the ultimate risk (i.e., the accumulated risk would not decrease with time). Regardless of whether smoking is largely a promotional pressure or not, it is a clear, avoidable health hazard, our single most preventable cause of illness and death, and one that can easily enhance the risk of exposure to many other carcinogens encountered in the workplace. The relationship of smoking to lung disease is detailed in Chapter 9.

Alcohol

Alcohol consumption is another clearly avoidable lifestyle-related cause of cancer. Epidemiologic evidence indicates that excessive alcohol consumption increases the risk of cancer of the mouth, larynx, esophagus, liver, and lung. Further, it appears that it acts additively or synergistically with tobacco in producing cancer of the upper digestive tract. It has been estimated that 76 percent of the human cancer at alcohol/tobacco-related sites could be eliminated if both addictions were avoided. Estimates of cancer associated with alcohol range as high as 5 percent, yet it is also estimated that there are some ten million problem drinkers in the United States, which suggests that the alcohol-influenced cancer rate should be higher. Research with pure alcohol (ethanol) in animals has failed to demonstrate that it is carcinogenic. Thus, some discussion has been raised about whether alcohol itself or the chemicals found in alcoholic beverages are carcinogenic. But metabolites of alcohol and alcoholic beverages are mutagenic, and the carcinogenic association with alcohol seems to be independent of the kind of beverage abused (i.e., wine versus beer, etc.). Thus, it must be concluded that alcohol abuse is a clear risk factor that may modulate or create associated workplace risks if not carefully controlled for in an epidemiologic study.

Diet

The relationship between diet and cancer is a complex and sometimes perplexing association. It is perplexing because the rate for stomach cancer in the United States decreased by about two-thirds between 1950 and the mid 1970s, a result believed to be related to a decreased consumption of salt-cured, pickled products or smoked foods, combined with the increased refrigeration of foods and increased consumption of antioxidants. Contrary to this, the increased consumption of fats, particularly of animal origin, is considered to be a promoter of breast and pancreatic cancer.

It is important to try to understand the relationship of diet and cancer because some estimates project that diet is an environmental factor associated with as much as 50 percent of human cancer. The complexity of the relationship arises because carcinogens and mutagens are natural products of many foodstuffs and can be produced by cooking; moreover, foods provide substrates for the formation of carcinogens in the body; diet alters the activation to or detoxification of carcinogens; and a number of nutrients (like antioxidants vitamins A, C, and E) act to prevent cancer.

Dietary fat is a problem that is faced by all of us. It is thought to promote cancer by a number of potential mechanisms, which include (1) alteration of hormone levels; (2) change in the composition of cellular membranes; (3) increase in fatty acids, which may inhibit immune responses or serve as precursors to prostaglandins, which then may act as promoters; and (4) stimulation of the production of bile acids, some of which act as promoters. Studies in animals have shown that decreasing fat or caloric intake lowers the cancer incidence, and this suggests that providing food to animals *ad libitum* (at will) in chronic bioassays may increase the background cancer incidence observed during studies.

Iron, zinc, and selenium deficiencies have all been associated with an increased cancer rate. Vitamin C has been shown to reduce carcinogen formation in animals and appears to inhibit the formation of certain initiating carcinogens; vitamin E appears to prevent promotion; and vitamin A appears to decrease the susceptibility of epithelial tissues to carcinogens. Nitrates and nitrites occur naturally and are salts that are introduced to our diet as food preservatives. Nitrites may form carcinogenic nitrosamines in the acid environment of the stomach by combining with amines from the protein in our diet. Broiling foods generates certain carcinogenic polyaromatic hydrocarbons and converts proteins to powerful mutagens. Lastly, chlorination of water during disinfection procedures creates a number of chlorinated hydrocarbons which are carcinogenic in animals.

From the preceding it is again clear that nutrition or diet can have a profound impact on cancer incidence. This may affect not only exposures in the workplace; it may also affect epidemiologic research because, combined with genetic makeup and smoking or drinking habits, the factor of diet gives strong reason to select control populations from the same area as the occupational (subject) group in epidemiologic studies. Ignorance of these influences may partially explain differences observed in similar factories located in different geographic regions.

Iatrogenic Cancer

Iatrogenic or drug-induced cancer is often overlooked. Some examples have gained widespread attention recently, such as problems associated with estrogens because of their widespread use in oral contraceptives, or the highly publicized problem of diethylstilbestrol- (DES-) caused cancer. Yet a number of other drugs have been linked with cancer as well, and the list is likely to continue to grow. Some of these drugs are rarely used, such as cancer therapeutic agents (e.g., cyclophosphamide, busulphan, etc.), but other less certain associations have been suggested for more widely used drugs, such as the sedative phenobarbital. Inasmuch as, when people age, medical problems increase and with them the use of medications increases, this use of drugs may also affect occupationally related cancer in a way that is difficult to factor into an epidemiologic assessment of risks.

15.9 Cancer Trends and Their Impact Upon the Evaluation of Cancer Causation

Human Cancer Trends

Having discussed occupational carcinogens and the environmental factors that influence cancer incidence, it seems logical to discuss cancer trends in the United States to gain some perspective as to how these two causative factors might be interacting and changing our cancer rates. Over the past decade or so three areas of concern have led some people to prophesy future doom: these are increased awareness of the number of carcinogens we are exposed to; the increasing number of industrial chemicals identified as carcinogens in animals; and increasing concern for environmental pollution from industrial sources.

As awareness of carcinogens, carcinogenesis, and pollution has increased, so has speculation that life in a "sea of carcinogens" will lead to a cancer epidemic. Formerly it was postulated that after some interval spanning the latency period for chemical-induced cancer, say 20 years or so, this epidemic would become evident. However, the data on human cancer trends do not support such speculation (see Figure 15-4).

Aside from the increase in lung cancer, which has been attributed to smoking, the rate of cancer mortality has remained fairly stable or has decreased during the past 50 years—a time of great industrialization in the United States. It seems that we have done remarkably well to date, therefore, and with the legislative steps taken in the past decade to identify and eliminate exposures to carcinogens, it now seems likely that the would-be future cancer epidemic will never occur. This is not to say that serious sources of occupational and environmental exposure have not been found, only that our current condition and future need not be considered dire.

Comparison of Populations

While the trends in cancer mortality are reassuring, the process for establishing the relative importance of ambient pollution, lifestyle factors, and, ultimately, occupationally related cancer has been complicated by technical difficulties. Any evaluation of these trends and their impact has been hindered by changes in the diagnostic criteria used for different types of cancer, reporting inade-

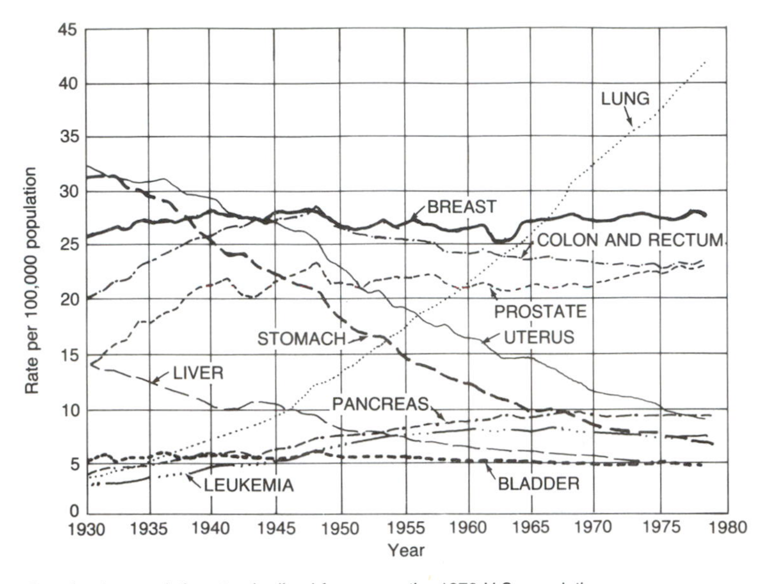

Rate for the population standardized for age on the 1970 U.S. population.
Sources of Data: National Center for Health Statistics and Bureau of the Census, United States.
Note: Rates are for both sexes combined except breast and uterus (female population only) and prostate (male population only).

(a)

Figure 15-4. Cancer death rates in the United States. (a) Cancer death rates by site, 1930–1978. (b) 25-year trends in age-adjusted cancer death rates per 100,000 population for the years 1951–1953 to 1976–1978. (Adapted from American Cancer Society, *Cancer Facts and Figures*, 1983.)

quacies, variations in population boundaries, changes in population makeup caused by the mobility of our society, and changes in those social habits that are confounding influences in epidemiologic studies. A consequence of these difficulties is that a number of artifacts are introduced into the trend analysis of such studies; sensitivity—ability to discriminate small changes—is reduced, and a number of biases, sometimes unknown or unsuspected, are introduced.

These problems hinder epidemiologic studies and emphasize the need for a careful, well-controlled comparative analysis between "identical" populations when attempting to identify carcinogens or factors influencing carcinogenesis. For example, in Table 15-9 the cancer mortality in the United States is listed by state. While we might expect heavily populated states like California and New York to have a larger number of cancer deaths than less heavily populated states like Rhode Island, Vermont, or Wyoming, that the rate—i.e., deaths per hundred thousand persons—varies as much as three- to four-fold between states is unexpected. It should be clear to the reader that there are problems inherent in discerning trends when local cancer rates fluctuate as

Sex	Sites	1951–1953	1976–1978	Percent Changes	Comments
Male	All sites	171.9	215.7	+ 25	Steady increase mainly due to lung cancer.
Female	All sites	146.4	136.1	− 7	Slight decrease.
Male	Bladder	7.2	7.2	*	Slight fluctuations; overall no change
Female	Bladder	3.1	2.1	− 32	Some fluctuations; noticeable decrease.
Male	Breast	0.3	0.3	*	Constant rate.
Female	Breast	26.0	27.1	+ 4	Slight fluctuations; overall no change.
Male	Colon and rectum	25.8	26.4	*	Slight fluctuations; overall no change
Female	Colon and rectum	24.8	20.0	− 19	Slight fluctuations; noticeable decrease.
Male	Esophagas	4.7	5.4	+ 15	Some fluctuations; slight increase.
Female	Esophagas	1.2	1.5	*	Slight fluctuations; overall no change in females
Male	Kidney	3.4	4.7	+ 38	Steady slight increase
Female	Kidney	2.1	2.2	*	Slight fluctuations; overall no change
Male	Leukemia	7.9	8.8	+ 11	Early increase, later leveling off.
Female	Leukemia	5.4	5.2	*	Slight early increase, later leveling off.
Male	Liver	6.7	4.8	− 28	Some fluctuations. Steady decrease in
Female	Liver	7.6	3.6	− 53	both sexes
Male	Lung	25.5	69.3	+ 172	Steady increase in both sexes due to
Female	Lung	5.0	17.8	+ 256	cigarette smoking
Male	Oral	5.9	5.8	*	Slight fluctuations; overall no change in
Female	Oral	1.5	2.0	*	both sexes.
Female	Ovary	8.1	8.6	+ 8	Steady increase, later leveling off.
Male	Pancreas	8.6	11.2	+ 30	Steady increase in both sexes, then leveling off.
Female	Pancreas	5.5	7.1	+ 29	Reasons unknown
Male	Prostate	21.0	22.6	+ 8	Fluctuations all through period; overall no change.
Male	Skin	3.1	3.4	*	Slight fluctuations; overall no change in both
Female	Skin	1.9	1.9	*	sexes.
Male	Stomach	22.8	9.3	− 59	Steady decrease in both sexes;
Female	Stomach	12.3	4.3	− 65	reasons unknown.
Female	Uterus	20.0	8.7	− 57	Steady decrease

*Percent changes not listed because they are not meaningful.

(b)

Figure 15-4. (*Continued*)

much as those in Table 15-8 do; the reader can also understand the difficulty in finding changes in the workplace by using the national average rate of cancer as the control population. (Note the number of states considerably higher or lower than the United States average figure for mortality per 100,000 persons; note also the considerable difference between neighboring northeastern states like Rhode Island and Massachusetts versus Vermont or New Hampshire, and the large differences even for rural states like Colorado and Wyoming versus Kansas and Nebraska.)

Overall Assessment of the Hazard

In conclusion, a review of the cancer trends in the United States for the past several decades indicates that there is no "cancer epidemic," nor is there likely to be an epidemic such as has been predicted from time to time. Likewise, there is little evidence to suggest that past pollution has had a major impact on cancer trends or incidence, or that ambient levels of industrial chemicals will have this result in the future. This does not imply that all industrial chemicals and pollution are harmless and should not be controlled. Instead, it appears to suggest that current vigilance and improvement will help curtail future prob-

Table 15-9. Estimated Total Cancer Mortality for 1983 for the United States.

	All sites			All sites	
State	Number of deaths	Death rate per 100,000 persons	State	Number of deaths	Death rate per 100,000 persons
Alabama	7,400	186	Montana	1,300	162
Alaska	300	69	Nebraska	3,200	201
Arizona	4,900	165	Nevada	1,300	136
Arkansas	4,700	203	New Hampshire	1,900	196
California	44,500	175	New Jersey	16,400	219
Colorado	3,900	124	New Mexico	1,800	129
Connecticut	6,700	210	New York	37,500	212
Delaware	1,200	198	North Carolina	10,400	170
Dist. of Columbia	1,600	260	North Dakota	1,200	179
Florida	26,600	237	Ohio	21,500	200
Georgia	9,000	155	Oklahoma	6,000	183
Hawaii	1,200	118	Oregon	5,000	186
Idaho	1,400	141	Pennsylvania	27,000	227
Illinois	22,500	195	Rhode Island	2,400	248
Indiana	10,400	192	South Carolina	4,900	150
Iowa	5,700	199	South Dakota	1,300	193
Kansas	4,600	190	Tennessee	8,600	184
Kentucky	7,000	191	Texas	22,400	140
Louisiana	7,600	167	Utah	1,500	91
Maine	2,500	217	Vermont	950	181
Maryland	8,600	197	Virginia	8,900	158
Massachusetts	12,600	215	Washington	7,300	165
Michigan	16,400	181	West Virginia	3,900	199
Minnesota	7,300	177	Wisconsin	9,000	186
Mississippi	4,600	180	Wyoming	650	120
Missouri	10,500	210	United States	440,000	187

Sources: Adapted from American Cancer Society (1983).

lems and that a general pollution-caused cancer epidemic is not a likely future prospect. The trends and wide differences also underscore how limited epidemiologic studies, with their associated variables and potential artifacts, are. In turn, this emphasizes the need for accurate interpretation and extrapolations of risk from animal data, since such data are the easiest and most basic approach to assessing human cancer risks.

Summary

Chemical-induced carcinogenesis represents a unique and complex area within toxicology. The difficulty in assessing the carcinogenic hazards and human risks of chemicals stems from the following characteristics of chemical carcinogenesis.

- It is a multistage process involving at least two distinct stages: initiation, which converts the genetic expression of the cell from a normal to aberrant cell line; and promotion, in which the aberrant cell is stimulated in some fashion to grow, thereby expressing its altered state.
- Since chemicals may increase cancer incidence at various stages and by different mechanisms, the term carcinogen by itself is somewhat limiting and a number of descriptive labels are applied to the chemical carcinogens that define or describe these differences; e.g., cocarcinogens, initiators, promoters, epigenetic, etc.
- Chemicals may produce or affect only a single stage or a single aspect of carcinogenesis which leads to a number of important differences and considerations about the potential health impacts of chemical carcinogens. Perhaps the most important considerations are the concept of thresholds and that qualitative differences do exist among carcinogens.
- Carcinogenicity testing raises many questions about interpretations of results. Considerations such as mechanism (genotoxic versus epigenetic), dose, relevant test species, etc., are important in determining probable human risk; thus, many additional toxicity test data are needed to improve the extrapolation of cancer bioassay data from test species to humans.
- A number of lifestyle-related factors influence carcinogenesis, altering the risks posed by carcinogenic chemicals and acting to confound epidemiologic evidence.

Considering the complexities involved in (1) determining the mechanism of cancer causation, (2) using animal and human data to identify carcinogenic substances, and (3) using these data to extrapolate risks with the aim of reducing or eliminating environmental risk factors, it should be clear to the reader that the best approach to occupational carcinogenesis is an interdisciplinary one. As depicted in Figure 15-5, identifying and reducing occupational cancer requires the interfacing of several scientific disciplines and several kinds of health professionals.

- The toxicologist is responsible for testing and identifying chemical carcinogens; through animal testing the toxicologist attempts to provide information about carcinogenetic mechanisms, and about species differences or similarities that can aid in assessing the human risk.
- Epidemiologists add human evidence to risk evaluations or ascertain if a chemical should or should not be considered a human carcinogen for various reasons (it may have weak or undetectable activity).

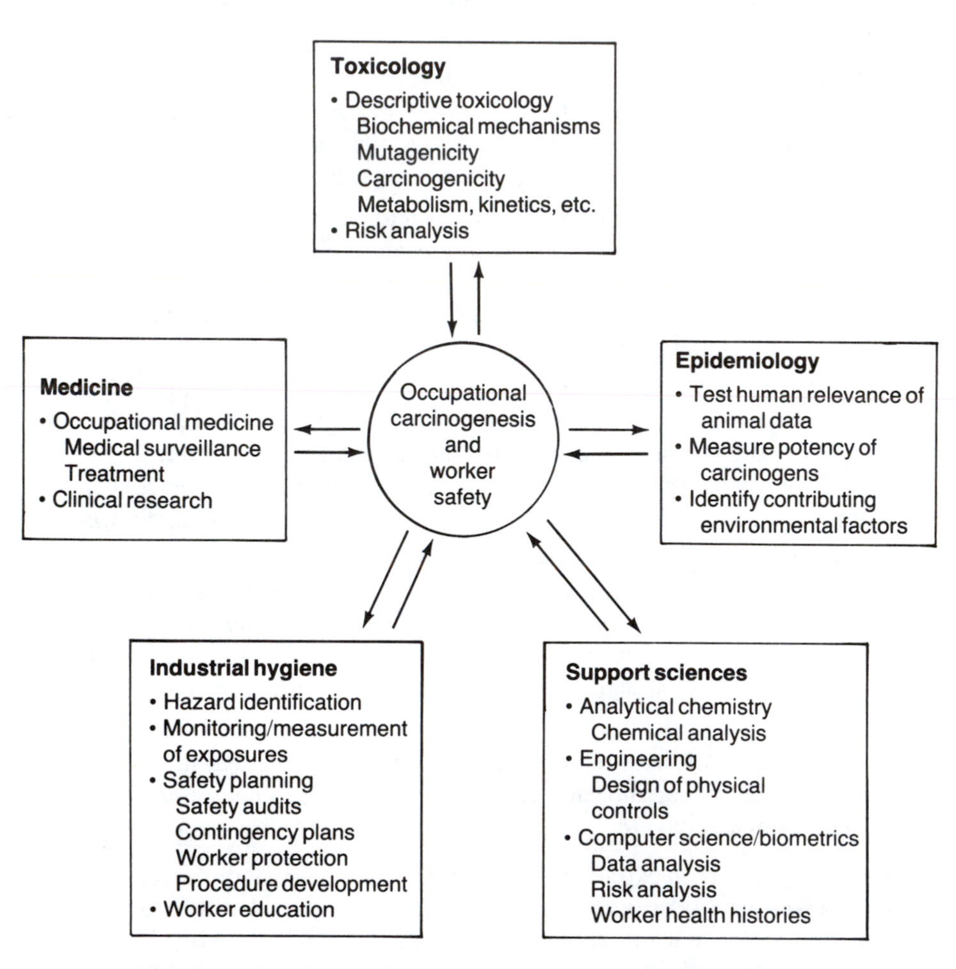

Figure 15-5. Identifying and reducing chemical carcinogens requires an interdisciplinary approach, in which health professions interface with other scientific disciplines.

- Specialists in occupational medicine provide health surveillance programs to protect the health status of the worker and attempt to prevent those exposures that could lead to serious, chronic health problems.
- Industrial hygienists help design better methods for evaluating and preventing worker exposures; and biometrists and computer scientists aid in risk analysis, data storage, and data analysis.

So long as these disciplines are utilized jointly and their relationship to the occupational carcinogenesis problem is understood, occupational health and safety professionals can have good reason to hope for improved success in the prevention of occupational carcinogenesis.

References and Suggested Reading

American Cancer Society. 1983. *Cancer Facts and Figures*. New York: American Cancer Society.

Ames, B. N. 1983. "Dietary Carcinogens and Anticarcinogens." *Science, 221*: 1256.

Bartsch, H., Tomatis, L., and Malaveille, C. 1982. "Mutagenicity and Carcinogenicity of Environmental Chemicals." *Regul. Toxicol. Pharmacol., 2*: 94.

Dinman, B. D. 1972. "'Non-Concept' of 'No-Threshold': Chemicals in the Environment." *Science, 175*: 495.

Doull, J. (Chairman). 1983. "The Relevance of Mouse Liver Hepatoma to Human Carcinogenic Risk." A Report of the International Expert Advisory Committee to the Nutrition Foundation. Washington, D.C.: The Nutrition Foundation.

Elmes, E. C. 1981. "Relative Importance of Cigarette Smoking in Occupational Lung Disease." *Brit. J. Ind. Med., 38*: 1.

Fishbein, L. 1979. *Potential Industrial Carcinogens and Mutagens*. New York: Elsevier Scientific Pub.

Griffin, A. C., and Shaw, C. R. (Eds.) 1979. *Carcinogens: Identification and Mechanisms of Action*. New York: Raven Press.

Henderson, R. E., Ross, R. K., Pike, M. C., and Casagrande, J. T. 1982. "Endogenous Hormones as a Major Factor in Human Cancer." *Cancer Res., 42*: 3232.

National Toxicology Program. 1982. *Third Annual Report on Carcinogens*. Washington, D.C.: Department of Health and Human Services.

National Toxicology Program. 1983. *Fiscal Year 1983 Annual Plan*. Washington, D.C.: Department of Health and Human Services.

Office of Technology Assessment. 1981. *Assessment of Technologies for Determining Cancer Risks from the Environment*. Washington, D.C.: OTA Publications.

Park, N. C., and Snee, R. D. 1984. "Quantitative Risk Assessment: State-of-the-Art for Carcinogenesis." *Fund. Appl. Toxicol., 3*: 320.

Peto, R. 1980. "Distorting the Epidemiology of Cancer: The Need for a More Balanced Overview." *Nature, 284*: 297.

Schaeffer, D. J. 1981. "Is 'No-Threshold' a 'Non-Concept'?" *Envir. Manag., 5*: 475.

Slaga, T. J., Sivak, A., and Boutwell, R. K. (Eds.) 1979. *Carcinogenesis—A Comprehensive Survey. Volume 2: Mechanisms of Tumor Promotion and Cocarcinogenesis*. New York: Raven Press.

Squire, R. A. 1981. "Ranking Animal Carcinogens: A Proposed Regulatory Approach." *Science, 214*: 877.

Squire, R. A., and Levitt, M. H. 1975. "Report of a Workshop on Classification of Specific Hepatocellular Lesions in Rats." *Cancer Research, 35*: 3214.

Stokinger, H. E. 1981. "Occupational Carcinogenesis." *In* Clayton, G. D., and Clayton, F. E. (Eds.) *Patty's Industrial Hygiene and Toxicology, Volume III, 3rd Edition*. New York: Wiley-Interscience Pub.

Stott, W. T., and Watanabe, P. G. 1982. "Differentiation of Genetic versus Epigenetic Mechanisms and Its Application to Risk Assessment." *Drug Metab. Rev. 13*: 853.

Tomatis, L., Agathe, C., Bartsch, H., Huff, J., Montesano, R., Saracci, R., Walker, E., and Wilbourn, J. 1978. "Evaluation of the Carcinogenicity of Chemicals: A Review of the Monograph Program of the International Agency for Research on Cancer." *Cancer Res., 38*: 877.

Truhaut, R. 1977. "Can Permissible Levels of Carcinogenic Compounds in the Environment Be Envisioned?" *Toxicol. Environ. Safety, 1*: 31.

Weisburger, E. K. 1978. "Mechanisms of Chemical Carcinogenesis." *Ann. Rev. Pharmacol. Toxicol., 18*: 395.

Weisburger, J. H., Cohen, L. A., and Wynder, E. L. 1977. "On the Etiology and Metabolic Epidemiology of the Main Human Cancers." *In* Hiatt, H. H., and Watson, J. D. (Eds.) *Origins of Human Cancer*. Cold Spring Harbor Laboratory.

Weisburger, J. H., and Williams, G. M. 1980. "Chemical Carcinogens." *In* Doull, J., Klaassen, C. D., and Amdur, M. O. (Eds.) *Casarett and Doull's Toxicology: The Basic Science of Poisons*. 2nd Edition. New York: Macmillan.

Weisburger, J. H., and Williams, G. M. 1981. "Carcinogen Testing: Current Problems and New Approaches." *Science, 214*: 401.

Williams, G. M. 1981. "Liver Carcinogenesis: The Role for Some Chemicals of an Epidemic Mechanism of Liver-Tumor Promotion Involving Modification of the Cell Membrane." *Fd. Cosmet. Toxicol., 19*: 577.

Chapter 16

Reproductive Toxicology

Martha Radike

Introduction

The purpose of this chapter is to present the range and types of effects that substances used in occupational settings can have on the reproductive systems of both men and women.

This chapter will:

- Review the nature and development of the male and female reproductive systems.
- Discuss substances that affect the human reproductive system.
- Discuss the potential for occupation-related birth defects and inherited anomalies.

The survival of a species depends on the integrity of its reproductive system. In human reproduction, genes in the chromosomes of germ cells, the sperm and the ovum, contain all the information necessary for the development of a single cell into a complex individual. Damage by physical or chemical agents to the sperm, ovum, or the fertilized ovum may cause infertility, spontaneous abortion, birth defects, or may result in mutations that are passed on to future generations. When considering occupational exposures that may damage reproductive processes, the hazard to both men and women must be examined.

16.1 Male Reproductive Toxicology

Development of the Male Reproductive System

The sexual development of a male begins at about seven weeks after conception, is completely determined by 85 days, and the male sex organs are fully developed, including testicular descent, by seven months of gestation. Male germ cells (sperm) are not produced until puberty, when internal and external stimuli trigger the release of specific hormones.

In the brain, hypothalamic releasing factors regulate the release of follicle stimulating hormone (FSH) from the pituitary into the blood stream. FSH stimulates specialized cells (Sertoli) in the testes, initiating a series of biochemical events that cause the hypothalamus-pituitary to release a second hormone, luteinizing hormone (LH). The target of LH is also specialized cells in the testes, Leydig cells, which synthesize and release the male steroid hormone, testosterone (Figure 16-1). It is interesting that the chemical structures of FSH and LH are identical in men and women; only the target cells and the

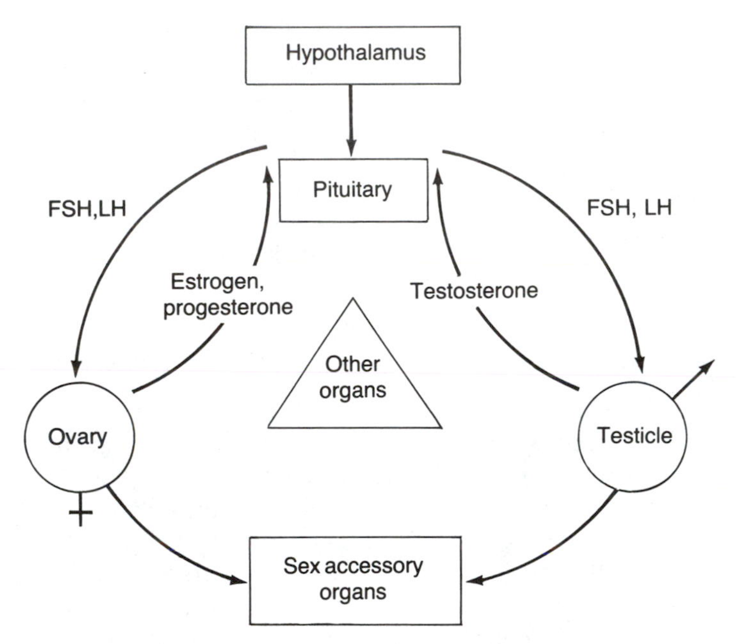

Figure 16-1. Normal hypothalamic-pituitary-gonadal relationships control reproductive functions in males and females. The synthesis and release of testosterone in the male and estrogen and progesterone in the female are controlled by the hypothalamic secretion of LHRH (luteinizing-hormone-releasing hormone), which stimulates the pituitary to release into the bloodstream the gonadotropic hormones LH and FSH (luteinizing hormone and follicle-stimulating hormone). Blood levels of the sex steroid hormones exert a feedback control on the hypothalamic–pituitary axis, thus assuring spermatogenesis in the male and menstruation and ovulation in the female. (Reproduced with permission from Smith, "Reproductive Toxicity: Hypothalamic–Pituitary Mechanisms." *American Journal of Industrial Medicine, 4*(1983). p. 108. Figure 1.)

biological responses differ. Leydig cells release about 7 mg testosterone per day, which is required for the production of sperm, sexual behavior, secondary sex characteristics, and accessory organ development. Both the Sertoli and Leydig cells are necessary for the proper functioning of the reproductive process. Physical or chemical agents that damage these cells can cause infertility.

The process of spermatogenesis (stem germ cell to mature sperm) takes about 70–80 days. This dynamic process of constant cell division makes the male reproductive system uniquely susceptible to chemical insult. Spermatogonia (stem germ cells), the only cells that can develop into mature sperm, undergo mitotic division (mitosis). Half of these cells (spermatocytes) undergo further division (meiosis) in which one spermatocyte produces four spermatids, each of which carries one-half of the genetic material (DNA) of the spermatocyte. These haploid cells cannot fertilize an ovum and are not motile.

There is a lengthy maturation process in which the nucleus of the spermatid is condensed, the cytoplasm is eliminated, the tail is developed, and the characteristic shape of the sperm is formed (Figure 16-2). The number of sperm produced per day ranges from 21 million to 374 million, lower in humans than in other mammalian species. A man is considered infertile if his ejaculate contains less than 20 million sperm per milliliter. The large number of sperm per ejaculate is necessary to overcome difficulties encountered in reaching

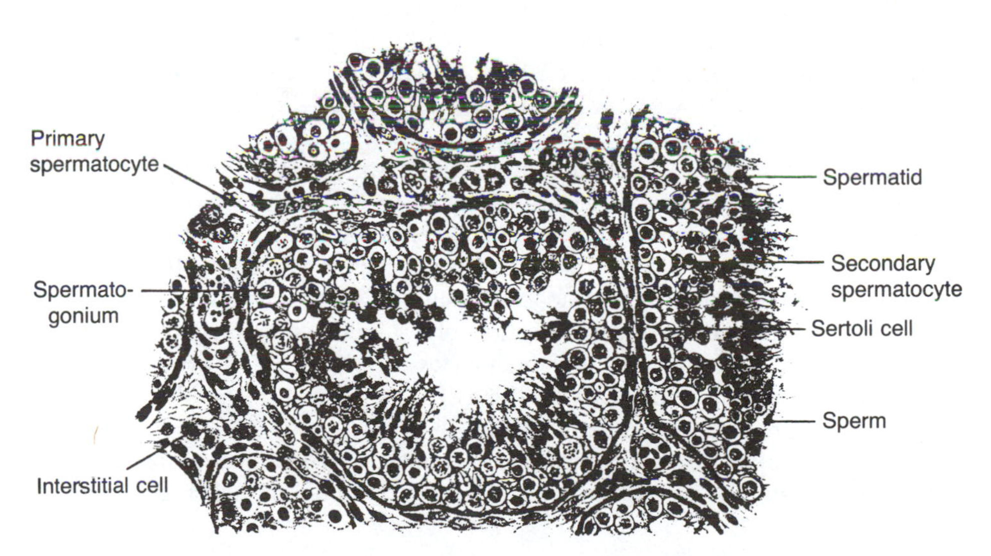

Figure 16-2. Section of human testis, illustrating spermatogenesis in seminiferous tubule. (Reproduced with permission from Ganog, *Review of Medical Physiology, Seventh Edition*. Los Altos (Calif.): Lange Medical Publications. 1975. Figure 23-14.)

and fertilizing the ovum. The average human sperm count is approximately 100 million per milliliter of ejaculate, with a range of 50 to 150 million per milliliter.

Throughout adult life, physical or chemical agents may cause hormonal imbalance by damaging testicular Sertoli or Leydig cells. Depression of testosterone synthesis or release induces lowered sperm counts and changes in sexual behavior. Sterility is the consequence of damage to spermatogonia or the developing sperm, and if spermatogonia are destroyed, sterility is permanent.

Foreign chemicals reach the testes by systemic distribution via the blood after ingestion, inhalation, or absorption through the skin. Substances may also accumulate in the prostatic secretions that make up part of the seminal fluids at the time of ejaculation.

Male Reproductive Toxins

Lead and Other Metal Compounds. Occupational exposure to lead compounds is known to decrease the fertility of male workers and increase the rate of spontaneous abortions in their wives. As long ago as the early 1900s, marital life records of men working in storage battery plants revealed that 24.7 percent had sterile marriages, compared to 14.8 percent in a control group. Another study indicated that wives of printers exposed to lead had an abortion rate of 14 percent, in comparison to 4 percent for the community. In a study of 442 pregnancies in women married to lead workers, 66 ended in spontaneous abortion and 241 in premature birth. Such data suggest that exposures were very high and that wives probably had considerable exposure from their husbands' clothing.

In a 1983 study, blood lead concentrations ≥ 53 μg lead/100 ml blood were associated with reduced fertility. Another recent study disclosed a significant increase in the number of abnormal sperm and a significant decrease in sperm count in men with a mean blood lead level of ≥ 41 μg/100 ml. The hypothesis that exposure to lead can result in mutations in the sperm is indirectly supported by the observation of a significantly increased rate of chromosomal aberrations in the lymphocytes of lead-exposed workers (mean blood-lead concentrations of 64 to 78 μg/100 ml). Lead in the blood reflects recent exposure rather than body burden.

Other metals and metal compounds are male reproductive toxins because they cause testicular damage in man or animals: cadmium, nickel, and methyl mercury.

Halogenated Pesticides. Chronic exposure to a number of halogenated pesticides has been linked to adverse responses of the male reproductive system. Factory workers who were exposed to the nematocide 1,2-dibromo-3-

chloropropane (DBCP) became sterile. Some men were sterile because of low sperm counts (oligospermia); others were sterile because the germ cell population was completely destroyed (germinal aplasia). Those men who suffered from germinal aplasia and had no sperm in their ejaculate (azoospermia) became permanently sterile. Normal spermatogenesis was recovered in men who were sterile because of insufficient numbers of sperm. Histologic examination of testes from permanently sterile men showed no damage to Sertoli and Leydig cells. The spermatogonia were apparently the only cells damaged by the pesticide. Other halogenated pesticides that have elicited toxic responses in men are kepone and DDT.

Organic Solvents. The solvent carbon disulfide (CS_2) has been reported in two U.S.S.R. studies to reduce sperm counts and induce abnormal morphology in the sperm of exposed workers. However, a recent NIOSH study of U.S. workers found no statistically significant differences in seminal abnormalities between CS_2-exposed and nonexposed workers; high exposure was > 10 ppm, moderate was 2–10 ppm, and low exposure was < 2 ppm. Only 50 percent of the workers initially contacted were willing to take part in the NIOSH study, however. Recent animal studies in which rats inhaled 600 ppm CS_2 suggest that decreased sperm counts in ejaculates are due to a central-nervous-system-mediated effect on ejaculation rather than testicular toxicity. Swedish workers who handle the organic solvents toluene, benzene, and xylene have been reported to have low sperm counts, abnormal sperm, and varying degrees of fertility.

Workers exposed to dinitrotoluene (DNT), used in the synthesis of toluenediamine (TDA), have been found to have reduced sperm counts, and several showed increased numbers of sperm with abnormal morphology. A NIOSH investigation of these workers was brought about by a young man who had fathered one child before he started working in the chemical industry. He and his wife wanted a second child but, ironically, before conceiving, they waited 90 days from the date of his employment so that their hospitalization benefits would apply to the wife's pregnancy, and this was sufficient time for exposure of all germ cell stages during spermatogenesis. The NIOSH investigation was requested after the worker's wife experienced three spontaneous abortions over several years. Animal studies indicate that DNT may be the chemical responsible for the toxic effects.

Other Agents. Fertility experts report that men whose work exposes them to extreme heat have fewer children than those who work in normal temperatures. For sperm to mature properly, the temperature of human testicles should remain about 2.2°C lower than the rest of the body. Infertility seen in welders and chefs and attributable to extreme heat is reversible if temperatures are lowered. The effects of long-term heat exposure on male reproductive processes are not known.

In the late 1960s and early 1970s, exposure to waste anesthetic gases was examined in relation to reproductive health. Of particular interest was an increase in congenital abnormalities in children born to wives of operating room-exposed male anesthetists. Since these women had no direct exposure to anesthetic gases, the findings suggested an effect on spermatogenesis. There was no increase in spontaneous abortions in this group of women. When male animals were exposed to anesthetic gases the effects were a decrease in sperm production and an increase in sperm abnormalities.

Based on reproductive histories, other agents known to decrease the number of human sperm include chloroprene, microwave exposure, x-irradiation, and occupational exposure to oral contraceptives (estrogens).

The best-known therapeutic drugs that cause infertility are those used for cancer chemotherapy, most of which are alkylating agents. Recreational drugs such as alcohol decrease the number of sperm, and chronic ingestion has produced testicular atrophy. Both the number of sperm and testosterone levels are decreased in marijuana smokers, and there is a small but statistically significant increase in morphologic abnormalities of sperm in cigarette smokers.

In addition to low sperm counts and infertility, reproductive failure may be manifested in other ways, as loss of libido or as impotence. Modes of toxic action may differ, too. Some agents may penetrate the blood-testes barrier and enter the fluids in which the sperm are stored; foreign substances in the semen can have a detrimental effect on implantation and the embryo.

When investigating possible reproductive toxins in the workplace, health professionals must examine not only exposures of workers, but also alcohol ingestion, drug intake, and smoking habits. Hobbies and activities in the home may also contribute to reproductive toxicity: stained glass manufacture is a source of lead exposure; gardening—pesticide exposure; furniture stripping—solvent exposure.

Agents known to be toxic to human male reproduction are listed in Table 16-1.

Table 16.1. Agents Toxic to the Human Male Reproductive System.

Physical agents	Chemicals
Microwaves	Alkylating agents
High-altitude exposure to ionizing radiation (14,000 feet or more)	Anesthetic gases
x-Irradiation	Carbon disulfide
High temperatures	Chloroprene
	Vinyl chloride
Social habits	Pesticides
Alcohol ingestion	Dibromochloropropane
Cigarette smoking	Kepone
Marijuana smoking	DDT
Metals	Therapeutic agents
Lead	Female oral contraceptives
Cadmium	Chemotherapeutic cancer agents
	Narcotics

Animal Test Methods for Evaluating Male Reproductive Toxins

Laboratory animal studies may not always predict the effect of a specific agent in humans; however, carefully controlled studies can indicate risk and, in some cases, the mechanism of toxicity. The male reproductive system of laboratory animals is similar to that of the human male, and chemically induced changes in hormones, testicular morphology, and sex accessory-gland responsiveness are frequently paralleled in man.

Dominant Lethal Assay. The dominant lethal assay has been a widely used method for determining if a chemical agent causes infertility in males or mutagenic damage to sperm (see Chapter 14). Male mice or rats are given high concentrations of the test compound for five days and are then sequentially mated with untreated females each week for ten weeks. Bred females are sacrificed before term and the number of early and late deaths and the number of live fetuses are recorded. Serial mating assesses sperm cell function and produces fertility patterns related in time to the stage of spermatogenesis damaged by the treatment. An indicator of lethal mutagenesis in sperm is a statistically significant increase in the number of early embryonic deaths in females mated to mutagen-treated males. In all animal studies the results from treated animals are compared to those obtained from an equal number of animals treated with the vehicle used for solubilizing the toxic agent under study.

Other Assessments. Useful information can be obtained by weighing and measuring the testes, prostate, seminal vesicles, and other structures from treated animals and comparing results to those from untreated animals. Light microscopy and transmission electron microscopy of the testes and pituitary aid in identifying the cell type that has been damaged. Many biochemical parameters can be investigated, such as enzyme activity indicative of normal or abnormal cell differentiation or function. Changes in blood levels of LH, FSH, and/or testosterone may indicate damage or suppression of activity in the testes, hypothalamus, or pituitary. Sperm analysis is valuable with viability, motility, abnormal morphology, and sperm count being scored. A more recent approach in male reproductive studies is to assess mating behavior. The use of various combinations of these methods can furnish a clue to the mechanism of a reproductive toxin's action.

Evaluation of Male Reproductive Toxins in Humans

Studies of human semen and microscopic analyses of sperm are well-established methods for assessing the reproductive potential of workers exposed to physical and chemical agents. For example, exposures to ionizing radiation and some chemicals have resulted in low sperm counts, decreased motility, and abnormal sperm morphology. Blood samples are frequently taken for the determination of circulating levels of FSH, LH, and testosterone.

Since between-male variability in semen characteristics is high among fertile and healthy men, rather large numbers of subjects are usually necessary to establish differences between control and exposure groups. After groups of men are selected, samples of semen and blood are collected and are coded on delivery to the laboratory so that analyses are carried out as blind studies. Each man is also asked to complete a questionnaire to determine his work history (including place and type of work, chemicals handled frequently, and exposure levels if available), medical history, personal habits, and fertility data.

Only a few epidemiologic studies have been directed towards occupationally related reproductive hazards. NIOSH recommends that a short occupational history form be developed for use in collecting data for studying the possible reproductive effects of specific hazardous substances or work environments.

16.2 Female Reproduction and Toxicology

Development of the Female Reproductive System

Sexual differentiation begins during the seventh week of embryonic development. If an ovum is fertilized by a sperm containing an X chromosome, the embryo develops into a female. Female sexuality is completely defined in about 100 days, and the sex organs are fully formed at about five months of gestation. In contrast to male development, in which no sperm are produced until puberty, the female fetus develops all the ova she will ever have before her birth. The proper term for the immature ovum is oocyte; for this discussion the terms *ovum* or *ova* will be used. At birth, the female infant has about 300,000 or 400,000 ova present in each ovary. At puberty, the number of ova per ovary has dimished to between 150,000 and 200,000, and at 30 years of age, about 25,000 remain; after menopause, remaining ova and follicles are resorbed. In a woman's lifetime, about 400 mature ova are released.

At the time of birth, all germ cells in the ovary are arrested in the first stage of meiosis and remain in that state until puberty, when FSH is secreted from the pituitary upon the release of specific factors from the hypothalamus (Figure 16-1). LH secretion from the hypothalamus-pituitary then acts with FSH to stimulate the developing ovarian follicle to secrete estrogen during the growth period of the ovum and to nurture and expel the mature ovum. FSH blood levels fall while LH and estrogen levels peak at the time of ovulation. The ovum does undergo meiosis and four haploid cells are produced from one diploid germ cell as in spermatogenesis; however, only one of the four cells develops into an ovum that can be fertilized by sperm.

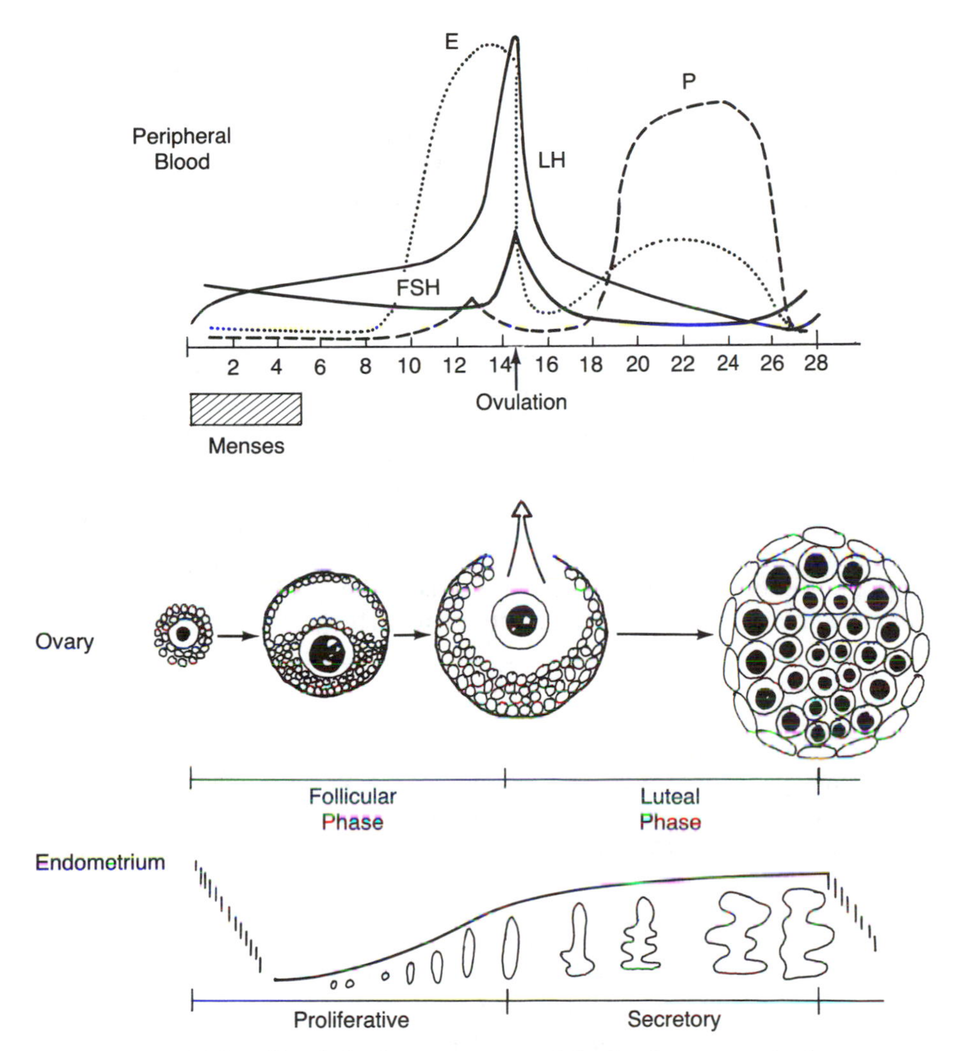

Figure 16-3. Circulating hormones control the development and release of the ovum, the synthesis and release of hormones during the luteal phase in the ovary, and the proliferation of the endometrium in the uterus prior to implantation of a fertilized ovum or menstruation: follicle stimulating hormone, FSH; estrogen, E; luteinizing hormone, LH; progesterone, P. If fertilization occurs, the blood levels of P and E do not drop, but remain high so that menstruation does not occur. (Reproduced with permission from Mattison (Ed.), "Reproductive Toxicology." *American Journal of Industrial Medicine*, 4(1983):1–2. p. 20.)

The ruptured follicle is transformed into a structure called the corpus luteum, which synthesizes and releases estrogen and a fourth hormone, progesterone. If the ovum is not fertilized, estrogen and progesterone levels fall and menstruation occurs; FSH and LH surges initiate another cycle of ovulation and menstruation (Figure 16-3). The steroid sex hormones, estrogen and progesterone, also have important regulatory roles in the growth, development, and maintenance of the female sex organs, and in the development of secondary sexual characteristics such as mammary glands, hair, fat distribution, and voice. The cyclic release of ova continues for 30 to 40 years after puberty.

There are no data indicating that immature ova before puberty or the maturing and mature ova are damaged by environmental or occupational agents other than irradiation. In addition to irradiation, drugs used in cancer chemotherapy offer the best evidence that chemical as well as physical agents can cause infertility in females. There is evidence in animal models that chemical agents can damage the ovary if administered in high doses. Of 77 compounds tested in juvenile mice, 21 caused significant ovum loss. All compounds positive for ovum-killing were agents known to have mutagenic or carcinogenic properties; however, 22 additional mutagens or carcinogens were negative in the assay.

Toxicology and Stages of Pregnancy

The term "perinatal toxicology" is one you will encounter often in the literature because it designates the study of toxic responses to occupational and environmental agents when exposure occurs from the time of conception through the neonatal period. It is important to recognize that exposures of the conceptus during specific periods of gestation elicit different responses. Figure 16-4 depicts the stages of gestation and the biological responses associated with exposures to reproductive toxins during specific periods.

In humans, fertilization of the ovum by the sperm occurs in the fallopian tube. From the time of fertilization to implantation on the wall of the uterus (Figure 16-5), the developing embryo is particularly susceptible to genetic abnormalities and toxic insults. Implantation can occur only during a limited period of time after which the endometrium loses its receptivity for the fertilized ovum. After implantation of the embryo (blastocyst), the placenta begins to develop from the outer cell layer of the embryo. The inner cell mass then differentiates into endodermal and ectodermal cells. Ionizing radiation and some chemicals and drugs have a toxic effect on the early embryo; alterations in membrane permeability, disruption of the mitotic spindle, inhibition of enzyme activities, and unrepaired mutations can lead to cell death, which is lethal to the embryo during these early stages. Before three weeks of gestation, differentiation has not progressed to the point that the embryo can survive cell death.

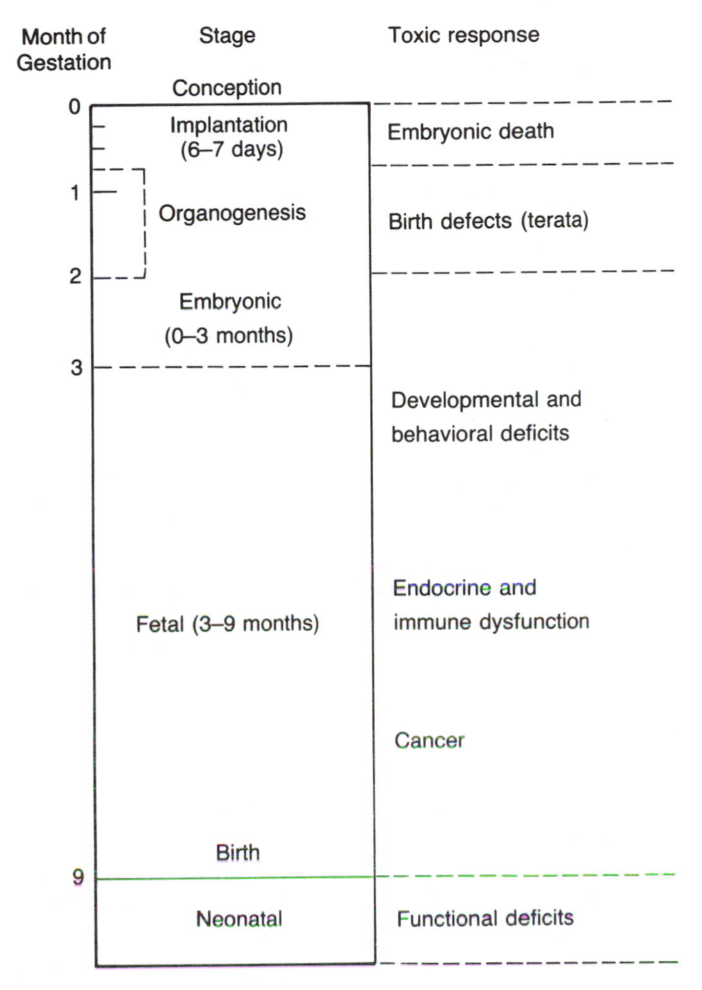

Figure 16-4. Relationship between time of exposure to toxic agents and response of human conceptus.

After organogenesis, the period of gestation from about day 20 to day 56 in humans when major organs are developed and the skeleton is formed, there is a long period of slower development and maturation, including the development of neurological tissues and sexual organs. Toxic exposures during the fetal stage of gestation (Figure 16-4) may induce neurological, behavioral, endocrine, and immune dysfunction, or cancer, in the offspring. There is little evidence that such effects occur often; however, if one agent can produce a specific defect, then it is possible that other agents may induce a similar response.

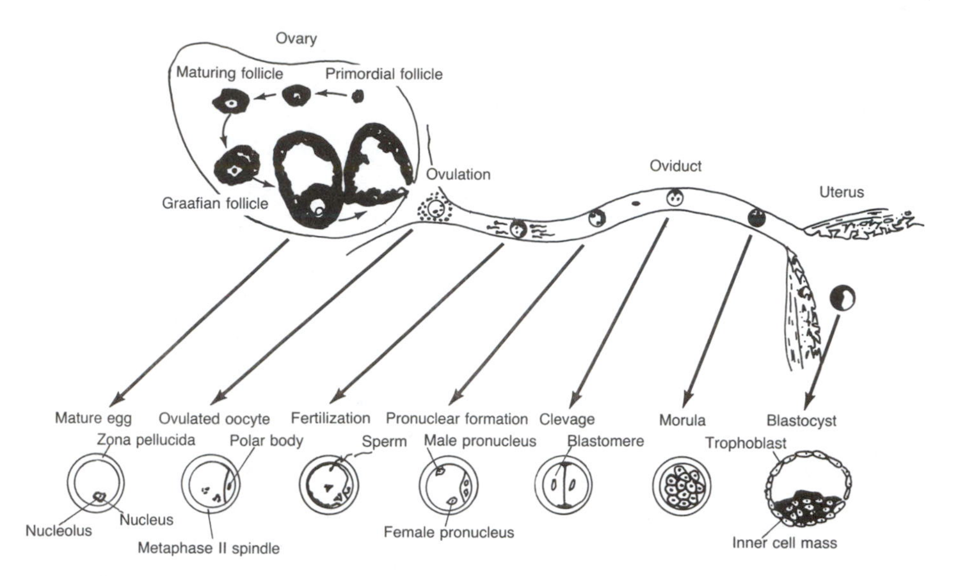

Figure 16-5. Mammalian preimplantation development: Diagrammatic representation of ovulation, fertilization, and development of the fertilized ovum to the preimplantation blastocyst. (Reproduced with permission from Dean, *American Journal of Industrial Medicine, 4*(1983): 3. pp. 31–39. Figure 1.)

Teratology. Teratology is the science dealing with the causes, mechanisms, and manifestations of developmental deviations of either structural or functional nature. Teratogens are agents that cause abnormal development resulting in congenital defects (birth defects), which are induced during the very early and short period of organogenesis. High doses of a teratogen result in embryolethality. In this critical period of embryonic development, cells undergo differentiation and mobilization into tissue groups that differentiate further into major organs and cartilage (pre-skeleton). The time span of organ development varies among species (Table 16-2), but in all species organogenesis includes the period between germ-layer differentiation (gastrulation) and completion of major organ formation.

The embryo is more susceptible to physical and chemical teratogens than is the mother. Overt toxicity and birth defects are not correlated, and, conversely, a number of chemicals producing toxic effects in the mother do not have an adverse effect on the embryo. Maternal and placental metabolism of a chemical and the ability of a chemical to move from maternal blood across the placenta into the embryo's blood influence the amount and chemical form of an agent interacting with the rapidly dividing and differentiating cells. Teratogenic effects of a chemical occur within a relatively narrow range of concentrations, between intermediate doses that kill the embryo and low doses that have

Table 16.2. Comparison of Gestational Events in Several Species.

	Number of days after conception		
Species	Implantation	Organogenesis[1]	Parturition
Human	6–7	20–56	260–280
Monkey	9–15	20–45	164–168
Rabbit	7–8	8–16$\frac{1}{2}$	30–32
Rat	5–6	9–16	22
Mouse	4$\frac{1}{2}$–6	7$\frac{1}{2}$–14	20

[1]Period of teratogenic risk.

no apparent effect. In rodents, a teratogenic dose of an agent produces some normal offspring, some dead and resorbed embryos, and a few fetuses with malformations. Different birth defects can be induced by administering a teratogen on different days of organogenesis. A cleft palate may be induced by exposure on one day and heart defects by exposure to the same agent on another day of gestation.

Mechanisms and Agents of Teratogenesis. Mechanisms of teratogenesis are not well understood and a number of mechanisms appear to be involved in the induction of birth defects. Many alkylating agents such as chemotherapeutic drugs are mutagens and teratogens and cause changes in the DNA and RNA of the rapidly dividing cells. Altered nucleic acids may be one mechanism of teratogenesis. The teratogen actinomycin D (in animals) specifically inhibits the formation of ribonucleic acid (RNA) by deoxyribonucleic acid- (DNA)-directed RNA polymerase. One can conclude that blocking the expression of DNA is a teratogenic event if timed properly. Colchicine, which inhibits cell division, is also a teratogen in animals. Heavy metals such as lead and cadmium are animal teratogens and probably inhibit enzymatic activity. Chemicals that inhibit energy production in embryonic cells are teratogenic in animal models, and osmolyte imbalance is known to result in damage to the embryo. In each case, rapidly dividing cells may die during a critical period of differentiation. Other evidence indicates that simply a delay in cell division during organogenesis induces birth defects. Therefore, physical or chemical agents that are lethal to rapidly differentiating cells, or which delay cell division by any mechanism, are teratogenic. It is not easy to predict which chemicals may be teratogenic. One would expect a high correlation between mutagens and teratogens, but many mutagens are not teratogens. Both deficiency and excess of vitamin A are teratogenic, as are deficiencies of other vitamins. Efforts have been made to relate the chemical structures of compounds to their activities as teratogens so that predictions can be made for the

Table 16.3. Incidence (per 1000) of Chromosomal Anomalies in Induced Abortions, Spontaneous Abortions, and Live Births.

	Abortions		
Chromosomal anomalies	Induced	Spontaneous	Live births
Total number	2782	1495	24,448
Number abnormal	48	921	126
Total incidence	17.25	614.82	5.15

teratogenicity of a new chemical. Such structure-activity studies have proved to be of little predictive value with regard to teratogenicity in animals.

Spontaneous Abortion. Spontaneous abortion, or miscarriage, is the loss of the products of conception from the uterus before the embryo or fetus is able to survive. During the first three weeks of a pregnancy, exposure to physical or chemical agents may cause the embryo to die (embryolethality), but will not cause birth defects. The greatest embryonic loss occurs near the time of implantation. The cause is unknown but chromosomal abnormalities account for a large proportion of the loss.

In many cases, early spontaneous abortions are not recognized because menstruation is only a week or two late. Fortunately, the embryo has a tremendous capacity for repair; if the repair capacity is exceeded in the first eight weeks of pregnancy the result will be spontaneous abortion, or birth defects. Because of the embryo's ability to repair cellular damage there is a "no effect level" for teratogens and agents causing spontaneous abortion. After two months of gestation, the incidence of spontaneous abortions drops sharply.

Actually, very few birth defects are seen (3–7%), and recent U.S. long-term studies have not revealed a relationship between birth defects and occupational exposures. In these studies, both induced and spontaneous abortions were recorded and tissues were examined for chromosomal abnormalities; abnormalities were hundreds of times higher in spontaneous than in induced abortions (Table 16-3). It is important to keep meticulous epidemiologic records on the current frequency of spontaneous abortions and birth defects for comparison with future generations.

Agents Inducing Spontaneous Abortion in Humans

The most common type of reproductive failure in humans appears to be spontaneous abortion during the first trimester. It has been estimated that as many as 60 percent of all conceptions end in spontaneous abortion. Such estimates suggest that routine surveillance of the male and female population be carried out by recording spontaneous abortions, occupational exposures, drug intake, health status, and personal and social habits. The loss of 60 percent of all conceptions represents a large unknown and it is important to determine how occupational and environmental factors contribute to the fetal wastage.

There is ample evidence that exposure to physical and chemical agents can cause spontaneous abortions. It has been known since the late 1800s that women who worked with lead were more likely to miscarry than nonexposed women. In one English factory, 90 miscarriages occurred in 212 pregnancies; 21 fetuses were full term, but were dead at birth.

Over the past eight years in a series of Swedish studies, a mixture of airborne metals was implicated in an increased incidence of spontaneous abortions. The workers who were studied were those exposed to copper, lead, arsenic, and cadmium. In a Finnish study, spontaneous abortions of female members of the Metal Workers' Union were analyzed. The ratio of spontaneous abortions to births was significantly higher than the average for all Finnish women. A high-risk industry appeared to be the production of radio and television sets. Exposure to solder fumes was thought to explain the increased risk.

A 1967 U.S.S.R. study of working conditions of anesthetists revealed that 18 of 31 pregnancies among female workers terminated in spontaneous abortion. In the early 1970s, the spontaneous miscarriage rate for female members of the American Society of Anesthesiologists actively working in operating rooms was found to be twice that of unexposed pediatricians. Exposed females included women who were in the operating room at least one year before pregnancy and during the first trimester of pregnancy.

Other occupational exposures that may carry a risk of spontaneous abortion include x-irradiation, laboratory work (virology, bacteriology, and chemistry) and product sterilization by ethylene oxide and glutaraldehyde. Personal habits linked to an increased risk of miscarriage are cigarette smoking and ingestion of alcohol.

Transplacental Carcinogenesis

Some chemicals can produce cancer in the offspring of experimental animals if exposure occurs during the fetal stage of development. Investigators call such compounds perinatal carcinogens. Transplacental carcinogenesis is associated with exposure during the fetal period of development, from the third to the ninth month of gestation.

In 1947, exposure of pregnant laboratory animals to urethane was reported to induce lung tumors in offspring. The first case of drug-induced transplacental carcinogenesis in humans was recognized in about 1970. Daughters of women treated during pregnancy approximately 14 to 22 years earlier with diethylstilbestrol (DES) to maintain pregnancy developed vaginal alterations, including a rare type of cancer, adenocarcinoma of the vagina. Two sons

(27–28 years of age) of DES-treated women are also known to have developed scrotal seminoma.

The association between DES and perinatal carcinogenesis was made because the tumors are very rare in the population. It is difficult to associate occupational exposure with perinatal carcinogenesis if the incidence of the type of cancer is common.

Agents Inducing Birth Defects in Humans

Before the twentieth century there was little evidence that birth defects were caused by external agents. In the 1920s, not long after the discovery of x-rays, it was observed that fetuses could be harmed *in utero* by transabdominal x-rays. Rather high doses of radiation produced a child with a small head circumference and mental retardation.

In 1941 it was recognized that blindness, deafness, and excess deaths occurred among children of women who contracted German measles (rubella) during the first trimester of pregnancy. The fact that a virus could damage an unborn child did not arouse public concern as much as the thalidomide disaster did in the late 1950s and early 1960s.

Thalidomide is one of the most potent teratogens known, with one in fifty women who used the drug bearing a child with congenital malformations, such as phocomelia or amelia (shortening or complete absence of limbs). This experience revealed that chemicals can cross the placenta and injure the embryo.

In the mid-1950s in Japan, a neurologic disease occurred as an epidemic among older children and adults who ate fish contaminated with methyl mercury. Inorganic mercury was being discharged in waste from a factory and was converted to the organic form in fish and other aquatic organisms. In 1965, a study of children born to women exposed 10 years earlier to methyl mercury revealed that 23 were severely affected by cerebral palsy; only one or two cases would be expected at normal rates.

Many other metals are suspected teratogens because birth defects can be induced in laboratory animals, but information on humans is limited. In a series of Swedish studies, employees of a copper smelter were exposed to a number of airborne metals, copper, lead, arsenic, and cadmium; an increased incidence of malformations was reported in the offspring of women employees. Three other Swedish studies reported an increase in malformations (gut atresias and oral clefts) in children of women who were working in chemical laboratories. Data were collected from company records and questionnaires. In a Finnish case-control study, an excess of central nervous system malformations was associated with exposure of women to solvents during pregnancy.

Maternal alcohol consumption during pregnancy was recently associated with a pattern of abnormalities called the "fetal alcohol syndrome." Such children have a characteristic facial appearance, growth deficiency, changes in central nervous system performance, and an increased frequency of malformations including eye and heart defects and cleft palate. It is estimated that 40 percent or more of infants born to chronic alcoholics will have structural malformations.

The cause of the majority of birth defects is still not known, and the role of environmental and occupational agents is not understood. Animal studies will continue to identify suspected teratogens; health professionals associated with industry must educate women about potential risks. As women enter the industrial workforce, programs must be set up that relate occupational exposure to chemicals and reproductive outcomes. Ideally, methods should be developed by which the health and reproductive function of individuals can be followed throughout their lives.

The science of reproductive toxicology developed in response to the need to evaluate the role of environmental and occupational agents in human reproductive failure, fetal wastage, and malformed children. There is no doubt that human reproductive wastage is significant, and that the cause of 60 percent of birth defects is unknown. A woman occupationally exposed to potential or known reproductive toxins who bears a child with a congenital defect will be more apt to believe that chemical exposure is the cause rather than an unknown agent. Nevertheless, there are health professionals who do not believe that occupational agents can cause reproductive toxicity. Therefore, you will encounter conflicting views. Clearly, occupation-related reproductive toxicology will be an important issue in the next 20 years.

Animal Test Methods for Evaluating Female Reproductive Toxins

The classical approach to the identification of teratogens in laboratory animals involves administration of an agent during the period of organogenesis (Table 16-2), sacrifice of dams before parturition, and examination of the fetuses for soft tissue and skeletal anomalies. Dose levels are selected so that overt toxicity is not produced in dams. The high rate of cell division and rapid differentiation in embryonic tissues account for the susceptibility of the conceptus. The embryo also lacks drug-metabolizing enzymes and an efficient excretory system. If birth defects are produced, the agent will be administered on individual days of organogenesis to pinpoint the day on which a specific defect is produced and to aid in investigations of teratogenic mechanisms.

After the drug thalidomide produced multiple defects in offspring, it became clear that the effects of therapeutic drugs on reproduction must be routinely investigated. In 1966, the Food and Drug Administration (FDA) issued guidelines for testing new drugs for their teratogenic potential. Three testing phases are now required before a drug can be authorized for use by childbearing women.

- Phase I. Male rodents are treated for 60 days, and females for 14 days, prior to mating. Treatment of pregnant females continues through gestation and weaning. Endpoints include breeding success, fertility, implantation, parturition, lactation, and neonatal effects.
- Phase II. Testing consists of treatment during the period of organogenesis as described above. Emphasis is on growth retardation, death, and structural and functional defects.
- Phase III. Pregnant rodents are treated during the final third of gestation. Endpoints are effects on late fetal development, labor, delivery, lactation, and neonatal viability.

Two dose levels are used in each phase with 20 rodents and 10 nonrodents per treatment group.

The FDA also devised a test for low-dosage, long exposures to chemicals such as food additives and pesticide residues on food. Recently this test has been used to evaluate the effects of other environmental chemicals on reproduction. Animals, usually rats, are treated continuously throughout three reproductive cycles and the study of three successive generations provides an opportunity to evaluate fertility and overall reproductive performance. This test is of relatively little value in identifying teratogenic risks because doses are generally lower than those inducing birth defects.

Transplacental carcinogenesis has not been investigated in relation to many chemicals except alkylating agents. To study suspected transplacental carcinogens, rather high doses are given to pregnant animals during the fetal period of development. Animals are allowed to litter, and litters are usually culled to equal numbers per litter and are then allowed to live a normal life span. Histologic examination of all tissues is performed on all animals. These are lengthy experiments, lasting two to four years, and are too expensive to be used for routine testing.

There are a number of short-term assays designed to test for teratogenic agents by using embryonic cells or tissue in laboratory culture systems. Fertilized mouse ova are used in culture to study the effects of chemicals on preimplantation embryos. In this system, the drastic effects of changes in osmolarity can be seen. The preimplantation embryo can be followed from the two-cell stage to the late blastocyst ready for implantation. This technique may become useful in detecting early reproductive toxins; however, it is a very sensitive system and achieving doses of a toxin as encountered *in utero* is extremely difficult. Chromosomal anomalies and somatic cell mutations are also detected in cell systems.

One of the difficulties in relating the findings from cell culture systems to humans is that many compounds will exhibit some toxicity in these assays, but there is not a satisfactory method of extrapolation to predict human reproductive failure or teratogenic outcomes.

Teratogenesis of occupational chemicals is the endpoint most often examined in relation to female reproduction. There is not an animal model used consistently to predict the embryolethal potential of occupational compounds.

Evaluation of Female Reproductive Toxins in Humans

The greatest number of female reproductive toxins discerned in the human population includes cytotoxic drugs such as chemotherapeutic agents used in the treatment of cancer, antinausea drugs, antibiotics, and other prescription drugs (Table 16-4). In many cases, the correlation between drug intake and adverse reproductive outcome was reported by astute physicians and then

Table 16.4. Agents Toxic to the Human Female Reproductive System[1].

Physical agents
- x Irradiation

Social and personal habits
- Alcohol ingestion
- Cigarette smoking
- Vitamin deficiency
- Mineral deficiency
- Hallucinogens

Metals
- Tellurium
- Lead

Therapeutic agents
- Chemotherapeutic cancer agents
- Large doses of steroid hormones
- Streptomycin
- Tetracyclines
- Sulfonamides
- Novobiocin
- Chloramphenicol
- Erythromycin
- Anticoagulants
- Antidiabetics
- Aminopterin
- Thiazide diuretics
- Quinine
- Prednisolone
- Antihistamine antiemetics
- Narcotics
- Thalidomide
- Excess vitamin K
- Chloroquine

Other
- Acute hypoxia

Chemicals
- Alkylating agents
- Anesthetic gases
- Methyl mercury

Biological agents
- Rubella virus
- Toxoplasmosis
- Syphilis

Pesticides
- Organophosphates
- Carbamates

[1]Human risk in many cases is not definitively known.

confirmed in animal studies. At this time, very few occupational chemicals have been associated with adverse reproductive effects in women.

It has been suggested that prenatal diagnosis using amniocentesis can help in monitoring for teratogenic agents. During development, the fetus is surrounded by amniotic fluid in which cells and macromolecules are of fetal origin and therefore reflect fetal properties. Amniotic fluid has been used to diagnose many genetic disorders such as inborn errors of metabolism. The risk associated with amniocentesis excludes the possibility of using the procedure for the monitoring of teratogenic defects in random populations.

A new technique for obtaining fetal cells for prenatal diagnosis is now being tested in the United States and Europe. The method permits prenatal testing only between the eighth and tenth weeks of pregnancy; chromosomal and biochemical disorders can be identified in a few days. The technique is called chorionic villus biopsy and preliminary evidence indicates that the procedure is safe, but the question of safety needs further study. Chinese and Soviet scientists have used the method in the last ten years, but only to determine sex of the fetus.

The epidemiologic approach helped to identify thalidomide as a teratogen. One case control study recognized there was an epidemic of birth defects. About two years later, case studies revealed the link between defects and thalidomide taken in early pregnancy. A cohort study recognized association of defects and thalidomide in less than a year. A cohort of women began to receive thalidomide as therapy for morning sickness. Nine months later three severely malformed infants were born to women in the study group. The physician recognized correctly that thalidomide was the causative agent.

Systematic, regular recording and analysis of spontaneously aborted specimens can indicate an increase in certain defects or the presence of unusual defects such as were seen with thalidomide. One objective of such monitoring is to reduce the time between the introduction of a teratogen in the human population and the discovery that a problem exists. Records should be made of occupation, chemical exposures, health status, hobbies, and personal habits as well as ingestion of drugs. To be effective, a standardized procedure would have to be adopted by a large number of hospitals.

Monitoring of the newborn should include the prompt and continuous reporting of birth defects, and so serve as a warning system by detecting clustering of defects or unusual trends suggesting new hazards in the environment or workplace. Efforts should be made to associate defects with heredity or other causes.

Another approach that can be useful in the industrial setting is to keep accurate records of menstrual dysfunction related to specific work and chemical exposure. Menstrual dysfunction caused by chemical agents suggests an effect on hormone levels either by direct or indirect action and may result in reproductive failure.

16.3 Regulation of Reproductive Hazards in the Workplace

There are statutes that address the problem of regulating hazards in the workplace, although none relate to both parents and offspring. Only the Toxic Substances Control Act specifically includes the significant risk of gene mutations and birth defects as a basis on which the Administrator of the Environmental Protection Agency can initiate appropriate action to prevent or reduce such risk.

The Secretary of Labor has the responsibility for setting health and safety standards for the workplace. In the regulation of reproductive hazards, medical removal protection (MRP) may prove to be a useful mechanism. Courts have approved of the MRP program as a valid exercise of the Secretary's authority in the case of lead exposure. Employees who are particularly susceptible to a hazard are reassigned to a work area where the specific hazard is not present. The MRP program is the only protective device in the Occupational Safety and Health Act for pregnant women or for male workers desiring to parent a child. The greatest difficulty in regulating reproductive hazards under the OSHA provisions is the lack of conclusive evidence that a specific chemical agent causes a particular reproductive injury.

Even though there is currently a 3–7 percent incidence of birth defects, there is not a valid method to determine if a congenital defect is induced by genetic diseases or environmental toxins.

As women of child-bearing age enter the workplace, management may choose to bar women from work involving real or perceived reproductive hazards. A woman usually does not know if she is pregnant at three weeks and may not be sure until about six weeks of gestation. If a woman's work exposes her to a suspected teratogen, should she be removed from the job if she is planning to conceive a child, or should all women of child-bearing age be barred from exposure to suspected teratogens? Some employers believe that women should be allowed to work in higher-paying jobs involving exposure to lead if they are not capable of bearing children. In a recent, widely publicized case, several women were sterilized voluntarily and released the employer from responsibility. More may have done the same in other, unpublicized instances. We do not believe that an employer has the right to require a woman to have a tubal ligation any more than an employer has the right to require a man to have a vasectomy in order to qualify for a particular position.

Clearly, in the future there will be difficult test cases to solve.

Summary

- In the male worker, the process of sperm production is uniquely susceptible to many occupational physical and chemical agents.

- Men can become sterile from chronic exposure to certain occupational and therapeutic agents.
- Records must be kept in order to relate male infertility to specific jobs and occupational chemicals.
- Test methods are developed for monitoring male reproductive toxins in humans and animal models.
- Reproductive toxins in the pregnant female can cause early spontaneous abortion, birth defects, and functional decrements in the offspring.
- Transplacental exposures can cause cancer in the offspring in later life.
- Agents cause birth defects in offspring at levels having no toxic effect on the mother.
- Physical and chemical agents can induce birth defects only during the period of organogenesis, from three weeks through eight weeks in humans.
- Safe test methods for detecting reproductive toxins in females are not developed for humans. Animal test methods are developed.
- Statutes regulating reproductive hazards in the workplace are not well defined.

References

Dean, J. 1983. "Preimplantation Development: Biology, Genetics, and Mutagenesis." *American Journal of Industrial Medicine, 4*:3, 31–49.

Dixon, R. L. 1980. "Toxic Responses of the Reproductive System." *In* Doull, J., Klaassen, C. D., and Amdur, M. O. (Eds.) *Casarett and Doull's Toxicology: The Basic Science of Poisons*. 2nd Edition. Philadelphia: Macmillan Publishing Company, Inc.

Dobson, R. L., and Felton, J. S. 1983. "Female Germ Cell Loss from Radiation and Chemical Exposures." *American Journal of Industrial Medicine, 4*: 175–190.

Ganong, W. F. 1975. *Review of Medical Physiology, Seventh Edition*. Los Altos (California): Lange Medical Publications.

Hamilton, W. J., Boyd, J. D., and Mossman, H. W. 1962. *Human Embryology*. Baltimore: Williams and Wilkins Company.

Harbison, R. D. 1980. "Teratogens." *In* Doull, J., Klaassen, C. D., and Amdur, M. O. (Eds.) *Casarett and Doull's Toxicology: The Basic Science of Poisons*. 2nd Edition. Philadelphia: Macmillan Publishing Company, Inc.

Infante, P. F., and Legator, M. S. (Eds.) 1980. *Proceedings of a Workshop on Methodology for Assessing Reproductive Hazards in the Workplace*. DHHS

(NIOSH) Publication No. 81–100. Washington, D. C.: U. S. Department of Health and Human Services, P. H. S., C. D. C., National Institute for Occupational Health and Safety. p. 423.

Johnson, A. D., and Gomes, W. R. (Eds.) 1977. *The Testis. Advances in Physiology, Biochemistry, and Function*. Volume IV. New York: Academic Press.

Kalter, H., and Warkany, J. 1983a. "Congenital Malformations, Part One: Etiologic Factors and Their Role in Prevention." *New England Journal of Medicine, 308*(8): 424–431.

Kalter, H., and Warkany, J. 1983b. "Congenital Malformations, Part Two: Etiologic Factors and Their Role in Prevention." *New England Journal of Medicine, 308*(9): 491–497.

Manson, J. M. 1978. "Human and Laboratory Animal Test Systems Available for Detection of Reproductive Failure." *Preventive Medicine, 7*: 322–331.

Mattison, D. R. (Ed.) 1983. "Reproductive Toxicology." *American Journal of Industrial Medicine, 4*: 1–2.

Mehlman, M. A., Shapiro, R. E., and Blumenthal, H. (Eds.) 1976. *New Concepts in Safety Evaluation. Advances in Modern Toxicology*. Volume I, Part I. Washington, D. C.: Hemisphere Publishing Company.

Norman, A. P. (Ed.) 1971. *Congenital Abnormalities in Infancy*. Oxford and Edinburgh: Blackwell Scientific Publications.

Rice, J. M. (Ed.) 1979. *Perinatal Carcinogenesis*. National Cancer Institute Monograph. Washington, D. C. 20402: Superintendent of Documents, U. S. Government Printing Office (Stock 017–042–00139–1). 282 pp.

Shepard, T. H., Miller, J. R., and Marois, M. (Eds.) 1975. *Methods for Detection of Environmental Agents That Produce Congenital Defects*. New York: American Elsevier Publishing Company, Inc.

Smith, C. G. 1983. "Reproductive Toxicity: Hypothalamic-Pituitary Mechanisms." *American Journal of Industrial Medicine, 4*: 108.

Tardiff, R. G. (Ed.) 1977. *Principles and Procedures for Evaluating the Toxicity of Household Substances*. Washington, D. C.: National Academy of Sciences, U. S. A.

Wilson, J. G., and Fraser, F. C. (Eds.) 1977. *Handbook of Teratology*. Volume I. New York: Raven Press.

Woollam, D. H. W. (Ed.) 1966. *Advances in Teratology*. Volume I. London: Logos Press.

PART III

Workplace Applications

Chapter 17

Risk Assessment

Robert C. James

Introduction

The purpose of this chapter is to explain how risks from chemical exposures can be assessed. The chapter will:

- Overview society's dependency on chemicals and, conversely, its need to assess the risks of their continued use.
- Describe the risk assessment process and the applications of models to quantitative risk/safety management.
- Discuss the use of various risk estimates.
- Explain the important factors to consider when evaluating toxicologic data.
- Provide examples of confounding factors for the evaluation process.

Chemical contamination of the workplace or the environment did not originate at any particular time but instead has always been with us. As man discovered and produced new materials, new chemicals, and new products, he also introduced new health hazards (in many instances unwittingly). As standards of living increased over the centuries and man found extensive uses for the various metals mined from ores, or discovered ways to control pests and increase agricultural yields, likewise serious human poisonings in the workplace or via contaminated food sources increased from useful but poisonous chemicals such as lead, mercury, cadmium, DDT, PCBs, benzene, methanol, etc. As was stated in Chapter 2, all chemicals can be hazardous at some dose; therefore, how we handle substances and at what level we are exposed to them determine whether they represent hazardous or safe substances.

17.1 Society Creates a Constant Need for Assessing the Risks Posed by Chemical Exposures

While there are many instances of natural processes that generate toxic contaminants, most of our current problems have arisen from anthropogenic sources. This will be so in the future, too. A summary of some of the problems or sources facing us are:

Chemicals. We are a chemical-based and chemical-dependent society. It is estimated that the chemical industry spends more than $100 million per year in production to meet consumer needs. The inventory mandated by the Toxic Substances Control Act identified some 100,000 different chemicals for potential use today, some 30,000 or more of which are routinely used. Chemical Abstracts Service lists 5 to 10 million different chemicals that have been synthesized, and it is estimated that some 100,000 new chemicals are synthesized each year for consideration of use. Of all these chemicals, threshold limit values (TLV) have been proposed by the American Council of Governmental Industrial Hygienists (ACGIH) for only about 400; even a smaller number have enforceable permissible exposure levels (PEL), which serve as the safe air standards under the Occupational Safety and Health Act (OSHA).

Drinking Water. There have been at least 1152 different organic chemicals found in the drinking water of five states by the U.S. Environmental Protection Agency. Yet there are only 20–30 primary or secondary drinking-water standards in the federal codes. Only 65 or so additional chemicals have been examined by the Environmental Protection Agency; this has resulted in a published water-quality-criteria guideline, yet these guidelines are not standards and are of unknown utility.

Pesticides. Hundreds of different pesticides have been formulated for use, several hundred million pounds of which are applied to crops each year. Although public concerns have been raised and will continue to be raised over the chemical contamination this introduces to our food supply, the U.S. Department of Agriculture has estimated that much of the world's population would starve to death within a single year without the use of modern fertilizers and pesticides. Similarly, even though DDT has been banned and has raised numerous environmental concerns, its use in the control of malaria has been an important aid to human health of citizens in a number of areas throughout the world.

Air Pollution. The major air contaminants from industrial or consumer sources include: ammonia, carbon monoxide, chlorine gas, oxides of nitrogen and sulfur, mercury and various metals, aldehydes and ketones, ozone, and hydrogen sulfide. While we might wish to remove these completely from the ambient air we breathe, our dependence upon such things as automobiles, paper products, and plastic products ensures our exposure to and generation of these contaminants.

The Pervasiveness of Exposures. With the awareness in the past decade or so of the extent of environmental contamination, and with the improvement of analytical methods for detecting chemicals, it has been demonstrated that there are no chemical-free environments or chemical-free persons in the United States. Persistent chemicals such as the heavy metals, chlorinated pesticides, and PCBs have been identified in almost all persons who have been tested for them, including persons not occupationally exposed to them. Many of these chemicals pass the "placental barrier" barrier or they are excreted in milk during nursing so that people may be exposed to chemicals almost from the time of conception.

Since it is clear that the use of chemicals and chemical products is pervasively and firmly entrenched in our modern lifestyle, attempts to determine acceptable levels of chemical exposure and the attendant risks is essential. And in an effort to respond to the myriad public health issues connected with chemical manufacturing, consumer product safety, food contamination or additives, industrial pollution, and the need for proper waste disposal, federal and state governments have begun to enact an array of statutes, regulations, and guidelines.

It is the aim and intent of these regulatory efforts to facilitate the safe handling of chemicals "from cradle to grave"; however, there is no straightforward method for setting acceptable limits to chemical exposure, nor is there always a consensus about how this should be done. Not surprisingly, most regulations have evolved from some definition or concept of an "acceptable risk" or "reasonable risk" written into a statute, rather than as the result of a specific logical decision or definitive procedure for determining allowable chemical exposures. Not only is the concept of determining the "reasonable risk" common among many diverse regulations governing chemical substances; it is the very cornerstone of much of the regulatory framework governing human exposure to chemicals. In a sense, that approach has become the policy of the Food and Drug Administration, the Consumer Product Safety Commission, the Environmental Protection Agency, and the Occupational Safety and Health Administration.

Although the laws establishing and governing the actions of those agencies differ in the amount and type of data required to establish exposure limits, assignment of liability, and relative influence to which cost/benefit consideration can be applied to the risk, the laws are remarkably consistent in that the critically important factor is the use of the term "acceptable risks." For example, Table 17-1 lists some of the sections dealing with risk analysis in the various federal acts governing chemical exposure.

Table 17-1 Review of Selected Federal Risk Assessment Regulations, Acts, and Laws (Courtesy of George Rusk, Ecology and Environment, Inc.).

Clean Air Act, Public Law (P.L.) 91-604, United States Environmental Protection Agency (EPA)

§109-Primary National Ambient Air Quality Standards must allow an "adequate margin of safety ... required to protect human health."

-Secondary National Ambient Air Quality Standards must be established at the level needed to protect the public welfare from any known or anticipated effects associated with the presence of air pollutants.

§111-New Source Performance Standards must prevent any emission that "causes, or contributes significantly to air pollution which may reasonably be anticipated to endanger public health or welfare."

§112-National Emissions Standards for Hazardous Air Pollutants must be established to prevent any emission of selected pollutants that "causes or contributes to air pollution which may reasonably be anticipated to result in serious irreversible, or incapacitating reversible, illness."

FWPCA, P.L. 95-217 (EPA)

§301-Requires the setting of effluent limitations for toxics according to the "Best Available Technology Economically Achievable."

§307-Requires an "ample margin of safety" in standards for toxic substances.

§311-Defines "hazardous substances" as those that, when discharged into the waters of the United States, "present an imminent and substantial danger to the public health or welfare."

§404-Prohibits "unacceptable adverse effects" of waste dumping in United States waters.

CERCLA, P.L. 96-510 (EPA)

§101-Defines "hazardous substances" as any that "present substantial danger to the public health or welfare or the environment."

Federal Insecticide, Fungicide, and Rodenticide Act, P.L. 92-516 (EPA)

§2-Requires EPA to regulate pesticides to protect against "unreasonable adverse effects on the environment."

Hazardous Materials Transportation Act, P.L. 93-633 (United States Department of Transportation)

§103-Defines a "hazardous material" as one that "may pose an unreasonable risk to health and safety, or property."

Occupational Safety and Health Act, P.L. 91-596 (Occupational Safety and Health Administration, Department of Labor)

§5(a)(1)-Requires employers to keep the workplace "free from recognized hazards that are causing or are likely to cause death or serious physical harm to employees."

§6(a)(5)-Requires that standards be established that "most adequately assure, to the extent feasible, on the basis of the best available evidence," that worker health will be protected.

RCRA, P.L. 94-580 (EPA)

§1004-Defines "hazardous waste" as waste that will either "significantly contribute to an increase in mortality or an increase in serious irreversible, or incapacitating reversible, illness" or "pose a substantial present or potential hazard to human health or the environment."

§7008-Authorizes the federal government to institute appropriate legal action to prevent situations involving "imminent and substantial endangerment" from hazardous wastes.

Safe Drinking Water Act, P.L. 95-110 (EPA)

§1401-Requires the establishment of primary standards for drinking water quality in order to prevent any "adverse effect on the health of persons" drinking the water.

TSCA, P.L. 94-469 (EPA)

§4-Requires premanufacture testing to determine whether chemicals will present an "unreasonable risk of injury to health or the environment."

§6-Requires that regulations be developed to prevent "unreasonable risk of injury to health or the environment."

Food, Drug, and Cosmetic Act, 52 Stat. 1040 (Food and Drug Administration), as amended by P.L. 85-929

§-Delaney clause states that *no* risk is acceptable when dealing with carcinogenic substances contained in food additives.

Since risk assessment methodologies are complex and their employment requires considerable judgment, Part III of this book discusses the procedures and problems inherent in this important process for setting safe exposure limits to the various chemicals in use today.

17.2 The Risk Assessment Process

Definitions

Since risk assessment is defined as "a determination of the probability that an adverse effect will be produced," risk assessment *per se* is the basic device by which we should arrive at our decisions governing accidental exposures, food tolerances, environmental pollution and cleanup, allowable workplace exposures, and the contamination represented by uncontrolled waste sites.

Risk assessment means different things to different people but the following definitions will be used in this chapter.

- Risk assessment: a methodologic approach in which the toxicities of a chemical are (1) identified, (2) characterized (species variations, mechanisms, etc.), and (3) analyzed for dose-response relationships. Finally, a mathematical model is applied to the data to generate a numerical estimate representing a guideline or decision concerning allowable exposures. Often, however, the process as defined is not completed and risk assessment is then often subcategorized, depending upon the depth of the analysis:
  - Quantitative risk assessment generates a numerical measure of the risk or safety of a chemical exposure. As such, it may ignore certain pertinent forms of decision logic when assessing the applicability of the numerical results to the human situation.
  - Qualitative risk assessment merely characterizes or compares the hazard of a chemical relative to others, or defines the hazards in only qualitative terms, such as mutagen or carcinogen, which connote certain risks or safety procedures and as such may not necessarily require a numerical assessment of risk.
- Safety: the inverse of risk; it is the probability or likelihood that an adverse effect will not occur.
- Risk management: the process of factoring the risk assessment against the possible alternative actions open to society, including cost/benefit considerations, consumer needs, or alternative chemicals, etc., to determine how best to regulate or manage exposure to the chemical. When value judgments and society's preferential but subjective bias modify the

risk assessment according to need, cost, or technical feasibility, this is risk management.

The Four Basic Steps of Risk Assessment

Risk assessments require a thorough evaluation of the toxicity data as well as a number of decisions or determinations at each step. The four basic steps are:

1. Hazard identification: a review of the toxicologic data base (described in detail in Chapter 18), or, if the data base is insufficient to make realistic decisions, toxicity testing to identify the range of toxicities the chemical is capable of. In this step the potential adverse human health effects are identified for further consideration.
2. Hazard evaluation: a determination of dose-response relationships, potency, species variation, mechanisms, etc. This is a thorough analytical review of data to determine their validity, their application to modeling, and to identify which dose-response values should be modeled.
3. Exposure evaluation: an estimation or modeling of the likely human exposures—routes of entry as well as levels of anticipated exposures.
4. Risk estimation: a mathematical modeling of the animal and/or human toxicity data, combined with human exposure evaluations in a manner that estimates the probability or incidence of the human health effects.

In a number of instances, step 3 is not required. Instead, the risk is computed for a range of hypothetical human exposures, or the exposure is determined for an acceptable level of risk and then the human exposure is regulated or modulated to meet this limit.

Unfortunately, many persons assume that the risk assessment procedure relies totally upon mathematical models, and that the goodness of the assessment is largely dependent upon the model. This impression is no doubt due in part to the considerable discussion and controversy surrounding the current mathematical approaches. And since we usually need a quantitative assessment (i.e., set some guidelines for exposure), we are forced to apply some mathematical model to the data. However, as is true for any mathematical model, which model to use to extrapolate human risks from animal data, the quality of the model, and utility of the model's outputs are all dependent upon the quality of data and the judgments that determine the inputs. Risk assessment is essentially a process of exercising good, scientific analysis and judgment.

The Application of Models to Quantitative Risk / Safety Assessments

In recent years the number of new chemicals entering the environment has grown at an impressive rate. Paralleling this growth, there has been increasing interest in the development of rational procedures for assessing human health risks associated with environmental exposures to potentially hazardous agents. Due to a lack of relevant human information, persons concerned with this assessment process have often been forced to rely solely on experimental animal data when evaluating possible human health effects. Therefore, the question of how to extrapolate the results of laboratory studies to man in a meaningful way has become a matter of growing concern in modern toxicology.

17.3 The Use of Risk Estimates

As discussed in Chapter 2 and in the introduction of this chapter, toxicity assumes a broad definition as it applies to all chemicals, since all chemicals are toxic at some dose. The key then is not necessarily establishing what toxicities a chemical possesses, but rather those prescribed conditions under which it elicits the undesirable response. To reiterate, risk is the probability that a substance will produce harm under specified conditions; therefore, risk assessment is a practical consideration to determine whether or not some harm will be elicited from a specific chemical exposure. Thus, when determining the risk or safety of a chemical, the critical factor is not necessarily the intrinsic toxicity of the chemical *per se* (e.g., carcinogenicity) but the likelihood that the level of exposure to the chemical is sufficient to express its intrinsic toxicity.

In general terms, then, the risk is qualitatively approximated by the equation

$$R = T \times E,$$

where R is risk, T is toxicity, and E is exposure. More accurately, the equation for risk is equal to the toxicity as some function (f) of the exposure, or

$$R = Tf(E).$$

This equation defines both qualitative and quantitative extrapolation, since under certain conditions the risk is not always linear over the entire dose-response curve; and since, depending on how one defines function (f), threshold or nonthreshold dose-response curves may be fitted to this equation. However, regardless of how one chooses to express the risks to chemical exposure, it is clear that: (a) the actual risk is dependent upon both the toxicity and the exposure, and (b) to change either alters the risk. Therefore, to make a risk assessment, one must compare the exposure probability to the various dose-response curves for each toxicity to determine whether or not a harmful response is likely to be induced.

In evaluating the risk associated with exposure to chemicals, one must attempt to answer two questions. First, is the level of exposure of a sufficient magnitude that a hazard exists? Second, if so, what is the hazard or undesirable response expected for that level of exposure? To answer both questions it is necessary first to evaluate the intrinsic toxicity associated with the chemical; i.e., what are the toxic responses and at what levels of exposure will they occur? Understanding and evaluating the intrinsic toxicity of a chemical will be discussed in subsequent sections of this chapter.

Safety Factor Calculations: Risk Estimates for Threshold Toxicities

As discussed in Chapter 2, some of the model(s) for extrapolating animal data to safe human exposures are relatively simple. For the toxicity of most concern, or for the toxicity occurring at the lowest exposure, the threshold for that toxicity or the dose closest to the threshold is chosen (see Figure 17-1). The calculation for toxicities demonstrating thresholds essentially converts the animal exposure to an equivalent human exposure based either on body weight or surface-area conversion. The equation given below incorporates some additional considerations often overlooked in most approaches—as, for example, in SNARL (suggested no-adverse-response level) or NOEL (no-observable-effect level) measurements. It "adjusts" the animal data to reflect differences in metabolism, absorption, and the exposure regimen for the test situation compared to the likely human conditions. Thus, the following calculation is similar to the one presented in Chapter 2; however, several new variables are

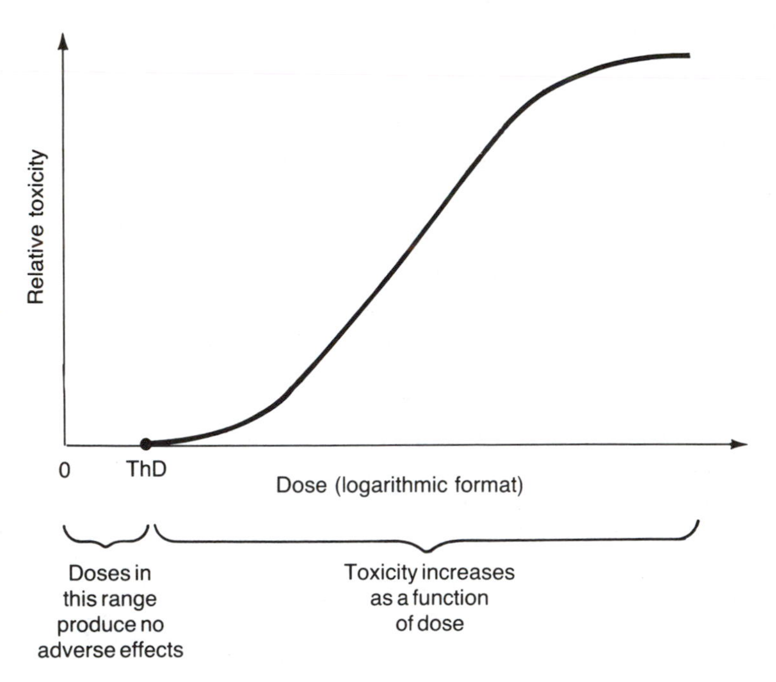

Figure 17-1. The relation between threshold dose and dose-response; ThD denotes a threshold dose.

introduced:

$$\text{Safe human dose, or SHD} = \frac{\text{ThD} \times \text{BW} \times \text{AF} \times \text{ER} \times t_{1/2}}{\text{SF}} = \text{mg/day},$$

where

ThD = threshold, or no-observable-effect doses in the test species;

AF = absorption factor, or ratio of animal absorption to human absorption (absorption in animal/absorption in human);

BW = body weight of the exposed individual (usually considered to be 70 kg);

ER = ratio of the exposure regimens (dosage regimen in test animals/exposure regimen in humans);

$t_{1/2}$ = ratio of elimination (half-life in animal/half-life in humans); and

SF = safety factor, which depends on the reliability of the data used for extrapolation.

As was discussed in Chapter 2, the number calculated should use chronic data where chronic exposures are expected. The SHD or safety factor type of model calculates one value. Exposure at or below this value is considered safe. A case has also been made for increasing the flexibility of the safety factor approach so as to better depict the quality of the underlying data base in any given determination of an acceptable daily intake (ADI). This sentiment was reflected to some degree in the 1977 report of the National Academy of Sciences Safe Drinking Water Committee, which proposed the following guidelines for selecting safety or, more appropriately, "uncertainty" factors to be used in combination with no-adverse-effect-level data.

1. A safety factor of 10 is suggested when valid human data based on chronic exposure are available.
2. A safety factor of 100 is suggested when human data are inconclusive (e.g., limited to acute exposure histories), or absent, but long-term, reliable animal data are available for several species.
3. A safety factor of 1000 should be utilized when no long-term or acute human data are available and the experimental animal data are scanty.

In addition, if the lowest-observable-effect dose is used rather than a true threshold dose, then an additional factor of—say—10 should be added to the SF.

This safety factor type of model does not generate a range of probabilities in the sense that we defined risk. Instead, it sets a ceiling limit on what is considered acceptable or safe and all doses or exposures below this level essentially carry the same probability of risk—i.e., zero. In a sense, exposures are judged in "yes" or "no" type of acceptability rather than with variable degree of acceptability or unacceptability.

Quantitative Risk Estimates for Carcinogens: Nonthreshold Toxicities

Quantitative estimates of the risks posed by carcinogenic substances can be done according to any of several models, most of which assume that there is no threshold dose. Essentially these models convert the dose-response curve for the cancer bioassay into an exponential curve, so that risk does not reach zero until exposure does (see Figure 17-2). However, all differ in their basic assumptions or in mathematical expression, as can be seen in Table 17-2. Hence, at the low exposures in question, the estimation of risk can vary dramatically depending upon which model is used. Currently, regulatory agencies usually use the multistage model. In this model, the risk is linearly proportional to the exposure at low doses. While each exposure is considered to carry some risk, the acceptable risk or safe dose is usually considered to be the exposure range for a 10^{-5} to 10^{-7} risk. This means that daily exposure to that dose for a lifetime would increase cancer by one person in 100,000 (i.e., 10^{-5}) or one in 10,000,000 persons (10^{-7}).

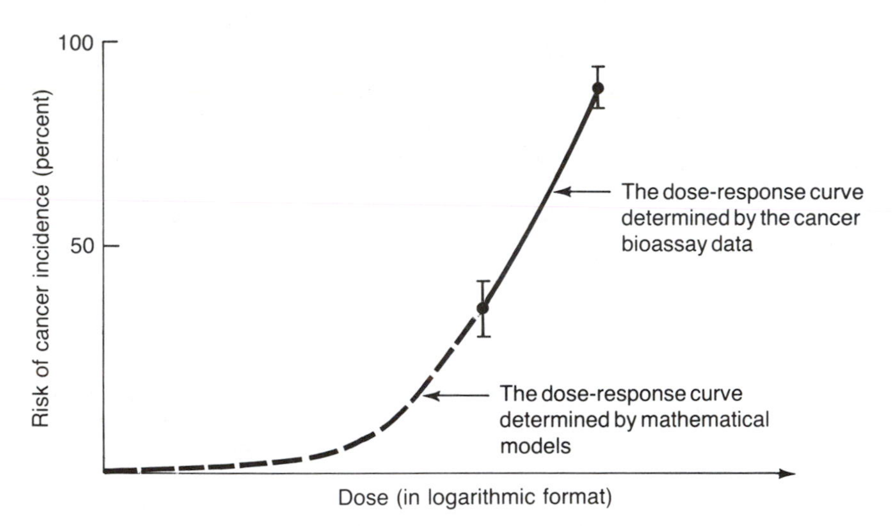

Figure 17-2. Because cancer bioassays are usually run at only two doses, there are few points with which to generate a dose-response curve describing the carcinogenic response of the test species. In addition, in risk assessment the response of interest is that which is so low it cannot be accurately measured by any study. The solution to this problem is to apply some mathematical model to the limited data set so as to describe an "assumed relationship" between dose and response (dashed curve). Since it is assumed that carcinogens do not have thresholds, the model typically generates an exponential curve. This type of curve predicts that risk (cancer incidence) will not be zero until the exposure (dose) is zero in the low-dose region of the curve, and yet it is still capable of merging with the actual data at higher doses.

Table 17-2. Examples of Extrapolation Models.

Model name	Mathematical expression
Probit function (point exposure)	$f_a = 1 \exp[-QD]$
Log-probit	$P_{(d)} = \Phi(\alpha + \beta \log_{10} d)$
Power law	$f_a = bD^k$
Doll	$I_t = b(t - v - w)^k$
Doll-Weibull	$I = bd^m(t - w)^k$
One-hit linear	$P_{(d)} = 1 - e^{(-\lambda d)}$
Extreme value	$P_{(d)} = 1 - \exp - [\exp(a + \beta \log d)]$
Linear interpolation	$P_c = \frac{\text{UCL}}{d_e} d$
Multistage	$P_{(d)} = 1 - \exp[(-a + b_1 d) \cdots (\lambda_k + \beta_k d)]$
Multihit (k-hit)	$P_{(d)} = 1 - \sum_{k-1} (\lambda d)^{ie^{-\lambda d/i}}$
Median effect principle of the mass-action law	$f_a/(1 - f_a) = (D/D_m)m$
Chou equation	$f_a = [1 + (D_m/D)^m]^{-1}$
Hartley-Seilken	$P(T, d) = 1 - e^{g(d)H(t)}$

Note: The models are provided to illustrate that a variety of choices are available when attempting to estimate risk posed by a carcinogen: thus, it is clear that the final answer one derives may depend not only on the choice of the model, but also on what data are entered.

Selection of the appropriate model for estimating risks at low doses would be made easier if some models clearly did not fit the observed data points. As mentioned above, hardly ever is it possible to select the best model or even to reject the worst on the basis of fit to observed data points. The low end of the dose-response curve would be the most informative for selecting the correct model but it is the part that is most difficult to measure and for which we have little or no information. In practice, incidence rates in animal tests much below 10 percent (five tumor-bearing animals in a test population of 50) can seldom be distinguished from the rate of spontaneous tumors. For this reason, the dose-response data that is modeled usually shows a fairly large response (i.e., greater than 10 percent incidence of cancer). At these higher responses, most equations are able to fit the animal data well but can generate relatively large differences in risk as dose is diminished. An example of the impact of choosing one model over another is given in Table 17-3. The table shows that two infralinear models and the one-hit model (which is essentially linear at doses that cause an incidence of 10 percent or less) are indistinguishable at high doses. For the table, a dose level of 1.0 was hypothetically set as a dose sufficient to cause a tumor incidence of 50 percent. As can be seen in the table,

Table 17-3. Expected Incidence of Tumors Calculated by Three Models When a Dose of 1.0 Caused Tumors in 50 Percent of the Tested Animals.

	Relative dose level	Model and incidence of tumors (in percent)		
		Log-normal	Log-logistic	One-hit
Basically similar incidence	16	98	96	100
	4	84	84	94
	1	50	50	50
	1/4	16	16	16
	1/16	2	4	4
	1/100	0.05	0.4	0.7
	1/1000	0.00035	0.026	0.07
	1/10,000	0.0000001	0.0016	0.007

(Adapted from Brown and Mantel, "Models for Carcinogenic Risk Assessment." *Science*, *202* (1978): 1105. Copyright © 1978 by the American Association for the Advancement of Science.)

the expected incidence at higher doses or at doses as low as 1/16 is nearly equal regardless of the model.

Brown and Mantel (1978) point out that no experiment of practical size can distinguish among the three models at high dose levels unless the incidence in control animals is very low. However, at the much lower dose levels of 1/100, 1/1000 and 1/10,000, the models diverge greatly in their projections of tumor incidence. These greatly lower doses and response levels are often the ones of most interest for estimating human risks, but these cannot be measured. Thus, it is clear that the incidence of cancer measured at higher doses in animals does not provide sufficient information to choose the appropriate model. This problem plagues most cancer risk-extrapolation efforts, and the correctness of the model chosen or its ability to reflect the biologic response is an often-debated issue.

As concluded by the Congressional Office of Technology Assessment, selection of the correct extrapolation model is important for only one of the three possible strategies by which regulatory agencies assess a carcinogen. The first strategy is to accept either human or animal evidence as sufficient to identify a carcinogen, and once the identification is made, try to eliminate the exposure. This approach requires no quantitative or numerical extrapolation. The second approach uses biologic and numerical extrapolation to rank substances in order, from that expected to be the most potent carcinogen to the weakest carcinogenic substance. This relative ranking or ordering process can be accomplished by consistently applying any model, and the numerical accuracy of the estimated incidence is not critical. The third approach, which includes a quantitative estimate of the human risk to be used in risk-benefit computations or for the consideration of levels representing an acceptable risk, requires the most accurate numerical estimates. Clearly, in the third approach (which is most often our purpose for modeling the data) the selection of models is important because the numbers produced by different models vary across a wide range, which in turn affects the level for the previously set "acceptable risk."

Gaylor's Approach. In lieu of arguments as to which model most accurately projects the actual risk involved, Dr. David Gaylor of the National Center for Toxicological Research has proposed a conservative but straightforward approach to the problem, in the form of a linear interpolation model. To simplify the discussion, let us assume that there is no spontaneous tumor rate at zero dose (no background incidence of cancer). Since it is often postulated that dose-response curves have a sigmoid shape, a line connecting zero dose with a point on the dose-response curve always lies above the true dose-response curve for the lower convex portion of the curve up to the inflection point (see Figure 17-3). Thus, Gaylor's linear interpolation model should be able to serve as a conservative upper bound for any sigmoid dose-response curves. However, Gaylor feels that his approach—to establish an upper bound on the proportion of tumor-bearing animals—may not be as conservative as might be thought, and he has cited examples to show that his linear interpolation model approximates quite well the dose-response curve resulting from the multistage model at low-dose levels. Thus, his method is in good agreement with the current model most often used by regulatory agencies.

An additional degree of conservatism is introduced into the risk estimate generated by Gaylor's model by interpolating from the upper confidence limit (UCL) of the animal data to zero. The linear interpolation model thus approximates the one-hit model for low doses and the probability of cancer is actually modeled as the estimate of the upper limit for the proportion of tumor-bearing animals. The equation for the linear interpolation model then can be written as follows:

$$P_c = \frac{\text{UCL}}{d_e} d,$$

where

P_c = the proportion of cancer bearing animals;

d = the given dose (the dose associated with the risk, P_c);

d_e = the dose used in the experiment; and

UCL = the upper confidence limit of the risk (i.e., percent cancer) observed in the experimental animals.

Conversely, the dose d for any given P_c is

$$d = \frac{P_c d_e}{\text{UCL}}.$$

Note that the probability of cancer generated in this manner is really the upper limit of the actual estimate.

An Example of Gaylor's Model. Let us illustrate the use of Gaylor's linear interpolation model. The following example was adapted from Gaylor and Shapiro (1979). A compound administered as 1 percent (10,000 ppm) of the diet results in 18 out of 100 animals having tumors, while 10 out of 100 control animals have tumors. The upper 99 percent confidence level for this example is 0.20. If one then desires to estimate an upper limit for the tumor incidence predicted for an exposure of 10 ppm based upon this data, it is calculated as

$$P_c = \frac{0.20}{10{,}000\ \text{ppm}} \times 10\ \text{ppm} = 2 \times 10^{-4}.$$

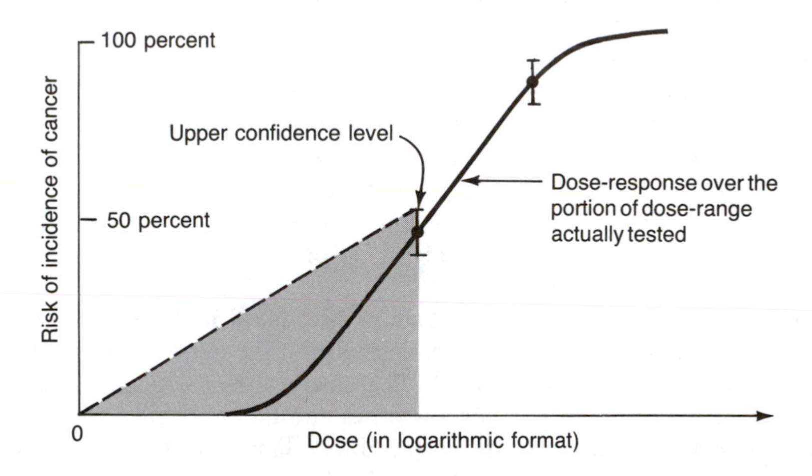

The solid line depicts the actual dose-response curve. Dashed lines represent estimated low-dose risks of cancer: Gaylor's model—the straight dashed line—is an extrapolation from the upper confidence limit of the lowest tested dose down to zero dose.

Figure 17-3. Gaylor's model simplifies the process of estimating human cancer risk. By extrapolating linearly from the upper confidence level of the lowest dose tested (or linearly from any dose higher than that on the solid portion of the dose-response plot) down to zero dose, Gaylor defines a region within which all risk will be found and thus obviates discussions concerning which model best fits the actual responses of humans to low doses of suspected carcinogens. While a number of different models might generate different estimated response curves, all would be somewhere within the shaded portion of the figure. Thus, Gaylor's model sets an upper bound on the estimated risk, and Gaylor is virtually certain that the actual risk of developing cancer from small doses of carcinogenic substances can be no greater than, and is probably less than, that bounded by the straight dashed line, which represents P_c.

This value of 2×10^{-4} means that the predicted rate of tumor incidence in the animal species tested is less than two tumors in 10,000 animals exposed to 10 ppm. Conversely, a dose could be calculated that represents a P_c of 10^{-5} (i.e., a 10^{-5} risk estimate); this means that animals exposed to this dose for a lifetime have only 1/100,000 chance of getting cancer from the chemical exposure:

$$d = \frac{10^{-5}}{0.2} \times 10{,}000 \text{ ppm} = 0.5 \text{ ppm}.$$

The linear interpolation model proposed by Gaylor has one advantage over most of the other models; it obviates a discussion of the "right" model. Essentially, the linear interpolation model states that the actual risk is somewhere below the generated line that connects the upper confidence limit with zero exposure (the shaded area in Figure 17-3). In other words, rather than argue over which model is more correct, Gaylor's model states that the actual risk should be no worse than the P_c at any given dose. It is a conservative estimation that derives the upper bound for safety. However, like other models, there is no evidence to prove that this model correctly estimates the risk, and the actual risk could be far less than what is predicted by this model at any given dose (note the actual dose-response curve in the shaded portion of Figure 17-3). Thus, while the linear interpolation model may serve regulatory agency purposes by erring on the side of safety, it may not reflect the actual risk at low exposures accurately and may not indicate the actual hazard when acceptable exposures derived by this model are exceeded.

Dealing with Overestimation of Risk. While numerous arguments can be made about the carcinogenic potential of specific chemicals or which model to use on carcinogenic substances, an estimate of the reliability of the computation can sometimes be found by a simple, direct method. Although the preceding extrapolations might provide useful guidelines for safe exposures when only animal data are available for consideration, they should be tempered by the human experience whenever possible. Furthermore, while human data may not be able to prove that the extrapolations from animal data are correct, in some instances they can be used to demonstrate that the cancer extrapolations are incorrect. Such is the case for a number of chemicals.

The models predicting cancer risks are generally linear in the low-dose (allowable human exposure) range. Therefore, extrapolating from the low acceptable exposure in a linear fashion backwards toward higher doses, so that the exposure approximating a 100 percent cancer incidence in humans is found, should provide an overestimation of the actual dose required for a high incidence of cancer in humans (see Figure 17-4).

Table 17-4 exemplifies this problem. The maximal daily occupational exposure was calculated by multiplying the permissible level established by OSHA (or the threshold limit value established by ACGIH) for the chemical by 10 (it was assumed that 10 m^3 of air would be inhaled per eight-hour shift and that 100 percent absorption would occur) to provide an estimate of the

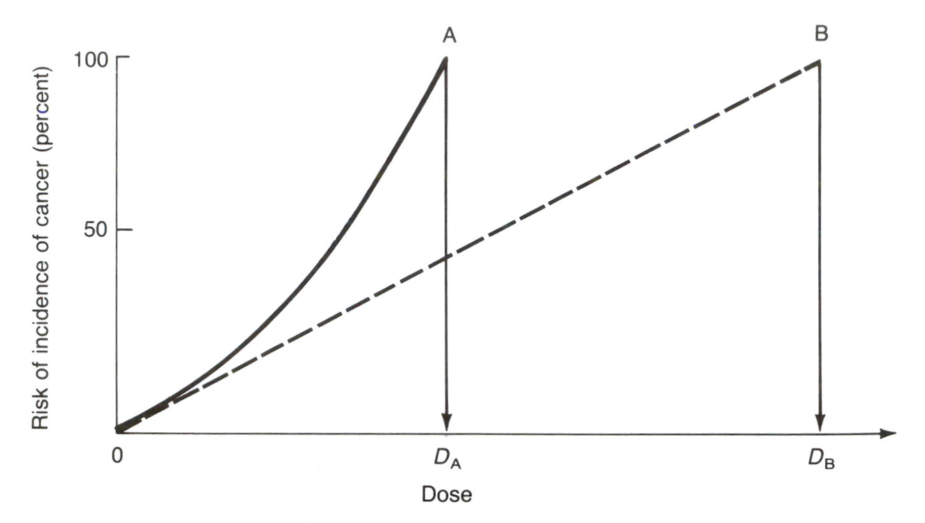

Figure 17-4. Extrapolation from exposures at which the risk of cancer is low to exposures at which the risk of cancer is high. A, actual risk as shown by data or by a model; B, risk as extrapolated from the low-dose region; D_A, actual dose representing a 100-percent risk of cancer; D_B, extrapolated risk representing a 100-percent risk of cancer.

amount of the chemical absorbed per work day. Each of these values was then compared to the level of the chemical allowed by the EPA water quality criteria, which are supposed to define acceptable chemical intake based upon a cancer risk estimate for the chemical calculated by using the multistage model. (This assumes two liters of water are consumed per day; or, 2 × the concentration for the 10^{-5} risk value ×100,000 = the daily dose expected to produce 100 percent cancer.)

Table 17-4. An Example of the Difference in Estimates of Risk Determined by Extrapolation From Daily Exposures (Low-Dose Exposures) to Lifetime Exposures.

Chemical	Estimated allowable occupational exposure (mg/day)	Exposure estimated by EPA to give a 100 percent cancer risk (mg · day/lifetime)
Carbon tetrachloride	600	800
Chloroform	1250	350
Chlordane	5	0.0046
DDT	10	0.0024
Dieldrin	2.5	0.0071
Perchloroethylene	6700	1600

Thus, it can be seen that if the EPA's 10^{-5}-risk estimates were correct, all of the chemicals in Table 17-4 would produce a large and easily identifiable cancer incidence in the worker populations, according to the guidelines set by OSHA. (For DDT and chlordane the allowable workroom air exposures would be 40,000–10,000 times that level expected to produce 100 percent cancer for a lifetime exposure.) Since the cancer incidence in workers exposed is nowhere near this estimated incidence, the mathematically estimated cancer risk must be wrong. It is conceded that workplace exposures are not for a lifetime and may be for only 1/3 to 1/2 of a lifetime or less; however, upward linear extrapolation by this method probably overestimates the exposure required by an order of magnitude or more (in Figure 17-4, notice the differences between dose *A*—the actual dose required to cause 100 percent cancer—and dose *B*—the dose calculated via linear extrapolation).

It is not my intent with the preceding example to undermine the use of models or their application to cancer risk estimates. This example does, however, underscore the need to use good judgment when assessing risks and to not become so inflexible that one becomes entirely dependent upon the numbers generated by such models and begins to believe that they represent the actual human risk. They do not represent the actual human risk; they are only a mathematical attempt to predict what it might be and are based on a variety of assumptions that may or may not be true.

17.4 Hazard Evaluation: Assessing the Intrinsic Toxicity of a Chemical

As stated earlier the second step in any risk assessment is an evaluation of hazard. This step derives the information to be modeled and may be used to determine which model should be chosen. While it might have been appropriate a few years ago to act on the premise that all potentially carcinogenic chemicals represented similar hazards, recent experience has shown that there are important theoretical and practical differences in the actual hazard each "carcinogen" represents to the public health. The same might be said for mutagens as well. In addition, there is a continual and substantial increase in the amount of data reflecting the toxicity of chemicals in animals and humans. Thus, as we continue to improve our understanding of the mechanisms of toxicants, we gain a better basis for extrapolating risks. Considering the national concern expressed about environmental contaminants and occupational hazards and the need for valid and appropriate judgments, it seems both appropriate and necessary to reevaluate the hazards and risks posed by chemicals as we obtain more information.

This section of the chapter will provide a framework for identifying the important toxicologic considerations used in interpreting the risks implied or identified by toxicity data. This is the necessary and important task of determining the intrinsic toxicity of a chemical (i.e., is it carcinogenic, hepatotoxic, nontoxic, etc?). In effect these steps (i.e., Hazard Identification and Hazard Evaluation) qualitatively establish the hazard and risk posed by a chemical and usually determines how the data ought to be modeled as well as what confidence can be applied to the numerical estimates that result.

Factors to Consider When Evaluating Toxicologic Data

Many aspects must be considered and weighed before the human hazard associated with any chemical exposure can be evaluated. An evaluation of the human hazard and subsequent risk estimation for any chemical must address the data base by utilizing thorough comparative toxicologic assessments. This means that the animal and human data available must be carefully considered in terms of:

- The breadth and variety of the toxic responses manifested.
- The degree of species variation or species consistency in the effects monitored.
- The possible and/or proposed mechanism of toxicity.
- The validity of the tests performed and their relevance for extrapolations to man.
- The dosage used in animal tests versus the expected level of human exposure.
- Finally, to the extent that the data are available, the outcome of serious poisonings and long-term occupational exposures as a guide to the expected human consequences and as a test of extrapolations from animal data.

Only when a consistent pattern of toxicity in animals coincides with human experiences can accurate and safe guidelines be promulgated. Moreover, the types and weight of the evidence often form an important part of determinations of which risk-assessment model to use, what safety factors are appropriate, and the probable accuracy with which the health risks can be estimated.

Breadth and Variety of Toxic Effects Reported per Species Variation

A review of the toxicologic data and the establishment of the variety and breadth (species consistency) of the toxic responses is an important initial step in any chemical evaluation, because this:

- Identifies potential effects that might be produced in humans and helps rank concerns according to priority.

- By comparing species we can estimate the likely consistency or variability in both the potency and types of toxicities to expect.
- A comparison of effects within species (e.g., sedation versus hepatotoxicity versus lethality) helps establish a rank order of the toxic effects manifested as the doses increase; this aids in identifying warning effects or symptoms as a means of predicting what outcomes to expect in humans who are affected.
- A comparison within a species may provide an estimation of the margin of safety between less harmful and more harmful effects.
- Variation or consistency among species is usually related to the mechanism of toxicity. This in turn may be useful for human extrapolations of the likely outcomes, and for attempting to identify the appropriate animal model to use when estimating the risk. Which species represents man is always a critical determination wherever it can be made.

In summary, comparing all reported effects is the initial step in identifying the toxicities and particular data base to use in the risk-assessment model chosen. It is an important step that should not be overlooked or minimized, as it can dramatically affect the validity of your extrapolation, whether it is only a qualitative risk assessment or a more definitive attempt at quantifying the risk.

Mechanism of Toxicity

Determining the mechanism of toxicity is extremely important and can be the key issue in many instances.

For one thing, mechanisms can define the differences between genetic and epigenetic carcinogens, which in turn may alter the risk estimate both qualitatively and quantitatively.

Second, understanding the mechanism of toxicity often identifies the conditions required to establish toxicity, which can help identify the hazard to be expected for each exposure. For example, acetaminophen, a synthetic analgesic used in many over-the-counter pain relief medications, can produce fatal liver injury in a number of animals, including humans. Yet, by determining that the mechanism of toxicity is the production of a toxic metabolite during the metabolism of high doses, it is possible to predict and establish its safe use in humans, determine the consequences of various doses, and develop and provide antidotal therapy.

Last, understanding the mechanism of action of a particular chemical helps establish the right animal species to use in assessing risk, and to determine whether the toxicity is likely to be caused in man. For example, squill is a rat pesticide that is relatively safe in humans because humans become nauseous and vomit the poison, while rats lack this reflex and are unable to vomit it, which proves fatal to this animal. Certain halogenated compounds are mutagenic and/or carcinogenic in test species, with the activity being dependent upon metabolism. Determining the route of metabolism for these compounds in humans therefore can dramatically alter the risk estimate. For example, certain halogenated hydrocarbons are metabolized in humans in the same way they are in animal species in which cancer is not produced. This may be a mechanistic basis for explaining why cancer has not been observed in humans even after extensive occupational use of a certain chemical or exposure to it.

Validity of a Test and Its Relevance to Human Extrapolations

The necessity of testing chemicals for toxicity in animal surrogates means that the validity of using the test to mimic human responses may always be questioned. While this is not always a major consideration and the assumption of correlation can often be accepted, it is a rule with important exceptions. As noted in the previous sections, species and mechanistic differences can essentially invalidate the use of some reported data.

Other guidelines to consider are:

- Has the test used an unusual, new, or unproved procedure?
- Does the test measure a toxicity directly, or is it a measure of a response purported to indicate an eventual change (i.e., a pretoxic manifestation, etc.)?
- Have the data and experiments been performed in a scientifically valid manner?
- Is it statistically significant against an appropriate control group?
- Has the test been reproduced by other researchers?
- Is the test considered more or less reliable than other types of tests in which the chemical has also been tested but has yielded different results?
- Is the species a relevant or reliable human surrogate or does this test conflict with other test data in species phylogenetically closer to humans?
- Are the conclusions of the experiment justified by the data included in the report of the experiment and consistent with the current scientific understanding of the test or of the area of toxicology being tested?
- Is the outcome of the reported experiment dependent on the test conditions or is it influenced by *competing toxicities*?
- Does the test indicate causality or merely suggest a possible chance correlation?

Dosages Tested

The dosages used in tests are dependent upon many of the preceding concerns. For example, increasing the dose may alter physiologic conditions such that changes observed at high doses are not consistent with or relevant to low-dose situations. Problems of dosage can largely be dealt with accurately by determining the threshold, mechanism of action, and other factors that can influence the toxicity of the chemicals, but the margin of safety provided by doses used in animal toxicity tests versus the expected human exposures might always be given some additional, separate consideration.

Epidemiologic Evidence

Known occupational exposures or accidental human poisonings can be used as an important test of the validity of the animal test data and as an indicator of a chemical's potency in man relative to animals. Where possible such evidence can be a useful adjunct or potentially determinant factor to be considered in risk estimations. However, like other toxicologic data, this evidence must be utilized appropriately and cannot always be given disproportionate attention.

In assessing epidemiologic evidence for any toxicity or chemically induced effects, it must be remembered that in any epidemiologic study it may be impossible to eliminate all of the unrelated but confounding variables in the observed populations (see Chapter 13). Moreover, it is not uncommon to see measurable increases in one type of cancer over what is mathematically expected since we cannot choose a subpopulation that reflects exactly the baseline cancer rates of the test population. This fact has been previously demonstrated by showing that different areas of the United States have different backgrounds for many of the specific types of cancer (see Chapter 15). All of this must be considered carefully, and no definitive conclusions should be made about any finding of excess cancer (or other type of toxicity) merely above expected values but not statistically significant, unless the same excess for the same types of cancer or toxicity is repeatedly demonstrated in subsequent studies.

A positive epidemiologic correlation between exposure to a chemical and a resultant adverse health effect relies on the following conditions:

- The positive association (correlation) must be seen in individuals with known exposures.
- The positive association cannot be explained by bias in recording, detection, or experimental design.
- The positive association must be statistically significant.
- The positive association should show both dose and exposure-period dependency.
- The positive association must be observed repeatedly in subsequent studies and cannot be a single, confounding, variable observation.

17.5 Factors That Limit or Confound the Hazard Evaluation

The preceding section outlined considerations that should be applied to the hazard evaluations. This section provides some of the factors that limit the use of the toxicity data for risk estimation purposes.

Acute versus Chronic Toxicities

Chemicals capable of chronic toxicities generally produce such effects via mechanisms and doses that differ from those conditions governing acute (short-term) effects. Likewise, the biologic conditions and changes produced by high acute exposures often differ from those produced by lower, chronic exposure. Thus, acute responses or exposure intervals have little or no relevance to chronic toxicities or exposure intervals and the outcomes related to chronic situations usually cannot be modeled with acute data.

Variations in Mechanisms between Species

Extrapolation made between species is always a "Catch-22" situation. To know whether it is a valid extrapolation essentially requires prior knowledge of the outcome and is similar to determining whether the chicken or the egg comes first. That is, we use animal data to predict human outcomes but often need to know the human response first to be able to pick the right animal model.

Since innumerable important, vast species differences have been documented (occurring for reasons such as differences in mechanism of action, metabolism, genetic disposition, etc.), we must always acknowledge the possibility of these when little or no human data are available. Along this line there are generally large species differences in the metabolism of a chemical, both in the rate of metabolism and type of metabolism, and this can always affect the outcome of chemical exposure. A partial list of some of the species- and mechanism-related factors affecting risk extrapolations using only animal data are:

- Age or sex.
- Genetic, sexual, or hormonal status.
- Metabolism and pharmacokinetics.
- Disease states and medical care.
- Diet, environment, and circadian rhythms.

Teratogenicity and Fetotoxicity Tests

Teratogenicity (generation of birth defects) is a specific toxic manifestation of a chemical that often occurs only at or near maternally toxic or embryolethal doses. This fact can confound the data in several ways. Lethality to the fetus or mother can severely limit or mask actual teratogenic effects by removing the endpoint (the deformed offspring) from the analysis (the teratogenic effects are, in a sense, resorbed).

Conversely, maternal toxicity may disrupt the delicate physiologic balance between mother and fetus such that teratogenic effects are produced as a secondary effect of the maternal toxicity. That is, by increasing the test dose, the chemical may eventually induce a maternal toxicity, which then affects the normal development of the fetus. This situation, however, is not particularly relevant to the human situation because we attempt to prevent toxic or lethal exposures of pregnant mothers in the first place; and the secondary nature of teratogenicity obviously cannot be inferred from maternally nontoxic doses.

This important consideration is sometimes referred to as Karnofsky's rule, which can be paraphased as "*any* compound administered at the proper dosage, at the proper stage of development, to embryos of the proper species, will be effective in disturbing normal embryonic development." What Karnofsky's rule calls attention to is the fact that doses capable of killing the fetus or producing teratogenic effects in the offspring secondary to maternal toxicity are of much less concern than those chemicals producing teratogenic effects in the offspring at doses *not* toxic to the mother. For example, disease, malnutrition, stress, excessive vitamin levels, hormonal balance, and even over-the-counter drugs such as aspirin have all produced teratogenic effects in animals. Yet, in these cases avoidance of the toxic effects in the adult animal prevents teratogenic effects in the offspring.

Carcinogenicity Tests

The importance of determining the mechanism of a carcinogen has already been discussed. Therefore, this discussion will begin by reviewing the qualitative measurement of carcinogenic risk as estimated from animal data.

Squire's Proposal. Recognizing the problems associated with accurately interpreting the relevance of animal carcinogenesis bioassays and the risk inferred from them, Robert Squire, former acting director of the Carcinogenesis Testing Program and head of the National Cancer Institute's Tumor Pathology Section, has proposed a ranking scheme for animal evidence. Squire proposes that the basis for determining the strength of a positive test resides in whether the "suspect carcinogen" is positive in a number of species, gives rise to histogenetically different types of neoplasms in one or more species, induces malignant rather than benign tumors, and is consistently positive in an appropriate battery of tests measuring genotoxicity. The dose required to induce tumors as well as the spontaneous incidence of neoplasms in the control animals are also criteria.

By ranking the test data for each criterion according to a point system developed by Squire, one obtains an overall score for assessing the strength of the evidence (see Tables 17-5 and 17-6). Squire's approach is appealing because it recognizes several features of an animal bioassay that should be considered when attempting to estimate the likelihood that human responses are similar. Squire classifies a chemical receiving a score of less than 41 as a class V

Table 17-5. Proposed System for Ranking Animal Carcinogens.

Factor	Score
A. Number of different species affected	
Two or more	15
One	5
B. Number of histogenetically different types of neoplasms in one or more species	
Three or more	15
Two	10
One	5
C. Spontaneous incidence in appropriate control groups of neoplasms induced in treated groups	
Less than 1 percent	15
1 to 10 percent	10
10 to 20 percent	5
More than 20 percent	1
D. Dose-response relationships (cumulative oral dose equivalents per kilogram of body weight per day for 2 years)*	
Less than 1 μg	15
1 microgram to 1 mg	10
1 milligram to 1 g	5
More than 1 g	1
E. Malignancy of induced neoplasms	
More than 50 percent	15
25 to 50 percent	10
Less than 25 percent	5
No malignancy	1
F. Genotoxicity, measured in an appropriate battery of tests	
Positive	25
Incompletely positive	10
Negative	0

*Based on estimated consumption of 100 g of diet per kg of body weight. Scoring could also be developed for inhalation or other appropriate routes.

(Reproduced with permission from Squire, "Ranking Animal Carcinogens: A Proposed Regulatory Approach." *Science*, *214* (1981): Table 1. Copyright© 1981 by the American Association for the Advancement of Science.)

Table 17-6. Classifying (or Categorizing) Animal Carcinogens According to Total Factor Score.

Total factor score	Carcinogen class	Regulatory options
86 to 100	I	Restrict or ban
71 to 85	II	↑
56 to 70	III	⁞ Increasing regulation
41 to 55	IV	↑
Less than 41	V	Several options (no action. limited use. labeling, public education)

(Adapted with permission from Squire, 1981. Table 2. Copyright © 1981 by the American Association for the Advancement of Science.)

carcinogen, a class for which he suggests that the regulatory options are many and far less restrictive than for chemicals scoring higher in the ranking scheme.

Thus, according to this scheme, even the "positive" evidence for chlorinated hydrocarbons such as chlordane or DDT does not rank them as likely to be a human cancer hazard compared to the probability expected for many other chemicals, some of which are still commonly used (see Table 17-7). This ranking scheme is not a quantitative measure of the risk, but is a method for assessing the qualitative strength of the cancer hazard identified by animal bioassays. Mathematical models might still be used to predict the risk or set allowable limits for the expected human exposures. What this scheme provides instead is some test as to whether human cancer is a likely result of exposure to

Table 17-7. Approximate Rank of Several Animal Carcinogens Based on the Proposed System.

Carcinogen	Score	Class
Aflatoxin	100	I
Dimethylnitrosamine	95	I
Vinyl chloride	90	I
2-Naphthylamine	81	II
Chloroform	65	III
Chlordane	40	V
Saccharin	36	V
DDT	31	V

(Adapted with permission from Squire, 1981. Table 3. Copyright © 1981 by the American Association for the Advancement of Science.)

that chemical. In a sense it is a test of the likelihood that the test data will mimic the human response for what might be a vastly different exposure regimen.

Liver Tumors in Cancer Bioassays. Rodents used in cancer tests have a high spontaneous incidence of liver tumors. Therefore, if a statistically significant increase is seen after chemical exposure, it is important to determine whether or not this phenomenon represents true initiation or merely promotion of the background incidence. Such is the problem for any chemical increasing liver tumors in rodents. To make the distinction between initiators and promoters, the mutagenicity data and other characteristics of the chemical or the carcinogenic response become critically determinant factors (see Chapter 15).

In addition, it seems that the criteria for establishing liver cancer may vary between animal and human studies; thus, if a nonmutagenic chemical induces only liver tumors, particularly if only in one species, the concern or likelihood that it can cause human cancer is probably far less than the hazard posed by a mutagenic chemical that induces cancer in a number of different tissues in several species. Liver tumors in rodents, particularly mice, are also of less concern when only this organ shows tumorigenicity; this effect may be caused by a number of reasons other than presence of a mutagenic chemical (Doull, 1983).

Cytotoxicity and Recurrent Tissue Injury as an Epigenetic Mechanism for Cancer. Recurrent injury (i.e., chronic cytotoxicity) has been recognized as a carcinogenic pressure for decades. It is particularly relevant for discussion as it is a threshold effect for which the risk extrapolation should be decidedly different from those currently used. If recurrent tissue injury causes the tumors observed in animals, then preventing this toxicity, which would be a goal in any event, renders the chemical a noncarcinogen at the allowable exposure level.

The lines of evidence for treating cytotoxicity as a tumor-promoting or inducing mechanism (particularly in the liver) are (adapted from Stott *et al.*, 1981):

1. Cytotoxicity increases all cell replication and, therefore, DNA synthesis, causing an increase in the spontaneous mutation rate.
   - It is estimated that, during DNA replication, misincorporated base pairs (i.e., those escaping proofreading-repair) occur in the range from 10^{-8} to 10^{-11} per base pair synthesized.
   - "Natural" sources of DNA base errors and mutations in humans are 10^{-5} to 10^{-6} mutations per gene per generation.
   - It is estimated that 10 percent of all human gametes contain at least one new mutation and several inherited ones.

2. Cytotoxicity shortens the cell cycle during which tissue is regenerated, thus leaving less time for DNA repair mechanisms to eliminate misincorporated bases or alkylated bases.
   - It has been reported that a fivefold increase in mutation frequency occurs in actively dividing liver cells versus nondividing liver cells exposed to an alkylating agent.
   - Normal cells have been observed to survive lethal, mutagenic doses of UV radiation if held in confluence (nondividing state), so that DNA repair can reverse the damage.
3. Cytotoxicity alters the levels of critical nucleotide pools, ions, and enzymes.
   - Changes in nucleotide pools and Mg^{++} concentrations have been observed to influence DNA polymerase error rates.
4. The increase in DNA synthesis in regenerating tissue means more genes are in a transcribable state and are more susceptible to the loss of regulatory control by a mutation in a regulatory gene locus.
5. DNA repair enzymes may become a significant target for protein binding chemicals that could compromise their activity.
6. Cytotoxicity may alter cellular methylase or polymerase activity.
   - Toxic doses of carbon tetrachloride increased the number of abnormal guanine bases (*O*-6-methyl- and *N*-7-methyl guanine) significantly; these bases were previously believed to be formed only as a result of alkylation by genotoxins.
7. There is empirical evidence for the role of enhanced cellular division and DNA synthesis in carcinogenesis. For example,
   - Tumors frequently develop in scar tissue (e.g., colon cancer occurs in persons with chronic colitis, skin cancer occurs in burn patients, and cirrhosis enhances the chance for liver cancer).
   - Repeated tissue damage with physical agents (such as dry ice) induces skin tumors in mice.
   - Repeated subcutaneous injections of nonreactive compounds such as saline, glucose, and distilled water have produced tumors at the injection site.
   - Partial hepatectomy increases the tumor incidence of initiating carcinogens.
   - Antiinflammatory steroids and retinoic acid (which inhibit DNA synthesis, reducing cell proliferation and inflammation) inhibit the tumor promotion of phorbol esters.

Example of a Chemical Producing Liver Tumors. As an example of a chemical that produces liver tumors in a chronic cancer bioassay in mice and for which the mechanism of cancer induction is probably recurrent liver injury,

Table 17-8. Perchloroethylene Data.

Treatment	Response	
	Mice	Rats
1. High doses	Liver cancer	No cancer
2. Percent metabolized	89.5	10.5
3. Total covalent binding	0.147 pmol/mg · protein	0.02 pmol/mg · protein
4. Liver weight increase	25%	5%
5. DNA synthesis	+82%	+3%
6. Histopathology positive for liver damage	++*	−*
7. DNA alkylation	−	−
8. Control incidence of tumors	High	Low

* ++ = Strong observation; − = weak observation.
(Adapted from Schuman *et al.*, "The Pharmacokinetics and Macromolecular Interactions of Perchloroethylene in Mice and Rats as Related to Oncogenicity." *Toxicol. Appl. Pharmacol.*, 55 (1980): 207.)

Table 17-8 lists various test data on perchloroethylene. Perchloroethylene induces liver tumors in mice but not rats, and only at high doses. It, like other chlorinated aliphatic chemicals, causes liver injury at high doses, and as a liver toxin ought to be capable of inducing liver cancer in certain animal species.

The data in Table 17-8 indicate that perchloroethylene enhances DNA synthesis by a cytotoxic rather than mutagenic mechanism in mice. These data also help explain why perchloroethylene does not induce tumors in rats. In rats, the toxicity of perchloroethylene as measured by covalent binding to protein and by histology sections of the liver is low to nonexistent. There is no increased liver weight or cell number caused by regenerative processes. DNA synthesis is normal; therefore, spontaneous tumors aren't likely. As there is no increased cellular division by recurrent toxicity, there is no stimulus for the low background incidence of liver tumors normally present. Interestingly, man metabolizes little perchloroethylene; thus, the rat, which does not generate toxic metabolites, seems to be a better animal than the mouse for modeling the human hazard of perchloroethylene exposure.

Readers interested in seeing how the ideas presented in this chapter are applied in a step-by-step assessment of risk for a hypothetical substance will find that the Appendix to this book provides such an example.

Summary

While numerous mathematical models may be used to calculate an acceptable risk, we must remember that the number generated is based upon many assumptions that are inherent in the model chosen and the test data used. We must never lose sight of the fact that these numbers are approximations at best, with an often unquantified amount of uncertainty in them. Thus, we must remember that, besides our extrapolated value of the risk, other factors such as those listed below may often help determine the "acceptable risk."

- The need and utility of the substance.
- The availability of alternative substances.
- Extent of public exposure.
- Economic impact of its use or disuse.
- Ecological impact.
- Use of natural resources.
- Availability of safety measures or engineering controls.
- Availability of appropriate waste disposal site and methods.

Moreover, for each "acceptable risk" calculated for carcinogens, the following questions may be raised regarding the accuracy of the estimation.

- Are there thresholds or no-effect exposure levels in man?
- Can "potential" or "real" be defined?
- To what extent should "lifestyle" be a factor?
- Can we identify and protect sensitive or susceptible individuals?
- How should negative results be evaluated?
- Should species differences in the metabolism and pharmacokinetics of the compound be considered?
- Should the risk of epigenetic carcinogens be estimated by a different method than those used for genotoxic carcinogens?

References and Suggested Reading

Brown, C. C., and Mantel, N. 1978. "Models for Carcinogenic Risk Assessment." *Science, 202*: 1105.

Clayson, D. B., Krewski, D., and Munro, I. C. 1983. "The Power and Interpretation of the Carcinogenicity Bioassay." *Reg. Toxicol. Pharmacol., 3*:329.

Committee on the Institutional Means for Assessment of Risks to Public Health, 1983. *Risk Assessment in the Federal Government: Managing the Process*. National Research Council. Washington, D.C.: National Academy Press.

Doull, J. (Chairman). 1983. "The Relevance of Mouse Liver Hepatoma to Human Carcinogenic Risk." A Report of the International Expert Advisory Committee to the Nutrition Foundation. Washington, D.C.: The Nutrition Foundation.

Dourson, M. L., and Stara, J. F. 1983. "Regulatory History and Experimental Support of Uncertainty (Safety) Factors." *Reg. Toxicol. Pharmacol., 3*:224.

Gaylor, D. W., and Shapiro, R. E. 1979. "Extrapolation and Risk Estimation for Carcinogenesis." *In* Mehlman, M. A., and Shapiro, R. E. (Eds.) *Advances in Modern Toxicology, Vol. 1: New Concepts in Safety Evaluation. Part II*. Washington, D.C.: Hemisphere Pub. Corp.

Hepatocarcinogenesis in Laboratory Rodents: Relevance for Man. Monograph No. 4. European Chemical Industry. Brussels, Belgium. October 1982.

Kolbye, A. C. 1981. "The Application of Fundamentals in Risk Assessment." *In* Bandal, S. K., Marco, G. J., Goldberg, L., and Leng, M. L. (Eds.) *The Pesticide Chemist and Modern Toxicology*. ACS Symposium Series No. 160. Washington, D.C.: American Chemical Society.

Krewski, D., and Brown, C. 1981. "Carcinogenic Risk Assessment: A Guide to the Literature." *Biometrics, 37*:353.

Office of Technology Assessment. 1981. *Assessment of Technologies for Determining Cancer Risks from the Environment*. Washington, D.C.: OTA Publications.

Park, C. N., and Snee, R. D. 1983. "Quantitative Risk Assessment: State-of-the-Art for Carcinogenesis." *Fund. Appl. Toxicol., 3*:320.

Schumann, A. M., Quast, J. F., and Watanabe, P. B. 1980. "The Pharmacokinetics and Molecular Interactions of Perchloroethylene in Mice and Rats as Related to Oncogenicity." *Toxicol. Appl. Pharmacol., 55*:207.

Squire, R. A. 1981. "Ranking Animal Carcinogens: A Proposed Regulatory Approach." *Science, 214*:877.

Stott, W. T., Reitz, R. H., Schumann, A. M., and Watanabe, P. G. 1981. "Genetic and Nongenetic Events in Neoplasia." *Food Cosmet. Toxicol., 19*: 567.

Task Force of Past Presidents, Society of Toxicology. 1982. "Animal Data in Hazard Evaluation: Paths and Pitfalls." *Fund. Appl. Toxicol., 2*:101.

Weil, C. S. 1972. "Statistics vs. Safety Factors and Scientific Judgment in the Evaluation of Safety for Man." *Toxicol. Appl. Pharmacol., 21*:454.

Chapter 18

Methods for Evaluating Chemical Exposures

Phillip L. Williams

Introduction

The purpose of this chapter is to explain how to develop guidelines for acceptable occupational exposure levels for chemicals for which no such standards exist, or in cases where the margin of protection provided by the existing standards is questionable. The material discussed is an extension of the process of risk assessment described in the previous chapter. This chapter will:

- Emphasize how toxicological data can be used by individuals in industry.
- Present a logical approach for the evaluaton of a chemical exposure.
- Provide a step-by-step procedure for using the information presented in the preceding chapters.

Figure 18-1 provides an overview of the evaluation process this chapter will present. The first step in evaluating an exposure is determining the specific chemical or chemicals to which the individual organism—animal or human—has been exposed. Typically, potential industry-related exposures affect humans. The difficult part of the procedure is determining whether or not the levels present are within acceptable limits. This can become a complicated problem when no published standards for acceptable exposure levels exist for the substance in question.

For an exposure of unknown cause to be of concern, those people who have been exposed must first show signs or symptoms of an adverse effect. This may occur where chemicals are not normally used in connection with work, such as in an office, or where new procedures or other changes have been put into effect that cause the release of toxins where none were present before. Obviously, an unknown exposure is the most difficult to evaluate because the first step—identifying the agent causing the adverse effects—can be extremely complex. Several cases of this type are presented in Chapter 19.

Once the source is determined, the evaluation process follows the pathway established for a known source (Figure 18-1). The exposure level is quantified through environmental sampling; a determination must be made about whether the exposure presents a hazard. In some cases this can be done by comparing the results to a standard or guideline, such as those established by OSHA, ACGIH, NIOSH, etc. But no established standards exist for most chemicals and, in some cases where standards do exist, new toxicological information may be available and established standards may need to be reexamined.

This chapter will deal primarily with the situation where standards for a chemical are thought to be inappropriate or have not been set at all. Discussion will focus on how to take published toxicity data, extrapolate them to an occupational exposure, and predict a guideline for an acceptable exposure level.

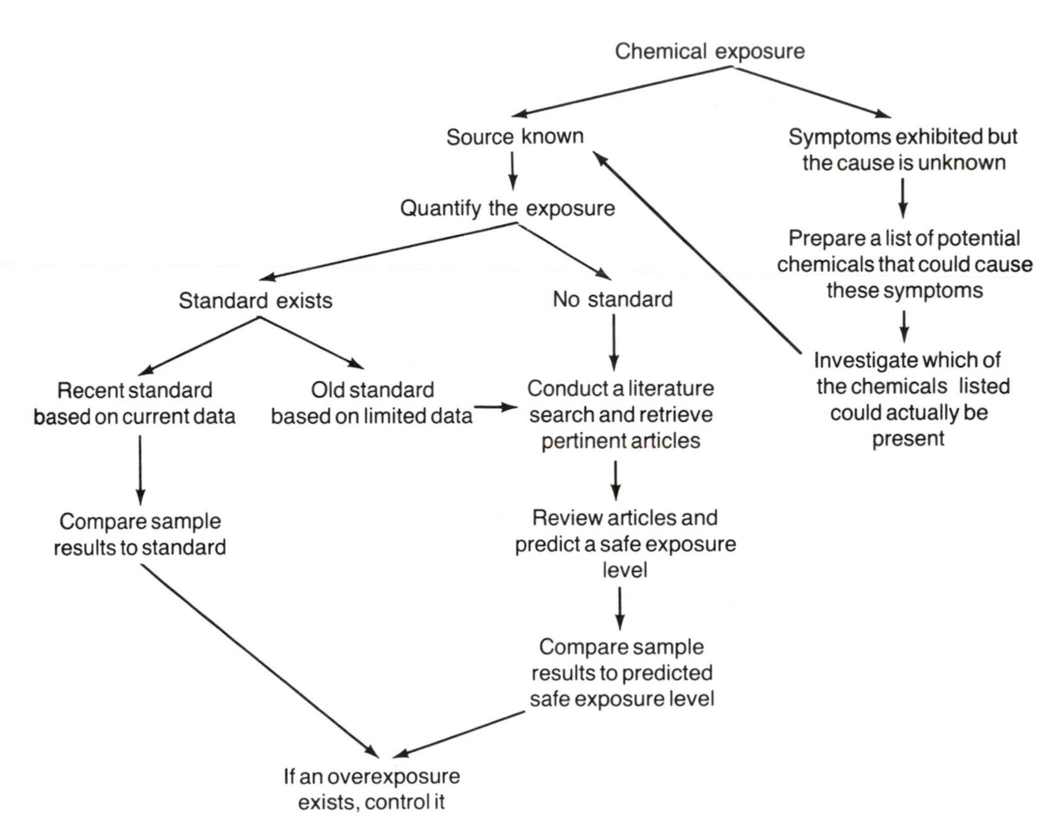

Figure 18-1. A flowchart for evaluating a chemical exposure.

18.1 Known Exposure Source

Determining Routes and Levels of Exposure

The chemicals used in industrial operations are usually known and, consequently, potential exposures can be predicted. Sometimes trade names, impurities, and reaction products can complicate the identification process; but with the proper research, usually the actual chemicals can be identified. For the purpose of our discussion, let us assume that the identity of a chemical toxin is not at issue.

The next step is to quantify the exposure level through appropriate sampling. Details of this activity are presented in Chapter 20. As discussed in Chapter 3, there are three major routes of entry into the human body for any occupational exposure: inhalation, skin absorption, and ingestion.

The oral route (or ingestion) is usually thought to be a minor pathway for workplace exposures and should be easily eliminated through strict observance of hygienic practices relating to eating, drinking, and smoking. The skin and inhalation routes can both be major exposure pathways and should always be evaluated. Of these two routes, inhalation is usually the simplest to evaluate. It is often relatively easy to quantify the air levels of contaminants present in the environment. When occupational health standards are established for a particular chemical, they denote an allowable air concentration for that chemical.

This does not mean that the dermal route should be ignored, because in some cases (for example, chemicals with low vapor pressures) skin absorption may be the major exposure pathway. A determination must first be made about how readily the chemical can be absorbed through the skin (detailed in Chapters 3 and 8) and what operations within the industrial process will result in a dermal exposure. Then, the appropriate controls to prevent direct contact with the skin should be implemented (discussed in Chapter 20).

Once the inhalation exposure is quantified, a determination must be made about whether the level of exposure is within acceptable limits. If a health standard exists, this can be as simple as comparing the air sampling data to the standard. However, if no standard exists, or if the adequacy of a standard may be questioned, an acceptable exposure level has to be determined.

Hazard Identification: Searching the Literature for Available Toxicologic Information

The next step in the evaluation process is to gather as much toxicological information on the chemical as possible. This can be done by conducting a thorough literature search. Such a procedure is time consuming, but it is necessary in developing an adequate exposure guideline.

Basically, there are two types of literature searches:

- Manual (or hand); and
- Computerized.

Both methods involve searching the indexes of various data bases, reviewing the abstracts, and retrieving pertinent articles. However, in computerized searches, the computer searches the indexes and provides a listing (and, in some cases, abstracts) of the citations. Obviously, the computerized search is based on information you provide, and the data bases rarely contain information published before the late 1960s. Because neither method is completely comprehensive, both are needed to perform a successful search.

To conduct a search, you need access to a library that has the specific journals, periodicals, and other necessary reference materials. Most major hospitals, colleges, and universities have such resources. Additionally, to conduct a computerized literature search, a computer terminal, phone interface, and access to a specific database are required. With these, the initial search can be performed in any office or even at home. However, to retrieve the actual citations and to perform the supplemental manual search, direct access to a library will be needed.

The first step is to determine the objective of the search. In the case of developing a guideline for exposure, this may require reviewing all published toxicological data on a specific chemical. However, in some cases, only specific areas of the toxicological profile may be of interest (for example, carcinogenesis or reproductive toxicology). Or, only information gathered after a certain date may be needed (for example, a standard may have been established in 1972, let us say, but one wishes to determine if new data suggest a need for reevaluating the acceptable exposure level). Whatever the need, the result that is wanted should be identified.

Next, the specific databases that may contain the material of interest should be identified. This will require familiarity with various databases and knowledge of how material is indexed, how subject matter is identified, how chemicals are listed, and which sources will provide what types of data. If a manual search is being performed, the indexes for various abstract services should be reviewed. Examples are *Chemical Abstracts*, *Biological Abstracts*, *Science Citation Index*, and *Index Medicus*.

When using an index of abstracts, first, determine how the chemical is listed in it (i.e., the specific type of chemical nomenclature used by the source); then, review the various yearly indexes, noting which abstracts for the chemical are listed with terms that appear related to the topic that interests you. Next, review the specific abstracts and, if they contain information of interest, retrieve the original source.

Conducting a Computerized Search. Although a computerized search does basically the same thing as a hand search, an important difference is that the computer will only do what it is told to do. With a manual search, decisions can be made, as one reviews the indexes, about the relevance of a subject listing or the possible usefulness of an article, on the basis of the information provided in the abstract. With the computerized search, the computer performs this function and it will only search for predetermined terms, making the preliminary work and design of a search strategy very critical.

Performing a computerized literature search is not computer programming. You work with an existing program that usually requires fewer than ten commands to conduct the search. The required commands can be found in the specific user's guide or reference manual for that particular system.

To access the system, simply pick up a phone, dial the number for the particular system, get a connection, connect the phone to the computer interface, provide an identification number, and begin the search.

Many systems offer files for conducting computerized *toxicologic* literature searches. Examples are:

- Systems Development Corporation has a system named ORBIT.
- The National Library of Medicine makes available various computer data bases related to toxicology.
- Lockheed has a system called DIALOGUE.

All of the systems have different files for different databases. To use these systems successfully, variations in specific commands, subject listings, file format, key words, etc. should be determined. The best approach is to become familiar with an example of an actual printed version of what the computer will provide before conducting a search. Usually the user's guide or reference manual for the system will provide this type of information.

Strategies for Computerized Searches. A computerized database consists of two main components:

- A branch of indexes (subject indexes, author indexes, title indexes, source indexes, etc.).
- Unit records.

A unit record is a bibliographic citation that provides a listing of the author, title, source, date of publication, listing of key words or identifiers for the material, and, with some databases, an abstract of the material. Although the computer will search both indexes and unit records, the unit record is what is provided to the user upon completion of the search. From this material a decision can be made about whether the article is pertinent to the search, and the source of the material (journal and volume or book publisher) can be identified for retrieval.

The cost of using these systems varies but is based on an hourly rate plus a fixed monthly rate per user. The hourly rate is another reason to do detailed preparation before beginning the search. You should know what files within the system you will want to access and what search terms will be used. Both of these require knowledge of the various computerized databases and knowledge of the subject to be searched. As a result, searches often require a team approach. A librarian trained in the use of the various databases and the individual requesting the information should work together to develop the best search strategy.

The principal components of the strategy are to:

- Outline what you want to search.
- Identify what system and which specific files you want to enter.
- Develop a list of synonyms for the chemical of interest.
- Develop a list of key words describing the type of information you wish to retrieve.

As stated previously, all of these concerns require detailed information about the databases. Different systems have different ways of listing chemicals and use different terms for subject headings.

You want the computerized search to provide you with pertinent data, not thousands of extraneous citations. For this reason, the terms and chemical names must be specific and properly entered into the system. Often truncated terms (that is, the root of a word) can be entered into the computer, with an appropriate command to produce a list of citations dealing with that word-root and its derivatives. For example, suppose you want to find material on the teratogenicity of Chemical A. The correct name for Chemical A and the proper conjunction (usually "and") are entered with the term "terato-" with the appropriate truncation symbol. This will result in the computer looking for all words that begin with "terato-" and appear in conjunction with Chemical A. These words would include teratology, teratogenic, teratogen, teratogenesis, and so on.

It is often necessary to search several databases. If you have a new chemical and little toxicologic data are anticipated, you may wish to search all the databases. This will result in some overlap, but it might prove useful.

If you want to conduct an exhaustive search on a substance, you may need to enter specific databases for specific reasons. For example, you can focus your search on chemical abstracts in the DIALOGUE or ORBIT data systems, to identify the manufacturing processes and any impurities in the chemical. The MEDLINE files of the National Library of Medicine can be searched to find information on metabolism and other biomedical data. Another of the NLM files, TOXLINE, could be searched for the toxicology literature. Here again, duplication will occur between the databases, but if the terms for each search are properly constructed, a successful search will result.

As each set of terms is entered into the computer, a number will be printed showing the number of "postings" or "hits" that can be located in the file with the requested combination of terms. At this point you can decide if the search should be continued, and whether to have the unit records printed on-line or off-line. If the computer shows 5000 postings, it will be obvious that the defining terms are too broad. If no listings are shown, the terms may be inappropriate for the database. If the data are needed immediately, they can be printed on-line, but this is more expensive than instructing the computer to print them off-line. If printed off-line, the material is usually received in less than ten days; it is advisable to have two or three of the unit records printed on-line to assure that the listings are indeed the sort you are searching for.

It is most effective to begin with a computerized search, and then perform a manual search as needed. The computer search provides a quick, relatively inexpensive source of data. The material provided in the unit records can be retrieved and reviewed. Then, from the bibliographies of the retrieved material, more data can be gathered to expand upon the computer search. A manual search can be conducted to further supplement areas not found in the computer search, such as material that existed prior to establishment of the databases, and the most recent literature that has not yet been added to the data bases.

Assembling Information Once It Is Obtained

Once the search is completed, the next step is to accumulate all the articles, books, and other pertinent publications. As mentioned in the preceding section, this requires reading the material and expanding on it from bibliographies and newly found material. It may seem like a never-ending chore, but it can be accomplished.

In order for a toxicologic evaluation of the literature to be performed, the material should be organized so that articles on similar topics are under a common heading for an overall review. The articles can be arranged similarly to the way this book is outlined.

For example, a category can be established for articles of general interest that cannot be assigned to a specific category. Next, there could be a category on the chemistry and any manufacturing impurities associated with the particular chemical. This category can be used to predict potential exposures to substances not generally associated with the chemical. A subsequent category could cover the broad field of metabolism (both in humans and laboratory animals) of the chemical; it would attempt to identify topics such as absorption, distribution, and elimination of toxic agents, concerns discussed in Chapter 3 of this book.

Once the chemistry and metabolism of a chemical are understood, its toxicology can be better studied. The next categories could therefore deal with acute and chronic toxicity and could be divided according to findings in humans and laboratory animals. The material should be reviewed and evaluated for specific areas of concern, as outlined in Chapters 4 through 9 of this book.

Aspects of acute toxicity would involve short-term human exposures or animal testing of generally less than 90 days. This issue can be divided into two broad categories: acute lethality, and irritation or local damage. Studies of chronic effects are those that evaluate the exposure for long periods of time. Usually, with a laboratory animal, this means the majority of the animal's lifespan (two years or longer with mice and rats).

The above categories should include effects of a general toxicologic nature. Specific areas of concern should then be treated in individual categories, too. These areas include studies on carcinogenesis, mutagenesis, and reproductive toxicology. Obviously, there will be some overlap especially of chronic general toxicity and carcinogenesis. As a result, you may want to refer to one study in more than one category. Also, epidemiologic data from human exposure should be identified.

Additionally, it may be appropriate to add a category for published levels of exposure found in various work settings or from your own experience. In this way, correlations can be drawn between various operations and control procedures may be found.

A summary should be made for each category, to present a cumulative gathering of the data. General conclusions can also be stated in each summary about whatever aspect of the chemical's toxicologic profile is dealt with in that category.

Hazard Evaluation and Risk Estimation

Now comes the key point in the evaluation process: integrating all the information and predicting an exposure guideline that can be realistically applied. Basically, this is accomplished by thoroughly evaluating the literature and then estimating the risk through the use of various mathematical models. In analyzing the test data all of the considerations discussed in Chapters 2 and 17 should be fully evaluated. These include such factors as dose-response relationships, species variations, mechanisms of toxicity, dosages tested, and the validation of the test and its relevance to human extrapolation.

This procedure may result in two exposure guidelines: an acute exposure level, such as a ceiling or peak level for a 15- or 30-min sampling period, and a chronic exposure level that will be acceptable for 8 to 10 hr/day, 40 hr/week. Also, in cases where the chemical has been proved a carcinogen (depending on its mechanism of action, see Chapters 15 and 17), the exposure level may need to be kept at the lowest feasible limits of exposure.

In any event, the extrapolation to humans should be made from the most sensitive animal species that has been tested, unless there is evidence that this species is not appropriate (e.g., detailed knowledge of metabolic differences). The most toxic response demonstrated in the published material and the dosage level that elicited that response should be identified. Ideally, the specific study that reported this effect would also have reported a no-observed-effect-level (NOEL) under the same testing procedures. If so, the NOEL can be taken (it will usually be given as a milligrams-per-kilogram dose), a safety or uncertainty factor can be applied, and the data can be plugged into one of the formulas presented in Chapters 2 and 17, to predict a safe exposure guideline.

As discussed in Chapter 17, the factors used are commonly either 100 or 10. A factor of 100 is used when extrapolating from animal data to humans (species to species). Generally, a factor of 10 is used if the data are from human exposures; this factor is to account for intraspecies variations. Sometimes a factor of 1000 is used, if there are very few data to work with. As the number of species tested increases, the confidence that the most sensitive species has been identified increases and the factor can be decreased.

Other conclusions can be drawn from the data. Such determinations as the significance of routes other than inhalation can made. Is the dermal route a significant exposure pathway? If so, controls such as those discussed in Chapter 20 can be implemented.

An obvious conclusion that can be reached regards areas that need further toxicologic testing. With nearly all chemicals, portions of the toxicologic profile will be incomplete, and by using the type of approach presented in this chapter, these areas will become readily apparent. As additional material is published about the chemical, categories can be added for these areas and conclusions can be altered as needed.

Case Study

I have used the approach I detailed in this chapter to evaluate various exposures. In some cases only portions of the method were needed, but in others the entire process was used. I have found it a useful tool in the effort to provide guideline information for evaluating employee chemical exposure.

One case that has been published involves occupational exposure to pentachlorophenol (PCP), a commonly used wood-preserving chemical (Williams, 1982; Williams, 1983). I was interested in seeing whether existing standards for human exposure were appropriate. However, the search for literature was not limited to that published within a specific time period.

I made a computerized literature search by using files in both DIALOGUE and the National Library of Medicine. The search was designed to cover a broad range, including occupational exposure cases, epidemiological reports, all types of toxic effects (acute, chronic, carcinogenic, mutagenic, and teratogenic), and general metabolism. This resulted in the retrieval of over 100 articles. Many of the bibliographies for these articles led to additional articles, and a supplemental manual search was conducted. (The identification of a recent review article may accelerate the search process; such a publication will usually have an extensive bibliography that may greatly assist in your project.)

As previously described in this chapter, the articles were then sorted into numerous categories. The specific headings and subheadings used were:

- General data
- Chemistry and manufacturing impurities in commercial products related to (PCP)
  - Manufacturing process
  - Impurities
- Metabolism
  - Absorption
  - Transport and distribution
  - Biotransformation
  - Excretion
- Toxicity to animals
  - Acute toxic effects
  - Subchronic and chronic effects
- Carcinogenesis
  - Carcinogenicity testing of PCP in test animals
  - Carcinogenicity of PCP impurities
- Mutagenesis
  - Mutagenicity testing of PCP
  - Mutagenicity testing of impurities
- Reproductive toxicology
  - Fetotoxicity
  - Teratology
  - Reproduction studies
- Occupational exposure
  - Industrial uses and processes
  - Cases of occupational exposures

Overviews and summaries were drawn from each section, and collectively an assessment of the occupational hazard of PCP began to take shape. It was learned that the toxicity of technical-grade PCP involves two issues: PCP as a toxic chemical, and the impurities of PCP. The various impurities were identified, as well as how they develop in the manufacturing process and the ranges of percentages found in commercial products. This then allowed a consideration of their place in the toxicologic evaluation. From the metabolism section it was concluded that both the dermal and the inhalation routes were primary routes of occupational exposures. Also, the biologic half-life was found and biologic monitoring procedures were identified. The acute toxicity data demonstrated that PCP causes local irritation to the eyes and nose, as well as systemic effects that result from its ability to uncouple mitochondrial oxidative phosphorylation. Some of the commercial impurities that contaminated PCP were found to be capable of causing chloracne and chronic liver damage. The carcinogenicity and mutagenicity studies were found to be negative with regard to both oncogenic effects and mutagenicity.

The literature made it apparent that the major area of concern was that PCP and possibly its hexachlordibenzo-*p*-dioxin impurity caused teratogenic and fetotoxic effects in test animals. The NOEL for this effect was reported to be 5.8 mg/kg · day for PCP.

I found many reports of industrial poisonings involving PCP, but little data on the long-term effects in workers. I was able to retrieve published exposure levels for occupational settings and identify various sources and controls of PCP exposure.

The resulting overall assessment of the hazard showed that fetotoxic and teratogenic effects were the most worrisome, as far as could be determined from the data available in 1981. When the NOEL of 5.8 mg/kg · day was used, and a safety factor of 100 was applied for extrapolation to human exposure, this resulted in a 0.058-mg/kg level. The next step was to use this finding in the following formula (discussed in Chapter 2):

$$\text{Dose} = \frac{\alpha(\text{BR})(\text{C})t}{\text{BW}} = \text{mg/kg},$$

where α is the absorption of the material into the body through the lungs (PCP was estimated to be 100%, based on published data); BR is the breathing rate, assumed to be 2 hr of heavy breathing (1.47 m^3/hr) and 6 hr of moderate breathing (0.98 m^3/hr), derived from published data; C is the concentration of PCP in the air; t is the time of exposure in hours; and BW is the body weight (estimated to be 60 kg for a female).

I wanted to make a comparison with the present OSHA standard and the present TLV set by the ACGIH. Both levels are 0.5 mg/m^3 for an 8-hr occupational exposure to PCP, so that concentration was chosen for use in the formula. Since the published data did reveal operations where such concentrations could be exceeded, the choice did appear realistic.

Solving the formula gave

$$\text{Dose} = \frac{1.0(0.98\ \text{m}^3/\text{hr})(0.5\ \text{mg/m}^3)(6\ \text{hr})}{60\ \text{kg}} + \frac{1.0(1.47\ \text{m}^3/\text{hr})(0.5\ \text{mg/m}^3)(2\ \text{hr})}{60\ \text{kg}};$$

$$\text{Dose} = 0.0735\ \text{mg/kg}.$$

The level of 0.0735 mg/kg was then compared with the 0.058 = mg/kg value of NOEL (allowing for a safety factor), and it was concluded that when the present TLV or OSHA standard is reached or exceeded, an occupationally exposed female may receive a dose that exceeds the NOEL of PCP for fetotoxicity or teratogenicity, allowing for a standard safety factor. Additionally, workers handling PCP would likely have an additional dermal exposure causing the total dosage to be even higher.

Another approach to the data would have been to rearrange the formula so that a predicted safe level of exposure could be determined:

$$C = \frac{\text{Dose(BW)}}{\alpha(\text{BR})t}$$

$$= \frac{(0.058\ \text{mg/kg})60\ \text{kg}}{1.0[(0.98\ \text{m}^3/\text{hr})6\ \text{hr} + (1.47\ \text{m}^3/\text{hr})2\ \text{hr}]}$$

$$= \frac{3.48\ \text{mg}}{8.82\ \text{m}^3};$$

$$C = 0.39\ \text{mg/m}^3.$$

This calculated level could then become a guideline for evaluating the occupational exposure of females to PCP. Here again, no consideration is given to dermal exposure, but it is expected that the proper precautions (e.g., personal protection and strict personal hygiene) could be used to limit these entry pathways.

Although a great deal of information was available on the toxicity of PCP, this review demonstrated many areas where additional work was needed. For example, the detailed data on the toxicity of the various impurities and the exact isomers present are not known. Additional chronic toxicity data are needed, as well as further testing on mutagenesis. Human epidemiological data are lacking, and testing should be conducted to show the toxicity of PCP in conjunction with other wood preservatives and the vehicles (solvents) commonly used with PCP in the occupational setting.

18.2 Unknown Exposure Source

An unknown source of exposure requires that several additional steps be added to the chemical exposure evaluation process shown in Figure 18-1. As mentioned at the beginning of this chapter, exposure can be suspected only after one or more persons display symptoms of an adverse effect, and the first step is to identify the cause of the problem.

The exposed individuals should be interviewed and their specific symptoms should be noted. Depending on the extensiveness of the illnesses, this step often requires a physician or an occupational health nurse.

The next step is to prepare a list of potential chemicals that could cause the symptoms. Then a well-conceived, methodical approach should be implemented to try to determine which chemical(s) actually caused the adverse effects. Chapter 19 presents several such cases. This process of identifying the source of the adverse effects can be extremely involved and time consuming. It requires the interaction of many disciplines—medicine, toxicology, industrial hygiene, chemistry, engineering, management, etc. The approach is something like that used by detectives to find clues for a crime.

Another added complication is that often the symptoms expressed by an individual are very subjective, and many chemicals may cause these effects in humans. Many of these cases occur in such places as office settings. Workers complain of headaches, upper respiratory irritation, and burning of the eyes. Many agents can cause these responses, and the process of elimination can be extensive. Once the list is narrowed to the substances most likely to cause the problem, sampling can be performed to determine which substances are present.

At this stage an additional problem is encountered: to assure that the substance identified is actually causing the problem, the exposure level has to be quantified and found to be in a hazardous range. For example, when workers complain of headaches, it is reasonable to think that carbon monoxide (CO) is the causative agent; sampling is conducted and the level is found to be 5.0 ppm of CO. Although CO is present, it then becomes clear that some other substance is causing the problem.

This chapter has emphasized chemical exposure, but in the case of an unknown exposure, other causes should also be considered. These include physical hazards such as radiation, biological hazards such as airborne micro-organisms, and physical factors such as humidity, heat, and cold. These types of agents can also be evaluated with the approach described in this chapter.

Once the source is identified and the exposure levels are quantified, evaluation follows a pathway identical to that used when the source of the exposure is known (Figure 18-1), as previously discussed in this chapter.

Case Study

Along with another industrial hygienist, a chemist, medical personnel, and company management representatives, I participated in the evaluation process summarized in this case study. A more detailed description of this case can be found in the published articles referenced at the end of the chapter (Williams, Spain, and Rubenstein, 1981; and Williams and Spain, 1982).

The case involved a group of artists working on the restoration of a famous painting, *The Battle of Atlanta*. The painting is 15 m tall by 115 m wide (the enormous size is accounted for by the fact that the painting is mounted around the inside of a round building). During the cleaning process, employees experienced headaches, nausea, weakness, and vomiting. One employee began vomiting almost immediately after beginning the cleaning work. The workers were removed from the operation and an evaluation was conducted.

The employees were interviewed and given medical examinations. In conjunction with the investigation, bulk samples were collected (paint from the canvas, settled dust, various residues, etc.) from all aspects of the operation, and the cleaning solutions were analyzed. Additionally, a list of chemicals that could potentially cause the symptoms was compiled from a literature search.

Upon further investigation, one of the artists mentioned that she had noticed a redness in her urine. Also, the bulk samples showed the presence of arsenic. Arsine was the chemical on our list that best fitted these facts and, since arsenic was present, there was a potential exposure to it.

This situation was complicated further, however, because arsine is usually generated under acidic conditions, and in this case the cleaning agents were determined to be mild alkaline solutions. In an attempt to confirm the suspected cause, employees were given NIOSH-approved air-supplied hoods to wear and were asked to resume work while air sampling was conducted. As a result of this sampling, airborne levels of arsine were found.

The next problem was that the levels found were well within the OSHA standard. A partial literature search was conducted, and a NIOSH publication, *Criteria for a Recommended Standard for Occupational Exposure to Inorganic Arsenic*, was retrieved. From this publication it was learned that NIOSH had recommended a much lower arsine exposure level, and our sample results were well above this recommended level.

Since the operation was only for a short duration, the exposure was controlled through the use of NIOSH-approved air-supplied hoods. The cause of the generation of the arsine was still not clear, but further work through the use of a literature search allowed us to postulate its source.

As one can see, many of the steps described in this process are performed concurrently and, depending on the urgency of the situation, interim controls are often put into use prior to a final conclusion.

Summary

This chapter has provided an approach for determining an exposure guideline for a given chemical.

- It outlined a model that can be followed to develop guidelines for employee exposure when no guidelines exist or when present standards are outdated.
- It discussed many of the factors that should be considered in the evaluation process.
- It explained how to conduct a review of toxicologic literature.
- It provided examples of case studies in which hazard identifications and evaluations were conducted and risk estimates were performed.

Once again, it should be noted that this is an approach for determining an exposure guideline. It does not take into account all the public, economic, and political complexities of regulatory standard-setting policy.

References and Suggested Reading

Department of Health, Education, and Welfare/National Institute for Occupational Safety and Health (DHEW/NIOSH). 1975. *Criteria for a Recommended Standard for Occupational Exposure to Inorganic Arsenic*. Publication No. 75–149. Washington, D. C.: DHEW/NIOSH.

National Academy of Sciences (NAS). 1977. *Principles and Procedures for Evaluating the Toxicity of Household Substances*. A Report Prepared by the Committee for the Revision of NAS Publication 1138, under the Auspices of the Committee on Toxicology, Assembly of Life Sciences, National Research Council. Washington, D. C.: NAS.

Teaf, C. M., and James, R. C. 1984. *Pentachlorophenol: A Toxicant Profile*. Contribution 1–84. Published by the Center for Biochemical and Toxicological Research, Florida State University. Tallahassee, Florida.

Wexler, P. 1981. *Toxicology: A Guide to Sources of Information*. Reference Information Series, National Library of Medicine. Bethesda, Maryland.

Williams, P. L. 1982. "Pentachlorophenol, An Assessment of the Occupational Hazard." *American Industrial Hygiene Association Journal, 43* (11): 799–810.

Williams, P. L. 1983. "Commercial PCP: Toxic Impurities." *Occupational Health and Safety, 52* (7): 14–16.

Williams, P. L., and Spain, W. H. 1982. "Industrial Hygiene and the Arts: Restoring *The Battle of Atlanta*." *Occupational Health and Safety, 51* (4): 34–38.

Williams, P. L., Spain, W. H., and Rubenstein, M. 1981. "Suspected Arsine Poisoning during the Restoration of a Large Cyclorama Painting." *American Industrial Hygiene Association Journal, 42* (12): 911–913.

Chapter 19

Case Studies

Renate D. Kimbrough

Introduction

This chapter will present overviews of six case studies. For each one, the following will be discussed:

- The symptoms of an illness that precipitated a toxicologic investigation.
- The methodology used to identify, evaluate, and control an exposure to a toxic agent.
- The interdisciplinary teamwork of such professions as medicine, toxicology, epidemiology, industrial hygiene, chemistry, and engineering, required when conducting these types of investigations.

For the reader interested in additional information, a reference for each case is provided at the end of the chapter.

19.1 A Case of Acute Arsine Poisoning

The Centers for Disease Control (CDC) received a call from an Atlanta hospital stating that a patient had just been admitted with what seemed to be hemolysis. The hospital also reported that a fellow employee of the patient had become ill at about the same time and had been admitted to another hospital. The CDC together with the Regional Office of the National Institute for Occupational Safety and Health (NIOSH) organized a visit to the plant where both employees worked. The concern was that other workers might have been exposed as well.

It was learned at the plant that the two ill employees had worked at cleaning a room on the preceding day. There had been a clogged floor drain in this room, and the workers used a drain cleaner that was one of the commercial products manufactured by the company. Once in the drain, the cleaner had bubbled and given off a gas with a pungent odor. Even though the employees developed headaches, they continued at their work for about four hours.

At home that night, the employees had become ill. They were nauseated, and vomited. One went to a hospital emergency room, was examined, and was sent home. The other went to another emergency room; his urine was sampled and it was bloody looking; he was admitted to the hospital. The first worker became progressively worse and was admitted to a hospital the next day. In the meantime it was determined in the hospitals that both workers had severe hemolysis, which caused the bloody-looking urine. The only gas that will cause hemolysis is arsine.

One of the necessary ingredients for the production of arsine is arsenic. The plant was therefore inspected for potential sources of arsenic, and samples from the floor drain were collected and analyzed. All the samples were analyzed at the Georgia Institute of Technology by neutron activation analysis, which is somewhat expensive but is a relatively easy way to analyze multiple matrices. In addition, urine and blood from the two patients, the drain cleaner, and air samples were also submitted for analysis.

Figure 19-1 is a general outline of the room in which the incident occurred. The plant consisted of a large work area and several offices. There were two loading docks, one small and the other large. One room was used for mixing the different products made there. In another area workers bagged products. The stopped-up floor drain from which the gases had bubbled was in the

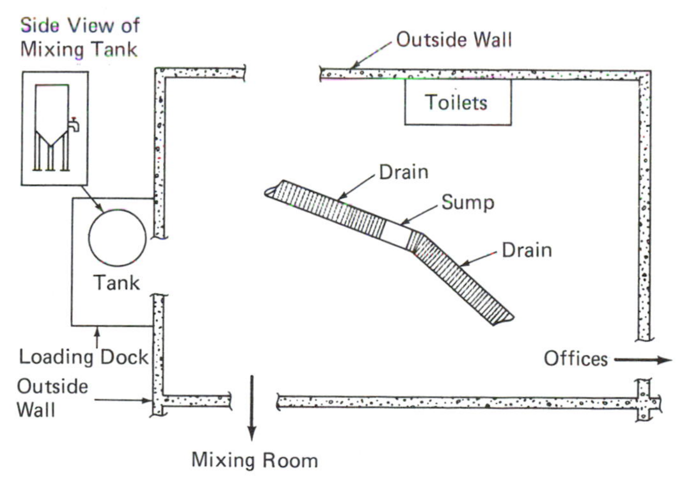

Figure 19-1. Floor plan of the area of the plant where the stopped-up drain was located.

center of this area. The loading dock outside this room was higher than the floor of the room; when it rained, water on the loading dock would run into the building.

The manager of the plant had worked there for only a few years and was unaware of any arsenic ever having been used in the plant. However, one of the maintenance men told investigators that before its present operation, the company had made arsenical herbicides, primarily arsenic trioxide; in fact, a tank located outside on the loading dock had previously been used for mixing arsenical herbicides. When the employees were conducting their general cleanup, they thought only water was in this tank and had drained it; the contents drained into the stopped-up drain. When the employees added their company's drain-cleaning product, primarily a mixture of sodium hydroxide, aluminum chips, and sodium, the gas had been emitted from the drain. Clearly the water and arsenic trioxide from the forgotten mixing tank had reacted to produce the gas arsine, as follows:

$$6\,NaOH + 6\,Al + 6\,H_2O + As_2O_3 \rightarrow 6\,NaAlO_2 + 3\,H_2O + 2\,AsH_3$$

Arsine is extremely toxic. Arsine poisoning does not reveal itself in elevated arsenic levels in those patients who have been poisoned by arsine; it does not produce the same type of symptomatology as arsenic poisoning. Indeed, arsenic poisoning is entirely different from arsine poisoning.

The most important consideration was to demonstrate that arsenic was present in the drain, that an arsenic-to-arsine reaction could have taken place, and that there was arsine in the gas the employees had inhaled. Drain samples, samples from the herbicide tank, samples from the drain cleaner, and the aluminum chips were all tested for arsenic. As shown in Table 19-1, there was a great deal of arsenic, comparatively speaking, in the drain and in the herbicide tank. There was also some antimony. If antimony were substituted for arsenic in the reaction shown above, the result would be formation of stibene. However, stibene is not as stable as arsine, and it is highly unlikely that it would have caused the illness in the patients. The drain cleaner and the aluminum chips had no arsenic. Analysis of the urine from a number of other workers demonstrated that the two people who were ill had higher arsenic levels in their urine than some of their fellow workers who had been in the vicinity of the drain but who had not gotten sick. A supervisor, who had never gone into the area where the drain was, showed no arsenic in his urine.

It was concluded that the presence of arsenic in the drain resulted in the formation of arsine when the drain cleaner was added. The arsine gas caused the hemolysis in the patients.

The two patients, after a very stormy medical course including renal failure and other problems, finally recovered, partly because such cases can now be treated with dialysis units. Before dialysis units were available patients with severe arsine poisoning usually died of renal failure. There are two effects of

Table 19-1. Arsenic and Antimony Levels in Environmental Samples Gathered at the Chemical Plant Where Arsine Poisoning Occurred.

Source	Concentration (ppm) Arsenic	Antimony
Drain		
Upper fraction		
Liquid	440	44
Solid	970	55
Lower fraction		
Liquid	410	43
Solid	4,460	170
Air sample*	13μg	
Abandoned herbicide tank		
With agitation		
Before	7,490	393
After	11,400	698
Drain cleaner		
NaOH and $NaNO_3$ solid	0.26	1.0
Aluminum chips	0.15	40.0

*Forty liters obtained on charcoal tubes minus background from unused tube.

arsine poisoning on the kidneys. One is the breakdown of the red blood cells with excretion of hemoglobin, which will damage the kidneys; in addition, arsine itself may directly damage the kidneys. Both employees had to be dialysized for several weeks.

19.2 A Case of Acute Dimethyl Nitrosamine Poisoning

In Omaha, Nebraska, the county health department became concerned because there seemed to be an outbreak of a new disease; five people had suddenly become ill.

The sick patients experienced vomiting, severe abdominal cramps, diarrhea, and two of them developed bruising. The first person had become ill on a Sunday morning while painting his house. Later in the day, his $2\frac{1}{2}$-year-old daughter became sick. During the afternoon, his wife's sister, her husband, and another child visited them. They stayed for a little while, drank some lemonade while there, and left. At six o'clock that night, they had also started feeling ill.

The man who had become ill early Sunday morning went to emergency rooms and then to a number of physicians, but he was repeatedly told he had the flu and should go home, rest, and take aspirin. Later, he developed severe nosebleeds; his nose was packed twice in emergency rooms early in the week. After each episode he was sent home again. He persisted in seeing physicians and, finally, on Thursday of that week, he walked into a physician's office and collapsed. At that point he was admitted to a hospital. On admission he showed abnormal liver function tests. The fact that his bilirubin was elevated also indicated a liver problem. The reason for the bleeding problems was that his platelets were severely decreased; there were about $6000/mm^3$ instead of the $150{,}000–300{,}000/mm^3$ or more that would be normal.

Because he became comatose, the child of the family who had visited the home on Sunday afternoon had already been admitted to another hospital. It took the hospitals a while to recognize that there might be a connection between the two cases. Once that was established, the county health department was called. The county health department then started looking at the other people who had visited the home and eventually established that of this group—a total of ten people who had been involved—five were ill and all of them had abnormal liver function tests. They all had low platelet counts. Not all of the people had platelet counts as low as those of the man or of the child who had visited the home.

After interviewing all who had visited the home on the Sunday when the illnesses were first noticed, the health department investigators concluded that the vehicle for the illness-causing agent must have been food or drink. The only food item all the ill people had consumed was lemonade. The parents of the man who had first become ill had visited the home, but while they were there, they did not consume any food or drink; they did not become ill. A sister living with the first family did not drink lemonade and did not become ill; and a baby did not eat any foods because it was being breastfed. The fact that the baby did not become ill eliminated to some extent the possibility of a highly contagious illness (except possibly a food-borne organism).

The lemonade therefore seemed implicated. By this time, of course, the family had discarded any remaining lemonade (always a problem in these types of cases).

The child died. The next day the man died. After that, the CDC was called. The county health department wanted to know what they should do, what kind of tests they should run, and what kind of samples they should take from the man. The CDC suggested that, in addition to taking tissues as is normally done for microscopic examination, they should also freeze tissue so that chemical analysis could be performed on the tissue if it became necessary to do so.

The autopsies of the father and child produced two common findings. One was that severe liver damage was the cause of death. In addition, there was extensive bleeding owing to the decrease in platelets. There were no other significant findings. No other organs were specifically affected. Review of the liver sections and other organs did not suggest an infection, because there was no inflammatory reaction in the tissues.

The question was, what type of chemical would produce these toxic effects? Quite a number of chemicals damage the liver, but they usually affect other organs as well. Because the investigation implicated the lemonade and the people had not complained about the lemonade's tasting peculiar, the chemical had to be something that did not taste strange and something that was water soluble. These observations somewhat limited the number of compounds. The substance also had to be relatively toxic so that a person would ingest a lethal dose by drinking a single or a half glass of lemonade in which this material was dissolved.

At the CDC we examined a number of the more common compounds that cause liver necrosis. We did not include in the list of compounds substances that, like carbon tetrachloride, severely affect the liver but also affect the sensorium. Carbon tetrachloride also has some effect on the kidneys but would be detected by people drinking lemonade. Portions of the liver tissues were examined for arsenic and yellow phosphorus.

None of this made very much sense. We searched further. Some people are sensitive to acetaminophen, a pain medicine. It can damage the liver, but typically not everyone in a family suddenly becomes ill. Usually there is a history of the patient's having taken this medication, but these patients had not taken acetaminophen. Aflatoxin, a mycotoxin and another hepatotoxin, was also a suspect, but it would be very difficult to have access to this type of compound. Furthermore, the microscopic appearance of the liver did not suggest aflatoxin. However, certain types of alkylating agents that are either used in cancer chemotherapy or cancer research might selectively damage the liver and could cause the severe centrilobular liver necrosis. This gave me an idea.

In talking to the county health department, I asked for additional information about the family to determine whether they would have had any contact with people in a cancer research institute or a hospital. The county health department thought this was a peculiar idea, but passed the request on to the police. A police officer in Omaha went to his files and found that this family did indeed know a man who was now working at the Eppley Institute in Omaha, which is a cancer research institute. Five years earlier this man had had a love affair with the wife of the man who later died, and had confronted the family with a gun; shots had been fired. He had then been sentenced to prison and was free on parole. Because he held a degree in biology, he was able to obtain a job at the Eppley Institute. His job was to mix diets for cancer research studies in animals, and one of the compounds he was working with was the alkylating agent dimethyl nitrosamine. The essential chemical structure of dimethyl nitrosamine is

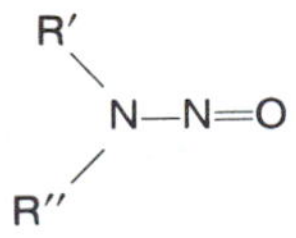

```
R′
  \
   N—N=O
  /
R″
```

Generalized structure of N-nitroso compounds

Dimethyl nitrosamine was eventually established as the compound that caused the illness and death in the family. While the family was away from their house, the suspect had climbed into a back window and added the poison to a pitcher of lemonade found in the refrigerator. His intention was to cause the people to have cancer; he wanted to watch them die. However, he picked a chemical that was also very acutely toxic. The oral LD_{50} in rats for dimethyl nitrosamine ranges between 27 and 41 mg/kg of body weight. The calculated lethal amount required for a child is less than one gram and for an adult roughly three grams.

Dimethyl nitrosamine is a yellow oil, so it mixes very well with lemonade. It is water soluble. It is somewhat more stable in a slightly acid environment. Other nitrosamines were examined to determine whether some other compound could have produced the effects evident in this case. They were ruled out for a number of reasons, and they were also not available at the institute. Many are not nearly as toxic, and the other more toxic nitrosamines do not specifically cause the selective centrilobular liver necrosis.

Dimethyl nitrosamine is very rapidly metabolized and excreted; therefore, five days after poisoning (the time that elapsed before investigation of this case began), all of the material would be metabolized and excreted and would not be found in the body. Nitrosamines and other alkylating agents do one thing by which they can be traced, however. They methylate nucleic acids, such as, for example, guanine.

There is now a test that has been developed, using high pressure liquid chromatography, by which methylation of guanine can actually be measured. Early in our investigation liver tissue from one of our patients was frozen. A total of eight tissue samples were then blindly submitted for analysis: seven controls and our sample. The chemist analyzed the eight samples and identified methylated guanine in the liver specimen that turned out to be our sample. Products like methylated guanine are also excreted in urine and would also probably be measurable in urine specimens. It is presently not possible to demonstrate methylation with acute yet nontoxic doses because the method is presently not sensitive enough. The amount of the concentration found in the liver of the adult male was as high as those demonstrated in rats that have been acutely poisoned with dimethyl nitrosamine, however. We were able to show this in court and the man who poisoned the family was convicted.

19.3 A Case of Industrial Exposure to Chlordimeform

This case occurred in Tennessee at a pesticide packaging plant. Apparently a group of people who were packaging a chemical called chlordimeform developed hematuria. In other words, their urine was bloody. The state health department in Tennessee was contacted first, and somebody from that department contacted the CDC. The chemical structure of chlordimeform is

$$Cl\text{—}C_6H_3(CH_3)\text{—}N{=}CH\text{—}N(CH_3)_2$$

Chlordimeform

A number of chlorinated aniline-type derivatives are currently used as pesticides. Many of them have not been studied very well; we know very little about their health effects, and little published information exists on animal toxicology data.

The medical problems among the workers were first discovered when they started going to physicians. Different people went to different physicians and it was not recognized at first that all workers had an occupational disease in common. It is at times difficult to recognize occupational illnesses, particularly if they resemble diseases common in the general population. One patient, for instance, was treated for gonorrhea, and another worker was treated for inflammation of the bladder. When three people from the pesticide packaging plant went to the same physician, however, this physician began to question whether or not the problems his patients were having might be associated with their work environment.

Questioning revealed that a similar outbreak of illness had occurred at the plant a year before when chlordimeform was being packaged. The affected workers displayed a number of symptoms at that time, including increased urinary frequency. Primarily, they all noticed that their urine was bloody. They also had such symptoms as skin rash and dizziness.

Limited animal studies had been conducted with chlordimeform, and CDC personnel reviewed these. There was no information on human exposures or toxic effects in humans. Dog, rat, and rabbit studies conducted with chlordimeform did not suggest that this material caused hematuria. However, the animal studies showed that chlordimeform is metabolized, and one of the compounds it is metabolized to is 2-methyl-4-chloroaniline. Sometimes a metabolite—not the parent compound—will cause toxic effects. Chlordimeform is metabolized to 2-methyl-4-chloroaniline by the removal of a side chain:

$$Cl\text{—}C_6H_3(CH_3)\text{—}NH_2$$

2-Methyl-4-Chloroaniline

A literature search concerning 2-methyl-4-chloroaniline produced several articles from Germany and England from the early 1930s. It was reported in these articles that workers who had worked with this material had developed hematuria. One of the authors had also studied 2-methyl-4-chloroaniline in cats. It is known that cats respond more like humans than do other animal species if they are exposed to anilines and chloroanilines.

At the CDC we performed limited animal experiments by exposing cats to chlordimeform and 2-methyl-4-chloroaniline, and found that toxic effects

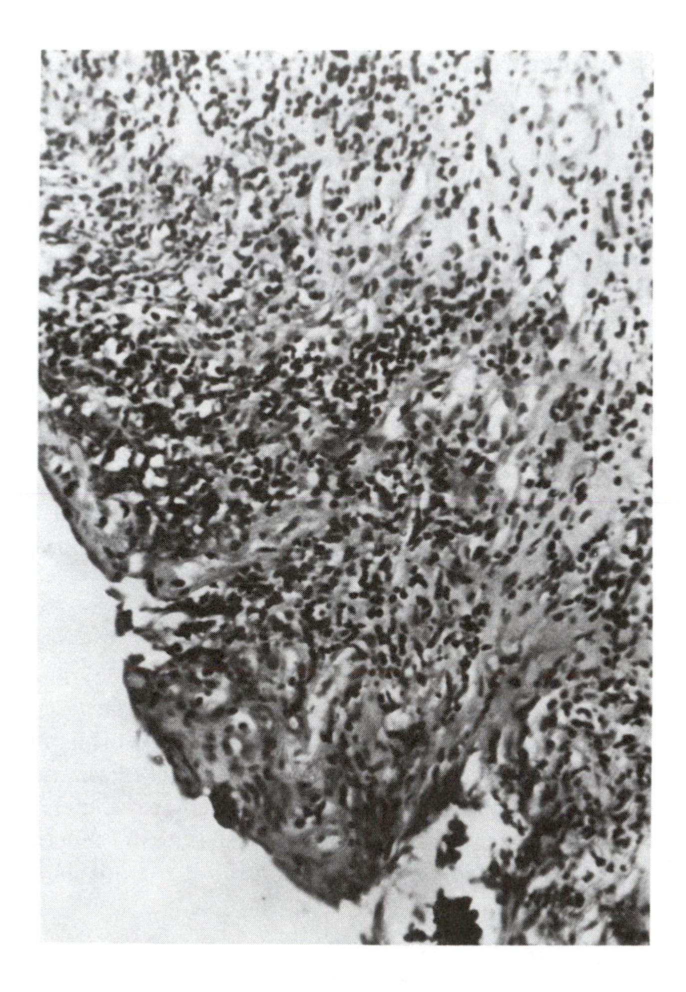

Figure 19-2. Section of bladder from a sick worker showing necrosis of the epithelium and an inflammatory reaction in the bladder wall.

could indeed be caused in the bladder of the cat, though these effects were not as pronounced as those that had been observed in the workers. Additionally, urine was collected from the workers to determine whether it contained the metabolite 2-methyl-4-chloroaniline, and whether chlordimeform could be identified. The toxic effect of chlordimeform can be seen in Figure 19-2, which shows a section of the bladder from one of the ill workers. The bladder mucosa is almost completely destroyed. The epithelium is sloughed off and many small blood vessels (capillaries) and other features of an inflammatory reaction are present.

The pesticide packaging plant was characterized by very poor work practices: workers were essentially "bathing" in chlordimeform. There were no facilities where the workers could change clothes or wash. The packaging process was a very dusty operation; we thought at first that the problem was simply very heavy overexposure in the packaging plant. But when we examined some farmers who used the material, we found that they also had red blood cells in their urine, though they did not have pronounced hematuria, and their symptoms subsided several weeks after they stopped using this particular pesticide.

Chlordimeform was subsequently taken off the market for a while. At present, the rules and regulations for using and packaging this material are much more stringent than they were, so there is no longer as much exposure to it. But marketing of the material is once again allowed and the pesticide is presently being used in the United States and other parts of the world.

19.4 A Carbamate and Propanil Pesticide Case

A plant in Arkansas manufactured several types of pesticides in rotation, manufacturing and packaging one pesticide for several weeks or months, and then doing the same with another pesticide, and then perhaps a third. The CDC became involved at this plant because there had been several hospitalizations of workers and also a number of complaints. The state health department called the CDC, which subsequently contacted the regional Occupational Safety and Health Administration (OSHA) office. OSHA inspected the plant and requested that a health hazard evaluation be done. Two physicians from the CDC then surveyed the plant. On investigation, they found that two types of compounds were produced.

One compound was a carbamate, methomyl (*N*-[(methylcarbamoyl)oxy] thioacetimidic acid methyl ester) which causes symptoms of toxicity similar to those caused by organophosphorous compounds—namely, nausea and vomiting, small contracted pupils, increased salivations, and muscle fasciculation. Methomyl is a highly toxic carbamate.

In addition to the methomyl, the plant was making and packaging an herbicide called propanil (3′,4′-dichloropropionanilide), made from 3,4-dichloroaniline. A number of employees had signs traceable to propanil: "acne," which was really chloracne (Figure 19-3), and also a nonspecific rash and skin irritation. These skin problems were more of a problem than the toxic effects of the methomyl. In trying to evaluate the problems, we divided the workers into different groups. There were some differences in symptoms, but because the chemicals were so pervasive, it was not possible to demonstrate that very well. Workers also moved freely from one area into another, and many had been exposed to both compounds.

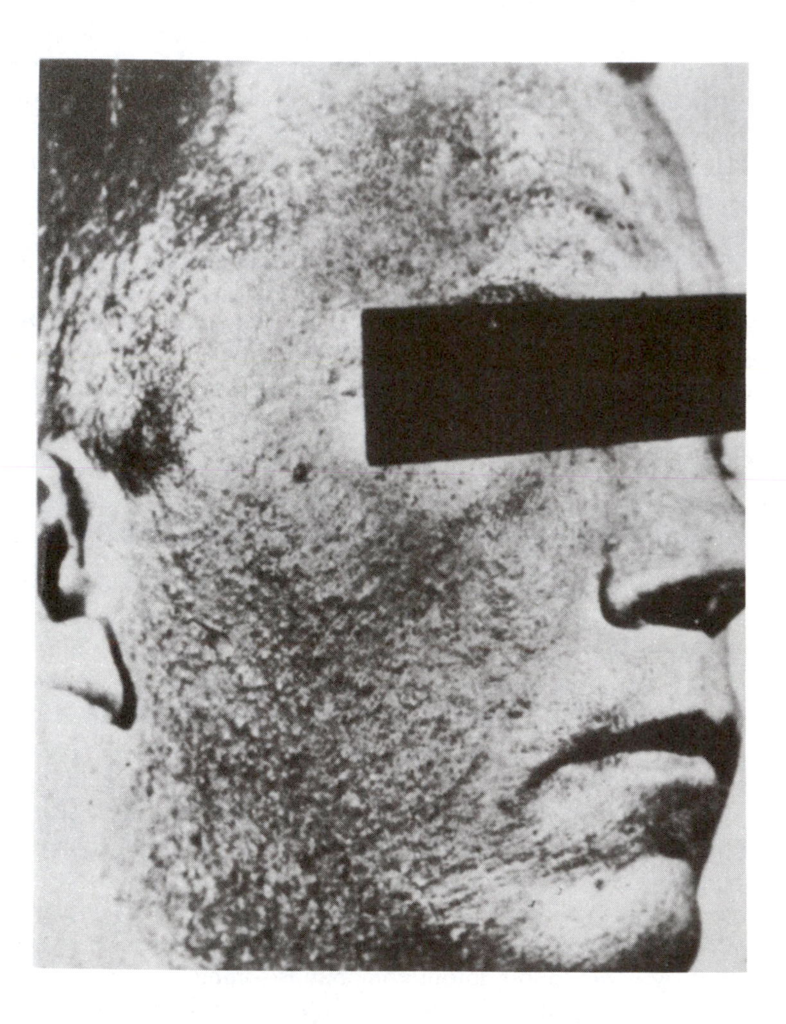

Figure 19-3. Illustration of the skin lesion chloracne.

The herbicide propanil, which is an aniline-type chlorinated compound, has the following structure:

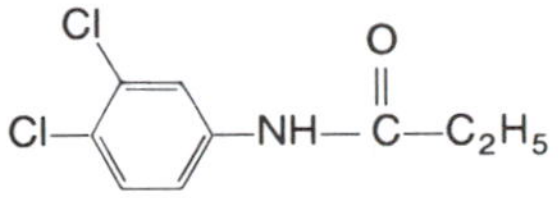

Propanil

Propanil has a number of different commercial names. It is not very toxic, but it does cause the skin disease chloracne, which can be quite persistent and disfiguring. Propanil has this effect because it is contaminated with 3,4,3′,4′-tetrachloroazoxybenzene and 3,4,3′,4′-tetrachloroazoxybenzene. These compounds are chloracnegic because of a peculiar chemical configuration, in which two chlorine atoms on one aromatic ring are connected either by another ring or by a double bond to two chlorines on another ring, reminiscent of TCDD (dioxin), 2,3,7,8-tetrachlorodibenzo-*p*-dioxin, which is extremely toxic. It has only recently been established that the chloroazobenzenes also cause chloracne. Other herbicides—diuron, linuron, neburon—may also be contaminated with tetrachloroazoxybenzene. Other chemicals that cause chloracne are the chlorinated naphthalenes, biphenyls, dibenzodioxins, and dibenzofurans.

In a few documented cases patients have had active chloracne for up to 20 years after they were exposed to one of these chemicals. In many instances chloracne gradually improves over a number of years following cessation of exposure. Until we made our investigation it had not been known that propanil caused chloracne. Our findings were substantiated by measuring chlorinated azobenzene in the propanil and by testing the chloracnegic effect of this material on the rabbit ear.

19.5 A Chlorinated Hydrocarbon Pesticide Case

A physician in Virginia saw a patient who worked in a plant where Kepone (chlordecone) was produced. The patient complained of weakness, nervousness, weight loss, shakiness and difficulty in reading and driving because he could not focus his eyes. The physician thought that the exposure to Kepone might have caused the patient's illness and sent a blood sample to the CDC for a determination of Kepone content. The full chemical name of Kepone is 1,1a,3,3a,4,4a,5b,6-decachlorooctahydro-1,3,4-metheno-2*H*-cyclobuta [*cd*] pentalen-2-one.

The CDC used to have a service for physicians in which, on request, we would analyze adipose tissues, blood samples, and other biologic samples for content of chlorinated hydrocarbon pesticides. The chemist at CDC who received the sample, not realizing the emergency, thought that the Virginia physician had previously arranged for such an analysis. He therefore accepted the sample, analyzed it, and found very high Kepone levels. He thought his analysis was wrong, so he repeated it three times before he said anything about it to anyone. In the meantime, the patient visited a neurologist at the Medical College of Virginia. The patient's symptoms and signs now included extreme nervousness, tremor, ataxia, skin rash, and an odd rolling of the eyes called opsoclonus. He also had lost weight. He couldn't concentrate even to read a newspaper, and he still had trouble driving.

Ultimately, the high Kepone blood level determined at CDC was reported to the neurologist at the medical college and also to the requesting physician.

The toxic effects of Kepone observed in animals include excitability, tremor, weight loss, and, in some cases, testicular atrophy. Kepone is a persistent compound. Once absorbed, Kepone will be stored in body tissues for a significant period of time. It is the only representative of the group of persistent halogenated pesticides that is not predominantly stored in adipose tissue. The ratio between blood and adipose tissue is not as large as for other chlorinated aromatic pesticides and industrial chemicals, and a great deal of Kepone is stored in the liver, because of its greater polarity.

After this index case was uncovered, we needed to determine some facts about the plant in which the man worked. Several phone calls were made in an attempt to gather information. The state health department did not know much about the plant. The neurologist who saw the index patient was a primary health care physician not usually involved with industrial cases. Nevertheless, the neurologist was asked to question the patient about other employees to determine if a larger problem existed. The patient reported that employees at the plant were unable to even drink much coffee during their breaks because they shook so severely they would spill the coffee, or would be unable to fill their cups, an indication that all probably had Kepone-caused tremor.

This was reported to the Virginia state health department; it became apparent that somebody needed to take direct action. The state epidemiologist inspected the plant and closed it, all within two days.

It was a small facility operating out of an old garage. In making Kepone, the firm used a completely open drying operation. During the drying process, dust was spread everywhere. There were no facilities in which the employees could change clothes or wash, and no separate eating facilities.

It was difficult to understand how people could be affected so profoundly—some people were unable to stand up from a chair—and still continue to work without realizing they were being affected by the chemical. In addition to the rash and nervousness noticed in the index case, many people complained of chest pains, but it was not possible to determine a cause for that complaint. It probably was a neurologic effect.

A divorced worker at the plant even seemed to have Kepone-related problems with alimony pay. He would begin to feel sick after working with

Kepone, would stay home, and then wouldn't be paid. His ex-wife, thinking this an intentional avoidance of work, would go to court, and the man would find himself in jail for failure to support her. While in jail, he would recover; and then he would resume work at the Kepone plant, become sick again, and the whole cycle would repeat. Other workers felt that perhaps they were being affected by Kepone, yet they had been told, "This compound doesn't really do anything. It's probably because you drink too much alcohol."

In addition to the workers, residents in the neighborhood were also examined, as well as employees' family members. Sometimes it is difficult to conduct these types of examinations and evaluations objectively because of publicity about the problem. As mentioned, the workers were unable to change clothes at the plant. Workers often carry dust of the chemicals they are working with—particularly persistent compounds, including chlorinated hydrocarbons, lead, and asbestos—home on their clothes and in their cars, thus exposing their families. Many family members of the employees had Kepone blood levels. They were lower than those of the plant employees, but they were higher than what might be expected in the general population. The Kepone produced at this plant was manufactured for export to other countries. Kepone is not registered for use in the United States, so normally Kepone blood levels are not observed in the general U.S. population.

The plant was closed and never reopened. Many political problems remained after the closure, however. The James River was found to be contaminated with Kepone and was closed to fishing for a while. The state of Virginia is still struggling with the problems resulting from this episode of contamination.

19.6 A Case of Tetrachlorodibenzodioxin Contamination

The final case in this chapter occurred in Missouri, and began in 1971. The CDC became involved when a child was admitted to the Children's Hospital in St. Louis. The child, a girl, had a hemorrhagic cystitis. She lived near a riding arena where a number of horses had become ill; some had, in fact, died. The physicians felt that her illness might somehow be related to the illness in the animals. At first an infectious disease, perhaps a viral infection, was considered.

An epidemiologist from the CDC was sent to Missouri to survey the riding arena. He collected soil samples and examined some of the sick horses. The epidemiologist was told that after the riding arena had been sprayed for dust control, birds and insects near the arena began to die. Perhaps something had been volatilized from the soil that was killing these animals; this might also have affected the little girl.

Because of this possibility, the chemists in the laboratory started looking for volatile substances, but were unable to come up with anything. Indeed, the search did not produce results until the winter of 1973, when one of the chemists, while trying to distill chemicals from the soil, obtained crystals that condensed on a cold finger of his distillation apparatus. He identified the crystals as 2,4,5-trichlorophenol. The chemist concluded that the illness of the girl in St. Louis might have been caused by 2,4,5-trichlorophenol.

However, 2,4,5-trichlorophenol is not extremely toxic; on the other hand, it is known that 2,4,5-trichlorophenol can be contaminated with an extremely toxic substance, 2,3,7,8-tetrachlorodibenzo-*p*-dioxin (TCDD), which is accidentally manufactured during production of 2,4,5-trichlorophenol. It was therefore decided to perform rabbit ear tests in order to establish whether the soil was also contaminated with TCDD; possibly TCDD might account for the illness.

The rabbit ear tests were done with an extract from the soil. Some of the rabbits died within a week. However, all rabbits developed hyperkeratosis on their ears, a typical reaction to TCDD. It could therefore be established that TCDD was most likely present, and additional chemical analysis confirmed this. The CDC informed the Missouri state health department that TCDD had caused the problem in the riding arena and that it was persistent and extremely toxic. A reinvestigation of the case was begun.

By now, three years had elapsed; the riding arena had been excavated twice and topsoil had been removed, because the soil caused continual problems. The removed soil had been used as fill beneath a road, in a back yard, and in various other locations.

Further investigation revealed other facts. Not only one riding arena had been affected, but three. One of the riding arenas stabled 125 horses, of which 65 had eventually died. Other animals also died. All horses exposed repeatedly to the contaminated arena surfaces suffered severe weight loss before they died. Investigation also revealed that not very many people had been consistently exposed to the soil of the riding arenas. Primarily, children showed most of the effects, probably because they played in the soil and had more opportunities for contact with it.

Symptoms in the index patient—the six-year-old girl who had been hospitalized in St. Louis—were hemorrhagic cystitis, headaches, and diarrhea. A ten-year-old girl had nosebleeding and headaches. People exposed to TCDD often complain of joint pains and pleuritic pains. Also, skin lesions are noticed. We assumed that the other children affected—two three-year-old boys—at one point had chloracne (Figure 19-3), which had subsided by the time the CDC reinvestigated the case.

The next question we attempted to answer in the investigation was where the TCDD came from and how it got onto the riding arena surface. We learned of an oil dealer who collected used oil and then sold it to refineries to be recycled. He had been hired to spray the riding arenas with oil for dust control, which is a customary procedure. We wondered if somehow he had picked up the material from a firm that made 2,4,5-trichlorophenol, the starting material for (2,4,5-trichlorophenoxy)acetic acid (known simply as 2,4,5-T) and for hexachlorophene.

The oil dealer had a very poor memory about where he collected the TCDD. What we then had to do was to find a company that was making either 2,4,5-T or 2,4,5-trichlorophenol. During the Vietnam War a number of factories produced 2,4,5-T as an ingredient for Agent Orange. Agent Orange was a mixture of 2,4,5-T and 2,4,-D used as a defoliant in Vietnam. Therefore we contacted the Department of Defense and asked them to review their records of all producers of 2,4,5-T. They identified a company located in the area of interest to us. This particular company had stopped its production of 2,4,5-T in 1969, before the contamination at the riding arenas had occurred. It had, however, subleased part of its facility to another company, which had made 2,4,5-trichlorophenol for the production of hexachlorophene. It was the second company that had produced the waste material. When 2,4,5-trichlorophenol is made, TCDD forms as a contaminant. When the company's factory was visited, a tank was found on the premises that was still partially filled with waste material containing TCDD.

The company's operators explained that at first they had sent the TCDD-containing waste material to a chemical disposal company; this had been expensive, and they later subcontracted disposal to the oil dealer we had interviewed earlier. The oil dealer had indeed mixed the waste with salvaged oil, and had sprayed the riding arenas with this mixture for dust control.

One of the oil dealer's drivers admitted, on questioning, that while hauling oil on one occasion, he found his truck was overloaded when he passed through a weighing station. Before reaching the next weighing station, he unloaded some of the excess material at a farm owned by the oil dealer. Chickens on that farm subsequently died. We later took samples on the farm and were able to find TCDD in soil there.

After establishing these facts, CDC personnel visited the other riding arenas and other locations and took samples for TCDD analysis. It was established that the other arenas were similarly contaminated. The oil dealer had several times picked up material from the plant where the TCDD-containing chemical tank was found and had also collected polychlorinated biphenyls from other companies; these were then mixed with the TCDD and the oil.

Recently it has been established that contamination by TCDD is much more extensive in Missouri than was earlier thought and that the waste-oil dealer sprayed many more sites with TCDD than those identified in 1974. The contaminated sites at present number 37; some are rural, others are urban, commercial, or residential. Many sites have soil levels above 100 μg/kg soil, and in a few areas the concentrations are above 1 mg/kg soil. The Environmental Protection Agency learned about these sites from former drivers employed by the waste-oil dealer and from the owner of one of the riding arenas who followed the trucks to their source. Human health and environmental studies are continuing in Missouri.

The structure of TCDD is

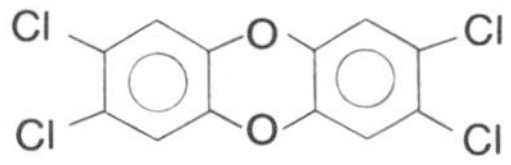

2,3,7,8-Tetrachlorodibenzo-*p*-dioxin

It is extremely toxic. In animals it causes severe weight loss, affects reproduction, causes liver damage and skin lesions in some species, depresses the cell-mediated immunity, and has been shown to be carcinogenic in rodents. Human illnesses following exposure to TCDD have not been well delineated. Chloracne, porphyria, cutanea tarda, sensory neuropathy, elevated serum cholesterol, and a neurasthenic syndrome have been described. Most chemicals that we are familiar with are toxic in the milligram or gram range. For example, parathion, which is extremely toxic, has a LD_{50} in rats of a few milligrams per kilogram body weight, but TCDD has an LD_{50} far lower than that: in rats, the oral LD_{50} is between 22 and 44 $\mu g/kg$. In addition to being acutely very toxic, TCDD also produces chronic toxic effects at much lower dosage levels. It accumulates in tissues and is persistent. Other extremely toxic compounds, such as some of the chemical warfare gases, break down very rapidly and the body is able to excrete them. TCDD is not easily eliminated from the body.

Summary

A total of six toxicologic case investigations have been presented in this chapter. Poisoning in each episode was caused by a different chemical: arsine, dimethyl nitrosamine, chlordimeform, mixed exposure to propanil and methomyl, Kepone, and dioxin (TCDD).

Each case was unique:

- The arsine poisoning case demonstrated that arsine gas can be generated under alkaline conditions and in environments unrelated to metal processing.
- In the dimethyl nitrosamine case, this substance was intentionally administered to humans in what turned out to be a matter of murder.
- The chlordimeform case arose when pesticide workers developed hemorrhagic cystitis, a condition not previously reported. It was established that the hemorrhagic cystitis was caused by a metabolite of chlordimeform, 2-methyl-4-chloroaniline.

- In the case of propanil and methomyl poisoning, pesticide workers were being simultaneously exposed to a variety of dangerous chemicals.
- The Kepone case illustrated problems with persistent chlorinated organic compounds and the inability of the exposed worker, in some cases, to recognize work-associated illness.
- The dioxin (TCDD) case involved environmental contamination with a highly toxic waste material. The material was improperly disposed of and caused illness in humans and death in animals. It was the first environmental case reported in the United States related to dioxin (TCDD) toxicity.

References and Suggested Reading

A Case of Acute Arsine Poisoning

Parish, G. G., Glass, R., and Kimbrough, R. D. 1979. "Acute Arsine Poisoning in Two Workers Cleaning a Clogged Drain." *Archives of Environmental Health, 34*:No. 4. July/Aug. pp. 224–227.

A Case of Dimethyl Nitrosamine Poisoning

Cooper, S. W., and Kimbrough, R. D. 1980. "Acute Dimethylnitrosamine Poisoning Outbreak. A Case Report." *Journal of Forensic Sciences, JFSCA, 4*:Oct. pp. 874–882.

A Case of Industrial Exposure to Chlordimeform

Folland, D. S., Kimbrough, R. D., Cline, R. E., Swiggart, R. C., and Schaffner, W. 1978. "Acute Hemorrhagic Cystitis—Industrial Exposure to the Pesticide Chlordimeform." *Journal of the American Medical Association, 239*: 1052–1055.

A Carbamate and Propanil Pesticide Case

Morse, D. L., Baker, E. L., Kimbrough, R. D., and Wisseman, C. L. 1979. "Propanil-Chloracne and Methomyl Toxicity in Workers of a Pesticide Manufacturing Plant." *Clinical Toxicology, 15*:1. 13–21.

A Chlorinated Hydrocarbon Pesticide Case

Cannon, S. B., Veazey, J. M., Jr., Jackson, R. S., Burse, V. W., Hayes, C., Straub, W. E., Landrigan, P. J., and Liddle, J. A. 1978. "Epidemic Kepone Poisoning in Chemical Workers." *American Journal of Epidemiology, 107*: 529–537.

A Case of Tetrachlorodibenzodioxin Contamination

Carter, C. D., Kimbrough, R. D., Liddle, J. A., Cline, R. E., Zack, M. M., Barthel, W. F., Koehler, R. E., and Phillips, P. E. 1975. "Tetrachlorodibenzodioxin: An Accidental Poisoning Episode in Horse Arenas." *Science, 188*: 738–740.

Chapter 20

Controls for Industrial Exposures

James L. Burson

Introduction

The purpose of this chapter is to develop the definitions of toxicity, hazard, and exposure in terms of the work environment, and to apply these definitions toward the establishment of an effective program designed to control industrial exposures. Such a program consists of three essential elements:

- Hazard identification and evaluation.
- Establishment of corrective and preventive measures.
- Evaluation of program effectiveness.

Many extremely toxic compounds are used daily in industry completely without hazard because adequate precautions are taken to limit contact with them to harmless amounts. The distinction between toxicity and hazard is important.

- The *toxicity* of a substance describes the nature, degree, and extent of its undesirable effects. It is the basic biological property of a material and reflects its inherent capacity to produce injury.
- Workplace *hazard* describes the likelihood that this toxicity will have its effect within an industrial setting. Thus, hazard in the workplace is the lack of effective controls to prevent injury, illness, or disease from occurring when a toxic substance is used in the quantity and manner proposed. Risk, discussed in detail in Chapter 17, describes the likelihood or probability that a toxic substance will constitute a hazard.
- These factors determine how much enters the body, by what route, how frequently, and for how long—all of which can be simply described as *exposure*. It is completely possible to have a high degree of hazard

associated with a low order of toxicity and a low degree of hazard associated with a high order of toxicity, depending on the exposure.

In this last chapter, we will examine one of the most important aspects of any discussion concerning toxicology—control of toxic substances. Again, we cannot control the toxicity of a given substance; that factor cannot be changed. However, we can control the degree of hazard a toxic substance poses to both people and the environment.

Hazard control involves more than the simple implementation of an engineering device, change in process or procedure, or the use of personal protective equipment. To be effective, hazard control must extend from an ongoing program of periodic hazard identification and evaluation to some method of evaluating the effectiveness of the control.

20.1 Hazard Identification and Evaluation

The first task in a hazard control program is to identify and evaluate potential toxic hazards in the workplace. One should develop a basic, systematic procedure that can be followed in the identification and evaluation of environmental health hazards. This procedure should be comprehensive, and should be applied to all processes, from the purchasing and receipt of all raw materials that will be used in manufacture, through all product-processing lines including intermediates and byproducts, and to the waste streams that will be generated for ultimate disposal.

One of the most important aspects of hazard identification is obtaining a complete list of raw materials, products, and byproducts used in the work environment. An excellent source for this information is the material safety data sheet (MSDS). The OSHA hazard communication standard (29 *CFR* 1910.1200) requires chemical manufacturers and importers to assess the hazards of chemicals they produce, distribute, and use. Hazard information concerning the product must be provided in the form of a warning label affixed to the product container and the material safety data sheet. The front side of a typical form (Figure 20-1) contains space in which to list manufacturer and product identification, hazardous ingredients, physical data, and fire and explosion data. On the reverse side (Figure 20-2) can be listed health hazard data, including emergency first aid procedures, reactivity data, spill and leak procedures, and any special protective information that might be required in handling the material.

Along with the basic physical, chemical, and toxicologic data, it is important to obtain accurate information about the industrial processes in which the material will be used. Plant engineering personnel are often a valuable source of information regarding many of the factors that could affect employee

Material Safety Data Sheet
Required under USDL Safety and Health Regulations for Shipyard Employment (29 CFR 1915)

U.S. Department of Labor
Occupational Safety and Health Administration

OMB No. 1218-0074
Expiration Date 05/31/86

Section I

Manufacturer's Name		Emergency Telephone Number
Address (Number, Street, City, State, and ZIP Code)	Chemical Name and Synonyms	
	Trade Name and Synonyms	
	Chemical Family	Formula

Section II - Hazardous Ingredients

Paints, Preservatives, and Solvents	%	TLV (Units)	Alloys and Metallic Coatings	%	TLV (Units)
Pigments			Base Metal		
Catalyst			Alloys		
Vehicle			Metallic Coatings		
Solvents			Filler Metal Plus Coating or Core Flux		
Additives			Others		
Others					

Hazardous Mixtures of Other Liquids, Solids or Gases	%	TLV (Units)

Section III - Physical Data

Boiling Point (°F)		Specific Gravity (H_2O=1)	
Vapor Pressure (mm Hg.)		Percent Volatile by Volume (%)	
Vapor Density (AIR=1)		Evaporation Rate (_______ =1)	
Solubility in Water			
Appearance and Odor			

Section IV - Fire and Explosion Hazard Data

Flash Point (Method Used)	Flammable Limits	Lel	Uel
Extinguishing Media			
Special Fire Fighting Procedures			
Unusual Fire and Explosion Hazards			

Page 1 (Continued on Reverse Side)

Form OSHA-20 (Rev. 3/84)

Figure 20-1. Material safety data sheet (front side).

Section V - Health Hazard Data

Threshold Limit Value

Effects of Overexposure

Emergency First Aid Procedures

Section VI - Reactivity Data

Stability	Unstable		Conditions to Avoid
	Stable		

Incompatability (Materials to Avoid)

Hazardous Decomposition Products

Hazardous Polymerization	May Occur		Conditions to Avoid
	Will Not Occur		

Section VII - Spill or Leak Procedures

Steps to be Taken in Case Material is Released or Spilled

Waste Disposal Method

Section VIII - Special Protection Information

Respiratory Protection (Specify Type)

Ventilation	Local Exhaust	Special
	Mechanical (General)	Other

Protective Gloves	Eye Protection

Other Protective Equipment

Section IX - Special Precautions

Precautions to be Taken in Handling and Storing

Other Precautions

Form OSHA-20
(Rev 3/84)

Figure 20-2. Material safety data sheet (back side).

exposure to toxic substances. An excellent aid in identifying and evaluating potential toxic exposures is the development of process flow sheets. Process flow sheets and standard operating procedures are often available and can aid greatly in plant surveys.

Evaluation of the work environment to determine the extent and degree of employee exposure to toxic substances usually involves sampling and analysis at numerous locations. It also involves the development of a sampling strategy in order to obtain representative information regarding the nature and intensity of exposures. Figure (20-3) shows a flow diagram of the NIOSH-recommended exposure-determination strategy.

An important component of the hazard evaluation process is the "walk-through" survey. Generally this survey is done prior to monitoring the work environment to better understand the process flow of the facility and to identify areas of potential exposure. The information gathered from the material safety data sheets, process flow sheets, and facility diagrams are an excellent source of help at this point. The walk-through survey is conducted to identify potential sources of toxic contaminants, new or changed processes, new or changed job functions, and control measures currently used to minimize exposures. Survey instruments (qualitative or semiquantitative) are often used during a walk-through survey as an aid in developing a strategy for the monitoring survey.

The "monitoring survey" or determination of employee exposures is usually comprehensive in scope and involves the use of a great deal of instrumentation—both areal and personal. These instruments may be direct-reading or they may be instrument systems that collect the contaminant from a measured quantity of air for subsequent laboratory analysis. The sampling strategy requires that numerous decisions be made in order for the survey to be representative of the actual exposures. These decisions include:

1. Sampling location.
    a. At breathing zone of employee.
    b. In the general room air.
    c. At point of operation.
2. Whom to sample.
    a. Employees directly exposed.
    b. Nearby employees.
    c. Remote employees.
3. Sample duration.
    a. Sensitivity of analytical procedure.
    b. Exposure limits for the substance.
    c. Estimated air concentration.

It is important that the proper evaluative instrumentation be selected for the particular hazard and that the instruments be properly calibrated. For

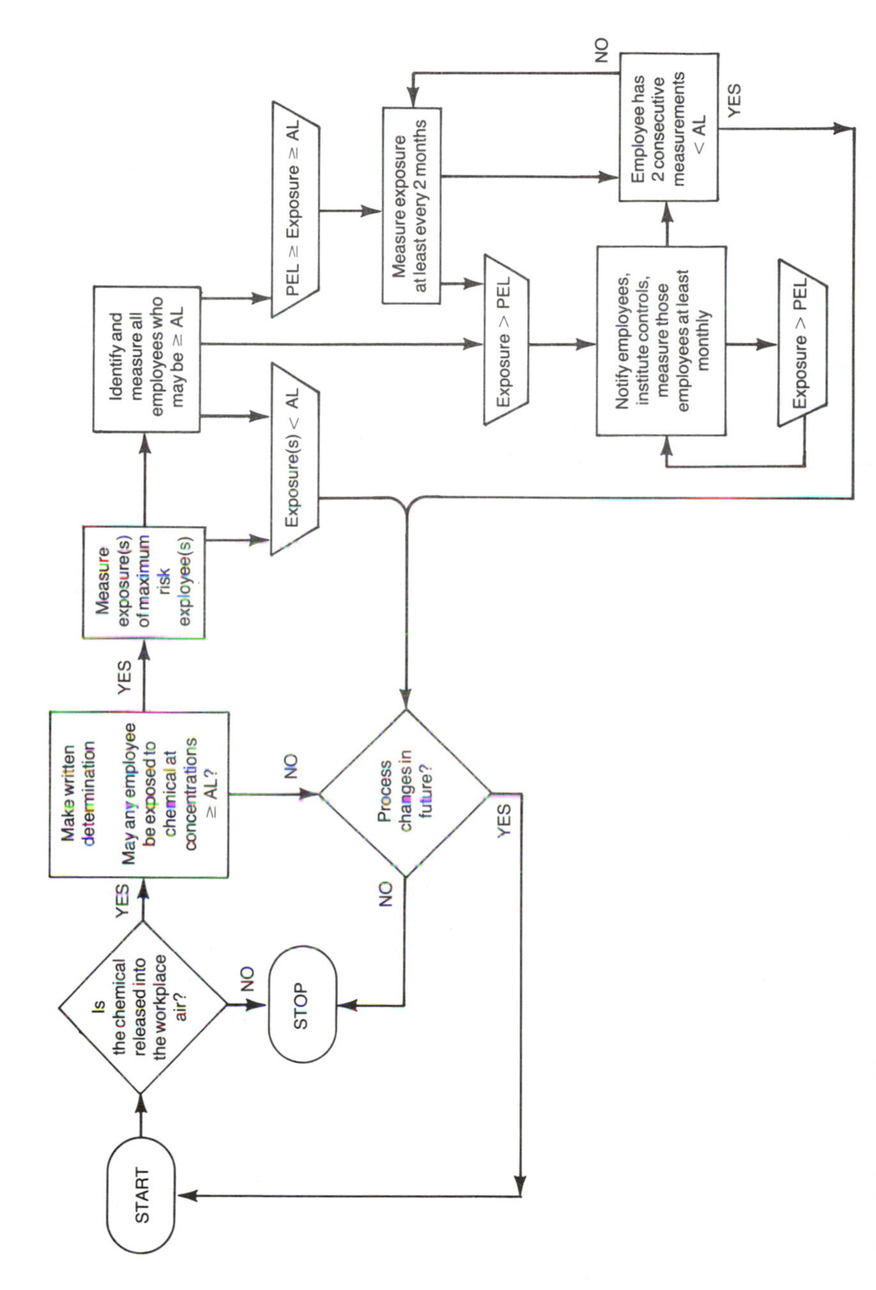

Figure 20-3. NIOSH-recommended employee exposure determination and measurement strategy. Health standard for each individual substance should be consulted for detailed requirements. AL = action level; PEL = permissible exposure limit.

those samples returned to the laboratory for analysis, it is equally important that the analytical method employed be accurate, sensitive, specific, and reproducible for the selected contaminant.

In evaluating employees' exposures to toxic substances, a sufficient number of samples of sufficient duration must be collected, and/or sufficient readings must be made with direct-reading instruments to permit the accurate assessment of daily, time-weighted average exposure and to evaluate peak exposure concentrations. The number and types of samples collected depend to a great extent upon the operations being studied and whether the threshold limit value (TLV) is a time-weighted average, a ceiling value, or both. For a time-weighted average, TLV-sampling done during most or all of the work day is preferable. For a ceiling TLV, grab samples (samples taken instantaneously for a short duration) should be taken to represent the worst potential conditions of exposure. If the results do not exceed the ceiling, then it can be reasonably assumed that the worker is not overexposed.

The time-weighted average (TWA) concentration of employees' exposures is the sum of the products of the duration of exposure multiplied by the concentration of air contamination of a specific substance during the time interval, divided by the total employee shift time.

$$\frac{C_1T_1 + C_2T_2 + \cdots C_NT_N}{\text{Total employee shift time}} = \text{TWA},$$

where C = indicates the concentration during each period of time T.

In the industrial environment, exposures often involve two or more hazardous substances, each of which has potentially harmful effects, rather than a single compound. When two or more toxic substances are present to act upon the same organ or organ system, their combined effect, rather than that of the individual substances, should be evaluated. Generally, unless there is evidence to the contrary, the effects of the separate toxic substances should be considered as additive. When the sum of exposures to the individual substances exceeds unity (1), then it should be considered that the threshold limit value of the mixture has been exceeded. Thus,

$$\frac{C_1}{T_1} + \frac{C_2}{T_2} + \cdots + \frac{C_n}{T_n} = 1,$$

where C = observed atmosphere concentration, and T = corresponding threshold limit value.

Examples of processes typically associated with exposures to two or more harmful atmospheric contaminants include spray painting, welding, diesel exhaust, and certain foundry operations.

After obtaining the sampling and analytical data, it is necessary to compare them with accepted occupational health standards (discussed further in

Chapters 17 and 18). It should be emphasized again that the samples collected during the study must be representative of the worker's daily, time-weighted average exposure before a comparison can be made with existing standards. There are numerous standards available for evaluation of data. Available standards include:

- Occupational Safety and Health Administration (OSHA) Regulations, published in the Code of Federal Regulations (CFR), Title 29, Part 1910. Permissible exposure limits (PEL) for many substances are listed in paragraph 1910.1000.
- Threshold limit values (TLV), published by the American Conference of Governmental Industrial Hygienists (ACGIH). These are reviewed annually.
- American National Standards Institute (ANSI) Z-37 Standards. The Z-37 series of standards includes recommendations for maximum exposures to several toxic substances.
- Hygienic Guides, published by the American Industrial Hygiene Association.

In addition, many states have adopted their own standards, some of which may be more stringent than those listed above.

The amount by which the standard or permissible level may be exceeded for short periods without injury to health depends upon a number of factors, such as the nature of the contaminant, whether the effects are cumulative, the frequency with which high concentrations occur, and the duration of such exposure. In many cases, adverse effects from exposure to toxic material do not appear until the exposure has existed for several years. The purpose of the TLVs and other standards is to protect against the future appearance of such symptoms.

20.2 Establishing Preventive and Corrective Measures

After hazards have been identified and evaluated, then the actual installation of control measures needs to be done. There are three major areas where the hazard posed by toxic substances can be controlled or eliminated.

The first and perhaps best control alternative is to limit a hazard at its source. The second alternative is to control the path of the hazard. The third alternative is to direct control efforts at the point where the hazard has its effect or is received.

Generally speaking, these three control alternatives can be grouped into nine classifications (designated a–i):

1. Control at the source
   a. Substitution of a less toxic material for one that is more toxic.
   b. Change or alteration of a process to minimize employee exposure.
2. Control of the path
   c. Isolation or enclosure of a process or work operation to reduce employee exposure.
   d. Localization of exhaust emissions at the point of generation, or localized dispersion of contaminants.
   e. Generalized or dilution ventilation with clean air to reduce concentration of contaminants.
   f. Dust-wetting methods to reduce generation of dust.
3. Control at the receiver
   g. Personal protective equipment to prevent exposure through inhalation, absorption, and ingestion.
   h. Training and education.
   i. Good housekeeping practices in the workplace, with adequate provisions for the personal hygiene of employees.

Control at the Source

Substitution. One of the most effective, and often least expensive, ways to control exposure to a toxic material is through substitution. Unfortunately, substitution is often one of the last alternatives considered when one thinks of controlling a hazard.

Replacement of a toxic material with a less toxic one is a very practical method of eliminating a health hazard. In many instances a solvent with a lower order of toxicity may be substituted for a more hazardous one. This principle can best be illustrated by citing a classic case from the drycleaning industry. For many years the principal cold-cleaning solvent was petroleum naphtha. Because of its flammability and potential for fire, a substitute solvent was sought. Carbon tetrachloride was chosen as a substitute because of its low flammability, cost, and effectiveness as a cleaning solvent.

Today we know that carbon tetrachloride is a highly toxic substance and that a serious fire hazard had been traded for an even more serious health hazard. Carbon tetrachloride has now been replaced by several other chlorinated hydrocarbons as a solvent of choice in several industries, including the cleaning industry. Solvents that have been successfully used as substitutes for carbon tetrachloride include perchloroethylene, methylene chloride, trichloroethylene, and 1,1,1-trichloroethane (see Chapter 12 for toxicologic data on these chemicals). Each of these solvents is less hazardous to handle and less toxic than carbon tetrachloride. However, even these substituted solvents are not without the potential for toxic exposures and have hazards of their own. More recently, the fluorinated hydrocarbons have begun to be used as substitutes for many of the more traditional solvents because of their low order of toxicity and flammability.

The preceding example of the successful substitution of one material for another also serves as a precautionary note. Substitution can indeed be used to reduce or eliminate an existing hazard; however, if it is not done carefully, one hazard can inadvertently be substituted for another. This method of control, like all others available, must be used carefully and only after a thorough analysis of the available alternatives.

Other examples of the substitution of a substance or material for a less toxic one include the substitution of water-based cleaners for solvents, the use of silica-free abrasives in abrasive blasting rather than silica sand, the use of aluminum oxide grinding wheels rather than those containing silica or other more toxic materials, and the use of toluene rather than benzene in numerous materials, including adhesives, paint removers, lacquers, and synthetic rubber products.

Changing or Altering a Process. Exposure to toxic materials can often be controlled or eliminated through the change of a process. This control alternative is often used and is not difficult to employ in many industrial processes. It has commonly been used in industry for other reasons, such as to reduce costs, increase productivity, and improve quality. It can also be an effective means of correcting a hazardous condition.

Substitution of equipment in the spray-coating industry in recent years illustrates the effectiveness of this control alternative. In this industry, the use of airless atomization and electrostatic attraction in place of conventional compressed-air spraying equipment has been shown to reduce both exposures and costs significantly in many applications. Compressed-air spraying equipment atomizes liquid coatings by directing a high-velocity air jet at the coating substance as it exits a nozzle. The air flow should convey the finely atomized droplets to the surface being coated. Numerous studies have shown that approximately 20 percent of a typical coating is of insufficient mass and fails to deposit on the object; instead it is deflected by a "bounce-back" air stream. These losses become sources of airborne contaminants for the employees operating the equipment. Total coating losses of 50 percent are not uncommon.

In airless spraying, the coating is atomized by forcing it through a small orifice under very high pressure. This method produces less fog than compressed-air spraying and reduces the "bounce-back" because the coating is transported to the surface by its own momentum rather than by a stream of air.

In electrostatic spraying, a coating can be atomized with compressed air, by high pressure as in airless spraying, or solely by electrostatic forces; the atomized particles are electrically charged as they leave the gun and are attracted to the workpiece, which is electrically grounded. This technique results in minimum dispersion of potential contaminants into the air and results in both an improved work environment and economy of operation. Electrostatic systems usually require the use of substantially less exhaust ventilation and make-up air for application of the coating than conventional compressed-air spraying.

Other examples of process changing include reducing exposure to the toxic dust generated by high-speed rotary sanders by changing to low-speed oscillating sanders, the use of vapor degreasers with adequate controls to replace manual cleaning of parts in open containers, and having machines install lead oxide battery grids in storage batteries instead of having employees perform this task manually.

Control of the Path

Isolation. Isolation is an effective control alternative that involves the interposition of a barrier between the source and the receiver. The barrier may be time, distance, or some type of physical component. This method of control can be particularly effective for equipment or processes that require little attention or "hands-on" activity by employees. Isolation can be accomplished by a physical barrier (enclosure of toxic dust process), time (use of semiautomatic equipment to reduce exposure time of employees), or by distance (use of remote controls/robotics).

In the chemical industry, total isolation of hazardous processes is commonly used. Similarly, total process enclosure is frequently utilized to reduce or eliminate exposure to toxic dusts or fumes in processes such as metal spraying or sandblasting.

A good example of the use of isolation can be seen in the use of ventilated storage cabinets in laboratories. Often highly toxic and flammable substances are stored in such cabinets. These cabinets are usually made of a fire resistant material and are constantly ventilated by a fan that discharges the air outside the laboratory. This type of isolation utilizes both a physical barrier and ventilation to isolate the contents of the cabinet from the employee environment.

Local Exhaust Ventilation. Local exhaust ventilation is an effective control method for many airborne contaminants. As indicated earlier, it is often used in conjunction with other control alternatives to reduce the potential for exposure to toxic substances. This control is widely used to control airborne contaminants produced by welding, cutting, and grinding.

Local exhaust ventilation reduces hazard by using directional air movement to capture the contaminant and transport it to a suitable collection device or to

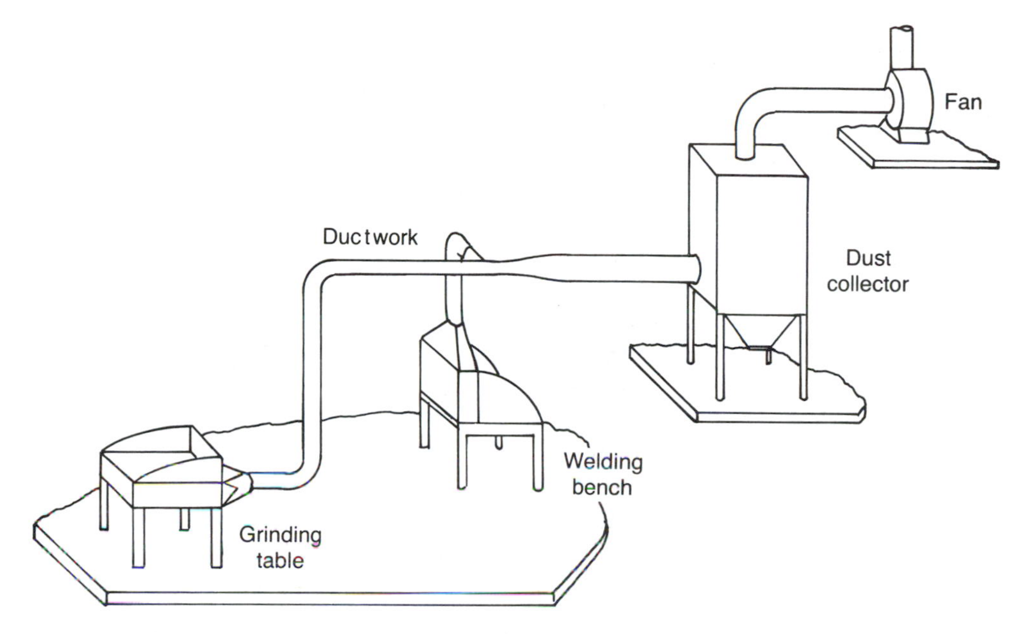

Figure 20-4. Components of a typical local exhaust ventilation system.

the outside of the building. The system is designed to capture air contaminants very near the source and to move them away from the breathing zone of employees performing the process or operation; this prevents them from being dispersed into the workroom air.

A typical local exhaust system consists of several components, including hood, duct(s), air cleaner, fan, and stack. A diagram of a typical system is shown in Fig. 20-4.

The *hood* is the entry point of the system where contaminants are either captured or entrained by air currents. Hood design is extremely important and is a function of the nature of the material to be captured and/or transported. Different hoods function in different ways—some simply catch and transport substances thrown into the hood while others reach out and capture contaminants before they are disposed into the work environment.

Ducts consists of a network of piping that connects the hoods with the other components of the system. Their design (sizing, turns, etc.) is extremely important in assuring transport of the contaminants from the point of capture to the exhaust.

Air Cleaners are devices designed to remove airborne contaminants from the exhaust air before it is recirculated or discharged into the community environment. Air cleaners are available to retain particulate or gaseous contaminants from exhaust air.

The *fan* is the air-moving device that provides the necessary energy to capture the contaminants and remove them from the workplace. The fan functions by inducing a negative pressure in the system and at the hoods.

Finally, the stack is the component through which the exhausted air is discharged outside the plant. Its design and location are important in the overall function and effectiveness of the local exhaust ventilation system.

More than most of the other control alternatives, the effectiveness of local exhaust ventilation is dependent upon proper design. All too often, ventilation systems have been designed by those lacking sufficient knowledge of proper air handling. Improperly designed systems frequently have included abrupt changes in duct sizing, 90° bends, and poorly conceived hoods. To correct problems created by poor design, they are often fitted with dampers or blast gates. Such poorly designed systems ultimately lead to great expense and frustration for management. Several excellent manuals now exist to provide guidance in the proper design of a local exhaust ventilation system.

Local exhaust, in conjunction with a local supply of make-up air, is a ventilation technique often used to control exposure to toxic substances used in open surface tanks. This technique, referred to as "push-pull" ventilation, "pushes" air from vents on one side of an open tank and "pulls" the air through vents on the opposite side of the tank. It is an effective means to control exposure to air contaminants in plating, etching, and cleaning operations. Properly engineered, the dual system operates more effectively and uses less air than systems that utilize only one or the other kind of exhaust flow.

As important as the proper design of a local exhaust ventilation system is the proper maintenance of the system. After installation, the performance of the system should be checked to determine if it meets the engineering specifications, including capture velocities, duct velocities, static and velocity pressures, etc. Performance should be rechecked periodically as a part of the maintenance program to insure proper operation of the system.

Results of initial tests should be recorded on data forms or on a drawing of the system to show the operating conditions, location of measurements, and measurement parameters. After a system is installed and is operating according to specifications, the system's effectiveness in capturing and removing contaminants may decrease for many reasons. These may include:

- Unauthorized damper adjustment.
- Accumulation of dust in ducts.
- Loose fan belt.
- Clogged air cleaners.
- Fan wear.

Occasionally, when new ducts are added to existing systems these may degrade or severely unbalance the entire system.

Periodic measurements can detect hardware problems and serve to maintain the system at or near design specifications. More important, though, is the measurement of how much protection the system is providing in terms of worker protection. A system may meet all design specifications in terms of capture velocity, air cleaner efficiency, fan speed, etc., but if air monitoring shows excessive employee exposures to an airborne toxin, the system is not performing adequately. Only by linking ventilation performance tests with the measurement of air contaminant levels can the overall system effectiveness be established.

Dilution Ventilation. General or dilution ventilation is an effective control alternative for areas generating low concentrations of hazardous substances. This control uses natural convection through open doors or windows, roof ventilators, or artificial air currents produced by fans or blowers. It maintains the concentration of a contaminant below hazardous levels by adding air to the workplace environment.

Dilution ventilation has limited application in the industrial environment and should only be used when certain criteria can be met, including the following:

- Contaminants should have a low order of toxicity.
- Only limited quantities of contaminants are released into the work environment at fairly uniform rates.
- Workers are located at sufficient distances from contaminant sources to permit adequate dilution to occur.
- Air cleaners or other cleaning devices are not needed before discharge of the exhaust air from the source into the external environment.

Dilution or general ventilation is often a more expensive control alternative than local exhaust ventilation primarily because of the large volumes of dilution air that may be required to maintain employee exposures at acceptable levels. Nevertheless, it can be a useful and effective control alternative when properly applied.

As with local exhaust ventilation, the proper design of general ventilation to control airborne contaminants is very important. Air moving and tempering equipment is expensive to purchase and maintain. Therefore, properly engineered systems, even though more costly in the beginning, can save money over time.

Recirculation of exhaust air is a growing occupational health problem, especially since many people are conscious of the need to conserve fossil-fuel energy. While recirculation of air is standard practice in some work environments (offices) and uses less energy, recirculation rarely is suitable in the industrial work environment because the exhaust air often cannot be adequately cleaned. Thus, once-through ventilation systems are commonly utilized in factories or industrial buildings, except where the exhausted contaminants have a low order of toxicity or where they can be easily and positively removed from the exhausted air.

Dust-wetting Methods. The use of wetting or dampening to control airborne dust and fiber concentrations is one of the simplest and most effective alternatives in many operations. Its effectiveness depends upon proper wetting of the dust. Often this requires adding a wetting agent to water or use of "amended water" and proper disposal of the wetted material before drying and redispersion of the material occurs.

This method is used extensively during removal of asbestos-containing insulation materials to control airborne concentrations of asbestos fibers. Foundries use pressurized water to clean castings and control dust exposures. The technique can be equally effective for molding and shakeout operations when properly used.

Control at Receiver

Personal Protective Equipment. The use of personal protective equipment is a legitimate and effective control alternative in many circumstances. However, its misuse by management and labor alike has caused this alternative to be viewed with suspicion. It should normally be considered a control alternative only until more positive controls can be implemented. When it is not feasible to control the source of the hazard, or its path, it then becomes necessary to protect the employee from the environment.

Technology today has provided us with equipment to protect all parts of the body from the harmful effects of many substances. Available equipment includes:

- Respiratory protective equipment, including devices to purify air through filtration and equipment to supply breathing air to employees through external sources.
- Eye and face protection, including goggles, face shields, and similar devices.
- Protective clothing, including gloves, aprons, boots, coveralls, and other items of impervious material to control or eliminate exposure to harmful materials.
- Protective creams and lotions, to help minimize skin contact with irritant chemicals.

Training and Education. In past years, little effort was made to educate employees about the toxic substances with which they work on a day-to-day basis, or about procedures or equipment they could use to minimize their exposure to potentially harmful agents. The main argument against providing employees with information about their work environment has been that such knowledge would cause apprehension among workers and create labor/management problems. More recently, safety and health professionals have demonstrated the benefits of educating employees about their work environments and how to avoid exposure to harmful agents. Employee education has not generated the feared responses. In most cases, employee morale has increased, labor/management relations have been more cooperative, and attendance has improved.

Proper training and education are essential to effective engineering controls. Employees must know how to use and operate engineering controls properly for the controls to perform as designed. If a welder performs his welding at some distance from a fume hood, which was designed to be placed 12–18 inches from the welding point, the exhaust hood cannot capture and remove the fumes as intended. The employee then is exposed to numerous substances that may be harmful to him and even his fellow workers.

Properly trained employees can also be instrumental parts of the overall hazard control program in industries. Because they usually work the closest to the harmful materials or substances, they can often anticipate and circumvent hazards before they become serious. In addition, once the source of a potential hazard has been identified, employees quite often have valuable ideas about the best way to eliminate or control the hazard.

Housekeeping and Personal Hygiene. Housekeeping and personal hygiene are key factors in occupational health protection. Where toxic substances are used, these factors are paramount. Accumulations of toxic dusts, such as lead, on floors tend to be pulverized under foot and further dispersed into the workroom air, contributing to the inhalation hazard. A regular cleanup schedule using vacuum cleaners with appropriate filters is the only effective method of removing dust from the work area. Sweeping and air hoses should not be used.

Good housekeeping is a day-by-day and even moment-by-moment control measure. It requires that refuse receptacles be conveniently located and that work flow be organized so as to prevent accumulations of raw materials or wastes on the floor. Immediate cleanup of spills of toxic material is very important. Plans should be established, in advance, of actions to be taken when such spills occur, and the proper equipment should be readily available to deal with such incidents.

Housekeeping practices should also address the proper disposal of all hazardous materials. Generally, this should be accomplished only by highly trained individuals. Procedures should be established for the proper disposal of unused dangerous chemicals, chemical containers, and contaminated waste in a timely manner.

Personal hygiene is an important part of the housekeeping program. Adequate provisions should be made for washing, toilet, and eating facilities. Change rooms with showers are often needed. In many operations, decontamination facilities are also required to control toxic agents effectively.

20.3 Evaluation of Program Effectiveness

The last essential element in the hazard control program deals with evaluating the effectiveness of control measures that have been implemented. This is an ongoing activity that helps management detect trends or spot deficiencies before they become problems.

An important component of this activity is periodic safety and health inspections of the work environment. This may be accomplished by trained staff personnel, safety committees, or some combination of both. Depending on the type of facility and toxic materials being used, it is recommended that a survey be conducted once a year by someone outside the facility, for example, a health professional from an insurance company, a corporate safety and health professional, or a consultant.

These surveys should provide the following assurances:

1. Hazard controls are performing according to specifications, are being used properly, and are maintaining contaminants in the work environment at acceptable concentrations.
2. Workplace additions and/or modifications have not altered hazard controls and reduced their effectiveness.
3. New or previously undetected hazards are not affecting the work environment.

In addition to periodic surveys of the work environment, a number of other "tools" are available to evaluate program effectiveness. These include:

1. Accident/Illness records.
2. Property damage reports.
3. Medical care costs.

Summary

The control of toxic substances requires a multidisciplined approach. It involves an ongoing program of hazard identification and evaluation, the selection of controls from among several available alternatives to reduce or eliminate exposure, and follow-up evaluation of control effectiveness.

- In the industrial environment, exposures to toxic materials must be studied and quantitatively evaluated. The data thus obtained must then be interpreted in the light of accepted standards, as well as according to other pertinent data about the potentially adverse effects of the exposure.
- After potential or actual exposures have been evaluated, then informed decisions can be made to implement control measures. These then must be periodically monitored to ensure their effectiveness.
- When a comprehensive program to control exposure to toxic substances is implemented, both the company and the employees will be served well, and the ultimate goal of such a program—to provide a workplace free from recognized hazards—will be realized.

References and Suggested Reading

Cralley, L. V., and Cralley, L. J. (Eds.) 1979. *Patty's Industrial Hygiene and Toxicology, Volume III. Theory and Rationale of Industrial Hygiene Practice*. New York: John Wiley and Sons.

Gleason, M. N., Gosselin, R. E., Hodge, H. C., and Smith, R. P. 1969. *Clinical Toxicology of Commercial Products*, 3rd Edition. Baltimore: The Williams and Wilkins Co.

Hemeon, W. C. L. 1963. *Plant and Process Ventilation*. New York: Industrial Press, Inc.

The Industrial Environment-Its Evaluation and Control. 1973. U.S. Department of Health, Education, and Welfare, Centers for Disease Control, National Institute for Occupational Safety and Health. Washington, D.C.

Industrial Ventilation: A Manual of Recommended Practice. 1984. American Conference of Governmental Industrial Hygienists. 18th Edition. Lansing, Michigan.

McDermott, H. J. 1979. *Handbook of Ventilation for Contaminant Control*. Ann Arbor (Michigan): Ann Arbor Science.

Olishifski, J. B., and McElroy, F. E. (Eds.) 1971. *Fundamentals of Industrial Hygiene*. Chicago: National Safety Council.

Rajhans, G. S., and Bragg, G. M. 1978. *Engineering Aspects of Asbestos Dust Control*. Ann Arbor (Michigan): Ann Arbor Science.

Sax, N. I. (Ed.) 1984. *Dangerous Properties of Industrial Materials*. 7th Edition. New York: Van Nostrand Reinhold.

Spain, W. H., and Burson, J. L. 1983. "Selecting Protective Clothing with Six Cs." *Occupational Health and Safety*, *52*(9): 17–23.

Threshold Limit Values for Chemical Substances and Physical Agents in the Work Environment and Biological Exposure Indices with Intended Changes. 1984–1985. Cincinnati. American Conference of Governmental Industrial Hygienists.

Williams, P. L., Luster, M. T., and Middendorf, P. J. 1983. "Area Sampling Methods: Selecting the Proper Sites." *Occupational Health and Safety, 52* (7): 14–16.

Appendix

A Typical Example of Risk Assessment

Robert C. James

For the purposes of providing the reader with an example, the chemical under consideration will be referred to by the abbreviation CE, which stands for chlorinated ethylene. The data provided here for CE are closely based, for the most part, on actual data presently available for a chlorinated aliphatic chemical found in common use today. Some of the actual data have been slightly altered, however, to present a better hypothetical problem.

For the sake of discussion and to set the problem up we will assume a state Division of Environmental Resources is reviewing the data in an attempt to set a statewide drinking-water standard for CE. As such, how they use the risk assessment (i.e., risk management) may differ from OSHA policies but in this manner the problem underscores a dilemma faced by regulators and the fact that different "allowable levels" may be generated depending upon the criteria used or the interpretation of certain parts of the database.

CE: Introduction and Historical Perspectives

CE, a volatile compound, was discovered in the 1800s and was subsequently used as a degreaser for machinery and as a solvent for organic products. CE is still widely used as an industrial solvent, particularly in metal degreasing and extraction processes, and has found uses in the boatbuilding, shoe, textile, chemical, painting, photography, optical, and dry cleaning industries. In addition, CE has been used at one time or another in a variety of consumer products. A partial listing of these products includes: glue, cleansers for automobiles, buffing solutions, rug cleaners, spot removers, disinfectants, deodorants, as a propellant for aerosol products, as a mildew preventive, as a cleaner for false eyelashes and wigs, and as an extraction solvent for decaffeinated coffee.

CE enjoyed a brief period in which it had several important medical uses. It was widely used as an inhalation anesthetic until the mid 1960s, when it began to be replaced by fluorinated halocarbons, which caused fewer postoperative complications. CE also has some analgesic properties and has been used for relief of pain in trigeminal neuralgia, proctoscopic examinations, early labor in pregnant women, postoperative pain, and angina pectoris.

Prior to the late 1970s CE was used as an extraction solvent in the manufacture of decaffeinated coffee and spice oleoresins. However, at that time the Food and Drug Administration (FDA) proposed regulations prohibiting its use as a direct or indirect food additive because of suspicions that CE was a carcinogen: food containing any detectable or added amount would be deemed adulterated. Prior to the FDA ban, CE was measured in food such as dairy products (1–10 ppm), meat (10–20 ppm), oils and fats (1–19 ppm), hot beverages (1–60 ppm), and on fruits and vegetables. CE levels reported in marine species include 0–250 ppb in molluscs and 0–479 ppb in fish. Besides food sources it was estimated that some 5000 medical, dental, and hospital personnel were exposed to CE and that about 60,000 additional persons were subjected to it annually when it was used as an anesthetic in medical situations. Concentrations in hospital rooms ranged from 0.3 to 103 ppm. Based on its widespread use and occurrence in our environment, it is not surprising that levels of 1–32 ppb were reported post mortem in human tissues.

The FDA therefore also moved against any medical uses, and banned all anesthetics containing CE. Furthermore, CE was disallowed in cosmetic products, animal and pet foods, animal drug products, and as an oil-extraction solvent for seed products.

The Environmental Protection Agency (EPA) quickly followed the precedent of the FDA and issued notice that CE was a candidate for denial of renewed registration (pending hearings) on the basis of its possible carcinogenicity.

The OSHA guidelines now limit air concentrations of CE in areas where employees work to 100 ppm (500 mg/m^3) during an eight-hour time-weighted average period. The NIOSH (National Institute for Occupational Safety and Health) guidelines, however, recommend that no occupational exposure to halogenated anesthetic agents be greater than 2 ppm (10 mg/m^3) for a one-hour sampling period.

Summary of the Toxicology of CE in Animal Species: Metabolism and Kinetics

Absorption. CE is readily absorbed by all routes of exposure, as would be predicted by its physical and chemical properties. Its historical use as an anesthetic indicates its rapid rate and high percent of absorption via inhalation. Measurements taken after oral administration of CE to rats indicate that at least 80 percent is absorbed from the GI tract.

Elimination. The elimination of CE in mammals is fairly rapid. After inhalation of CE at 300 ppm for 8 hr, CE was undetectable in the expired air of rats in one study. However, another study indicated approximately 80 percent would be eliminated via the lungs. The estimated half-life for CE for most tissues is about 100 min while for fat tissues the half-life is 200 min. Thus, eight hours after administration some 95 percent of the CE absorbed should be eliminated.

Metabolism. Much of the early toxicity research with CE was thought to support the idea that metabolism generated a reactive metabolite, probably the epoxide form, which was hepatotoxic and possibly capable of binding DNA and initiating cancer. However, recent work has changed much of the current thought on CE metabolism by demonstrating that the alcohol metabolite is not dependent upon nor derived from the epoxide metabolite form. This in turn suggests that species differences in hepatotoxicity and carcinogenicity are not related to species differences in the conversion rate of the epoxide to the less reactive alcohol metabolite. In addition, it has been shown that covalent protein binding of CE intermediates does not correlate with the formation of the epoxide metabolite. However, research has demonstrated that species differences do exist for covalent binding of reactive metabolites in rats and mice and that these differences in covalent binding parallel the different carcinogenic responses observed in these species.

Mechanistic Differences between Species. Interestingly, human microsomes form DNA adducts at a low rate similar to that found in the noncarcinogenic model, Osborne-Mendel rats. Both rats and humans appear to have substantially lower rates of microsomally generated DNA-adduct formation when compared to the B6C3F1 mouse, a species that is positive for liver cancer after chronic CE exposure. Researchers have also reported a species difference in microsomally generated covalent protein binding, which was likewise consistent with the results from the carcinogenicity tests. A sex difference in covalent binding in the B6C3F1 mouse has also been demonstrated which is consistent with the higher liver cancer incidence observed in male mice.

Differences between Rats and Mice. Observations of rats versus mice show differences in metabolism, both quantitative and qualitative. It has been found that the B6C3F1 mouse metabolized CE to a greater extent and generated even more macromolecular binding than expected when compared to the Osborne-Mendel rat. Also noted in the mouse was a lack of DNA alkylation, an increased hepatotoxicity, and an increase in DNA synthesis. This led to the proposal that CE was probably not initiating liver cancer in the mice through alkylation of DNA, but was increasing tumor formation through a recurrent injury (i.e., a cytotoxic type of mechanism). The importance of this proposed mechanism is that chronic, hepatotoxic doses would be required to induce cancer in any species.

Acute and Chronic Toxicities of CE. The acute and chronic toxicities of CE have been adequately summarized in a number of reviews and the data used here are drawn largely from them. The acute lethal effects of CE occur at moderately high concentrations. The oral LD_{50} in rats is 5–7 g/kg and about 6 g/kg in dogs. For intraperitoneal injections of CE the LD_{50} in mice is 3 g/kg while that of dogs is 2.8 g/kg. The maximum vapor concentrations that produce no toxic effects after an exposure of 7 hr/day, for 5 days/week, for 6 months have been determined to be:

- 100 ppm (500 mg/m^3) in guinea pigs.
- 200 ppm (1 g/m^3) in rats and rabbits.
- 400 ppm (2 g/m^3) in monkeys.

Likewise, 30 exposures each of 8-hr duration to 700 ppm (3.5 g/m^3) or continuous exposure to 200 mg/m^3 for 90 days produced no visible signs of toxicity in rats, dogs, monkeys, guinea pigs, or rabbits.

Like other chlorinated aliphatics, when administered at acutely toxic levels CE is principally a central nervous system depressant. It also produces some liver and kidney damage but at doses far higher than those at which carbon tetrachloride produces similar damage. In rodents liver injury occurs after a single dose at about 2 g/kg. CE, like many chlorinated solvents, may sensitize the heart to epinephrine such that stress or excitement may produce cardiac arrhythmias.

Teratogenicity and Reproductive Toxicity. CE produced no teratogenic effects in either rats or mice exposed for 7 hr/day at 300 ppm (1.5 g/m^3) on days 6–15 of gestation. In this study the indices monitored for adverse effects were the number of implantations, litter size, incidence of resorptions, fetal sex ratios, fetal body measurements, and morphologic anomalies. It was concluded that CE was not significantly maternally toxic, embryotoxic, fetotoxic, nor teratogenic in this study. Another study has reported that male mice exposed to 2000 ppm (0.2 percent in air) for 4 hr/day for 5 days had 2.4 percent abnormal spermatozoa compared to a 1.4 percent abnormal sperm count in controls. While this was a statistically significant difference, no change was observed at the 200 ppm exposure and the relevance of the 2000 ppm exposure in mice to expected human exposures, if any, is minimal.

Mutagenicity. CE has been reported as weakly positive or positive in only a few mutagenicity tests but negative in a number of others. The role that might have been played by contaminants in some of the positive tests has been suggested but not proved. Therefore, evidence of the mutagenic potential of CE is considered to be inconclusive and confounded by the presence of mutagenic contaminants in a number of studies using the technical grade compound. Pure CE is apparently not mutagenic.

Carcinogenicity. The National Cancer Institute published results of its bioassay in 1978. B6C3F1 mice had been administered CE 5 days/week for 78 weeks. Male animals received either 1.2 g/kg · day (low dose) or 2.4 g/kg · day (high dose) while female mice had received 0.9 g/kg · day (low dose) or 1.8 g/kg · day (high dose). In low-dose groups the rate of hepatocellular carcinoma was 4/50 in females and 26/50 in males; for the high-dose groups it was 11/49 for female animals and 31/48 in males. In control animals the rate was 1/40 when both sexes were combined. In Osborne-Mendel rats exposed to doses of either 0.6 or 1.2 g/kg · day, no liver cell tumors occurred and there was no statistically significant increase for any other type of tumor. The rat data are complicated somewhat by the poor survival of all groups during the course of this particular bioassay. Likewise, the positive result in the B6C3F1 mouse is somewhat undermined by the fact that about 0.3 percent of the CE consisted of the mutagenic contaminants common to technical grade preparations. Subsequent studies in these strains of rodents yielded essentially the same results. An inhalation study performed by an independent laboratory reported that CE again produced positive results in the B6C3F1 mouse but was not carcinogenic in the Osborne-Mendel rat. A third study using rats also yielded negative results. Dermal and oral administration of CE to ICR/Ha Swiss mice has also yielded negative results. Finally, no increased incidence of tumors was found as the result of an 18-month inhalation exposure to 100 ppm (500 mg/m^3) or 500 ppm (2.5 g/m^3) of nonepoxide-stabilized CE in hamsters, rats, and mice. Thus, even when using technical grade CE with mutagenic contaminants present, CE was found to be positive in only one strain of mice.

Summary of the Toxicology of CE in Humans

The acute toxicity of CE in humans is remarkably similar to that observed in animals and will not be discussed here for the sake of brevity.

Chronic Toxicity Produced in Humans. Of two epidemiologic studies examining the cancer mortality in occupationally exposed persons, one study reported a lower-than-expected total mortality (49 versus 62) and cancer mortality (11 versus 14.5) for 600 male workers exposed to 50 ppm of CE. The other study reported a cohort study of 2117 CE-exposed workers. Again, both total mortality and cancer mortality were lower than expected based on rates established for the general population. Out of these two studies some 1050 workers could be identified who had been exposed to CE for at least 15 years at an exposure level of about 50 ppm or greater.

Extrapolation of the Animal and Human Data

The method used to extrapolate a risk estimate for CE is dependent upon whether or not it is concluded from the animal data that CE poses a sufficient cancer hazard to man such that the cancer hazard warrants exclusion of other considerations when modeling the risk. Since there is conflicting, or at least inconclusive, evidence suggesting that CE may exert mutagenic or carcinogenic effects, both threshold and nonthreshold models might be considered for the animal data. In addition, since a significant human database is available, risk estimates can be generated from these data as well and then used in comparison with the animal extrapolations and as a guideline for final recommendations.

For noncancer toxicity, or threshold situations, models for extrapolating to no risk are reasonably straightforward and basically are similar to the suggested no-adverse-response level (SNARL) developed by the National Research Council of the National Academy of Sciences or the no-observable-effect limit (NOEL) of the EPA. Calculations are applied that make the conversion from a no-effect level in animals to a safe upper limit of exposure to humans (i.e., the human threshold dose). Although some investigators have proposed that an animal-surface-area comparison better reflects species-to-species conversion, typically an extrapolation on a basis of animal weight is adequate. The conversion must also take into consideration such factors influencing species response as absorption, metabolism, and excretion. Finally, a safety or uncertainty factor is applied, which is derived from an evaluation of the extent, reliability, reproducibility, and interspecies variability of the available data from animal and, when available, human studies. An equation may then be constructed.

Constructing a Formula for Estimating the Acceptable Daily Intake of CE Based on Animal Data (Noncarcinogenic Considerations)

A review of the animal data on CE toxicity, listed earlier, shows that the best estimate for an animal-derived no-observable-effect level (TD_0) is 35 ppm. This can then be used as a threshold dose in a risk assessment formula for human exposure to CE. The particular formula we will use here is

$$\text{ADI} = \frac{\text{ThD} \times \text{BW} \times \text{AF} \times \text{ER} \times t_{1/2}}{\text{SF}},$$

where

ADI = acceptable daily intake;

ThD = the threshold dose for which no observable effect was produced in the animal species;

BW = body weight of human (average assumed to be 70 kg);

AF = absorption factor, the percentage of the dose absorbed via the designated route of exposure in the animal species divided by the percentage of the dose absorbed by humans via the designated expected route;

ER = exposure ratio, the dosage regimen (i.e., rate of exposure) in the animal species divided by the anticipated exposure in humans;

$t_{1/2}$ = a ratio of the elimination half-life in the animal species to the elimination half-life for humans;

SF = a safety factor included to dampen individual variability in responses to chemicals; SF depends upon the reliability of the database used for the extrapolation. A value of 10, 100, or 1000 is selected as shown below:

10 = When the extrapolation is derived from a good database of human origin and there exists a substantial animal data base.

100 = Human data are scanty, inconclusive, or absent; long-term, reliable animal data are available.

1000 = No long-term human data are available; animal data are scanty, or interspecies sensitivity varies greatly.

The BW factor is given a value of 1.0 rather than 70 because it is assumed that for an inhalation exposure at steady state the concentrations in critical tissues will be similar for all species and there will be no size adjustment needed.

The AF is assumed to be 0.70 even though this value is probably similar for all species and would therefore be 1.0 when making animal-to-man comparisons. It is included as 0.70 in order to provide an additional margin of safety in this calculation.

The ER is 1.0 in as much as exposure situations are set to be the same in animals and humans.

The ratio $t_{1/2}$ is assumed to be 1.0.

The SF is set at 100 and is derived from a factor of 10 because good human data exist that suggest a ThD of 35 ppm does not lead to chronic illness in humans, but since chronic animal data were not utilized, and as the actual ratio for $t_{1/2}$ cannot be calculated because the human half-life is not known, a second factor of 10 is included to insure safety for a heterogeneous human population.

Thus, substituting in the equation:

$$\text{ADI}_{\text{air-continuous exposure}} = \frac{35 \text{ ppm} \times 1.0 \times 0.7 \times 1.0 \times 1.0}{100},$$

or,

$$\text{ADI} = 0.245 \text{ ppm in air, or } 1.225 \text{ mg/m}^3.$$

To convert this value to a safe water concentration it is assumed that a 70-kg man will inhale about 24 m^3 of air per day and consume 2 liters of water per day. Thus the ADI becomes

$$\text{ADI}_{\text{water}} = \frac{\text{ADI}_{\text{air}} \times 24 \text{ m}^3}{2 \text{ l/day}} = \frac{(1.225 \text{ mg/m}^3)(24 \text{ m}^3)}{2 \text{ l/day}}$$

$$= 14.7 \text{ mg/l} = 14.7 \text{ ppm}.$$

Constructing a Formula for Estimating the Permissible Concentration of CE Based on Human Data (Noncarcinogenic Considerations)

A second approach can also be utilized that uses human data. An estimated permissible concentration (EPC) can be calculated by using safe occupational air concentrations as a guideline. This approach has been outlined in the document entitled *Multi-Media Environmental Goals* (U.S. EPA report #600/7-77-136; 1977). Essentially, the EPC for air is calculated by converting the amount of CE a worker theoretically can be exposed to safely during a 40-hr week to the same total amount of CE but derived from a continuous daily air exposure (i.e., 40 hr/168 hr per week). A safety factor of 100 is recommended in calculating an EPC. The EPC for water is then derived by applying the same conversion as was made for the ADI calculation in the preceding section. The occupational threshold limit value (TLV) for CE is 100 ppm (500 mg/m^3) for an 8-hr shift; thus the EPC for air is calculated to be:

$$\text{EPC}_{\text{air}} = \frac{(500 \text{ mg/m}^3)(40/168)}{100} = 1.19 \text{ mg/m}^3.$$

Using the same extrapolation as before to convert safe air exposures to safe drinking-water exposures, we get

$$\text{EPC}_{\text{water}} = \frac{(1.19 \text{ mg/m}^3)(24 \text{ m}^3)}{2 \text{ l/day}}$$

$$= 14.3 \text{ mg/l} = 14.3 \text{ ppm}.$$

Note that the techniques used to extrapolate risk from the animal and human data produce very similar values.

Estimating a Permissible Concentration of CE Based on Animal Data (Cancer Risk Estimation)

For the nonthreshold (i.e., cancer) risk estimates, several different models can be chosen and each will derive a slightly different number. The linear interpolation model proposed by Gaylor is selected here because it probably represents the most conservative approach and essentially derives the upper bound of the risk estimates that would be arrived at in other ways in other models. The formula for deriving the risk estimate is

$$R = \frac{\text{UCL}}{d_e} d,$$

where

R = the probability that cancer will occur as extrapolated from the animal data;

UCL = the upper confidence limit of the cancer incidence observed in animals;

d_e = the experimental dose used in the animal studies;

d = the expected human dose.

The equation is rearranged to solve for the safe dose by setting the acceptable risk (R) at 10^{-5} or 1/100,000:

$$d = \frac{Rd_e}{\text{UCL}}.$$

Using the lowest positive dose for male mice from the NCI cancer bioassay data and the upper bound of the 95 percent confidence interval (0.50), the safe human dose (SHD) can be calculated as

$$\text{SHD} = d = \frac{(10^{-5})(1200 \text{ mg/kg} \cdot \text{day})}{0.50} = 0.024 \text{ mg/kg} \cdot \text{day}.$$

Extrapolating to a 70-kg man, and assuming a human lifetime exposure of 70 years versus 2 for a rat, the safe human daily intake (ADI) becomes:

$$\text{ADI} = \frac{(24\ \mu\text{g/kg} \cdot \text{day})(70 \text{ kg})}{35}$$

or

$$\text{ADI} = 48\ \mu\text{g/day}.$$

If humans are assumed to consume 2 liters of water per day, then the acceptable water concentration, based on animal data for the extrapolation purposes, is:

$$\frac{\text{ADI}}{2 \text{ l/day}} = \frac{48\ \mu\text{g/day}}{2 \text{ l/day}} = 24\ \mu\text{g/l}$$

or 24 ppb.

Estimating a Permissible Concentration of CE Based on Human Data (Cancer Risk Estimation)

As human data are available, a relatively simple but straightforward risk estimate can also be obtained by another means. If the human data do not show an observable increase in the cancer rate—i.e., the epidemiologically derived cancer rate is considered to be zero—then the 99 percent upper confidence interval for a test population of 1000 individuals is 0.45 percent. That is, the actual cancer rate could be as high as 0.45 percent (0.0045) or 4.5/1000 persons exposed and, statistically, once in a hundred times it will be mistakenly identified as zero. Thus, the assumed worst case by statistical inference when no cancer risk is observed in 1000 exposed individuals is that the risk could be as high as 4.5×10^{-3} but has not been detected.

There exist two epidemiologic studies based on occupational exposures for CE. In both, the total mortality and cancer mortality were less than expected, so the risk observed can be considered to be zero. If we combine the populations from both studies, which had approximately the same CE exposures, we may make the following assumptions and calculations:

- There were 1050 persons exposed to about 50 ppm in an occupational setting.
- The average daily exposure was 1.4 g/day (about 2.0 g/day for a 5-day work-week; conversion to a 7-day week will need to be made for daily environmental exposure).
- The 99 percent upper confidence limit on the risk for this population is approximately 4.5/1000 or 4.5×10^{-3}.
- It was assumed that the average exposure interval was only 15 years; thus, this estimate will need to be converted to a lifetime or 70-yr exposure period.
- The ADI is contained in 2 liters of water, which is the amount of liquid a person consumes each day.

Since the risk for the 1050 persons exposed to 500 ppm of CE is equivalent to

$$4.5 \times 10^{-3} = 1.428 \text{ g/day},$$

a 10^{-3} risk will be

$$\frac{1428}{4.5} = 317 \text{ mg/day}.$$

Since the 10^{-3} risk = 317 mg/day after only 15 years of exposure to CE, a 10^{-3} risk for a lifetime exposure (70 yr), given the same dose, would be

$$317 \text{ mg/day} \times (15/70) = 70 \text{ mg/day};$$

and a 10^{-5} risk, with daily exposure for a lifetime, would be 700 μg/day. Therefore, assuming 2 l are consumed per day, the safe water concentration for a 10^{-5} risk estimate is 350 μg/l (or ppb) if human epidemiologic data are used for extrapolation purposes.

Note that our hypothetical risk has been calculated from a substantial body of negative data by making worst-case assumptions that reflect the actual data. A hypothetical risk has therefore been calculated to help set an exposure guideline even though no such risk has been identified in humans.

Setting a Recommended Groundwater Standard for CE

The final recommended water standard for CE, based on a threshold model, is listed below:

Extrapolated standards for exposure to CE, using threshold model

	Safety factor of 100	Safety factor of 1000
1. Threshold derived from animal data	14.7 mg/l	1.47 mg/l
2. Threshold derived from human data	14.3 mg/l	1.43 mg/l

The values are very similar; the similarity stems from coincidental use of nearly equivalent inhalation levels obtained from animal experiments and time-established human occupational guidelines. These rather high values for the CE content of water reflect the fact that CE is not a potent toxicant at acute exposures. The few studies that reflect its noncancerous chronic toxicity likewise suggest that it is a relatively unremarkable toxicant at other than high concentrations. As was seen earlier, while there is animal evidence demonstrating that CE increases the tumor incidence in the B6C3F1 mouse, these studies are mitigated by the fact that CE is not carcinogenic in the rat, or hamster, or in another strain of mice. The positive studies are also confounded by the fact that contaminants in technical-grade CE are mutagenic and may have contributed significantly to the response observed in the susceptible mouse.

Studies comparing metabolic and biochemical differences between susceptible and nonsusceptible species for CE-induced oncogenicity suggest that the susceptible strain of mouse is not a relevant animal model for predicting human risks. In addition, a ranking of CE using the system proposed by R. A. Squire (explained in Chapter 17) indicates that CE would score approximately 31 points, which is equivalent to the score for saccharin. This score suggests that CE is not a likely human carcinogen, especially at low doses. Epidemiologic evidence, while perhaps not yet of sufficient strength to eliminate all concern for the carcinogenic potential of CE, indicates that even for the worst possible case it is a relatively weak human carcinogen. Thus, there is much evidence to suggest the values in the preceding table are adequate for the protection of health in the circumstances of chronic ingestion of CE and would adequately protect human health from all adverse effects. However, cancer data were also considered so as to calculate levels in the event that additional evidence better defining the carcinogenicity of CE is produced.

The values for the estimated safe water concentration, based on carcinogenic potential, are listed below.

Extrapolated standards assuming CE is carcinogenic

	10^{-5} Risk	10^{-6} Risk	10^{-7} Risk
1. Estimates derived from animal data	24.0 μg/l	2.40 μg/l	0.24 μg/l
2. Estimates derived from human data	350 μg/l	35 μg/l	3.5 μg/l

In both of the tables of extrapolated standards, the risk estimates generated from the animal data can be considered reasonably conservative estimates. Gaylor's model has been applied to the second table, which is based on the assumption that CE is carcinogenic; the animal data in that table were considered to give a worst-case estimate of the carcinogenic potential of CE. That estimate was then tested for its relevance to the human situation by extrapolating to, and identifying for comparison, the exposures that would be predicted to produce measurable cancer rates in humans if the base estimates of risk were correct. This test of the data is an upward linear extrapolation from the dose representing the 10^{-5}-risk estimate. When this is done, the additional increase in cancer for persons occupationally exposed should be about 17 percent for an exposure, over a 15-yr work interval, to 50 ppm of CE. This incidence is very high and would have been observed in the two actual epidemiologic studies that were made of workers with similar exposure levels and exposure intervals. It can be concluded then that the risk estimates generated here from the animal data *do overestimate* the human risk. A similar test of the extrapolation made from the human data in the two actual epidemiologic studies cannot, of course, be done, but the true risk of CE exposure can be no worse than the estimate derived from the human data in which a cancer incidence was assumed for negative data. Therefore, if it is assumed on the basis of the animal data that CE might carry some carcinogenic potential and risk, a final standard for groundwater could be reasonably chosen as that value corresponding to a previously determined "acceptable risk" level (i.e., 10^{-5}, 10^{-6}, or 10^{-7}) using the extrapolation of the human epidemiologic data.

Summary of the Estimates of Risk Posed by CE

It is recognized that risk management is a process in which environmental standards should reflect a variety of important considerations when they are established for contaminant levels in potential human exposure sources such as water. Of primary importance is the protection of human health. Once this criterion is satisfied a regulatory standard may reflect other societal concerns as well. For example, while it could be argued that 1 mg/l might adequately protect human health from CE exposure, if subsequent human or animal studies removed concerns for its carcinogenic potential, the value chosen for a standard might be lower and reflect society's desire for a chemical-free drinking water source. For comparative purposes the EPA has established federal primary drinking water standards for a number of chlorinated organic solvents and chlorinated pesticides. These standards range from the 1-ppb to the 100-ppb level. Since the 10^{-5}-risk estimates calculated from the animal and human data for our example are reasonably conservative, and generally bracket this range, the state might argue to set the standard somewhere between 100 and 350 ppb. Such a standard should reasonably and safely protect human health within the limits of the uncertainty that are inherent in the database and the models used to calculate the values from it. It would also be in keeping with the intent of the EPA and the guidelines proposed by that agency for other chlorinated aliphatics. Reiterating that much data are available to greatly lessen concern for the human cancer risk posed by CE, a state Division of Environmental Resources might reasonably select the cancer-risk estimate for a drinking-water standard as high as 350 ppb and at the same time choose to adopt the threshold-derived values in the 1–2-ppm range as a reasonably safe guideline to protect aquatic life and human health under short-term, unexpected situations such as spills into rivers acting as drinking-water supplies. It is hoped that the reader will appreciate the fact that should the state finally pick either level as a standard, various interest groups could argue against one or the other of them by reasoning either that (1) risk of carcinogenic potential must be assumed until better and more definitive evidence proves there is none, or (2) since the animal and human data indicate no cancer risk, the higher value is safe because apparently-safe occupational exposures are far higher.

Conclusion

It is hoped that this sample problem, which reflects experience with an actual compound, has provided some useful insight into the problem of risk assessments for the reader. It may interest the reader that currently OSHA allows up to a 500-mg/day-exposure to CE-type compounds in the workplace, while the EPA, which has not proposed a formal standard, has recommended about a 60-μg/day-exposure. These are clearly very different exposures. This difference is caused because the EPA criteria are based upon the extrapolation of animal carcinogenicity data, and typically the EPA does not consider modifying factors such as mechanistic and species variations in the carcinogenic response; nor does it use human data as a test or modifier of the animal extrapolation. Thus, while the OSHA standard is designed to protect workers occupationally exposed for perhaps most of their lifetime, the EPA by virtue of its risk assessment methodologies advocates an exposure that is some 60–70,000 times lower. Such a disparity between standards or recommendations for federal regulatory agencies is common today, because each agency assesses and manages risk by a different process.

Glossary

Absorption. The movement of a chemical from the site of initial contact with the biologic system across a biologic barrier and into either the bloodstream or the lymphatic system.

Accumulative effect of a chemical. Describes the effect of a chemical on a biologic system when the chemical has been administered at a rate that exceeds its elimination from the system. Sufficient accumulation of the chemical in the system can lead to toxicity.

Acetylcholine. An acetic acid ester of choline normally present in many parts of the body and having important physiologic functions, such as playing a role in the transmission of an impulse from one nerve fiber to another across a synaptic junction.

Acetylcholinesterase. An enzyme present in nervous tissue and muscle that catalyzes the hydrolysis of acetylcholine to choline and acetic acid.

Acidosis. A pathologic condition resulting from accumulation of acid in, or loss of base from, the body.

Acute toxicity. Adverse effects caused by a toxic agent and occurring within a short period of time following exposure.

Adduct. A chemical addition product (i.e., a chemical bound to an important cellular macromolecule *like DNA or protein*).

Administrative control. A method of controlling employee exposures to contaminants by job rotation or work assignment within a single work shift.

Aflatoxins. Toxic metabolites produced by some strains of the fungus *Aspergillus flavus*. They are widely distributed in foodstuffs, especially peanut meals.

Albuminuria. Presence of serum albumin in the urine; proteinuria.

Alcohol. An organic compound in which a hydrogen atom attached to a carbon atom in a hydrocarbon is replaced by a hydroxyl group (OH). Depending on the environment of the —C—OH grouping, they may be classified as primary, secondary, or tertiary alcohols.

Aldehyde. A broad class of organic compounds having the generic formula RCHO.

Alicyclic. Organic compounds characterized by arrangement of the carbon atoms in closed ring structures.

Aliphatic. Organic compounds characterized by a straight- or branched-chain arrangement of the constituent carbon atoms.

Alkane. *See* Paraffin.

Alkyl. A chemical group obtained by removing a hydrogen atom from an alkane or other aliphatic hydrocarbon.

Alkylation. The introduction of one or more alkyl radicals (e.g., methyl, CH_3—; ethyl, C_2H_5—; propyl, $CH_3CH_2CH_2$—; etc.) by addition or substitution into an organic compound.

Allele. Either of the pair of alternative characters or genes found at a designated locus on a chromosome. Chromosome pairing results in expression of a single allele at each locus.

Allergy. General or local hypersensitive reactions of body tissues of certain persons to certain substances (allergens) that, in similar amounts and circumstances, are innocuous to other persons. Allergens can affect the skin (producing urticaria), the respiratory tract (asthma), the gastrointestinal tract (vomiting and nausea), or may result from injections into the bloodstream (anaphylactic reaction). *See also* Anaphylactic type reaction.

Alveolar macrophages. Actively mobile, phagocytic cells that process particles ingested into the lung. They originate outside the lungs from precursor cells (promonocytes) in the bone marrow and from peripheral blood monocytes. They enter the alveolar interstices from the blood stream and are able to migrate to terminal bronchioles and lymphatic vessels.

Alveolus (pl., alveoli). In the lungs, small outpouchings along the walls of the alveolar sacs, alveolar ducts, and terminal bronchioles, through the walls of which gas exchange takes place between alveolar air and pulmonary capillary blood.

Amelia. The congenital absence of a limb or limbs. *See also* Phocomelia.

Ames assay. A screening test capable of revealing mutagenic activity through reverse mutation in *Salmonella typhimurium*. Mammalian metabolism can be simulated by addition of S9 liver enzyme to the bacterial growth medium.

Amide. A nitrogenous compound with the general formula $RNH_2C{=}O$, related to or derived from ammonia. Reaction of an alkali metal with ammonia yields inorganic amides (e.g., sodium amide, $NaNH_2$). Organic amides are closely related to organic acids and are often characterized by the substitution of one or more acyl groups (RCO) for an H atom of the ammonia molecule (NH_3).

Amine. An organic compound formed from ammonia (NH_3) by replacement of one or more of the H atoms by hydrocarbon radicals.

Amyotrophic lateral sclerosis (ALS). A disease marked by progressive degeneration of the neurons that give rise to the corticospinal tract and of the motor cells of the brain stem and spinal cord, resulting in a deficit of upper and lower motor neurons; the disease is usually fatal within two to three years.

Anaphylatic type reaction. One of four types of allergic reaction. A violent allergic reaction to a second dose of a foreign protein or other antigen to which the body has previously been hypersensitized. Symptoms include severe vasodilation, urticaria or edema, choking, shock, and loss of consciousness. Can be fatal.

Angiosarcoma. A malignant tumor formed by proliferation of endothelial and fibroblastic tissue; usually in a blood vessel.

Anoxia. A complete reduction in the oxygen concentration supplied to cells or tissues.

Antibody. An immunoglobulin molecule that has a specific amino acid sequence by virtue of which it interacts only with the antigen that induced its synthesis, or with antigens closely related to it.

Antigen. A substance that, when introduced into the body, is capable of inducing the formation of antibodies and, subsequently, of reacting in a recognizable fashion with the specific induced antibodies.

Antipyretic. An agent that relieves or reduces fever.

Aplasia. Lack of development of an organ or tissue, or of the cellular products of an organ or tissue.

Aplastic anemia. A form of anemia generally unresponsive to specific antianemia therapy, in which the bone marrow may not necessarily be acellular or hypoplastic but fails to produce adequate numbers of peripheral blood elements; term is all-inclusive and probably encompasses several chemical syndromes.

Apnea. Cessation of breathing; asphyxia.

Aromatic. A major group of unsaturated cyclic hydrocarbons containing one or more rings. These are typified by benzene, which has a six-carbon ring containing three double bonds. These are also known as arene compounds.

Arrhythmia. Any variation from the normal rhythm of heart beat, including sinus arrhythmia, premature beat, heart block, atrial fibrillation, atrial flutter, pulsus alternans, and paroxysmal tachycardia.

Arthro-osteolysis. Dissolution of bone; the term is applied especially to the removal or loss of calcium from the bone; the condition is attributable to the action of phagocytic kinds of cells.

Asbestosis. A bilateral, diffuse, interstitial pulmonary fibrosis caused by fibrous dust of the mineral asbestos; also referred to as asbestos pneumoconiosis.

Asphyxiant. A substance capable of producing a lack of oxygen in respired air, resulting in impending or actual cessation of apparent life.

Asthmatic response. Condition marked by recurrent attacks of paroxysmal dyspnea, with wheezing caused by spasmodic contractions of the bronchi; the response is a reaction in sensitized persons.

Ataxia. Failure of muscular coordination; irregularity of muscular action.

Atrophy. A decrease in the size and activity of cells, resulting from such factors as hypoxia, decreased work, and decreased hormonal stimulation.

Atropine. An alkaloid in the form of white crystals, $C_{17}H_{23}NO_3$, soluble in alcohol and glycerine; it is usually derived from belladona, *Hyoscyamus*, or strammonium, or is produced synthetically. Used as an anticholinergic for relaxation of smooth muscles in various organs, to increase the heart rate by blocking the vagus nerve, to treat Parkinsonism, and as a local application to the eye to dilate the pupil and paralyze ciliary muscle during medical procedures.

Autonomic nervous system. The part of the nervous system that regulates the activity of cardiac muscle, smooth muscle, and glands.

Autosome. Any chromosome that is not a sex chromosome.

B cell. An immunocyte produced in the bone marrow. They are responsible for the production of immunoglobulins but do not play a role in cell-mediated immunity. They are short-lived.

Bactericidal. Destructive to bacteria.

Basophil. A granular leukocyte with an irregularly shaped, relatively pale-staining nucleus that is partially constricted into two lobes; cytoplasm contains coarse, bluish black granules of variable size.

Benign tumor. A new tissue growth (tumor) composed of cells that, though proliferating in an abnormal manner, are not invasive—i.e., do not spread to surrounding, normal tissue; benign tumors are contained within fibrous enclosures.

Bilirubin. A bile pigment; it is a breakdown product of heme mainly formed from the degradation of erythrocyte hemoglobin in reticuloendothelial cells, but is also formed by the breakdown of other heme pigments. Normally bilirubin circulates in plasma as a complex with albumin, and is taken up by the liver cells and conjugated to form bilirubin deglucuronide, which is the water-soluble pigment excreted in bile.

Biologic half-life. The time required to eliminate one-half of the quantity of a particular chemical that is in the system at the time the measurement is begun.

Biotransformation. The series of chemical alterations of a foreign compound that occur within the body, as by enzymatic action. Some biotransformations result in less toxic products while others result in products more toxic than the parent compound.

Bradycardia. A slowness of the heart beat, as evidenced by a slowing of the pulse rate to less than 60 beats per minute.

Bronchitis. Inflammation of one or more bronchi, the larger air passages of the lungs.

Byssinosis. Respiratory symptoms resulting from exposure to the dust of cotton, flax, and soft hemp. Symptoms range from acute dyspnea with cough and reversible breathlessness and chest tightness on one or more days of a work-week to permanent respiratory disability owing to irreversible obstruction of air passages.

Cancer. A process in which cells undergo some change that renders them abnormal. They begin a phase of uncontrolled growth and spread. *See also* Malignant tumor.

Carbamate. A compound based on carbamic acid, NH_2COOH, which is used only in the form of its numerous derivatives and salts; as pesticides, carbamates are reversible inhibitors of cholinesterase. They may be direct or delayed in action. Inhibition of the enzyme is reversed largely by hydrolysis of the carbamylated enzyme and to a lesser extent by synthesis of a new enzyme.

Carcinogen. Any cancer-producing substance.

Carcinoma. A malignant tumor that arises from embryonic ectodermal or endodermal tissue.

Cardiomyopathy. General diagnostic term designating primary myocardial disease, often of obscure or unknown etiology.

Cell-mediated immunity. Specific acquired immunity in which the role of small lymphocytes of thymic origin is predominant; the kind of immunity that is responsible for resistance to infectious diseases caused by certain bacteria and viruses, certain aspects of resistance to cancer, delayed hypersensitivity reactions, certain autoimmune disease, and allograft rejections, and which plays a part in certain allergies. *See also* T cell.

Cephalosporidine. A broad-spectrum antibiotic of the cephalosporin group, which are penicillinase-resistant antibiotics derived from *Cephalosporium*.

Chelate. A chemical compound in which a metallic ion is sequestered and firmly bound into a ring with the chelating molecule; used in chemotherapeutic treatments for metal poisoning.

Chloracne. An acneiform eruption caused by exposure to halogenated compounds, especially the polyhalogenated naphthalenes, biphenyls, dibenzofurans, and dioxins.

Cholestasis. Stoppage or suppression of the flow of bile.

Cholestatic. Pertaining to or characterized by cholestasis.

Chromhidrosis. The secretion of colored sweat.

Chromosome aberrations. Structural mutations (breaks and rearrangements of chromosomes) or changes in number of chromosomes (additions and deletions).

Chronic toxicity. Adverse effects occurring after a long period of exposure to a toxic agent (with animal testing this is considered to be the majority of the animal's life). These effects are considered to be permanent or irreversible.

Cocarcinogen. Any chemical capable of increasing the observed incidence of cancer if applied with a carcinogen, but not itself carcinogenic.

Comedo. A plug in an excretory duct of the skin, containing microorganisms and desquamated keratin; a blackhead.

Competitive inhibition. Inhibition of enzyme activity in which the inhibitor (substrate analog) competes with the substrate for binding sites on the enzymes; such inhibition is reversible since it can be overcome by increasing the substrate concentration.

Conformational change. A change in the particular shape of a molecule.

Conjunctivitis. Inflammation of the conjunctiva, the delicate mucous membrane that lines the eyelids and covers the exposed surface of the eye.

Contact dermatitis. *See* Dermatitis.

Contraindication. Any condition, especially one of disease, that renders some particular line of treatment improper or undesirable.

Corpus luteum. A yellow glandular mass in the ovary formed by an ovarian follicle that has matured and discharged its ovum.

Cutaneous sensitization. Immune reaction characterized by local skin rashes, urticaria (hives), erythema, edema, and itching. Cutaneous sensitization is thought to be initiated by the release of histamine.

Cyanosis. A bluish discoloration, especially of skin and mucous membranes, owing to excessive concentration of reduced hemoglobin in the blood.

Cytochrome P-450 enzymes. *See* Mixed-function oxidase system (MFO).

Cytoplasm. The protoplasm of a cell exclusive of the nucleus, consisting of a continuous aqueous solution (cytosol) and the organelles and inclusions suspended in it (phaneroplasm); the site of most of the chemical activities of the cell.

Cytosol. The liquid medium of the cytoplasm (i.e., cytoplasm minus organelles and nonmembranous insoluble components).

Dalton. A unit of mass, 1/12 the mass of the carbon-12 atom. Carbon-12 has a mass of 12.011, and thus the dalton is equivalent to 1.0009 mass units, or 1.66×10^{-24} grams. Also called the atomic mass unit (amu).

Denaturation. The destruction of the usual nature of a substance, usually the change in the physical properties of proteins caused by heat or certain chemicals.

Dermatitis. Inflammation of the skin. Contact dermatitis is a delayed allergic skin reaction resulting from contact with an allergen. Irritant dermatitis describes irritation of the skin accompanying exposure to a toxic substance.

Diethylstilbestrol (DES). A synthetic estrogenic compound, $C_{18}H_{20}O_2$, prepared as a white odorless crystalline powder.

Dimethyl sulfoxide (DMSO). An alkyl sulfoxide, C_2H_6OS, practically colorless in its purified form. As a highly polar organic liquid, it is a powerful solvent, dissolving most aromatic and unsaturated hydrocarbons, organic compounds, and many other substances.

Direct carcinogen. *See* Primary carcinogen.

Dissociation constant. The equilibrium constant for the reaction by which a weak acid compound is dissociated into hydrogen ions and a conjugate base, in solution. *See also* pK_a.

Distal alveolar region. The part of the lung composed of the alveoli, or tiny air sacs, through which gas exchange between alveolar air and blood takes place.

DMSO. *See* Dimethyl sulfoxide.

Dose. The amount of a drug needed at a given time to produce a particular biologic effect. In toxicity studies it is the quantity of a chemical administered to experimental animals at specific time intervals. The quantity can be further defined in terms of quantity per unit weight or per body-surface-area of the test animal. Sometimes the interval of time over which the dose is administered is part of the dose terminology. Examples are: grams (or milligrams) per kilogram of body weight (or per square meter of body-surface area).

Dose-response relationship. One of the most basic principles of both pharmacology and toxicology. It states that the intensity of responses elicited by a chemical is a function of the administered dose (i.e., a larger dose produces a greater effect than a smaller dose, up to the limit of the capacity of the biologic system to respond).

Drug-induced toxicity. Toxicities that are "side effects" to the intended beneficial effect of a drug. They represent pharmacologic effects that are undesirable but that are known to accompany therapeutic doses of the drug.

Dyspnea. Difficult or labored breathing.

Dysrhythmia. Disturbances of rhythm—e.g., speech, brain waves, heart beat, etc.

Eczema. A superficial inflammatory process involving primarily the epidermis; characterized early by redness, itching, minute papules and vesicles, weeping of the skin, oozing, and crusting, and later by scaling, lichenification, and often pigmentation.

ED_{50}. The dose of a particular substance that, administered to all animals in a test, elicits an observable response in 50 percent of the animals.

Edema. The presence of abnormally large amounts of fluid in intercellular spaces within a tissue.

Elimination. The removal of a chemical substance from the body. The rate of elimination depends on the nature of the chemical and the mechanisms that are used to remove the chemical from the organism. Examples of mechanisms include expiration from the lungs, excretion by the kidneys by way of the urinary system, excretion in the sweat or saliva, and chemical alteration by the organism and subsequent excretion by any of these mechanisms. *See* Excretion.

Emphysema. Literally, an inflation or puffing up; a condition of the lung characterized by an increase, beyond the normal, in the size of air spaces distal to the terminal bronchiolus.

Encephalopathy. Any degenerative disease of the brain.

Endoplasmic reticulum. An ultramicroscopic organelle of nearly all cells of higher plants and animals, consisting of a more or less continuous system of membrane-bound cavities that ramify throughout the cell cytoplasm.

Endothelial. Pertaining to the layer of flat cells lining blood and lymphatic vessels.

Engineering control. A method of controlling employee exposures to contaminants by modifying the source or reducing the quantity of contaminants released into the work environment.

Enterohepatic circulation. The recurrent cycle in which the bile salts and other substances excreted by the liver pass through the intestinal mucosa and become reabsorbed by the hepatic cells, and then are reexcreted.

Environmental toxicology. That branch of toxicology that deals with incidental exposure of biologic tissue (more specifically, human life) to chemicals that are basically contaminants of the biologic environment, or of food, or of water. It is the study of the causes, conditions, effects, and limits of safe exposure to such chemicals.

Eosinophil. A structural cell or histologic element readily stained by eosin; especially, a granular leukocyte containing a nucleus usually with two lobes connected by a slender thread of chromatin, and having cytoplasm containing coarse, round granules that are uniform in size.

Epidermis. The outermost and nonvascular layer of skin. It derives from embryonic ectoderm.

Epithelioma. Any tumor developing in the epithelium, which is the kind of tissue that covers internal and external surfaces of the body, including the linings of vessels and other small cavities.

Epoxide. An organic compound containing a reactive group comprising a ring formed by an oxygen atom joined to two carbon atoms, having the structure

```
      O
     / \
  —C—C—
```

Erethism. Excessive irritability or sensitivity to stimulation, particularly with reference to the sexual organs, but including any body parts. Also a psychic disturbance marked by irritability, emotional instability, depression, shyness, and fatigue, which are observed in chronic mercury poisoning.

Erythema. The redness of the skin produced by congestion of the capillaries.

Erythropoiesis. The production of erythrocytes (red blood cells).

Erythropoietic stimulating factor (ESF). A factor or substance that stimulates the production of erythrocytes; may be the same as erythropoietin.

Erythropoietin. A protein that enhances erythropoiesis.

Ester. A compound formed from an alcohol and an acid by removal of water.

Ether. A colorless, transparent, mobile, very volatile liquid, highly inflammable, and with a characteristic odor; many ethers are used by inhalation as general anesthetics; the usual anesthetic forms are diethyl ether or ethyl ether.

Excretion. The process whereby materials are removed from the body to the external environment. If a chemical is in solution as a gas at body temperature, it will appear in the air expired from the animal; if it is a nonvolatile substance, it may be eliminated by the kidney via the urinary system, or it may be chemically altered by the animal and then excreted by means of any of the mechanisms available to the animal, such as excretion in the urine, in the sweat, or in the saliva. *See* Elimination.

Follicle-stimulating hormone. One of the gonadotropic hormones of the anterior pituitary, which stimulates the growth and maturation of graafian follicles in the ovary, and stimulates spermatogenesis in the male.

Forensic toxicology. The medical aspects of the diagnosis and treatment of poisoning and the legal aspects of the relationships between exposure to and harmful effects of a chemical substance. It is concerned with both intentional and accidental exposures to chemicals.

Gastritis. Inflammation of the stomach.

Gene. The basic unit of inheritance, recognized through its variant alleles; a segment of DNA coding a designated function (or related functions).

Genotype. The entire allelic composition of an individual (or genome), or of a certain gene or set of genes.

Gingivitis. Inflammation of the gums of the mouth.

Granulocyte. Any cell containing granules, especially a leukocyte containing neutrophil, basophil, or eosinophil granules in its cytoplasm.

Halogenation. The incorporation of one of the halogen elements, usually chlorine or bromine, into a chemical compound.

Hematopoietic. Pertaining to or affecting the formation of blood cells; an agent that promotes the formation of blood cells.

Hematuria. Blood in the urine.

Hemolytic. Pertaining to, characterized by, or producing hemolysis. The liberation of hemoglobin; the separation of hemoglobin from the red cells and its appearance in the plasma.

Hemolytic anemia. Anemia owing to shortened *in vivo* survival of mature red blood cells, and inability of the bone marrow to compensate for their decreased life span.

Hemorrhagic cystitis. Urinary bladder inflammation compounded with bleeding.

Hepatomegaly. Enlargement of the liver.

Hepatotoxin. A toxin destructive of liver cells.

Homolog. One of a series of compounds, each of which is formed from the one before it by the addition of a constant element or a constant group of elements, as in the homologous series C_nH_{2n+2}, compounds of which would be CH_4, C_2H_6, C_3H_8, etc.

Hydrocarbon. An organic compound consisting exclusively of the elements carbon and hydrogen. Derived principally from vegetable sources, petroleum, and coal tar.

Hydrophilic. Readily absorbing water; hygroscopic.

Hyperemia. An excess of blood in some part of the body.

Hyperkeratosis. Overgrowth of the corneous layer of the skin, or any disease characterized by that.

Hyperpigmentation. Abnormally increased pigmentation.

Hyperplasia. Abnormal multiplication or increase in the number of normal cells in normal arrangement in a tissue.

Hypersensitivity. A state of extreme sensitivity to an action of a chemical; for example, the individuals of a test population who fit into the "low end" of an ED_{50} or LD_{50} curve (i.e., those individuals who react to a very low dose as opposed to the median effective dose).

Hypokinesis. Abnormally decreased mobility; abnormally decreased motor function or activity.

Hyposensitivity. The state of decreased sensitivity; for example, the individuals of a test population who fit into the "high end" of an ED_{50} or LD_{50}

curve (i.e., those individuals who respond only to a very high dose as compared to the median effective dose).

Hypoxia. A partial reduction in the oxygen concentration supplied to cells or tissues.

Immune response. *See* Sensitization reaction.

Incidence. An expression of the rate at which a certain event occurs, as the number of new cases of a specific disease occurring during a certain period.

Infarct. An area of necrosis in a tissue caused by local lack of blood resulting from obstruction of circulation to the area.

Inhalation route. The movement of a chemical from the breathing zone, through the air passageways of the lung, into the alveolar area, across the epithelial cell layer of the alveoli and the endothelial cell layer of the capillary wall, and into the blood system.

Intraperitoneal. Within the peritoneal cavity; an intraperitoneal injection is one in which a chemical is injected into the abdominal fluid of an animal.

Ionization. The dissociation of a substance in solution into ions.

Irritant dermatitis. *See* Dermatitis.

Ischemia. Deficiency of blood owing to a functional constriction or actual obstruction of a blood vessel.

Kepone. Insecticide and fungicide having the formula $C_{10}H_{10}O$; causes excitability, tremor, skin rash, opsoclonus, weight loss, and in some cases (in animals) testicular atrophy.

Keratoacanthoma. A rapidly growing papular lesion, with a crater filled with a keratin plug, which reaches maximum size and then resolves spontaneously within 4–6 months from onset.

Keratosis. Any horny growth, such as a wart or callosity.

Ketone. Any compound containing the carbonyl group C=O and having hydrocarbon groups attached to its carbonyl carbon.

LD_{50}. That dose of a particular substance that, administered to all animals in a test, is lethal to 50 percent of the animals. It is that dose of a compound which will produce death in 50 percent of the animals—hence, the median lethal dose. The values of LD_{50} should be reported in terms of the duration over which the animals were observed. If a time is not given, it is assumed they were observed for 24 hours.

Lacrimation. The secretion and discharge of tears.

Laryngitis. Inflammation of the larynx, a condition attended with dryness and soreness of the throat, hoarseness, cough, and dysphagia (difficulty in swallowing).

Leukocyte. A white blood cell or corpuscle; classified as either granular or nongranular.

Leukocytosis. A transient increase in the number of leukocytes in the blood, resulting from various causes, such as hemorrhage, fever, infection, inflammation, etc.

Leukopenia. Lower-than-normal number of leukocytes in the blood; the normal range in adults is from 4 to 11×10^3/ml.

Leydig's cells. The interstitial cells of the testes (between the seminiferous tubules), believed to furnish the male sex hormone.

Lichen planus. An inflammatory skin disease characterized by the appearance of wide, flat, violaceous, itchy, polygonal papules, occurring in circumscribed patches, and often very persistent. The hair follicles and nails may become involved, and the buccal mucosa may be affected.

Lipid peroxidation. Interaction of free radicals with the lipid constituents of a membrane, resulting in alterations of structure and function of the membrane.

Lipophilicity. Having an affinity for fats.

Locus of action (site of action). The part of the body (organ, tissue, or cell) where a chemical acts to initiate the chain of events leading to a particular effect.

Luteinizing hormone. A gonadotropic hormone of the anterior pituitary, which acts with the follicle-stimulating hormone to cause ovulation of mature follicles and secretion of estrogen by thecal and granulosa cells.

Lymphocyte. A mononuclear leukocyte with a deep-staining nucleus containing dense chromatin and a pale blue-staining cytoplasm. Chiefly a product of lymphoid tissue. Participates in humoral and cell-mediated immunity. *See also* B cell; T cell.

Makeup air. In workplace ventilation, air introduced into an area to replace the air that has been removed.

Malignant tumor. Relatively autonomous growth of cells or tissue. Each type of malignant tumor has a different etiology and arises from a different origin. The condition tends to become progressively worse and to result ultimately in death. There are many common properties of malignant tumors but the invasion of surrounding tissue and the ability to metastasize are considered the most characteristic.

Margin of safety. The magnitude of the range of doses involved in progressing from a noneffective dose to a lethal dose. Consequently, the slope of the dose-response curve is an index of the margin of safety of a compound.

Megakaryocyte. A giant cell found in bone marrow, containing a greatly lobulated nucleus from which mature blood platelets originate.

Mesenchymal cells (tissue). The meshwork of embryonic connective cells or tissue in the mesoderm from which are formed the connective tissues of the body, the blood vessels, and the lymphatic vessels.

Mesothelioma. A tumor developed from the mesothelial tissue—the simple squamous-celled layer of the epithelium, which covers the surface of all true serous membranes (lining the abdominal cavity, covering the heart, and enveloping the lungs).

Metabolism. The biochemical reactions by which energy is made available for the use of an organism (catabolism and anabolism).

Metastasis. The establishment of a secondary growth site, distant from the primary site. One of the primary characteristics of a malignant tumor.

Methemoglobin. A compound formed from hemoglobin by oxidation of iron in the ferrous state to the ferric state. Methemoglobin does not combine with oxygen.

Methemoglobinemia. Presence of methemoglobin in the blood, resulting in cyanosis.

Microsomes. The fragments of the smooth reticular endothelium. This is the source of the microsomal enzymes that are capable of catalyzing a variety of biotransformation reactions, including hydroxylation, dealkylation, deamination, alkyl side-chain oxidation, hydrolysis, and reduction.

Miosis. Contraction of the pupil of the eye.

Mitochondria. Small spherical or rod-shaped components (organelles) found in the cytoplasm of cells, enclosed in a double membrane. They are the principal sites of energy generation (ATP) resulting from the oxidation of foodstuffs, and they contain the enzymes of the Krebs and fatty acid cycles and the respiratory pathways. Mitochondria contain an extranuclear source of DNA and have genetic continuity.

Mixed-function oxidase system (MFO). A nonspecific, multienzyme complex on the smooth endoplasmic reticulum of cells in the liver and various other tissues. These enzymes constitute the important enzyme system involved in phase I reactions (i.e., oxidation/reduction reactions). Also called cytochrome P-450 enzymes.

Monocyte. A mononuclear phagocytic leukocyte with an ovoid or kidney-shaped nucleus, containing lacy, linear chromatin, and abundant gray-blue cytoplasm fitted with fine, reddish and azure granules.

Morbidity. The rate of sickness or ratio of sick persons to well persons in a community.

Mutagen. Any substance causing genetic mutation.

Mutagenesis. The induction of those alterations in the information content (DNA) of an organism or cell that are not due to the normal process of recombination. Mutagenesis is irreversible and it is cumulative in the event of increased mutation rates or decreased selection pressures. The genetic damage can occur in both somatic and germinal cell lines.

Mutagenic tests. Test of an agent to determine effects on the faithful replication of genes.

Mutation. A permanent transmissible change in the genetic material. Abnormalities that manifest themselves as altered morphology or altered ability to direct the synthesis of proteins originate as mutations.

Mycotoxin. A fungal toxin.

Myeloid leukemia. Leukemia arising from myeloid tissue (bone marrow) in which the granular, polymorphonuclear leukocytes and their precursors predominate.

Myelotoxin. A cytotoxin that causes destruction of the bone marrow cells.

Myopathy. Any disease of muscle tissue.

Nasopharyngeal region. The part of the pharynx lying above the level of the soft palate.

Necrosis. Death of one or more cells, or of part of a tissue or organ, owing to irreversible damage.

Necrotic. Pertaining to or characterized by necrosis.

Neoplasm. Literally, new growth, usually of an abnormally fast-growing tissue.

Nephritis. Inflammation of the kidney; a focal or diffuse proliferative or destructive process, which may involve the glomerulus, tubule, or interstitial renal tissue.

Nephropathy. Disease of the kidneys.

Neurofibril. One of the delicate threads running in every direction through the cytoplasm of the body of a nerve cell and extending into the axon and dendrites of the cell.

Neuromuscular endplate. A flattened discoid expansion at the neuromuscular junction, where a myelinated motor nerve fiber joins a skeletal muscle fiber.

Neuropathy. General term denoting functional disturbances and/or pathologic changes in the peripheral nervous system.

Neutropenia. A decrease in the number of neutrophilic leukocytes in the blood.

Neutrophil. A granular leukocyte having a nucleus with three to five lobes connected by slender threads of chromatin, and cytoplasm containing fine, inconspicuous granules.

Nicotinic effect. Poisoning by nicotine or a compound related in structure or action, characterized by stimulation (low doses) and depression (high doses) of the central and autonomic nervous systems. In extreme cases, death results from respiratory paralysis. Also referred to as nicotinism.

Nitrosamine. Any of a group of *n*-nitroso derivatives of secondary amines. Some show carcinogenic activity.

NOEL. *See* No-observable-effect level.

Noncompetitive inhibition. Inhibition of enzyme activity by inhibitors that combine with the enzyme on a site other than that utilized by the substrate; such inhibition may be irreversible or reversible.

Nonspecific chemical action. The action of a chemical, such as a strong acid or base or concentrated solution of organic solvent, which occurs in all cells in direct proportion to the concentration in contact with the tissue. This is a nonselective effect and its intensity is directly related to the concentration of the chemical.

Nonspecific receptor. Secondary receptor within the body, which combine with or react with a chemical; however, the function of the cell is not influenced by the product that is formed. Such receptor are usually combining sites on proteins.

No-observable-effect level (NOEL). A measure of the toxicity of a substance, established by the U.S. Environmental Protection Agency; the level of a substance that, when administered to a group of experimental animals, does not produce those effects observed at higher levels, and at which no significant differences between the exposed animals and the unexposed or control animals are observed.

Olefin. A class of unsaturated aliphatic hydrocarbons having one or more double bonds. Also called alkene.

Oncogenic. Giving rise to tumors or causing tumor formation.

Opsoclonus. A condition characterized by rapid, irregular, nonrhythmic horizontal and vertical oscillations of the eyes, observed in various disorders of the brain stem or cerebellum.

Optic neuritis. Inflammation of the optic nerve; it may affect the part of the nerve within the eyeball, or the portion behind the eyeball.

Oral route. The entry of a chemical into the body by way of the gastrointestinal tract. Although absorption to some extent takes place throughout the tract, the majority of the absorption takes place in the area of the villi of the small intestine.

Organic acid. Any acid the radical of which is a carbon derivative; a compound in which a hydrocarbon radical is joined to COOH (carboxylic acid) or to SO_3H (sulfonic acid).

Organochlorine pesticide. These compounds are extremely stable and persistent in the environment. They are efficiently absorbed by ingestion, and act on the central nervous system to stimulate or depress it. Signs and symptoms of toxicity vary with the specific chemical. In general, mild poisoning cases cause symptoms such as dizziness, nausea, abdominal pain, and vomiting. In chronic poisoning, weight loss and loss of appetite, temporary deafness, and disorientation can occur.

Organophosphate pesticide. These are irreversible inhibitors of cholinesterase, thus allowing accumulating of acetylcholine at nerve endings. They are rapidly absorbed into the body by ingestion, through intact skin, including the eye, and by inhalation. Poisoning symptoms range from headache, fatigue, dizziness, vomiting, and cramps in mild cases, to the rapid onset of unconsciousness, local or generalized seizure, and other manifestations of a cholinergic crisis in severe cases.

Osteomalacia. A condition of softening of the bones characterized by pain, tenderness, loss of weight, and muscular weakness.

Osteoporosis. Abnormal rarefaction of bone, seen most commonly in the elderly.

Osteosclerosis. Hardening or abnormal density of bone.

Paraffin. A class of aliphatic hydrocarbons characterized by a straight or branched carbon chain and having the generic formula C_nH_{2n+2}; also called alkane.

Paranoid schizophrenia. A psychotic state characterized by delusions of grandeur or persecution, often accompanied by hallucinations.

Parkinsonism. A group of neurologic disorders characterized by hypokinesia, tremor, and muscular rigidity.

PEL. *See* Permissible exposure limit.

Percutaneous absorption. The transfer of a chemical from the outer surface of the skin through the horny layer (dead cells), through the epidermis, and into the systemic circulation. A variety of factors, such as pH, extent of ionization, molecular size, and water- and lipid-solubility govern transfer of chemicals through the skin.

Perinatal toxicology. The study of toxic responses to occupationally or environmentally encountered substances when a woman's exposure to them occurs from the time of conception through the neonatal period.

Peripheral neuritis. Inflammation of the nerve ending or of terminal nerves.

Permissible exposure limit (PEL). A measure of the toxicity of a substance, established by the U.S. Department of Labor, Occupational Safety and Health Administration (OSHA); an eight-hour, time-weighted average limit of exposure is assumed. The limit is commonly expressed as the concentration of a substance per unit of air volume (e.g., mg/m^3, ppm, $fibers/cm^3$, etc.)

Personal protective equipment. Any devices worn by workers as protection against hazards in the environment of the workplace, including respirators, gloves, goggles, ear muffs, etc.

Pesticide. Any substance used to destroy or inhibit the action of plant or animal pests. *See* Carbamate; Organochlorine pesticide; Organophosphate pesticide.

Petechiae. Tiny, nonraised, perfectly round, purplish red spots caused by intradermal or submucosal hemorrhaging.

pH. A value taken to represent the acidity or alkalinity of an aqueous solution. It is defined as the logarithm of the reciprocal of the hydrogen-ion concentration of a solution:

$$pH = \ln \frac{1}{[H^+]}$$

Pharmacokinetics. The field of study concerned with the techniques used to quantify the absorption, distribution, metabolism, and excretion of drugs or chemicals in animals, as a function of time.

Pharmacology. The unified study of the properties of chemical agents (drugs) and living organisms and all aspects of their interactions. An expansive science encompassing areas of interest germane to many other disciplines.

Phenothiazine. A green, tasteless compound with the formula $C_{12}H_9NS$, prepared by fusing diphenylamine with sulfur; also, a group of tranquilizers resembling phenothiazine in molecular structure.

Phocomelia. A developmental anomaly characterized by the absence of the proximal portion of a limb or limbs, such that hands or feet are attached to the trunk of the body by a single small, irregularly shaped bone.

Photophobia. Abnormal sensitivity, usually of the eyes, to light.

Photosensitivity reactions. Undesirable reactions in the skin of persons exposed to certain chemicals when the skin is also exposed to sunlight (in some cases, to artificial light). Dermatologic lesions form, which vary from sunburnlike responses to edematous, vesiculated lesions or bullae.

pK_a. The acidic dissociation constant of a compound; the pH of an aqueous solution of an acid or base at which equal concentrations of each are present, at the point at which dissociation is half-complete. The negative logarithm of the ionization constant K_a.

Pneumoconiosis. Accumulation of dusts in the lungs and the tissue reaction to the presence of such dust.

Point mutation. An alteration in a single nucleotide pair in the DNA molecule, usually leading to a change in only one biochemical function.

Poison. The term used to describe those materials or chemicals that are distinctly harmful to the body.

Polymorphonuclear. Having a nucleus deeply lobed or so divided as to appear to be multiple.

Polyneuropathy. A disease involving several nerves.

Porphyrin. Any of a group of iron-free or magnesium-free cyclic tetrapyrrole derivatives occurring universally in protoplasm. They form the basis of the respiratory pigments of animals and plants.

Potency. A comparative expression of chemical or drug activity measured in terms of the dose required to produce a particular effect of given intensity relative to a given or implied standard of reference. If two chemicals are not both capable of producing an effect of equal magnitude, they cannot be compared with respect to potency.

Potentiation. A condition whereby one substance is made more potent in the presence of another chemical that alone produces no response.

Pressure, static. The potential pressure exerted in all directions by a fluid at rest.

Pressure, total. The algebraic sum of static and velocity pressures, representing the total energy in the system.

Pressure, velocity. The kinetic presure exerted in the direction of flow necessary to cause a fluid at rest to flow at a given velocity.

Primary carcinogen. Chemicals that act directly and without biotransformation. Also called direct carcinogen.

Primary irritants. Chemicals that induce local minor to severe inflammatory response, or even extreme necrosis, of cells of a tissue, in direct relation to the concentration available to the tissue. This is termed a "nonspecific chemical action," the toxicity of which may be manifested at the site of exposure (e.g., skin or in the respiratory tract). Examples of these types of chemicals are strong acids or bases, ammonia, and acrolein.

Probenecid. A white, odorless crystalline powder, with the formula $C_{13}H_{19}NO_4S$, soluble in dilute alkali, alcohol, and acetone; used to increase serum concentrations of certain antibiotics, as well as being an agent to promote uric acid secretion in the urine.

Procarcinogen. Chemicals that require metabolism to another, more reactive or toxic chemical form before their carcinogenic action can be expressed.

Proerythropoietin. A precursor of erythropoietin.

Psoriasis. A chronic, hereditary, recurrent, papulosquamous dermatitis, the distinctive lesion of which is a vivid red macula, papule, or plaque covered almost to its edge by silvery lamellated scales. It usually involves the scalp and extensor surfaces of the limbs, especially the elbows, knees, and shins.

Raynaud's phenomenon. Intermittent attacks of severe pallor of the fingers or toes and sometimes of the ears and nose, brought on characteristically by cold and sometimes by emotion.

Receptors. *See* Specific receptors; Nonspecific receptors.

Renal osteodystrophy. A condition resulting from chronic kidney disease. The onset early in childhood is characterized by impaired renal function, elevated serum phosphorus and low or normal serum calcium levels, and stimulation of parathyroid function. The resultant bone disease includes a variety of symptoms including osteitis fibrosa cystica, osteomalacia, osteoporosis, and osteosclerosis. Renal dwarfism may result from childhood onset.

Reproduction tests. Tests that determine (or estimate) the effects of an agent on fertility, gestation, and offspring; usually conducted on more than one generation of test animals. Toxicity in either parent may affect fertility as the direct result of altered gonadal function, estrus cycle, mating behavior, and conception rates. Effects on gestation concern the development of the fetus. Effects on offspring concern growth, development, and sexual maturation; and effects on the mother concern lactation and acceptance of the offspring.

Resorption. The loss of substance in the mucous lining of the uterus.

Reticuloendothelial system. Phagocytic macrophages present in linings of sinuses and in reticulum of various organs and tissues. A functionally important bodily defense mechanism; the phagocytic cells have both endothelial and reticular attributes and the ability to take up particles of colloidal dyes.

Risk assessment. A methodologic approach in which the toxicities of a chemical are identified, characterized, and analyzed for dose-response relationships, and a mathematical model is applied to the data to generate a numerical estimate that can serve as a guide to allowable exposures.

Risk estimation. Mathematical modeling of the animal and/or human toxicity data, combined with evaluation of human exposures, so as to estimate the probability or incidence of effects on human health.

Risk management. The process of applying a risk assessment to the conditions that exist in society, so as to balance exposures to toxic agents against needs for products and processes that may be inherently hazardous.

Safety factor. A factor that presumably reflects the uncertainties inherent in the process of extrapolating data about toxic exposures (i.e., intraspecies and interspecies variations). With this approach, an allowable human exposure to a compound can be determined by dividing the no-observable-effect level (NOEL) established in chronic animal toxicity studies by some safety factor.

Sarcoma. A cancer that arises from mesodermal tissue (supporting or connective tissue).

Sclerodermatous skin change. A chronic hardening and shrinking of the connective tissues of any part of the body, including the skin, heart, esophagus, kidney, and lung. The skin may become thickened, hard, and rigid, and pigmented patches may occur. The condition may be generalized, limited to the distal parts of the extremities and face, or to the digits, or localized to oval or linear areas a few centimeters in diameter.

Sensitization reaction. An immunologic response to a chemical. The mechanism of immunization involves the following events: initial exposure of an animal to a chemical substance; an induction period in the animal; and the production of a new protein termed an antibody. The initial exposure does not result in cellular damage but causes the animal to be "sensitized" to subsequent exposures to the chemical. Exposure of the animal to the same chemical on a subsequent occasion will lead to the formation of sensitized antigen, which will react with the preformed antibodies and lead to a response in the tissues in the form of cellular damage.

SGOT. Serum glutamic-oxaloacetic transaminase. An enzyme found in the liver and muscle tissue and used to detect early membrane permeability as part of a test of the activity of enzymes present in liver cells.

SGPT. Serum glutamic-pyruvic transaminase. An enzyme used in the identification and measurement of the activity of enzymes present in liver cells. SGPT is found in the liver and heart tissues. It is an indicator of early membrane permeability, as is SGOT.

Silicosis. A type of pneumoconiosis that is due to the inhalation of the dust of stone, sand, or flint containing silicon dioxide. It results in the formation of generalized nodular fibrotic changes in both lungs.

Site of action. *See* Locus of action.

SNARL. Suggested no-adverse-response level. A measure of toxicity established by the National Research Council.

Specific receptors. Macromolecular constituents of tissue capable of combining reversibly with a compound by means of chemical bonds; the tissue element with which a compound interacts to provide its characteristic biologic effect.

Spirometer. An instrument for measuring the air taken into and exhaled from the lungs.

Spirometry. The measurement of the breathing capacity of the lungs.

Squamous cell carcinoma. Carcinoa developing from squamous epithelium (composed of flattened, platelike cells) and characterized by cuboid cells.

Stereoisomers. Two substances of the same composition differing only in the relative spatial positions of their constituent atoms and/or groups.

Steric hindrance. The nonoccurrence of an expected chemical reaction owing to inhibition by a particular atomic grouping.

Sympathomimetic. Mimicking the effects of impulses conveyed by adrenergic postganglionic fibers in the sympathetic nervous system.

Synapse. The anatomical relation of one nerve cell to another; the region of junction between processes of two adjacent neurons, forming the place where a nervous impulse is transmitted from one neuron to another.

Synergism. The situation in which the combined effects on a biologic system of two chemicals acting simultaneously is greater than the algebraic sum of the individual effects of these chemicals.

T cell. Thymus-dependent lymphocytes; these pass through the thymus or are influenced by it on their way to the tissues; they can be suppliers or assist the stimulation of antibody production in B cells in the presence of antigen, and can kill such cells as tumor and transplant-tissue cells. T cells are responsible for all cell-mediated immunity and immunologic memory.

TD_{50}. That dose of a substance that, administered to all animals in a test, produces a toxic response in 50 percent of them. The toxic response may be any adverse effect other than death.

Teratogen. Any substance capable of causing malformation during development of the fetus.

Thalidomide. A sedative and hypnotic drug commonly used in Europe in the early 1960s. It was discovered to be the cause of serious congenital anomalies in the fetus, notably amelia and phocomelia.

Threshold dose (ThD). The minimal dose effective in prompting an all-or-none response.

Threshold limit value (TLV). A term for exposure limits established by the American Conference of Governmental Industrial Hygienists. That concentration of any airborne substance to which it is believed, through animal toxicity-testing and human exposure data, workers can be exposed to eight hours per day, 40 hours per week for a working lifetime, without suffering adverse health effects or significant discomfort. TLV measurements are usually based on eight-hour time-weighted average (TWA) exposures but may be expressed as ceiling values.

Time-weighted average (TWA). A method of combining multiple air-sample results collected on one individual during a workshift, so as to derive the overall average exposure for the entire shift (or exposure period). Measurements of the chemical exposure can be made in each phase, and the exposure estimate is calculated according to the formula

$$E = \frac{C_1T_1 + C_2T_2 + \cdots + C_NT_N}{T_1 + T_2 + \cdots + T_N}$$

where E = exposure, C_N = concentration measured in phase N, and T_N = duration of phase N.

TLV. *See* Threshold limit value.

Tolerance. The ability of an organism to show less response to a specific dose of a chemical than it showed on a prior occasion when subjected to the same dose.

Toxicity. A relative term generally used in comparing the harmful effect of one chemical on some biologic mechanism with the effect of another chemical.

Toxicology. The scientific study of poisons, their actions, their detection, and the treatment of the condition produced by them. Also the study of the effects of chemicals on biologic systems, with emphasis on the mechanisms of harmful effects of chemicals and the conditions under which harmful effects occur. Thus, toxicology is a multidisciplinary science.

Tracheitis. Inflammation of the trachea.

Tumor. An abnormal mass of tissue, the growth of which exceeds and is uncoordinated with that of normal tissue. The basic types are benign and malignant. *See also* Benign tumor; Malignant tumor.

Uncertainty factor. *See* Safety factor.

Unction. An ointment; the application of an ointment or salve.

Uropathy. Any pathologic change in the urinary tract.

Ventricular fibrillation. Arrhythmia characterized by fibrillary contractions of the ventricular muscle owing to rapid repetitive excitation of myocardial fibers without coordinated contraction of the ventricle.

Xenobiotic. A chemical foreign to the biologic system (i.e., chemicals that are not normal endogenous compounds for the biologic system).

Index